STATISTICAL THERMODYNAMICS
UNDERSTANDING THE PROPERTIES
OF MACROSCOPIC SYSTEMS

T0203626

STATISTICAL THERMODYNAMICS
UNDERSTANDING THE PROPERTIES
OF MACROSCOPIC SYSTEMS

LUKONG CORNELIUS FAI
GARY MATTHEW WYSIN

CRC Press
Taylor & Francis Group
Boca Raton London New York

CRC Press is an imprint of the
Taylor & Francis Group, an **informa** business

CRC Press
Taylor & Francis Group
6000 Broken Sound Parkway NW, Suite 300
Boca Raton, FL 33487-2742

First issued in paperback 2019

© 2013 by Taylor & Francis Group, LLC
CRC Press is an imprint of Taylor & Francis Group, an Informa business

No claim to original U.S. Government works

ISBN-13: 978-1-4665-1067-8 (hbk)
ISBN-13: 978-0-367-38076-2 (pbk)

Visit the Taylor & Francis Web site at
http://www.taylorandfrancis.com

and the CRC Press Web site at
http://www.crcpress.com

Contents

Preface

This book on statistical thermodynamics has been written with professional theoretical physicists in mind. The text is appropriate for training of graduate and postgraduate students in theoretical physics. The authors present the modern level of evolution of statistical thermodynamics. In presenting the material, the pivot is the application of the general and fundamental principles of theoretical physics. The student should be acquainted with the basics of statistical thermodynamics and in particular, its experimental basis. The chapters go beyond classical thermodynamics and describe how it relates to statistical mechanics. Knowledge of the microscopic constituents of a system enables the evaluation of their macroscopic parameters, via the approach of statistical thermodynamics.

Although statistical physics differs from thermodynamics, both complement each other. Thermodynamics is the macroscopic phenomenological theory of heat. The physical system in thermodynamics is mapped onto some large totality of macroscopic measurable quantities (temperature, pressure, heat capacity, etc.). The link between them and the general laws they obey are developed from axioms in a logical manner and postulated based on experimental facts.

Statistical physics deviates from thermodynamics in that it describes the atomic system through a model or theory for the microscopic motions of matter in the presence of statistical fluctuations due generally to heat. The effects of heat are random and must be calculated through a statistical approach, based on the ideas of probability theory. All relations between macroscopic measurable quantities that arise from thermodynamics are contained in the general statistical theory as relations between corresponding statistical expectations. In this sense thermodynamics is a corollary of statistical physics. However, thermodynamics represents just a general theory. This is because on its basis the general axioms are inherent and do not contain concrete atomic representations and definite dynamic models of matter.

Both statistical physics and thermodynamics are partitioned into the theory of equilibrium states and the theory of nonequilibrium processes. In the former, the theory is based on the probabilities and expectation values of macroscopic parameters that are not dependent on time. In the latter, the probabilities and expectations of measurable parameters depend on time, but generally have a tendency to move toward the equilibrium conditions.

Statistical physics is partitioned into classical and quantum statistical physics, dependent on the physical situation of the matter. Some physical phenomena may be described correctly assuming that the atoms or molecules in the system move according to the laws of classical mechanics. When statistical physics is constructed on this model we call it classical statistics. The quantum model is required for the motion of atoms, molecules, field oscillators, and other matter that when the discrete energy states are perceptible, or when the wave functions of quantum particles overlap. In this case the needed theory is called quantum statistics. We establish relationships between macroscopic parameters of the system when an equilibrium system is considered. This is done using classical thermodynamics. Valid conclusions are drawn based on a minimum number of postulates.

Statistical thermodynamics and the related domains of statistical physics and quantum mechanics nowadays are very important in many fields of research, such as plasmas, rarefied gas dynamics, nuclear systems, lasers, semiconductors, superconductivity, ortho- and para-hydrogen, liquid helium, and so on. To study these systems a statistical approach at microscopic as well as macroscopic scales is needed. Also, in order to understand the foundations, laws, properties and applications of thermodynamics, a study of statistical physics is extremely helpful.

This book starts with the fundamental principles of statistical physics, and then goes to thermodynamics, where the criterion for equilibrium is that the entropy be a maximum. Statistical physics of equilibrium systems are constructed using the Gibbs formalism that is more compatible with quantum mechanics and fits very well with ensemble statistics. From the Gibbs formalism the Maxwell–Boltzmann, Bose–Einstein, and Fermi–Dirac statistics follow in a more logical manner. The procedure in the book presents thermodynamics and statistical physics in a fundamentally cohesive and simple manner. Lattice dynamics are treated in detail and clearly bring out understanding of condensed bodies. The chapters aid very well in the understanding of condensed matter and other areas of physics. The book contains applications on the polaron theory, the electronic gas in a magnetic field, thermodynamics of dielectrics in an electric field, and thermodynamics of magnetic materials in a magnetic field. The last chapter describes statistical thermodynamics with the approach of functional integration and Feynman path integrals. The book has a wide range of problems with solutions that explain the theory very well. The authors do not claim originality neither of the problems nor of their solutions. Their goal is that the problems may help in investigating, illustrating, and concretizing various aspects of the theoretical course of statistical thermodynamics.

We would like to acknowledge those who have helped us at various stages of the elaboration and writing, through discussions, criticism, and especially, encouragement and support. We single out the contributions of Professor Dr. Wolfgang Dieterich of Universitaet Konstanz Fachbereich Physik and Professor Victor Fleurov of Tel-Aviv University for the comments, suggestions, and corrections on the original manuscript that have certainly contributed to improve the quality of this book. We are grateful to Malcolm and Julie Eastoe of Norwich, UK for proofreading the first manuscript in English. Fai Cornelius is very thankful to his wife, Dr. Fai Patricia Bi, for all her support and encouragement and to his three sons (Fai Fanyuy Nyuydze, Fai Fondzeyuv Nyuytari, and Fai Ntumfon Tiysiy) for their understanding and moral support during the writing of this book. Gary Wysin expresses his appreciation for the encouragement of his wife, Marcilene Sousa Wysin, and the understanding of his daughter, Nayara Sousa Wysin, while working on the book.

Fai Cornelius wants to acknowledge with gratitude the financial and library support under the associate scheme received from the Abdus Salam International Centre for Theoretical Physics, Trieste, Italy and the hospitality during his visits there.

While all the above persons have helped to improve the quality of the book, we alone are responsible for all errors and inadequacies that remain. We will be grateful to readers to bring errors to our attention so that corrections can be made in the subsequent edition. The typing of the manuscript has been done by the authors.

Authors

Lukong Cornelius Fai is a professor of condensed matter physics, and head of Condensed Matter Division as well as head of the Mesoscopic and Multilayer Structures Laboratory in the Department of Physics, Faculty of Sciences, at the University of Dschang, Cameroon. He also holds the position of chief of Division for Cooperation at the same university, and delivers lectures on quantum mechanics, statistical physics, and Feynman path integrals. Professor Fai performs research in condensed matter and statistical physics, and also is a senior associate at the Abdus Salam International Centre for Theoretical Physics, Trieste, Italy. Some of his current research involves studies on the interactions between electrons and phonons in metals, nanocrystals, and quantum dots—in which there are the so-called polaron and bipolaron excitations. He was awarded a Doctor of Science in physics and mathematics in 1997 from Moldova State University, based on studies of optical absorption due to polarons. He obtained Master of Science from the same university in 1991, which involved research on the quantum excitations in metal–dielectric–semiconductor structures.

Gary Wysin is a professor in the Physics Department at Kansas State University (KSU) since 1988. He obtained a bachelor's degree in electrical engineering in 1978 at University of Toledo, Ohio, and in 1980, a master's degree in physics, for research on optical bistability at the boundary of a metal and a dielectric, also at the same institution. He was awarded a doctorate in theoretical physics from Cornell University in 1985 while working on the theory and Monte Carlo and spin-dynamics simulations for magnetic models with solitons and vortices. While at KSU, he has lectured on classical electrodynamics, quantum mechanics, computational physics, statistical physics, and introductory physics and astronomy. He carries out research in condensed matter physics, including phase transitions, magnetization reversal, spin waves, solitons and vortices in low-dimensional magnetism, and optical effects in dielectrics and metals. Some of the research was carried out in collaborations with scientists at Los Alamos National Laboratory, New Mexico, USA, University of Bayreuth, Germany, Universidade Federal de Minas Gerais, Belo Horizonte, Brazil, Universidade Federal de Viçosa, Minas Gerais, Brazil, and Universidade Federal de Santa Catarina, Florianópolis, Brazil.

1 Basic Principles of Statistical Physics

1.1 MICROSCOPIC AND MACROSCOPIC DESCRIPTION OF STATES

Statistical thermodynamics is the science that describes the effect of heat interacting with matter, on a microscopic and statistical basis. This heading constitutes the most essential part of this course: to introduce the reader to the foundations of the subject and prepare him/her for subsequent headings. In order to complete the structure of statistical thermodynamics, we create a basis for relating macroscopic properties of thermodynamic systems to the microscopic properties of their mechanical models at the atomic level. The detailed studies of individual mechanical systems are carried out by methods of classical and quantum mechanics [1–38], to obtain the physical properties of their individual microstates.

Let us examine such microscopic and macroscopic systems. A system is *microscopic (small scale)* if it is roughly of an atomic scale or of the order of $10\,\text{Å}$ or less. An example of such a system is a molecule. Typically, a microscopic system has very few degrees of freedom, in the mechanical sense, and so a description in terms of statistical physics is not applicable. A system is *macroscopic (large scale)* when it is large enough to be visible in the ordinary sense, say for example, it may be greater than 1 micron such that it can at least be observed with a microscope with the use of ordinary light. A macroscopic system usually possesses very many degrees of freedom; hence, a statistical description of its properties is not only applicable but also necessary. In dealing with macroscopic systems, one is interested in *macroscopic parameters* of the system, such as volume, pressure, magnetic moment, thermal conductivity, and so on, that can be measured in experiments, and are representative of the average behavior of all its molecules. For an isolated system, if the macroscopic parameters do not vary with time, the system is in *equilibrium*. If the isolated system is not at equilibrium, the parameters of the given system change with time so as to approach their equilibrium values, possibly through some very complex processes. Generally, any system in equilibrium will return to that state or condition even if affected by some temporary outside force that is later removed. A great part of statistical physics considers only systems in equilibrium. Unless otherwise stated further in the chapters, we deal with statistical physics of equilibrium systems.

Macroscopic systems such as gases, liquids, or solids were first investigated from a macroscopic phenomenological point of view in the nineteenth century and the laws thereof formed the subject of *thermodynamics*. In the second half of the twentieth century, the atomic theory of matter gained general acceptance and macroscopic systems began to be analyzed from the fundamental microscopic point of view as

TABLE 1.1

Object, Goal, and Method of Thermodynamics and Statistical Physics

	Object	Goal	Method
Thermodynamics	Macroscopic body in the thermodynamic equilibrium state	Link between macroscopic parameters	Phenomenological
Statistical	Macroscopic body in the thermodynamic equilibrium state	From microscopic description follows macroscopic properties	Probabilistic Microscopic

systems consisting of very many atoms or molecules. The development of quantum mechanics (QM) after 1926 provided an adequate theory for the description of atoms and thus opened the way for an analysis of systems composed from very many atoms on the basis of realistic microscopic concepts. Thus, the basic concept applied in statistical physics is the idea that macroscopically observed variables can be determined from the average behavior of many individual atoms. In this chapter, the main objective is to derive the laws of thermodynamics from statistical physics. We will derive statistical physics formulas for the evaluation of thermodynamic functions of a system, giving their values in terms of various controlled variables such as the system's temperature.

The *object of statistical physics* is a study of particular types of laws that are found in the behavior and properties of macroscopic states, that is, the states of a large number of separate particles such as atoms and molecules. QM provides the fundamental laws for evaluating the properties of individual microscopic particles and their interactions. Statistical physics begin with these results and lays down the statistical hypothesis about systems with a large number of particles. This hypothesis includes the principle of *equal a priori probabilities* for a set of classical or quantum states, that we see further in the chapters. In general, statistical physics is strongly based on QM. Here, almost all the important results of statistical physics are developed using only certain basic concepts and results of QM.

Let us represent the above in Table 1.1.

1.2   BASIC POSTULATES

The goal of thermodynamics is an investigation of the link between macroscopic parameters, and especially, the temperature dependence of measured quantities. Statistical physics intends to find these links from a viewpoint of the microscopic dynamics of the constituent atoms and molecules that make up the system. Although a quantum mechanical description is the fundamental one, it is possible to describe much of the problem in the language of classical mechanics and classical statistics. It will become apparent that this point of view is only slightly modified when the constraints of QM are included. Indeed, we can follow a system's dynamics as though it obeys classical mechanics, and then develop the corresponding formalism for quantum statistics based on that. In the end, it must be insured that the quantum formalism satisfies the Heisenberg uncertainty principle, as well as fundamental conservation laws. Further, the QM formalism must be able to describe the subtle ways in which the presence of indistinguishable identical particles affects the QM states and thus the thermodynamic properties.

In formulating the basic problem of classical statistics we first have to introduce the notion of phase space that we will use frequently. We use the notion suggested by Gibbs [38]. Suppose the given macroscopic mechanical system has s degrees of freedom. This implies that there are s coordinates that we denote by q_i, $i = 1, 2, ...,$ s, which characterize the positions (or other generalized coordinates) of the microscopic entities (atoms, molecules) that make up the system.

The state of a classical system at a given moment will be defined by the values of s coordinates q_i and s corresponding velocities $\dot{q}_i$. In statistics, we use for the characteristics of the system its coordinates q_i and canonical momenta p_i and not the velocities, as the momenta are most closely connected to the coordinates, in both classical and quantum mechanics. For one degree of freedom, the specification of the position coordinate q and its corresponding momentum p is sufficient, as the laws of classical mechanics are such that knowledge of q and p at any one time permits prediction of the values of q and p at any other time. The specification of q and p is equivalent to specifying a point in a two-dimensional space or phase space. The coordinate and momentum of a particle change with time. This representative point (q, p) (phase point) moves through this phase space, describing a curve called the phase trajectory. Each point in the phase space corresponds to a microscopic state, and then the phase trajectory shows exactly what states the dynamical system evolves through. As we really have a thermodynamic system described by s coordinates q_i, $i = 1, 2, ..., s$ and s corresponding momenta p_i, $i = 1, 2, ..., s$, we then have a total of $2s$ parameters and consequently a phase space of $2s$ dimensions. In general we will use the symbol (q, p) to represent the full set of s coordinates and their corresponding momenta.

Thus, microscopic states in classical statistical physics make a continuous set of points in phase space. In quantum statistical physics, the Heisenberg uncertainty principle does not allow the coordinates and momenta to be known simultaneously to such perfect precision. Instead, we might think of any point in the phase space as being smeared out over an area $\Delta q_i \Delta p_i \approx h$ per degree of freedom, where h is Planck's constant. Classically or quantum mechanically, we cannot possibly follow all the microscopic dynamics, especially of a macroscopic sample that might have on the order of Avogadro's number of particles. The basic idea of statistical physics is the existence of a probability distribution for the relative frequency that the different microscopic states are visited by the system. If we only know the probability that each state will be visited, then statistics can be used to predict real physical measurements on the system.

1.3   GIBBS ERGODIC ASSUMPTION

It is impossible to calculate the full time evolution of a complex macroscopic system, from which one could, in principle, calculate averages over long time. The foundations of statistical physics and the success of its methods are based on the fact that theoretical values for physical quantities may be found using statistical averages in the phase space. This is to be contrasted with experimental measurements, which are typically long-time averages. According to Boltzmann, a system left in free evolution for a sufficiently long time interval will pass through all of its *accessible states*, namely the states with the given value of the total energy if the system is isolated. This hypothesis is the so called the *ergodic hypothesis*. The ergodic hypothesis and the use of the method of statistical methods assumed a central role indicated in the work of J. Willard Gibbs of 1902 [38]. That approach is strongly based on using phase space averages, with an appropriate type of probability distribution, to replace long-time averages.

To make theoretical progress, it is necessary to introduce some postulate on the relative probability of finding a system in any of its accessible states. For example, let the system under consideration be *isolated*. Thus it cannot exchange energy with its surroundings. From here, the laws of mechanics then imply that the total energy of the system is conserved. With a given fixed energy, only a limited set of the possible microscopic states can be visited; these are the accessible states referred to above.

Suppose the isolated system is in equilibrium, characterized by the fact that the probability of finding the system in any one state is independent of time. All the macroscopic parameters describing the isolated system are then also time independent. From here, we may conclude that whenever the system is observed, it must be in one of its accessible states consistent with the constant value of its energy. On the other hand, this says nothing about exactly which of the accessible states it should be found in, or more importantly, how often it would be found in a particular state. If the same system is observed a short time later, it would be expected to be found in a different one of the accessible states, as the phase point of the system moves with time.

An arbitrary system has complex dynamics in the phase space of $2s$ dimensions, so that practically speaking, the future motion of the phase point cannot be obtained. Instead, we use Boltzmann's ergodic assumption: "An isolated system in equilibrium is equally likely to be in any of its accessible states." There is no way to prove this statement for arbitrary systems. In fact, one can find some simple mechanical systems where it is not true. For very many realistic problems, however, it describes the average long time behavior of the system exceedingly well, leading to the great success of statistical physics in many areas.

What do we mean, most generally, by *accessible states*? There are generally some specified physical conditions that a system is supposed to satisfy, even for a nonisolated system that may not have constant energy. These conditions act as constraints that limit the number of states in which the system can be found. The *accessible states* are the states consistent with these physical constraints.

1.4 GIBBSIAN ENSEMBLES

The ergodic hypothesis is necessary to justify and give physical meaning to this *statistical ensemble method*, which we describe in more detail. Consider a macroscopic system of bodies. We suppose that this system is closed, that is, it does not interact with any other body. Select a small macroscopic system from within the given system. That selected small macroscopic system is called a *subsystem* of the given system. This subsystem is still mechanical but not closed as it is subjected to all possible interactions from the rest of the other particles. These interactions have a complex and confusing character and as a consequence the state of a given subsystem changes with time in a very complex and confusing manner. Due to the complicated external influence of the rest of the particles for an exceedingly long interval of time, the subsystem could be expected to attain all of its possible states.

Some physical quantity, say, the magnetization, could be measured by using observations of a subsystem over a very long time. On the other hand, one could imagine doing a different kind of measurement, by using observations of all of the subsystems simultaneously, without a long-time average. If each subsystem is under physical conditions the same as all the other subsystems, then they all move through the same accessible region of phase space. At any given instant, however, they are distributed with some likelihood or probability distribution over the accessible phase space. We can use the basic methods of statistical averaging to calculate what should be measured in the real system, especially, the quantities referred to as *expectation values*. The idea is that averaging over the subsystems should give the same result as a long-time averaging of one subsystem. Besides, one finds that this statistical method can be carried out more readily than trying to calculate the time evolution of a complex dynamical system.

Now, we can think about the terminology in a slightly different fashion. Conceptually, the subsystems are themselves really nothing different than small copies of the original macroscopic system. Instead of thinking of averages over subsystems, we can simply imagine doing averages over many identical copies of the original system of interest. This is the idea of an ensemble and ensemble averaging. The term *ensemble* was coined by Gibbs. A certain number N of copies of the system are followed or averaged over, as a way to think about how to calculate expectation values of macroscopic observables. We refer to these copies now not as subsystems but as members of the ensemble. This is a useful concept especially because the averages we need are made over the phase space. As long as we know the probability for the members of the ensemble to be in the different allowed parts of phase space, any kind of averages can be predicted.

1.5 EXPERIMENTAL BASIS OF STATISTICAL MECHANICS

Statistical thermodynamics is a general theory designed to apply to a very broad range of material systems and physical situations. It is based strongly on the idea that all states of a system under given or desired physical conditions (e.g., constant total energy as in the microcanonical ensemble that is discussed below) are equally probable. Whether this idea is correct or not can only be verified by comparison of the theoretical results with many experiments. The basic ideas, such as the ergodic hypothesis, are not necessarily directly provable in general for all possible systems. Indeed, even in a system that does nearly explore all of its available phase space, it can only do so if observed over a sufficiently long time interval. This minimum observation time depends on the fundamental physical time scales in the problem, over which the system scans the phase space. As a simple example, an individual harmonic oscillator must be observed, at the very least, over a time greater than its period, in order that averages over time would be close to the averages obtained by integration over the accessible phase space.

Of course, any physical system of interest is never truly isolated, and besides that, we never know the true Hamiltonian. There are always additional interactions that we cannot account for or even know about. Generally, they might be considered as due to the "thermal reservoir" or "heat bath," which in some vague sense can be expected to mix the system more than we can possibly account for. Further, quantum fluctuations will also be present due to interactions which we may not take into account. We hope that these additional perturbations that we are not totally aware of lead the system to behave as ergodic. But again, the best test of the validity of statistical mechanics is the entire realm of experiments for which it is found to apply, in a wide variety of physical systems and physical measurements.

1.6 DEFINITION OF EXPECTATION VALUES

Any measurable physical quantity must be determined by the state of the system; hence, it must be a function like $F(q, p)$. As time passes, the observable F changes, but when viewed over a long time, an average value will result, which is the expectation value. We imagine first how it would be calculated based on an averaging process over long time.

The state of any subsystem or system in an ensemble may be specified approximately by saying that its coordinate lies in some interval between q and $q + dq$ and its momentum lies in some interval between p and $p + dp$, that is, by stating that the representative point (q, p) lies in a particular part of a phase space. We take q and p to represent all s degrees of freedom, so that each of these has a certain small range. The Gibbs space has $2s$ dimensions, where the number of degrees of freedom is s. The coordinates are q_i, $i = 1, 2, ..., s$ and the corresponding momenta are p_i, $i = 1, 2, ..., s$. Thus, $q \equiv \{q_i\}$, $p \equiv \{p_i\}$, $dq = dq_1 dq_2 ... dq_s$, $dp = dp_1 dp_2 ... dp_s$, and the infinitesimal region surrounding a phase space point is a hypervolume $d\Gamma = dpdq$, with

$$dqdp = dq_1 dq_2 \ ... \ dq_s dp_1 dp_2 \ ... \ dp_s \tag{1.1}$$

Typically, we want to imagine observing when the system passes near a desired point (q, p) in phase space, that is, in the small volume $d\Gamma = dpdq$. Suppose that for an exceedingly long interval of time T, the confused phase trajectory passes through an arbitrary part of phase space a very large number of times. Let $dt = dt(q, p; d\Gamma)$ be the part of the total time T for which the subsystem is found in that part of phase space $d\Gamma$ near (q, p). For an infinitely long time T, the ratio dt/T tends to a limit dW, which is a number less than unity

$$dW(q, p; d\Gamma) = \lim_{T \to \infty} \frac{dt(q, p; d\Gamma)}{T} \tag{1.2}$$

The value dW in Equation 1.2 is the probability that if we observe the subsystem at an arbitrary moment in time, we find it in a desired infinitesimal volume $d\Gamma = dpdq$

of phase space surrounding the point (q, p). This means that dW can be used to define a probability density $\rho(q, p)$ in phase space, which gives probability when multiplied by a phase space volume:

$$dW(q, p; d\Gamma) = \rho(q, p)d\Gamma \qquad (1.3)$$

Thus, $\rho(q, p)$ is the probability per unit phase space volume to find the system in a small region of phase space $d\Gamma = dpdq$ surrounding the point (q, p). Further, being a probability density means that a sum (or integral) over all the possible microstates gives unity:

$$\int \rho(q, p)\, dqdp = 1 \qquad (1.4)$$

Usually we would state this relationship by saying that the probability density is normalized to unity. This relation just represents the fact that the sum of all the times spent in different parts of phase space must add up to the total observation time, T. Or, it can be interpreted to mean that the sum of probabilities for the system to be in any state of phase space is unity.

Now suppose the observable $F(q, p)$ is measured. The idea of measurement for a macroscopic system, involves observing the system over long enough times that the system evolves through a large part of phase space. During that time, the dynamically measured quantity $F(t) = F(q(t), p(t))$ can be changing; however, the experimental apparatus will only produce a time-averaged value because it cannot follow the microscopically rapid motions. For an instantaneous measurement of $F = F(t)$ to give a desired value $F(q, p) = F(t)$ requires that the system pass through the particular points $(q, p)_{F=F(t)}$ satisfying this constraint, $F(q, p) = F(t)$. In fact, this constraint could be visualized as contours in the phase space. Alternatively, we can look for the observable to lie in some range, say, from F to $F + dF$. There will be some limited region $d\Gamma$ of phase space, bounded by contours at F and $F + dF$, possibly made from disjoint sections, where $F(q, p)$ lies in the desired range, with a width dF. As time evolves, $F(t)$ takes on values in some desired range from F to $F + dF$ only during some total time interval dt, out of a total observation time T, which is a kind of mapping:

$$F \leq F(t) \leq F + dF \quad \Rightarrow \quad t \in dt(F; dF) \qquad (1.5)$$

Thought of in this way, Equation 1.2 can be interpreted as the probability for finding F in a desired range:

$$dW(F; dF) = \lim_{T \to \infty} \frac{dt(F; dF)}{T} \qquad (1.6)$$

Here, of course, $dt(F; dF)$ means the *total time* the system spent within the correct region of phase space to give $F(q, p)$ in the range from F to $F + dF$; the system may be dynamically going into and out of that region.

The expectation value of the observable $F(q, p)$, denoted as $\overline{F}$ or as $\langle F \rangle$, could be found as its average using this probability:

$$\overline{F} = \int F \, dW(F; dF) \tag{1.7}$$

In fact, this allows the definition of the corresponding unit-normalized *probability density* for F, similar to the probability density per unit phase space $\rho(q, p)$. We can denote it as $\rho_F(F)$, such that

$$dW(F; dF) = \rho_F(F) \, dF \tag{1.8}$$

$$\overline{F} = \int F \rho_F(F) \, dF \tag{1.9}$$

where
$$\int \rho_F(F) \, dF = 1 \tag{1.10}$$

These results show how the time averaging of F leads to a probability distribution for F. However, the expectation value of F can be viewed differently, using the Gibbs phase space and the idea of ensemble averaging. Let us examine the quantity $F(t)$ of a dynamical state of a system. As time evolves, $F(t)$ is changing ultimately because the coordinates and momenta are evolving. Instead of thinking of a probability for F, it is better to think of the probability for a system to be in a desired infinitesimal element of phase space, $d\Gamma = dpdq$, surrounding some point (q, p). Thinking this way means that we imagine a very large number N of identical copies of the system, all evolving together in phase space. This is the ensemble. When the ensemble is observed at some instant of time, its coordinates and momenta will be distributed in phase space, according to a probability distribution. At any instant, there can be expected to be a number $dN(q, p; d\Gamma)$ of members of the ensemble within a desired volume element $d\Gamma = dpdq$. Indeed, that probability distribution has the probability density $\rho(q, p)$ introduced earlier. As time evolves, the "positions" $(q, p)_m$ of each member m of the ensemble change, following trajectories in phase space. However, how they are distributed in phase space should remain the same, if all are subjected to the same conditions and the system is in equilibrium. Therefore, this probability distribution can be used to compute averages, that is, expectation values. Further, one probability distribution will apply to averages for all dynamical observables.

We already saw that the probability of observing any system in the volume $d\Gamma = dpdq$ is related to the time spent by a system within $d\Gamma$. Alternatively, we simply look at what number dN of the N systems in the ensemble are in the desired phase space element to define the probability:

$$dW(q, p; d\Gamma) = \frac{dt(q, p; d\Gamma)}{T} = \frac{dN(q, p; d\Gamma)}{N} \tag{1.11}$$

The different systems occupy the different microstates with the probability dW. Hence, this can be reexpressed in terms of the probability density in phase space:

$$dW = \frac{dN(q,p)}{N} = \rho(q,p)\,dqdp \qquad (1.12)$$

where $\rho(q, p)$ is the probability density (or distribution function) of phase points in the given phase space. Here, the argument $d\Gamma$ has been suppressed. The number of macroscopically identical systems distributed among accessible microstates by the distribution function $\rho(q, p)$ defines a statistical ensemble. Thus, this statistical ensemble is defined and characterized by $\rho(q, p)$. The quantity dW or $\rho(q, p)\,dqdp$ is the probability that the dynamical state of any system in the ensemble is in $dqdp$ about the point (q, p).

With the probability of a given microstate being $\rho(q, p)\,dqdp$, the *ensemble average* (*statistical average*) of any function $F(q, p)$ depending on the dynamical state of a system is

$$\overline{F} = \int F(q,p)\,\rho(q,p)\,dqdp \qquad (1.13)$$

The single integral sign means integration over all the p and q (the integral is taken over all of phase space). Although we used the idea of an ensemble, there is no explicit sum over its members. That has been accomplished by the use of the probability density. What has not been fully specified, however, is the actual probability distribution $\rho(q, p)$. Its definition depends on the conditions that determine the accessible states.

We should also mention one further advantage of the ensemble method over the concept of long-time averages. A physical system could be subjected to time-dependent external forces. In that case, long-time averages would smear out possibly important dynamical effects. The only way to make a sensible theoretical calculation of macroscopic averages that depend on time is to imagine using an ensemble average.

1.7 ERGODIC PRINCIPLE AND EXPECTATION VALUES

In this section, we discuss the correlation between the *two types of expectation values* that are of interest, that is, *statistical expectation values* and *chronological expectation values*. The statistical expectation value is the ensemble average of a thermodynamic quantity $F(q, p)$, $q = q(t)$, $p = p(t)$, say, at a given time t over all systems of the ensemble. The chronological expectation value is the time average of the thermodynamic quantity $F(q, p)$, say, for the given system of the ensemble over an exceedingly long time interval $2T$ (i.e., where $T \to \infty$). For simplicity, it is assumed that the only time dependence of F is that implicitly due to the time variation in $(q(t), p(t))$.

The correlation between the statistical and the chronological expectations values will aid in choosing the specific form of the distribution function that will be used in computing statistical expectations. In particular, for an equilibrium situation, we will find that the distribution function ρ will be independent of time t and dependent only on q and p.

The *statistical expectation value* of the quantity $F(q, p)$ is denoted interchangeably by $\overline{F}$ or $\langle F \rangle$ and is defined by using an average over the ensemble:

$$\overline{F(t)} \equiv \langle F(t) \rangle \equiv \frac{1}{N} \sum_i F_i(t) = \frac{1}{N} \sum_i F\left(\{q(t)\}_i, \{p(t)\}_i\right) \rightarrow \int \rho(q, p; t) \, F(q, p) \, dqdp$$

(1.14)

or

$$\overline{F(t)} = \int \rho(q, p; t) \, F(q, p) \, dqdp \tag{1.15}$$

Here, $F_i(t)$ is the value assumed by $F(q, p)$ in the ith system of the ensemble, N is exceedingly large and is the total number of systems in the ensemble, and the coordinates and momenta in the ith system, $\{q(t)\}_i$, $\{p(t)\}_i$, are evaluated at a particular time t. The result could depend on the time t through $\rho(q, p; t)$, if the number of members of the ensemble within $d\Gamma = dqdp$ depends on time, that is, $dN(q, p; t)$.

The *chronological expectation value* of the quantity $F(q, p)$ is denoted interchangeably by $\{F\}$ or $\tilde{F}$. It is defined for the ith system (perfectly isolated) of the ensemble by a symmetrical time average

$$\tilde{F}_i(t) = \{F_i(t)\} = \frac{1}{2T} \int_{-T}^{T} F_i(t + t') dt' = \frac{1}{2T} \int_{-T}^{T} F\left(\{q(t + t')\}_i, \{p(t + t')\}_i\right) dt' \quad (1.16)$$

The chronological and statistical expectations commute because the time and phase space integration commute. To see this, consider the statistical expectation of the time average, $\tilde{F}(t)$:

$$\overline{\tilde{F}(t)} = \frac{1}{N} \sum_i \tilde{F}_i(t) = \frac{1}{N} \sum_i \left[\frac{1}{2T} \int_{-T}^{T} F_i(t + t') dt' \right] = \frac{1}{2T} \int_{-T}^{T} \left[\frac{1}{N} \sum_i F_i(t + t') \right] dt'$$

$$\rightarrow \frac{1}{2T} \int_{-T}^{T} \left[\int \rho(q, p; t + t') F(q, p) dqdp \right] dt' = \frac{1}{2T} \int_{-T}^{T} \overline{F(t + t')} \, dt' = \tilde{\overline{F}}(t)$$

(1.17)

It follows that

$$\overline{\tilde{F}(t)} = \tilde{\overline{F}}(t) \tag{1.18}$$

The quantity T is exceedingly long ("microscopically long") so that $\tilde{F}$ becomes independent of T through the smoothing out of microscopic fluctuations.

We consider now a *stationary* situation for the quantity F. This implies that there exists no preferred origin in time for the statistical description of F. That is to say, we ensure the same ensemble when all member functions F of the ensemble are shifted by arbitrary amounts in time. This is true for all statistical quantities for an equilibrium situation. Here, there is an intimate union between the statistical and chronological expectations of the ensemble. This is on condition that the function $F(t)$ for an exceedingly long time will pass through all values accessible to it (the *ergodic assumption*). Here, we do not consider some exceptional systems in the ensemble.

Consider again the function $F(t)$ as observed in a system of the ensemble. We partition the time scale into exceedingly long intervals of magnitude $2T$. As T is exceedingly large, the behavior of $F(t)$ in each partition is independent of its behavior in any other partition. Consider some large number N for such partitions. Then, this is a representation of the statistical behavior of F in the original ensemble. Let us view a system in equilibrium. A way to define equilibrium is to say that chronological expectation values do not depend on the choice of the origin of the long time interval. Then, the chronological and statistical expectations should be equivalent. It follows that

$$\tilde{F}(t) = \overline{F} \tag{1.19}$$

The chronological expectation is independent of the partition, say the ith. Similarly, it must be true that for such a stationary ensemble the statistical expectation of F must be independent of time, because ρ is independent of time:

$$\overline{F}(t) = \overline{F} \tag{1.20}$$

Consider again Equation 1.18. Examine the statistical expectation of Equation 1.19; then

$$\overline{\tilde{F}}(t) = \overline{F} \tag{1.21}$$

Thus, if we consider Equation 1.18, then for a stationary ergodic ensemble, we have the important result

$$\tilde{F} = \overline{F} \tag{1.22}$$

Let us again examine the phase space probability density ρ. We may get an exact integral of motion for ρ, if a system is isolated. Gibbs proposed that the first integral of motion of ρ is the energy E, and as a consequence E can be considered as the independent variable:

$$\rho = \rho(E), \quad E = E(q, p) \tag{1.23}$$

This form for ρ corresponds to what is called the microcanonical ensemble (system at fixed energy). Let us examine a constant energy surface

$$E = E(q, p) = E_0 = \text{const} \tag{1.24}$$

in the $2s$-dimensional phase space (q, p), represented abstractly in Figure 1.1.

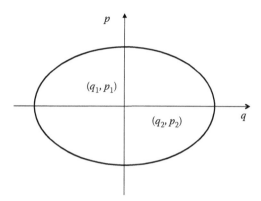

FIGURE 1.1   An energy surface in a 2s-dimensional phase space.

On the given constant energy surface, consider the points (q_1, p_1) and (q_2, p_2), corresponding to two different accessible states, which must satisfy the requirements

$$E_1 = E(q_1, p_1) = E_0, \quad E_2 = E(q_2, p_2) = E_0 \tag{1.25}$$

Then, under Gibbs' assumption that the density can be determined from the energy alone

$$\rho(q_1, p_1) = \rho(E_0), \quad \rho(q_2, p_2) = \rho(E_0) \tag{1.26}$$

It follows that on the constant energy surface, the densities are equal

$$\rho(q_1, p_1) = \rho(q_2, p_2) \tag{1.27}$$

This is for an isolated equilibrium system.

From the definition

$$dW(q, p) = \rho(q, p)dqdp \tag{1.28}$$

we conclude that the probabilities of all microstates with the given value of the energy are the same. This is one of the most important fundamental principles in statistical mechanics. It serves as a guiding principle for all ensemble averages: "The probabilities of all accessible microstates are the same."

In classical mechanics, q and p are continuous variables and define a microstate. Mathematically, however, the probability is only defined for finding a system in a microstate within some phase space region $dqdp$. In QM, the state of one degree of freedom cannot be defined more precisely than $\Delta q \Delta p = 2\pi\hbar$. It follows from the uncertainty principle that one cannot completely localize a system in phase space. In other words, each quantum state occupies a phase volume $\Delta q \Delta p = 2\pi\hbar$ per degree of freedom, or a total hypervolume $(2\pi\hbar)^s$ for a system with s degrees of freedom. In a quantum system, Gibbs' assumption still holds: all quantum states consistent with the conservation laws have equal probability.

Based on Gibbs' assumption, the average (mean) statistical value $\overline{F}$ for the system at chosen energy E_0 is

$$\overline{F} = \langle F \rangle = \int F(q,p)\, dW(q,p) = \int F(q,p)\,\rho(q,p)\, dqdp \qquad (1.29)$$

In the simplest case of an isolated system, the chronological expectation coincides with the statistical expectation value

$$\tilde{F} = \overline{F} \qquad (1.30)$$

We see that if the chronological expectation coincides with the expectation of the statistical value, then statistical physics has a firm foundation. The ergodic problem consists precisely of this justification and finds the correspondence between particular values of phase space functions and the experimental measurements for the physical quantity. Relation (1.30) shows that time averaging is equivalent to averaging over microcanonical ensembles and the property is often called *ergodicity*.

We see here that ρ really represents the probability density of points in phase space and that is why the expectation value of the quantity F is properly computed by relation (1.29).

Let us examine relation (1.18) from the light of the Schrödinger and the Heisenberg pictures. We rewrite Equation 1.18 again in the form

$$\langle F(t) \rangle = \int dqdp\, F(q,p,t)\,\rho(q,p,t) = \int dqdp\, F(q,p,t_0)\exp\{-i\hat{L}^s t\}\,\rho(q,p,0) \quad (1.31)$$

The operator $\hat{L}^s$ evolves the density or other physical quantities forward in time. The quantity F at time t_0 corresponds to the *Schrödinger picture* in QM where $F(q, p, t_0) = F(q, p)$ is fixed in time and $\rho(q, p, t)$ evolves with time. If we perform integration by parts of the relation in Equation 1.31, then we have as follows:

$$\langle F(t) \rangle = \int dqdp\, F(q,p,t)\,\rho(q,p,0) = \int dqdp\, \rho(q,p,0)\exp\{-i\hat{L}^s t\}\, F(q,p,t) \quad (1.32)$$

This corresponds to the *Heisenberg picture* in QM where

$$i\frac{\partial F(q,p,t)}{\partial t} = \hat{L}^s F(q,p,t) \qquad (1.33)$$

We see that the probability density $\rho(q, p)$ for any given configuration (q, p) may be interpreted as representing a particular realization of a system of ensembles.

1.8   PROPERTIES OF DISTRIBUTION FUNCTIONS

A probability distribution function $\rho(q, p)$ relates to the likelihood of finding a system of the ensemble in a region of phase space surrounding (q, p). As a probability distribution, it satisfies some basic properties.

1.8.1   ABOUT PROBABILITIES

Let us review some principles of the probability, related to events or situations denoted by symbols A and B, within some space of possible events:

1. *Addition*: If the two events can both occur but if they are mutually exclusive (they occupy different nonoverlapping spaces in a Venn diagram), the probability that one or the other occurs is

$$P(A \cup B) = P(A) + P(B) \tag{1.34}$$

   The symbol $P(A \cup B)$ is the probability of the *union* of A and B. This only applies if the events A and B cannot take place together. Otherwise, they have some overlapping part in the Venn diagram, which has the probability of *intersection*, represented with the symbol $P(A \cap B)$. Then, the probability that either A or B occurs, or both occur, is

$$P(A \cup B) = P(A) + P(B) - P(A \cap B) \tag{1.35}$$

   The last term corrects the overcounting of probability in the first two terms by the intersection of the spaces of A and B.
2. *Multiplication*: If two events are mutually independent, the probability that they both occur together is

$$P(A \cap B) = P(A)P(B) \tag{1.36}$$

   Further, if there are more than just two events, all mutually independent, then the probability of all of them happening together is just the product of all of their separate probabilities.
3. *Conditional Probability*: One may be interested in the probability that B occurs, given that A is already present. This is the conditional probability, denoted by the symbol, $P(B|A)$. It is equal to

$$P(B|A) = \frac{P(A \cap B)}{P(B)} \tag{1.37}$$

   If the events are independent, then Equation 1.36 can be applied and there results only

$$P(B|A) = P(A) \tag{1.38}$$

1.8.2 NORMALIZATION REQUIREMENT OF DISTRIBUTION FUNCTIONS

A probability density integrates to unity:

$$\int \rho(q, p)dqdp = \int \frac{dN(q, p)}{N} = \frac{1}{N}\int dN = 1 \tag{1.39}$$

This expresses the fact that the sum of the probabilities of all possible states must equal unity.

1.8.3 PROPERTY OF MULTIPLICITY OF DISTRIBUTION FUNCTIONS

The fact that different subsystems may be considered weakly interacting with each other leads to the fact that they may also be considered independent in the statistical sense. The statistical independence implies that the state in which one of the subsystems is found does not influence in any way the probabilities of different states of the other subsystems. The relevant observables are said to be statistically independent or uncorrelated.

Let us examine any two subsystems (Figure 1.2) and let $dq_I dp_I$ and $dq_{II} dp_{II}$ be their phase volumes.

If we consider the two subsystems taken together as a single system, then from the mathematical point of view, the statistical independence of the subsystems implies that the probability of the system to be found in the element of its phase volume

$$dq_{I+II} dp_{I+II} = dq_I dp_I dq_{II} dp_{II} \tag{1.40}$$

factors into the product of the probabilities for each of the subsystems correspondingly in $dq_I dp_I$ and $dq_{II} dp_{II}$. Each of these probabilities is dependent only on the coordinates and momenta of the given subsystem. Consequently, we may write

$$\rho_{I+II} dq_{I+II} dp_{I+II} = \rho_I dq_I dp_I \; \rho_{II} dq_{II} dp_{II} \tag{1.41}$$

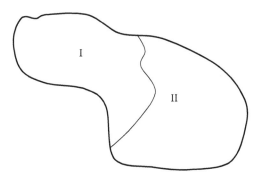

FIGURE 1.2   Two subsystems I and II of a larger system.

from where it follows that

$$\rho_{I+II} = \rho_I \rho_{II} \tag{1.42}$$

This is for statistically independent or uncorrelated subsystems. Here, ρ_{I+II} is the joint statistical distribution function of both subsystems and ρ_I and ρ_{II} are the distribution functions of separate subsystems. Similarly, we may generalize for the totality of a higher number of subsystems.

1.9 RELATIVE FLUCTUATION OF AN ADDITIVE MACROSCOPIC PARAMETER

Based on the factoring of the distribution function, one can consider measurements of observables more generally. We limit ourselves to the evaluation of expectation values. Consider the observables F_I and F_{II} defined in two subsystems, assumed to be weakly interacting and statistically independent. From Equations 1.13 and 1.42, it follows that the expectation value of their product is

$$
\begin{aligned}
\overline{F_I F_{II}} &= \int F_I(q_1, p_1) F_{II}(q_{II}, p_{II}) \rho_{I+II} dq_1 dp_1 dq_{II} dp_{II} \\
&= \int F_I(q_1, p_1) \rho_I(q_1, p_1) F_{II}(q_{II}, p_{II}) \rho_{II}(q_{II}, p_{II}) dq_1 dp_1 dq_{II} dp_{II} \\
&= \overline{F_I} \cdot \overline{F_{II}}
\end{aligned}
\tag{1.43}
$$

Thus

$$\overline{F_I F_{II}} = \overline{F_I} \cdot \overline{F_{II}} \tag{1.44}$$

This implies that the average of a product is equal to the products of the averages of these observables F_I and F_{II} that are statistically independent. If F_I and F_{II} are statistically not independent, then Equation 1.44 is generally not true. Similarly, it may be shown that

$$\overline{F_I + F_{II}} = \overline{F_I} + \overline{F_{II}} \tag{1.45}$$

Furthermore, if $C = $ const, then it follows that

$$\overline{CF} = C\overline{F} \tag{1.46}$$

Let us examine the value F related to some macroscopic body. This value within a certain time interval t fluctuates around its expectation value $\overline{F}$. Recalling that the expectation value represents the long-time average in an experiment, a short time observation of F would not necessarily give $\overline{F}$. We can investigate what could be expected to happen in experiments that measure F. There could be a certain spread in instantaneous measurements of F, especially if the system is small (few degrees of freedom) or the measurement time is short.

Consider the dispersion $\overline{(\Delta F)^n}$ of F or second moment of F, where $\Delta F = F - \overline{F}$ is the deviation of F from its mean. The value $\overline{(\Delta F)^n}$ is never negative since $(\Delta F)^2 \geq 0$ and each term in the sum makes a nonnegative contribution. If $F = \overline{F}$, there is no fluctuation at all in F, and the dispersion vanishes. The larger the spread of the values of F about its expectation value $\overline{F}$, the larger the dispersion. It follows that the dispersion $\overline{(\Delta F)^n}$ measures the degree of scatter of values of the variable about its expectation value $\overline{F}$. It is easy to show that the dispersion is a difference of the mean of F^2 and the square of the mean of F:

$$\overline{(\Delta F)^2} \equiv \overline{(F - \overline{F})^2} = \overline{(F^2 - 2F\overline{F} + \overline{F}^2)} = \overline{F^2} - 2\overline{F}\,\overline{F} + \overline{F}^2$$
$$= \overline{F^2} - \overline{F}^2 \geq 0 \tag{1.47}$$

and consequently

$$\overline{F^2} \geq \overline{F}^2 \tag{1.48}$$

In most practical applications, F does fluctuate about the mean, and then this is really an inequality. As in any experimental measurements, it is convenient to express the deviation from the mean in terms of the root-mean-square (RMS) deviation of F, also known as the standard deviation, defined as

$$(\Delta F)_{rms} = \sqrt{\overline{(\Delta F)^2}} \equiv \sqrt{\overline{(F - \overline{F})^2}} \tag{1.49}$$

The ratio

$$\frac{(\Delta F)_{rms}}{\overline{F}} = \frac{\sqrt{\overline{(\Delta F)^2}}}{\overline{F}} \tag{1.50}$$

is called the relative fluctuation of F.

It should be noted that knowledge of $\rho(q, p)$ implies complete information about the actual distribution of values of the parameter F. Knowledge of quantities like $\overline{F}$ and $\overline{(\Delta F)^n}$ implies only partial knowledge of the characteristics of the given distribution although this knowledge is useful. If we consider Equations 1.47 through 1.50, then in order to evaluate the relative fluctuation, one has to evaluate $\overline{F}$ and $\overline{F^2}$.

We may define higher moments such as $\overline{(\Delta F)^n}$, called the nth moment of F about its expectation value, for integers $n > 2$. They are, however, less commonly used.

We can show that the relative fluctuation of a physical observable decreases with an increase in the number of particles in the bodies in question. It should be noted that most parameters of physical interest are additive. That is a result of the quasi-closeness of the separate parts of the given body. For example, the internal energy of the separate parts is usually greater than their energy of interaction. Consequently, the energy of the entire body may be considered equal to the sum of the energy of its parts, to a very good approximation.

We have been evaluating expectation values for very general situations using very simple and transparent methods. We consider now a general case. Let F be an additive parameter, meaning, it is composed from different parts of the system. Let the given body be partitioned into N equal parts so that

$$F = \sum_{i=1}^{N} F_i \qquad (1.51)$$

where the parameter F_i is related to each part. Thus

$$\overline{F} = \overline{\sum_{i=1}^{N} F_i} = \sum_{i=1}^{N} \overline{F_i} \qquad (1.52)$$

Here, we use property (1.45). Since ρ is the same and independent of i, each expectation value $\overline{F_i}$ is the same. Then, Equation 1.52 is simply the sum of N equal terms:

$$\overline{F} = N\overline{F_1} \qquad (1.53)$$

It is obvious that when the number of partitions increases, $\overline{F}$ increases proportionately to N (but only if $\overline{F_1}$ does not change). Let us determine the dispersion of F:

$$\overline{(\Delta F)^2} = \overline{\left(\sum_i \Delta F_i \right)\left(\sum_j \Delta F_j \right)} \qquad (1.54)$$

Considering the statistical independence of the different parts of the given body, we have

$$\overline{\Delta F_i \Delta F_j} = \overline{\Delta F_i}\ \overline{\Delta F_j} = 0, \quad i \neq j \qquad (1.55)$$

because

$$\overline{\Delta F_j} = \overline{(F_j - \overline{F_j})} = \overline{F_j} - \overline{\overline{F_j}} = \overline{F_j} - \overline{F_j} = 0 \qquad (1.56)$$

Consequently

$$\overline{(\Delta F)^2} = \overline{\left[\sum_j \left(F_j - \overline{F_j}\right)\right]\left[\sum_{j'} \left(F_{j'} - \overline{F_{j'}}\right)\right]} = \sum_j \overline{\left(F_j - \overline{F_j}\right)^2}$$

$$+ \sum_j \sum_{j' \neq j} \overline{\left(F_j - \overline{F_j}\right)\left(F_{j'} - \overline{F_{j'}}\right)}$$

(1.57)

Here, the first term on the right-hand side (RHS) represents all square terms and the second one represents all cross terms that originate from multiplication of the sum by itself. The second term is zero, however, because subsystems j and j' are statistically independent (see previous equations). This leaves only the square terms, which give

$$\overline{(\Delta F)^2} = \sum_{j=1}^{N} \overline{\left(F_j - \overline{F_j}\right)^2} = \sum_{j=1}^{N} \overline{(\Delta F_j)^2}$$

(1.58)

Since ρ is the same for each subsystem and independent of j, then $\overline{(\Delta F_j)^2}$ is the same for each subsystem. Then the dispersion consists merely of N equal terms:

$$\overline{(\Delta F)^2} = N \overline{(\Delta F_1)^2}$$

(1.59)

The dispersion is a measure of the squared width of the distribution of F about its expectation value.

The quantity

$$(\Delta F)_{\mathrm{rms}} = \sqrt{\overline{(\Delta F)^2}}$$

(1.60)

that is, the RMS deviation from the expectation value, provides a direct measure of the width of the distribution about the expectation value $\overline{F}$. It follows that as N increases, the dispersion $\overline{(\Delta F)^2}$ also increases proportionately to N, provided the subsystems are held fixed in size. The RMS width $(\Delta F)_{\mathrm{rms}}$ of the distribution increases proportionately to $\sqrt{N}$. Thus, the relative RMS width decreases as $1/\sqrt{N}$:

$$\frac{\sqrt{\overline{(\Delta F)^2}}}{\overline{F}} = \frac{(\Delta F)_{\mathrm{rms}}}{\overline{F}} \propto \frac{1}{\sqrt{N}}$$

(1.61)

The relative deviation of the distribution of values of F about its expectation value $\overline{F}$ becomes increasingly negligible as N becomes larger. This is a characteristic feature of statistical distributions that involve large N.

1.9.1 QUESTIONS AND ANSWERS

Q1.1 Consider a particle of mass m that moves in a one-dimensional parabolic potential well characterized by a spring constant K. Find the probability that for trial measurements of its position it will be found within the interval $[x, x + dx]$.

A1.1 The parabolic potential is selected in the form

$$U(x) = \frac{Kx^2}{2} \tag{1.62}$$

and the Hamiltonian of the particle in the form

$$H(x, p_x) = \frac{p_x^2}{2m} + \frac{Kx^2}{2} \tag{1.63}$$

The corresponding equation of motion is

$$\ddot{x} + \omega^2 x = 0, \quad \omega^2 = \frac{K}{m} \tag{1.64}$$

This is an equation of an oscillator that has the solution

$$x = a\sin(\omega t + \phi) \tag{1.65}$$

If the initial phase $\phi = 0$, then

$$x = a\sin \omega t \tag{1.66}$$

where a is the amplitude of oscillations.

As the particle executes a harmonic oscillation, then the period

$$T = \frac{2\pi}{\omega} \tag{1.67}$$

is taken as the total time of observation. If dt is the time during which the mass is found in the interval $[x, x + dx]$ moving toward positive x, then the required probability will be

$$dW = \frac{2dt}{T} \tag{1.68}$$

The factor of 2 is due to the fact that the mass moves within the given interval twice during one period T.

From Equation 1.66, it follows that

$$dt = \frac{1}{a\omega} \frac{dx}{\cos\omega t} = \frac{dx}{\omega\sqrt{a^2 - x^2}} \tag{1.69}$$

and

$$dW = \begin{cases} \dfrac{dx}{\pi\sqrt{a^2 - x^2}}, & |x| \leq a \\ 0, & \text{otherwise} \end{cases} \tag{1.70}$$

The probability density $\rho(x) = dW/dx$ is minimum at the point $x = 0$, where the velocity of the particle is maximum. The probability density dW/dx achieves its maximum as $|x| \to a$, where the velocity of the particle is minimum.

Q1.2 An ideal gas in equilibrium has N molecules enclosed in a container of volume V. Find the probability that n molecules are located in a subvolume v of this container. Each molecule is equally likely to be located anywhere within the container. What is the mean number $\bar{n}$ of molecules located within the volume v? Find the dispersion $(n - \bar{n})^2$ in the number of molecules located within v. Show that as $\lim_{V \to \infty}(v/V) \to 0$ (for $N/V = \text{const}$) the distribution in the number of molecules is approximated by a Poisson distribution. Find the asymptotic expression of the Poisson formula for the case when the mean number of molecules in the selected volume is exceedingly high, such that $n \approx \bar{n} \gg 1$ and $\Delta n = n - \bar{n} \ll n$.

A1.2 We assume that in the state of thermal equilibrium, the probability that we find a certain molecule within the volume v is equal to $p = v/V$. The probability that we find it within the volume $V - v$ is obviously equal to $1 - p$. Then the probability that within the volume v we find n molecules and the remaining molecules within the volume $V - v$ is $p^n(1 - p)^{N-n}$. As the molecules in the volume v are identical, we multiply the probability by the number of ways in which n arbitrary molecules may be selected from the total number of N molecules, that is, by the number of combinations of n elements from N.

The total probability that we find n arbitrarily selected molecules within the volume v is

$$W_N(n) = \binom{N}{n} p^n(1 - p)^{N-n} \tag{1.71}$$

Here

$$\binom{N}{n} \equiv \frac{N!}{n!(N - n)!} \tag{1.72}$$

gives the number of arrangements of N molecules taken n at a time (combination).

We have useful identities like

$$\binom{N}{n} + \binom{N}{n+1} = \binom{N+1}{n+1}, \quad \binom{N}{n} = \binom{N}{N-n} \tag{1.73}$$

The distribution (1.71) is called the binomial law. We evaluate the mean $\bar{n}$:

$$\bar{n} = \sum_{n=1}^{N} n W_N(n) = Np \sum_{n=1}^{N} \binom{N-1}{n-1} p^{n-1}(1-p)^{N-1-(n-1)} \tag{1.74}$$

We can see that in Equation 1.74, the sum may be reduced to a binomial law by letting $n - 1 \equiv l$. This results in

$$\bar{n} = Np \sum_{l=0}^{N-1} \binom{N-1}{l} p^l (1-p)^{N-1-l} = Np \left[p + (1-p) \right]^{N-1} = Np \tag{1.75}$$

The dispersion may be found from the relation

$$\overline{(\Delta n)^2} = \overline{(n - \bar{n})^2} = \overline{n^2} - \bar{n}^2 = \overline{n(n-1)} + \bar{n} - \bar{n}^2 \tag{1.76}$$

Then, similar to Equation 1.74, the mean value $\overline{n(n-1)}$ may be found as follows:

$$\overline{n(n-1)} = \sum_{n=2}^{N} n(n-1) W_n(n) = N(N-1)p^2 \sum_{n=2}^{N} \binom{N-2}{n-2} p^{n-2}(1-p)$$

$$= N(N-1)p^2 \left[p + (1-p) \right]^{N-2} = N(N-1)p^2 \tag{1.77}$$

In the limit $p = v/V \rightarrow 0$, as $V \rightarrow \infty$, with $n/N \ll 1$ or with N exceedingly large, formula (1.71) may be written in the form

$$W_N(n) = \frac{N!}{n!(N-n)!} \left(\frac{\bar{n}}{N}\right)^n \left(1 - \frac{\bar{n}}{N}\right)^{N-n} = \frac{\bar{n}^n}{n!} \left(1 - \frac{\bar{n}}{N}\right)^N \frac{N!}{(N-n)!\left(1 - \frac{\bar{n}}{N}\right)^N N^n}$$

$$= \frac{\bar{n}^n}{n!} \left(1 - \frac{\bar{n}}{N}\right)^N \frac{N(N-1)(N-2) \cdots (N-n+1)}{N^n \left(1 - \frac{\bar{n}}{N}\right)^n}$$

$$\tag{1.78}$$

Since N is exceedingly large, the quantity $(1 - \bar{n}/N)^N$ to a good approximation follows the limit

$$\lim_{N\to\infty}\left(1-\frac{\bar{n}}{N}\right)^{N} = \exp\{-\bar{n}\}, \quad \frac{n}{N} \ll 1 \tag{1.79}$$

Then, the n factors

$$1, 1-\frac{1}{N}, 1-\frac{2}{N}, \cdots \tag{1.80}$$

cancel out n factors of

$$\left(1-\frac{\bar{n}}{N}\right) \tag{1.81}$$

Thus

$$W(n) = \frac{\bar{n}^{n}}{n!}\exp\{-\bar{n}\} \tag{1.82}$$

Expression 1.82 is called the Poisson distribution. It is the (asymptotic) probability of finding n outcomes (molecules inside volume v) after many trials.

Let us find the asymptotic expression for the Poisson formula for the case when the mean number of molecules within the selected volume v is large, such that $n \approx \bar{n} \gg 1$ and $\Delta n = n - \bar{n} \ll \bar{n}$. Using the Stirling formula for very large n:

$$\ln n! \cong n\ln n - n \tag{1.83}$$

From Equation 1.82, we get

$$\ln W(n) \cong n\ln\bar{n} - n\ln n + n - \bar{n} = -(\bar{n}+\Delta n)\ln\left(1+\frac{\Delta n}{\bar{n}}\right) + \Delta n \tag{1.84}$$

Using the expansion

$$\ln\left(1+\frac{\Delta n}{\bar{n}}\right) = \frac{\Delta n}{\bar{n}} - \frac{1}{2}\left(\frac{\Delta n}{\bar{n}}\right)^{2} + \cdots \tag{1.85}$$

we obtain

$$W(n) = C\exp\left\{-\frac{1}{2}\frac{(n-\bar{n})^{2}}{\bar{n}}\right\} \tag{1.86}$$

The normalization condition

$$\int_{-\infty}^{\infty} W(n)\, dn = 1 \tag{1.87}$$

results in

$$C = \frac{1}{\sqrt{2\pi\,\bar{n}}} \tag{1.88}$$

so that

$$W(n) = \frac{1}{\sqrt{2\pi\,\bar{n}}} \exp\left\{ -\frac{1}{2} \frac{(n - \bar{n})^2}{\bar{n}} \right\} \tag{1.89}$$

This result is called the Gaussian distribution after the German mathematician Carl Friedrich Gauss, who discovered it while investigating the distribution of errors in measurements. Considering Equation 1.77, the dispersion of the distribution is

$$\overline{(\Delta n)^2} = Np = \bar{n} \tag{1.90}$$

Q1.3 Suppose that a release of electrons from the surface of a body during thermo-electric emission is a statistically independent event and that the probability of the release of one electron during an infinitesimal time interval dt is equal to $W_1 = \lambda dt$, where λ is a constant. Evaluate the probability of a release of n electrons during time t. Also find $\bar{n}$ and the dispersion $\overline{(\Delta n)^2}$.

A1.3 Let us first find the differential equation for the function $W_n(t)$, which is the probability that n electrons have been released during a macroscopic time interval t. The release of the electrons from the surface of the body is statistically independent. Then the probability of release of n electrons during the time $t + dt$ is

$$W_n(t + dt) = W_n(t)(1 - W_1) + W_{n-1}(t)W_1, \quad n \neq 0 \tag{1.91}$$

where the first term on the RHS corresponds to having had n electrons released up to time t, and no extra one emitted during dt, and the second term corresponds to having had $n - 1$ electrons released up to time t, with another one emitted during dt. The probability that there are still no electrons released is

$$W_0(t + dt) = W_0(t)(1 - W_1) \tag{1.92}$$

If we expand the left-hand side (LHS) of these equations in series with respect to dt, then at $dt \to 0$, we have

$$\frac{dW_n(t)}{dt} = \lambda\left[W_{n-1}(t) - W_n(t)\right], \quad n \neq 0 \tag{1.93}$$

$$\frac{dW_0(t)}{dt} = -\lambda W_0(t) \tag{1.94}$$

The differential Equation 1.94 is solved with the initial condition $W_0(0) = 1$

$$W_0(t) = \exp\{-\lambda t\} \tag{1.95}$$

If we substitute this expression into Equation 1.93 for $n = 1$, then we obtain a non-homogeneous differential equation relative to the function $W_1(t)$. The solution of this equation is found in the form:

$$W_1(t) = C_1(t)\exp\{-\lambda t\} \tag{1.96}$$

with the initial condition

$$W_0(0) = 0, \quad n \neq 0 \tag{1.97}$$

For this we obtain

$$C_1(t) = \lambda t \tag{1.98}$$

We continue the process for $n = 2, 3, \ldots$ and finally we find that

$$W_n(t) = C_n(t)\exp\{-\lambda t\} = \frac{(\lambda t)^n}{n!}\exp\{-\lambda t\} \tag{1.99}$$

that is, the desired probability is described by a Poisson distribution.
   Now we evaluate

$$\bar{n} = \sum_{n=0}^{\infty} nW_n(t) = \lambda t\exp\{-\lambda t\}\sum_{n=1}^{\infty}\frac{(\lambda t)^{n-1}}{(n-1)!} = \lambda t\exp\{-\lambda t\}\sum_{l=0}^{\infty}\frac{(\lambda t)^l}{l!} = \lambda t$$

$$\tag{1.100}$$

From here it follows that λ is the mean number of electrons that escape from the surface of the body per unit time. Further

$$\overline{n^2} = \sum_{n=0}^{\infty} n^2 W_n(t) = \lambda t\exp\{-\lambda t\}\sum_{n=1}^{\infty}\frac{n(\lambda t)^{n-1}}{(n-1)!}$$

$$= \lambda t\exp\{-\lambda t\}\left[\lambda t\sum_{n=2}^{\infty}\frac{(\lambda t)^{n-2}}{(n-2)!} + \sum_{n=1}^{\infty}\frac{(\lambda t)^{n-1}}{(n-1)!}\right] = (\lambda t)^2 + \lambda t \tag{1.101}$$

Thus the dispersion can be found:

$$\overline{(\Delta n)^2} = \overline{n^2} - \overline{n}^2 = \lambda t \tag{1.102}$$

1.10   LIOUVILLE THEOREM

Before stating the hypothesis that determines the density function, we must establish sufficient conditions that are imposed on stationary ensembles by the equations of motion. These can be provided by Liouville's theorem where we show that the density function $\rho(\ldots, q_i, \ldots p_i, \ldots)$ is an integral of motion. Let us prove the Liouville theorem: If motion of a system obeys the laws of mechanics, the phase volume occupied by the ensemble is invariant (it does not change with time).

Let us consider an isolated system specified by s generalized coordinates $q_1, \ldots, q_s$ and s generalized momenta $p_1, \ldots, p_s$. A set of points in phase space represents the ensemble for the system at the moment $t = 0$ in phase volume Γ_0.

At a time moment t, these points move according to the equations of motion and occupy a volume Γ_t (see Figure 1.3). We show that $\Gamma_t = \Gamma_0$.

For convenience, let us denote the ensemble of the coordinates $q_1, \ldots, q_s$ and momenta $p_1, \ldots, p_s$ by $Q_1, \ldots, Q_{2s}$. Let $Q_1^0, \ldots, Q_{2s}^0$ be the variables in the volume Γ_0:

$$\Gamma_0 = \int dQ_1^0 \ldots dQ_{2s}^0, \quad \Gamma_t = \int dQ_1 \ldots dQ_{2s} \tag{1.103}$$

The phase points for the ensemble members are in motion; their coordinates and momenta change with time. The motion of the phase points can be represented as

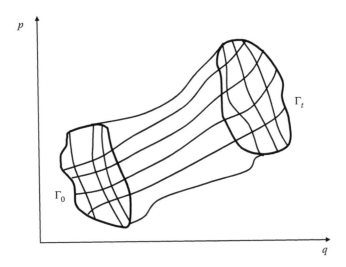

FIGURE 1.3   Phase points initially occupying phase volume Γ_0 at time $t = 0$ that move so as to occupy a phase volume Γ_t at an arbitrary time t.

$$Q_i = Q_i(Q_1^0, \ldots, Q_{2s}^0, t), \quad i = 1, 2, \ldots, 2s \tag{1.104}$$

Here, $Q_1^0, \ldots, Q_{2s}^0$ are constants of integration and have the sense of the coordinates and momenta at the initial moment of time $t = 0$.

If in Equation 1.103 we carry out the change of variables for the quantities $Q_1^0, \ldots, Q_{2s}^0$ in Equation 1.104, then

$$\Gamma_t = \int |J| \, dQ_1^0 \ldots dQ_{2s}^0 \tag{1.105}$$

where J is the Jacobian, defined as the determinant of the Jacobian matrix,

$$[J] = \frac{\partial(Q_1, \ldots, Q_{2s})}{\partial(Q_1^0, \ldots, Q_{2s}^0)}, \quad \frac{\partial Q_i}{\partial Q_k^0} \equiv J_{ik}, \quad i, k = 1, 2, \ldots, 2s \tag{1.106}$$

It should be noted that the transformation Jacobian J is a function of time t. It is equal to unity when $t = 0$, since $Q_i = Q_i^0$. We show that in general $(\partial J/\partial t) = 0$ and thus the Jacobian is always equal to unity. In this way, Γ_t is transformed into Γ_0 and the theorem will be proven.

We consider $J_{ik} = \partial Q_i / \partial Q_k^0$ to be the elements of the Jacobian matrix in Equation 1.106 and a_{ik} its minors. Then

$$\sum_{k=1}^{2s} a_{ik} J_{lk} = J\delta_{il} \tag{1.107}$$

Differentiating the determinant

$$\frac{dJ}{dt} = \sum_{i,k=1}^{2s} \frac{\partial J}{\partial J_{ik}} \frac{dJ_{ik}}{dt} = \sum_{i,k=1}^{2s} a_{ik} \frac{dJ_{ik}}{dt} = \sum_{i,k=1}^{2s} a_{ik} \frac{d}{dt} \frac{\partial Q_i}{\partial Q_k^0} = \sum_{i,k=1}^{2s} a_{ik} \frac{\partial \dot{Q}_i}{\partial Q_k^0}$$

$$= \sum_{i,k,l=1}^{2s} a_{ik} \frac{\partial \dot{Q}_i}{\partial Q_l} \frac{\partial Q_l}{\partial Q_k^0} = \sum_{i,k,l=1}^{2s} a_{ik} J_{lk} \frac{\partial \dot{Q}_i}{\partial Q_l} = \sum_{i,l=1}^{2s} J \frac{\partial \dot{Q}_i}{\partial Q_l} \delta_{il} = J \sum_{i=1}^{2s} \frac{\partial \dot{Q}_i}{\partial Q_i} \tag{1.108}$$

If the Hamiltonian of the system is denoted by $H = H(q, p)$, then the motion of a phase point is along the phase trajectory and is determined by the canonical equations of motion

$$\dot{q}_i = \frac{\partial H(q, p)}{\partial p_i}, \quad \dot{p}_i = -\frac{\partial H(q, p)}{\partial q_i} \tag{1.109}$$

It should be noted that as $H(q, p) = E = \text{const}$, the phase trajectory must lie on a surface of constant energy (ergodic surface). Equation 1.109 allows one to write Equation 1.108 as follows:

$$\frac{dJ}{dt} = J\sum_{i=1}^{2s} \frac{\partial \dot{Q}_i}{\partial Q_i} = J\sum_{i=1}^{s}\left[\frac{\partial \dot{q}_i}{\partial q_i} + \frac{\partial \dot{p}_i}{\partial p_i}\right] = J\sum_{i=1}^{s}\left[\frac{\partial^2 H}{\partial q_i \partial p_i} - \frac{\partial^2 H}{\partial p_i \partial q_i}\right] = 0 \quad (1.110)$$

It follows that $J = \text{const} = 1$ and $\Gamma_t = \Gamma_0$. Thus, the theorem is proven. It is seen that the volume of phase space occupied by an ensemble of systems is constant. It only changes its shape.

Let us represent the set of possible microstates of a system by a continuous set of phase points. Each phase point can move by itself along its own phase trajectory. The overall picture of this movement may be best described by the so-called distribution function $\rho(q, p, t)$, already introduced earlier, but considered as a function of time. We define this function such that at any time moment t, the number of representative points in the volume element $d\Gamma$ around the point (q, p) of phase space is given by the product $\rho(q, p, t)d\Gamma$. The distribution function $\rho(q, p, t)d\Gamma$ here symbolizes the manner in which the members of the ensemble are distributed over various possible microstates at various time moments.

Let $d\Gamma_0$ and $d\Gamma_t$ be the elements of phase space at the moments t_0 and an arbitrary moment t, respectively. Then, considering the Liouville theorem, the number of points $dN_0 = \rho^0 d\Gamma_0$ in the phase volume $d\Gamma_0$ is equal to the number of points $\rho_t d\Gamma_t$ in the phase volume $d\Gamma_t$. It follows from the Liouville theorem that

$$\rho_t^0 = \rho_{t+\tau} \quad (1.111)$$

Then

$$\rho^0(q(t), p(t)) = \rho(q(t + \tau), p(t + \tau)) \quad (1.112)$$

and for a small time displacement τ,

$$\rho^0(q(t), p(t)) = \rho^0(q(t), p(t)) + \frac{\partial \rho}{\partial \tau}\tau + \sum_{i=1}^{s}\left[\frac{\partial \rho}{\partial q_i}\dot{q}_i + \frac{\partial \rho}{\partial p_i}\dot{p}_i\right]\tau$$

$$\equiv \rho^0(q(t), p(t)) + \frac{d\rho}{dt}\tau \quad (1.113)$$

We limit ourselves only to the first order of τ because we consider τ as an infinitesimal time interval. The operator

$$\frac{d}{dt} \equiv \frac{\partial}{\partial t} + \sum_{i=1}^{s}\left[\dot{q}_i \frac{\partial}{\partial q_i} + \dot{p}_i \frac{\partial}{\partial p_i}\right] \quad (1.114)$$

obtained from Equation 1.113 is called the convective time derivative.

Thus, by virtue of the equations of motion in Equation 1.109, Equation 1.113 now is simplified as follows:

$$\frac{d\rho}{dt} = \frac{\partial \rho}{\partial t} + \sum_{i=1}^{s} \left[\frac{\partial \rho}{\partial q_i} \frac{\partial H}{\partial p_i} - \frac{\partial \rho}{\partial p_i} \frac{\partial H}{\partial q_i} \right] = 0 \qquad (1.115)$$

This measures the rate of change of ρ if we move along the phase trajectory of the system. Relation (1.115) is also known as the Liouville theorem (the Liouville equation is the foundation on which statistical physics rests). Relation (1.115) shows explicitly how the value of the distribution function at any fixed point (q, p) changes with time t. It should be noted that, in particular, the masses and forces associated with molecules of the system determine $H(q, p)$. Thus, $\partial H/\partial q_i$ and $\partial H/\partial p_i$, and the initially chosen ρ at t_0 determine $\partial H/\partial q_i$ and $\partial H/\partial p_i$ at t. It follows that Equation 1.115 may be integrated, in principle, starting at t_0. This yields the value of ρ at the point (q, p) and time moment t, given $H(q, p)$ and the form $H(q, p)$ of ρ at t_0.

Considering the second summand of Equation 1.115 to be the Poisson bracket

$$\{\rho, H\} = \sum_{i=1}^{s} \left[\frac{\partial \rho}{\partial q_i} \frac{\partial H}{\partial p_i} - \frac{\partial \rho}{\partial p_i} \frac{\partial H}{\partial q_i} \right] \qquad (1.116)$$

then we may write Equation 1.115 as an alternative form of the convective time derivative

$$\frac{d\rho}{dt} \equiv \frac{\partial \rho}{\partial t} + \{\rho, H\} \qquad (1.117)$$

and here $d\rho/dt = 0$. Considering Equation 1.116 we may write the Poisson operator as follows:

$$\hat{H}^s = \sum_{i=1}^{s} \left[\frac{\partial H}{\partial p_i} \frac{\partial}{\partial q_i} - \frac{\partial H}{\partial q_i} \frac{\partial}{\partial p_i} \right] = \sum_{i=1}^{s} \left[\dot{q}_i \frac{\partial}{\partial q_i} + \dot{p}_i \frac{\partial}{\partial p_i} \right] = \{\dots, H\} \qquad (1.118)$$

This may be used to define the Liouville operator as follows:

$$\hat{L}^s \equiv -i\hat{H}^s \qquad (1.119)$$

The i is placed here as a matter of convention and has the effect of making $\hat{L}^s$ a Hermitian operator. It follows from above that the probability function ρ obeys a dynamics determined by the Liouville operator:

$$i \frac{\partial \rho}{\partial t} = \hat{L}^s \rho \qquad (1.120)$$

This may be formally solved to give

$$\rho(q,p,t) = \exp\{-i\hat{L}^s t\}\rho\{q,p,0\} \tag{1.121}$$

The operator $\exp\{-i\hat{L}^s t\}$ is defined by the Taylor expansion as follows:

$$\exp\{-i\hat{L}^s t\}\rho(q,p,0) \equiv \rho(q,p,0) + t\{H,\rho(q,p,0)\} + \frac{1}{2!}t^2\{H,\{H,\rho(q,p,0)\}\} + \cdots$$

$$\tag{1.122}$$

Let us suppose that during some interval of time near time t the probability function ρ remains constant in time. This implies that the systems are uniformly distributed over all of phase space, or, more generally we suppose that at equilibrium, ρ is a function of the energy E of the system. This energy is of course a constant of motion. Since E does not depend on time explicitly, then the system with s degrees of freedom has $2s - 1$ integrals of motion that are independent of time. Consequently, ρ should be a function of these integrals of motion, or in particular, could be a function of one of those integrals of motion. Further, if we consider that ρ is a single-valued and continuous function, then it depends only on the "uniform" integrals of motion. These "uniform" integrals of motion should be single-valued and continuous functions of q and p. Such integrals of motion are usually few. They may be energy and perhaps certain momenta or angular momenta. It should be noted that in most isolated closed systems, the energy is usually the only uniform integral of motion. It follows from the equation that, in equilibrium, the distribution of systems over states remains unchanged in time. Thus, $\rho = $ const and $\partial\rho/\partial t = 0$ or $\{\rho, H\} = 0$. In the neighborhood of a phase point following its trajectory, the probability function ρ remains constant (the distribution function is an integral of motion). That is to say, the "probability fluid" occupying phase space is incompressible. The relation $\partial\rho/\partial t = 0$ is also the condition for a stationary ensemble. It is obvious that for such an ensemble, the average value $\bar{F}$ of any physical quantity $F(q, p)$ is independent of time. Therefore, a stationary ensemble qualifies to represent a system in equilibrium.

We shall therefore apply the postulate formulated by Gibbs that the density function ρ is dependent only on the energy E in an equilibrium ensemble:

$$\rho = \rho(E) \tag{1.123}$$

To proceed, we will examine two types of the function (1.123) (microcanonical and canonical distribution functions) playing very important roles in statistical physics.

1.10.1 QUESTIONS AND ANSWERS

Q1.4 A particle of mass m executes a one-dimensional motion within the interval $0 \leq x \leq l$ and elastically bounces off walls at $x = 0$ and $x = l$. Draw the phase path of the particle. Evaluate the volume of phase space $\Gamma_0(E)$ corresponding to the energy less than E and show that the quantity $\Gamma_0(E)$ remains invariant for a slow motion of the wall $x = l$ (adiabatic invariance).

A1.4 As the momentum of the particle is conserved within the interval $0 \le x \le l$, the phase path has the form illustrated in Figure 1.4. The momentum reverses each time the particle collides elastically with a wall.

The energy of the particle is equal to $E = p^2/2m$, where p is the momentum of the particle. The energy E increases monotonically as the momentum p increases. Consequently

$$\Gamma_0(E) = l \times 2|p| = 2l\sqrt{2mE} \tag{1.124}$$

If the wall $x = l$ moves with the velocity v, then, as the particle bounces off the wall elastically, it simply reverses its velocity in a frame moving with the wall. The original relative velocity is $(p/m) - v$, and after the collision it is $(-p'/m) - v$, where p and p' are the magnitudes of initial and final momentum. These relative velocities then satisfy

$$\frac{p}{m} - v = \frac{p'}{m} + v \tag{1.125}$$

Consequently, for each collision, the magnitude of the particle's momentum changes from p to $p' = p - 2mv$. During a time interval δt, the position of the wall changes by $\delta l = v\,\delta t$. The time between subsequent collisions with the wall at $x = l$ is $t_{coll} = 2l/v$. During δt, the particle experiences $\delta t/t_{coll} = (\delta t/2l/(p/m)) = p\,\delta l/2mvl$ collisions and its momentum changes by

$$\delta p = (p' - p)\frac{p\delta l}{2mvl} = (-2mv)\frac{p\delta l}{2mvl} = -p\frac{\delta l}{l} \tag{1.126}$$

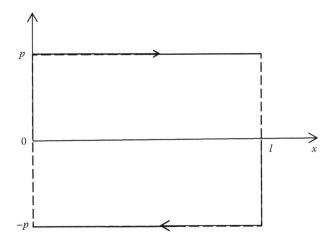

FIGURE 1.4 Phase path of the momentum of a particle constrained within the interval $0 \le x \le l$.

From here

$$\delta\Gamma_0(E) = \delta(2lp) = 2(l + \delta l)(p + \delta p) - 2lp = 2(l\,\delta p + p\,\delta l) = 0 \quad (1.127)$$

Q1.5 Construct the phase path of a linear oscillator of mass m if the coefficient of the elastic force is equal to K. Evaluate the volume of phase space $\Gamma_0(E)$ corresponding to the energy less than E.

A1.5 For a motion influenced by an elastic force, the total energy of the particle is conserved with time, that is, $E = $ const. From here it follows that

$$\frac{p_x^2}{2mE} + \frac{m\omega_0^2 x^2}{2E} = 1 \qquad (1.128)$$

where $\omega_0 = \sqrt{K/m}$ is the angular frequency of the oscillator. Thus, the phase path of a linear oscillator represents an ellipse with semimajor axes

$$a = \sqrt{\frac{2E}{m\omega_0^2}}, \quad b = \sqrt{2mE} \qquad (1.129)$$

The volume of the phase space corresponding to the energy not larger than E maps onto the area of an ellipse and is equal to

$$\Gamma_0(E) = \oint p_x dx = \pi\sqrt{\frac{2E}{m\omega_0^2}}\sqrt{2mE} = TE \qquad (1.130)$$

where $T = 2\pi/\omega_0$ is the period of the oscillator.

Relation (1.130) played a very important role during the creation of the quantum theory. It should be noted that it does not contain the mass m. In principle, with the help of the canonical transformation

$$p_x = \sqrt{m}\,p, \quad x = \frac{Q}{\sqrt{m}}, \quad (p_x\delta x - p\delta Q) = 0 \qquad (1.131)$$

we can obtain the Hamiltonian function $H(p, Q)$ that does not contain the mass m in explicit form. Although the form of the energy surface for such a canonical transformation changes, it retains its elliptical form and the volume $\Gamma_0(E)$ does not change.

Q1.6 Find the phase paths and evaluate the change with time of the phase volume for

   a. A free particle subjected to a frictional force that is proportional to the velocity
   b. A linear harmonic oscillator with a small frictional force

A1.6 a. From the equation of motion

$$m\ddot{x} = -\alpha\dot{x} \tag{1.132}$$

where α is the coefficient due to friction, we have

$$\dot{x} = v_{x_0}\exp\left\{-\frac{t}{\tau}\right\}, \quad \tau = \frac{m}{\alpha}, \quad x = v_{x_0}\int_0^t \exp\left\{-\frac{t}{\tau}\right\}dt + x_0$$

$$= x_0 + v_{x_0}\ \tau\left(1 - \exp\left\{-\frac{t}{\tau}\right\}\right) \tag{1.133}$$

The momentum is

$$p_x = mv_{x_0}\exp\left(-\frac{t}{\tau}\right) = mv_{x_0} - \alpha(x - x_0) \tag{1.134}$$

or

$$p_x - p_{x_0} = -\alpha(x - x_0) \tag{1.135}$$

Thus, the phase paths form a family of lines with the slope $-\alpha$. The time scales the momentum with the factor $\beta = \exp(-t/\tau)$. An individual phase path will terminate at the point $p_x = 0$, $x = x_0 + v_{x_0}\tau$, in the limit of infinite time. In the absence of the frictional force, $\alpha = 0$, and the phase paths would be parallel to the abscissa with $p_x = \text{const}$. The motion is not periodic, so there is no enclosed path for which to find an area. Instead, one can look at how an infinitesimal volume element $d\Gamma(t) = dx(t)\,dp_x(t)$ changes with time. Imagine the phase points within an original rectangular volume between coordinates $x_1(0)$ and $x_2(0)$ and between momenta $p_1(0)$ and $p_2(0)$. The beginning phase volume is $d\Gamma(0) = dx(0)dp_x(0)$, where $dx(0) = x_2(0) - x_1(0)$ and $dp_x(0) = p_2(0) - p_1(0)$. At a later time, both $x_1(t) = x_1(0) + v_0\tau(1 - \beta)$ and $x_2(t) = x_2(0) + v_0\tau(1 - \beta)$ have been simply shifted by the same amount, so that $dx(t) = dx(0)$. The momentum infinitesimal becomes $dp_x(t) = p_2(t) - p_1(t) = dp_x(0)$ β, that is, the volume element gets flattened only along the momentum axis by the exponential factor. Therefore, any volume element simply decays exponentially with time

$$d\Gamma(t) = dx(0)\,dp_x(0)\beta = d\Gamma(0)\exp\left(-\frac{t}{\tau}\right) \tag{1.136}$$

This can be generalized to any arbitrary phase space volume, and alternatively, using the Jacobian, one gets

$$\Gamma(t) = \iint_{(G)} dp_x dx = \iint_{(G_0)} \frac{\partial(p_x, x)}{\partial(p_{x_0}, x_0)} dp_{x_0} dx_0 = \Gamma(0)\exp\left(-\frac{t}{\tau}\right) \qquad (1.137)$$

That is, the phase volume for any set of particles influenced by this frictional force decreases exponentially with time.

b. Representing the equation for a damped harmonic oscillator

$$m\ddot{x} + \alpha\dot{x} + Kx = 0 \qquad (1.138)$$

in the form

$$\ddot{x} + \frac{1}{\tau}\dot{x} + \omega_0^2 x = 0, \quad \frac{1}{\tau} = \frac{\alpha}{m}, \quad \omega_0^2 = \frac{K}{m} \qquad (1.139)$$

we suppose that $x = \exp\{rt\}$ and we obtain for r a characteristic equation

$$r^2 + \frac{1}{\tau}r + \omega_0^2 = 0 \qquad (1.140)$$

The roots to this equation are

$$r_{1,2} = -\frac{1}{2\tau} \mp \sqrt{\frac{1}{4\tau^2} - \omega_0^2} = -\frac{1}{2\tau} \mp i\omega, \quad \omega = \sqrt{\omega_0^2 - \frac{1}{4\tau^2}} \qquad (1.141)$$

Usually, the friction is weak, with $\omega_0\tau \gg 2$, implying that ω is real. In this case

$$x = a_1 \exp\{r_1 t\} + a_2 \exp\{r_2 t\} = a\exp\left\{-\frac{t}{2\tau}\right\}\cos(\omega t + \phi) \qquad (1.142)$$

where a_1, a_2, and a are amplitudes. The motion represents a decaying oscillation. The corresponding generalized momentum is equal to

$$p_x = m\dot{x} \cong -Cm\exp\left\{-\frac{t}{2\tau}\right\}\omega\sin(\omega t + \phi) \qquad (1.143)$$

where we neglect a term due to the exponential decay, because $1/2\tau \ll \omega_0$.

Introducing an exponentially decreasing amplitude $a(t) = a_0\exp\{-(t/2\tau)\}$:

$$\left(\frac{p_x}{\omega a(t)}\right)^2 + \left(\frac{x}{a(t)}\right)^2 = 1 \qquad (1.144)$$

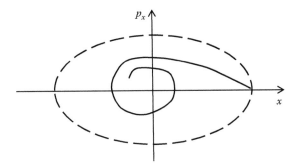

FIGURE 1.5 Elliptical spiral phase path of a linear harmonic oscillator subjected to a small frictional force.

or

$$\left(\frac{p_x}{\omega a}\right)^2 + \left(\frac{x}{a}\right)^2 = \left[\left(\frac{p_{x_0}}{\omega a_0}\right)^2 + \left(\frac{x_0}{a_0}\right)^2\right] \frac{1}{\exp\{-t/\tau\}} \qquad (1.145)$$

Thus, the phase point describes an ellipse; the semimajor axes decrease with time proportional to $\exp\{-t/2\tau\}$. In other words, the phase path represents an elliptical spiral (see Figure 1.5) with semimajor axes

$$A(t) = \omega a(t) = \omega a_0 \exp\{-t/2\tau\}, \quad B(t) = a(t) = a_0 \exp\{-t/2\tau\} \qquad (1.146)$$

As the friction is small, the phase point makes many orbits while the size of the ellipse changes slowly. So the phase volume can be approximated using the changing area of the ellipse. The change of the phase volume with time is equal to

$$\Gamma(t) = S = \pi A(t)B(t) = S_0 \exp\left\{-\frac{t}{\tau}\right\} = \Gamma(0)\exp\left\{-\frac{t}{\tau}\right\} \qquad (1.147)$$

where $S_0 = \pi\omega a_0^2$ is the initial area of the ellipse.

Q1.7 Find the phase path of a physical pendulum of mass m, length l, and moment of inertia J.

A1.7 We denote by ϕ the angle of inclination of the pendulum from its equilibrium (vertical) position. We write the kinetic energy of the pendulum in the form

$$T = \frac{J\dot{\phi}^2}{2} = \frac{p_\phi^2}{2J}, \quad p_\phi = \frac{\partial T}{\partial \dot{\phi}} = J\dot{\phi} \qquad (1.148)$$

During the motion of the pendulum, kinetic energy is periodically transformed to gravitational potential energy U, while the total energy E is conserved with time. Thus

$$T + U = \frac{p_\phi^2}{2J} + mgl(1 - \cos\phi) = E \qquad (1.149)$$

and the momentum is found to depend on ϕ according to

$$p_\phi = \pm\sqrt{2J(E - mgl) + 2Jmgl\cos\phi} \qquad (1.150)$$

The graph of $p_\phi(\phi)$ depends strongly on the sign of the minimal value of the quantity under the square root sign, $Q_{min} = 2J(E - 2mgl)$. This value corresponds to the pendulum being at its highest angular position, $\phi = \pi$ and highest potential energy $U_{max} = 2mgl$.

If $E > 2\ mgl$, then the pendulum never reaches a state of zero kinetic energy, and $Q_{min} > 0$. The positive value of p_ϕ oscillates about the value $p_0 = \sqrt{2J(E - mgl)}$, reaching the extremes $\left[\sqrt{2J(E - 2mgl)}, \sqrt{2JE}\right]$. This branch corresponds to the positive values $\dot{\phi} > 0$, that is, the pendulum rotates counterclockwise about its support, never pausing as it goes around. The phase path in this case is illustrated as the upper curve in Figure 1.6. The lower branch on this figure corresponds to $\dot{\phi} < 0$, that is, the rotation of the pendulum clockwise.

If $E = 2mgl$, the pendulum has just the minimum energy necessary to reach a vertical orientation. Then, $Q_{min} = 0$ and at the points $\phi = (2n + 1)\pi$, $n = 1, 2, \ldots$, both branches of the phase path overlap. There arises a nonunivalency as the pendulum at these positions may reverse its direction of rotation. For this case, the phase point changes from one path to another.

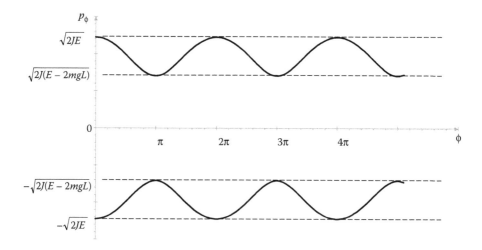

FIGURE 1.6   Variation of the momentum $p_\phi(\phi)$ of a physical pendulum with the angle ϕ.

It should be noted that technically, the pendulum would reach $\dot{\phi} = 0$ and stop at a point of unstable equilibrium. In the real world, any slight deviation from $E = 2mgl$ would make this an impossibility.

It is interesting to note that the time necessary to attain the point $\phi = \pi$ is

$$T = \int_0^\pi \frac{Jd\phi}{p_\phi} = \sqrt{\frac{J}{2mgl}} \int_0^\pi \frac{d\phi}{\sqrt{1 + \cos\phi}} = \sqrt{\frac{J}{2mgl}} K(1) = \infty \qquad (1.151)$$

where $K(x)$ is the total elliptical integral of the first kind.

It should be noted that practically speaking, there could be no true reversal of the pendulum's rotation. This case is a singular example that cannot actually take place because there are always external influences on the pendulum.

If $E < 2mgl$, with $Q_{min} < 0$, the pendulum does not have sufficient energy to rotate over the vertical position; it will reverse its rotation after reaching a maximum angle ϕ_0. The maximum angle occurs when $p_\phi = 0$, and all the energy is potential energy:

$$E = mgl(1 - \cos\phi_0) \qquad (1.152)$$

so that the dynamical angle satisfies

$$\cos\phi \geq 1 - \frac{E}{mgl} \equiv \cos\phi_0 \qquad (1.153)$$

that is, it lies on the interval $-\phi_0 \leq \phi \leq \phi_0$, where $\phi_0 = \cos^{-1}(1 - (E/mgl))$. The pendulum in this case executes a back and forth oscillatory motion. The phase path in this case is represented by an oval as in Figure 1.7. For low enough energy, which ensures $\phi \ll 1$, using $1 - \cos\phi \approx (\phi^2/2)$, the oval figure is approximately transformed into an ellipse. Thus, it would be easy to find its total phase volume as in previous examples.

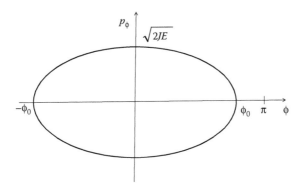

FIGURE 1.7 Phase path of a physical pendulum.

Q1.8 Define the phase path for a charged particle of mass m in motion. It has an electric charge of $-e$. This particle is subjected to the action of a Coulomb force of attraction to a fixed charge $+e_1$. The initial distance between the charges is r_0 and the initial velocity of the particle is $v_0 = 0$.

A1.8 The Coulomb field is a potential field with the force (with suppressed units)

$$F = -\frac{ee_1}{r^2} \tag{1.154}$$

Thus, the total energy of the particle is an integral of motion:

$$E = T + V = \text{const} \tag{1.155}$$

where

$$V(r) = -\int_{r_0}^{r} F(r)dr = -ee_1\left(\frac{1}{r} - \frac{1}{r_0}\right) \tag{1.156}$$

From here, considering the initial conditions (both the kinetic and potential energies start at zero)

$$E = \frac{p^2}{2m} - ee_1\left(\frac{1}{r} - \frac{1}{r_0}\right) = 0, \quad p = \pm\sqrt{2mee_1\left(\frac{1}{r} - \frac{1}{r_0}\right)} \tag{1.157}$$

Only one of the roots actually applies, which is that root that gives an attractive motion.

Q1.9 Verify the truthfulness of the Liouville theorem for the case of an elastic collision of two particles with masses m_1 and m_2 moving on the same line.

A1.9 Let us denote by q_i, p_i and q'_i, p'_i, $i = 1, 2$, the coordinates and momenta of the particles before and after the collision, respectively. From the laws of conservation of momentum and energy

$$p'_1 + p'_2 = p_1 + p_2, \quad \frac{p'^2_1}{2m_1} + \frac{p'^2_2}{2m_2} = \frac{p^2_1}{2m_1} + \frac{p^2_2}{2m_2} \tag{1.158}$$

and we find

$$p'_1 = \frac{m_1 - m_2}{m_1 + m_2}p_1 + \frac{2m_1}{m_1 + m_2}, \quad p'_2 = \frac{2m_2}{m_1 + m_2}p_1 + \frac{m_2 - m_1}{m_1 + m_2}p_2 \tag{1.159}$$

This could be obtained most readily by analyzing the collision in the frame of reference moving with the center of mass.

Now, let us evaluate the outgoing phase volume

$$\int d\Gamma' = \int \left|\frac{\partial(q_1', q_2', p_1', p_2')}{\partial(q_1, q_2, p_1, p_2)}\right| d\Gamma \tag{1.160}$$

Considering the relation $\partial q_i / \partial q_j = \delta_{ij}$, we have $q_1' = q_1, q_2' = q_2$ (at the instant of the collision, the particle positions do not change). However, a Jacobian in which the same quantities appear in the numerator and the denominator leads to a Jacobian with fewer variables. Thus

$$\frac{\partial(q_1', q_2', p_1', p_2')}{\partial(q_1, q_2, p_1, p_2)} = \frac{\partial(p_1', p_2')}{\partial(p_1, p_2)} \tag{1.161}$$

After evaluation, we find that $|J| = 1$, that is, the phase volume is constant.

Q1.10 Verify the truthfulness of the Liouville theorem for a completely inelastic collision of two spheres.

A1.10 After the completely inelastic collision, the spheres begin to move together as a whole, that is, $p_2' = p_1'$. The dimension of the phase space of the system for this reduces by half. Thus the Jacobian of the transformation is equal to zero as the minor in the determinant has two equal rows. Thus

$$d\Gamma' = |J| d\Gamma = 0 \tag{1.162}$$

and the phase volume is conserved.

1.11 GIBBS MICROCANONICAL ENSEMBLE

The most fundamental way to look at a system in equilibrium is to consider an iso-lated system at a fixed value of total energy. When considering an ensemble of sys-tems, all at the same fixed energy, their distribution in phase space will be called the microcanonical ensemble. Let us examine an isolated equilibrium state with energy $E(q, p) = E_0$ for which $\rho = \text{const}$ for a certain set of q and p. Let us examine two constant energy surfaces (see Figure 1.8) separated by an infinitesimal amount, dE.

Let us select some volume element $d\Gamma = dqdp$ and find the distribution of all sys-tems in that volume, that is, the probability is

$$dW(q, p) = \rho(q, p) \, dqdp = \rho(q, p) \frac{dqdp}{dE} dE \equiv \rho(q, p) \, d\Omega dE \tag{1.163}$$

Here

$$d\Omega(E) \equiv \frac{dqdp}{dE} = \frac{d\Gamma}{dE} \tag{1.164}$$

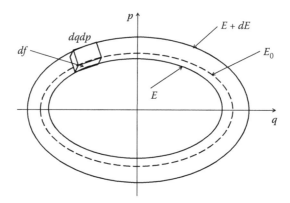

FIGURE 1.8   Constant energy surfaces in phase space.

is the element of the constant energy surface with "area" $\Omega(E)$ (see Figure 1.8):

$$\Omega(E) = \int d\Omega(E) \tag{1.165}$$

This can be considered an integral over the elements of phase space at constant energy. The product $\Omega(E)dE = \Gamma(dE)$ represented gives the amount of phase space within the desired energy interval dE.

Thus far we assumed a set of identical systems, with a small width in energy dE. From the practical viewpoint, even a set of classical systems could not be prepared all with exactly the identical energy E_0. This is because there is always some small but finite experimental error present. Also, any real experiment has uncontrolled external influences, all of which cannot be completely eliminated. Thus, it makes sense initially to include a tiny width in energy; let us call that width for the ensemble ΔE. We expect the probability distribution in phase space to be strongly peaked around some energy E_0. A particular sample distribution with these properties is the function (an example only)

$$\rho(E) = \frac{C_0}{(E - E_0)^2 + (\Delta E)^2} \tag{1.166}$$

The constant C_0 is a normalization constant, which is selected to make the total probability for all energies sum to unity:

$$\int \rho(E)dE = \int_{-\infty}^{+\infty} \frac{C_0}{(E - E_0)^2 + (\Delta E)^2} \, dE = \frac{\pi C_0}{\Delta E} = 1 \tag{1.167}$$

The energy was assumed to be unbounded, making the evaluation of the integral possible. Then the normalization constant is $C_0 = \Delta E/\pi$. Now, this makes this distribution a very interesting function. Ultimately, we wish to consider a distribution very narrow in energy, that is, $\Delta E \to 0$. Even before this limit is applied, one sees that $\rho(E)$ is dominated by energies within ΔE of E_0. In addition to this, its maximum value is

$$\rho(E_0) = \frac{C_0}{(\Delta E)^2} = \frac{1}{\pi \Delta E} \tag{1.168}$$

which diverges if the width is allowed to vanish. It means that this distribution becomes of infinite height and infinitesimal width, but with finite area equal to unity, as $\Delta E \to 0$. These are the conditions that define a Dirac delta function (i.e., delta distribution), denoted as

$$\lim_{\Delta E \to 0} \rho(E) \to \delta(E - E_0) \tag{1.169}$$

It is essentially the kind of probability distribution we are seeking, to be applied to an ensemble at fixed (or closely defined) energy. The example function used above is just one of many possible distributions that can go over to this Dirac delta function in an appropriate limit.

Now, in actual application, we need the probability distribution to be expressed in the phase variables (q, p), rather than just in the energy variable. This modifies the way in which the probability distribution is normalized.

We examine a particular case of the function $\rho = \rho(E(q, p))$ for which $\rho \neq 0$ when $E(q, p) = E_0$ and $\rho = 0$ when $E(q, p) \neq E_0$. From the geometrical point of view the equation $E(q, p) = E_0$ maps the equation of a constant energy surface in phase space. This implies that the distribution function ρ is different from zero only on the constant energy surface $E(q, p) = E_0$. Such a distribution function is called microcanonical. As $dE \to 0$, then $d\Omega = \text{const}$, $\rho = \text{const}$ and from here it follows that $\rho(q, p)$ has a delta function form:

$$\rho(q, p) = \begin{cases} \infty, & E(q, p) = E_0 \\ 0, & E(q, p) \neq E_0 \end{cases} \tag{1.170}$$

or

$$\rho(q, p) = C\delta(E(q, p) - E_0) \tag{1.171}$$

where C is a constant needed to get the correct normalization in phase space. This distribution is infinitely narrow in energy width, but with finite area, as assumed. This gives a probability

$$dW(q, p) = C\delta(E(q, p) - E_0)dqdp \tag{1.172}$$

This is the probability of the microstate with coordinate functions given in the interval $q, q + dq$ and $p, p + dp$ for the equilibrium-isolated system. The function (1.172) is called the Gibbs microcanonical distribution. Equations 1.170 and 1.172 that represent the microcanonical ensemble take into consideration that it is impossible for an actual system to be perfectly isolated. It follows that the energy can only be specified within some very small range dE. Notwithstanding this, in an idealized conservative classical mechanical system, there is no fault in letting $dE \rightarrow 0$.

For a quantum mechanical system, the situation with regard to this limit is rather different. In this case, some uncertainty dE in E will arise since the measurement of the thermodynamic property E implies interaction of the experimental system with the measuring instrument over an exceedingly long microscopic period of measurement. It follows that even if the system is prepared to have exactly the energy E at the initial moment of the start of the experiment, a small variation, dE, still needs to be introduced. It should be noted that due to the uncertainty principle, at the start of the experiment there will still be an uncertainty in the energy. The limit $dE \rightarrow 0$ may be achieved only under the physically impossible condition of perfect isolation of the experimental system for an exceeding long time prior to the start of the measurement.

Let us normalize this function over the phase space:

$$1 = \int dW(q,p) = C \int \delta(E(q,p) - E_0) \frac{dqdp}{dE} dE$$
$$= C \int \delta(E(q,p) - E_0) d\Omega(E) dE \tag{1.173}$$

If we consider $\Omega(E)$ as being a continuous function, then we have

$$1 = C \int \delta(E(q,p) - E_0) \left[\int d\Omega(E) \right] dE$$
$$= C \int \delta(E(q,p) - E_0) \Omega(E) dE = C\Omega(E_0) \tag{1.174}$$

where the last step is due to the infinitely narrow, unit-valued area of the delta function. Hence

$$dW(q,p) = \frac{1}{\Omega(E_0)} \delta(E(q,p) - E_0) dqdp \tag{1.175}$$

This is the correctly normalized microcanonical distribution, which can be used to find ensemble averages in the usual way:

$$\bar{F} = \bar{F}(E_0) = \int F(q,p) dW(q,p) = \int F(q,p) \frac{1}{\Omega(E_0)} \delta(E(q,p) - E_0) dqdp \tag{1.176}$$

From here

$$\bar{F} = \int F(q,p) dW(q,p) = \bar{F}(E_0) \tag{1.177}$$

This shows that the value of $\overline{F}$ is always a function of the selected energy, E_0.

It is obvious that $\rho(q,p)$ in the general case is dependent on $2s-1$ stationary integrals of motion, one of which is the energy E.

Now consider the meaning of the area $\Omega(E)$ needed to normalize the microcanonical probability distribution for a sample problem. Imagine a free nonrelativistic classical particle of mass m, coordinate $\vec{r}=(x,y,z)$, and momentum $\vec{p}=(p_x,p_y,p_z)$. Suppose it is constrained within a rectangular box of dimensions $L_x \times L_y \times L_z$. The free-particle energy expressed in terms of momentum magnitude, $E=p^2/2m$, leads us to

$$d\Omega = \frac{d\Gamma}{dE} = \frac{dxdydzdp_xdp_ydp_z}{dE} = \frac{dxdydzp^2dpd\Omega_{\theta,\varphi}}{pdp/m} \tag{1.178}$$

Here, we used spherical coordinates for momentum, $\vec{p}=(p,\theta,\varphi)$, with the solid angle element being $d\Omega_{\theta,\varphi}=-d(\cos\theta)d\varphi = \sin\theta\,d\theta\,d\varphi$. The integration over these remaining coordinates gives the result

$$\Omega(E) = \int dxdydzd\Omega_{\theta\varphi}mp = 4\pi L_xL_yL_zmp = 4\pi L_xL_yL_zm\sqrt{2mE} \tag{1.179}$$

Therefore, the normalized microcanonical distribution for a single free particle at fixed energy E_0 is

$$\rho(\vec{r},\vec{p}) = \frac{1}{4\pi Vm\sqrt{2mE_0}}\delta\left(\frac{p^2}{2m}-E_0\right) \tag{1.180}$$

where V is the volume of the system.

These results suggest a simpler way to get $\Omega(E_0)$. Instead of finding the differential $d\Omega$ and then integrating over $2s-1$ phase space coordinates, one can first find the volume of phase space out to energy E_0, denoted $\Gamma_0(E_0)$:

$$\Gamma_0(E_0) = \int\limits_{E(q,p)=0}^{E(q,p)=E_0} dqdp \tag{1.181}$$

and then differentiate with respect to the upper energy limit

$$\Omega(E_0) = \int \frac{d\Gamma}{dE}\bigg|_{E=E_0} = \frac{d}{dE_0}\int\limits_{E=0}^{E=E_0} d\Gamma = \frac{d}{dE_0}\Gamma_0(E_0) \tag{1.182}$$

This explains why it was useful to find the total volume of phase space out to some desired energy limit as in earlier examples.

1.12   MICROCANONICAL DISTRIBUTION IN QUANTUM MECHANICS

Here, we consider the use of the microcanonical ensemble in quantum systems. Application of phase space to quantum problems leads to an exact counting of quantum microstates, but it also has conceptual problems. Let us examine two examples:

1. A free particle. Consider the possible quantum states Ψ_n and corresponding energies E_n of a single noninteracting free particle. We consider this particle to be nonrelativistic and denote its mass by m, its position vector by $\vec{r}$, and its momentum by $\vec{p}$. Suppose this particle exists within a rectangular box of size $L_x \times L_y \times L_z$, within which it is subjected to no forces. If we neglect the effect of the boundary walls, the wave function $\Psi(\vec{r},t)$ of the particle may be represented by a plane wave at fixed energy $E = \hbar\omega$:

$$\Psi(\vec{r},t) = A\exp\{i(\vec{K}\cdot\vec{r} - \omega t)\} = \Psi(\vec{r})\exp\{-i\omega t\} \tag{1.183}$$

This wave propagates in the direction of the wave vector $\vec{K} = (K_x, K_y, K_z)$. The de Broglie relation relates the wave vector $\vec{K}$ to the momentum by $\vec{p} = \hbar\vec{K}$. Further, for a nonrelativistic particle, the energy determines the magnitude of $\vec{K}$ according to

$$E = \frac{\vec{p}^2}{2m} = \frac{\hbar^2\vec{K}^2}{2m} = \hbar\omega \tag{1.184}$$

The space wave function $\Psi(\vec{r})$ is expressed as

$$\Psi(\vec{r}) = A\exp\{i\vec{K}\cdot\vec{r}\} = A\exp\{i(K_x x + K_y y + K_z z)\} \tag{1.185}$$

To make progress, we apply "periodic boundary conditions." Although unphysical, we imagine that the wave function at one wall of the box must match its value at a point on the opposite wall of the box. If we consider the x-axis direction, then the wave function must satisfy the condition

$$\Psi(x + L_x, y, z) = \Psi(x, y, z) \tag{1.186}$$

Thus, to satisfy the above condition requires

$$K_x(x + L_x) = K_x x + 2\pi n_x \tag{1.187}$$

or

$$K_x = \frac{2\pi n_x}{L_x} \tag{1.188}$$

Here, n_x is an integer. Thus, the momentum p_x of the particle is restricted to the quantized values:

$$p_x = \hbar K_x = \frac{2\pi \hbar n_x}{L_x} \tag{1.189}$$

Let us represent this simply as

$$p_{n_x} = \frac{2\pi \hbar}{L_x} n_x \tag{1.190}$$

Then it follows that the momentum cannot be specified more precisely than the amount

$$\Delta p_x = p_{n_x+1} - p_{n_x} = \frac{2\pi \hbar}{L_x} \tag{1.191}$$

As the coordinate of the particle is spread over $\Delta x = L_x$, it follows that

$$\Delta p_x \Delta x = \frac{2\pi \hbar}{L_x} L_x = 2\pi \hbar \tag{1.192}$$

Not too surprisingly, this just reproduces the Heisenberg uncertainty principle. Obviously there is a similar relation for the other coordinate axes. For quantum statistics, however, Equation 1.192 has interesting consequences. It shows that for each degree of freedom, there is a fixed phase space volume per degree of freedom

$$\Delta \Gamma = \Delta p_x \Delta x = 2\pi \hbar \tag{1.193}$$

The full momentum of the particle in three dimensions can be expressed using an integer for each axis

$$\vec{p} = 2\pi \hbar \left(\frac{n_x}{L_x}, \frac{n_y}{L_y}, \frac{n_z}{L_z} \right) \tag{1.194}$$

Thus, for one quantum state in three dimensions, we have a constant phase space volume per state given by

$$\Delta\Gamma_\psi = (\Delta p_x \Delta x)(\Delta p_y \Delta y)(\Delta p_z \Delta z) = (2\pi\hbar)^3 \tag{1.195}$$

But now the energy is a discrete function of the quantum numbers (n_x, n_y, n_z). Thus, one must consider more carefully how to define, if at all, the distribution in phase space. Even more importantly, the way to calculate expectation values for a quantum system must be carefully reanalyzed.

2. Let us examine the phase space volume of the quantum state of a harmonic oscillator, with energy levels $E_n = (n + 1/2)\,\hbar\omega$, obtained from

$$\frac{p^2}{2m} + \frac{m\omega^2 x^2}{2} = E = \text{const} \tag{1.196}$$

This can be rearranged as

$$\frac{p^2}{2mE} + \frac{m\omega^2 x^2}{2E} = 1 \tag{1.197}$$

It follows that (x, p) forms an ellipse, with the total area

$$S = \pi ab = \pi\sqrt{2mE}\sqrt{\frac{2E}{m\omega^2}} = \pi\frac{2E_n}{\omega} \tag{1.198}$$

Here, a and b are the minor and major axes of the ellipse. Thus, the area between two ellipses corresponding to an energy difference of $\Delta E = E_{n+1} - E_n = \hbar\omega$ is (see Figure 1.9):

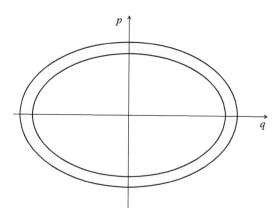

FIGURE 1.9 The area between two ellipses in phase space.

$$\Delta q \Delta p = \Delta S = S_{n+1} - S_n = \frac{2\pi}{\omega}(E_{n+1} - E_n) = \frac{2\pi}{\omega}\hbar\omega = 2\pi\hbar \qquad (1.199)$$

It follows that if a system has s degrees of freedom (e.g., a collection of independent oscillators), the total phase space volume per state is

$$\Delta\Gamma_\psi = \Delta p \Delta q = (2\pi\hbar)^s \qquad (1.200)$$

It is obvious that for an isolated equilibrium system, the probability is a *discrete* function dependent on $E(q, p)$, that is,

$$W_n = W(E_n(q, p)), \quad E = E_n(q, p) \qquad (1.201)$$

Let us reexamine the quantum phase space, in light of the above examples and the Heisenberg uncertainty principle. For some arbitrary volume of phase space $dqdp$ (Figure 1.10), the number of quantum states contained in that volume is $dqdp/(2\pi\hbar)^s = d\Gamma$. This can be called the statistical weight. It is simply the number of microstates within that volume. It is proportional to the phase volume of the given region. As all the allowed microstates have equal probabilities, the probabilities of alternative microstates are proportional to their statistical weights.

Now

$$dW_n = W(E_n(q, p))\frac{dqdp}{(2\pi\hbar)^s} = W\left(E_n(q, p)\right)d\Gamma \qquad (1.202)$$

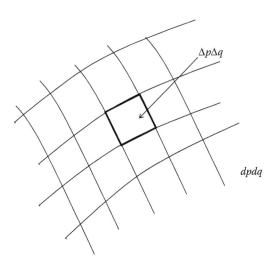

FIGURE 1.10 Some arbitrary volume of phase space $dqdp$.

If we consider Equation 1.172, then we have

$$dW_n = W(E_n(q,p))d\Gamma = C\delta(E_n(q,p) - E_0)d\Gamma \qquad (1.203)$$

which differs from the classical formula in that here we have considered the number of states. Note, however, that calculation of expectation values based on energy eigenstates must be more carefully considered. The formalism for quantum expectation values, known as the *density matrix* approach, is developed next.

1.13   DENSITY MATRIX

For dynamical systems on atomic scales, classical mechanics is totally replaced by QM. Quantum mechanical language is a proper formulation of statistical physics as a molecular theory. Notwithstanding this, there are very many problems that may be correctly treated classically. However, it should be noted that classical mechanics is a limiting case of QM. There are particular conditions under which classical statistical physics is a limiting case of QM.

The density matrix is a quantum mechanical distribution function. It plays a role analogous to the distribution function ρ in classical phase space. Conceptually, the use of continuous phase space for quantum problems has difficulties. It follows that we need a different approach to define expectation values in QM, which is provided by the density matrix.

For the study of the properties of systems we need to solve the time-dependent Schrödinger equation:

$$i\hbar \frac{\partial \Psi}{\partial t} = \hat{H}\Psi \qquad (1.204)$$

The operator $\hat{H}$ is the Hamiltonian of the system, usually postulated based on classical mechanics ideas. The wave function may also be the solution of the stationary equation $\hat{H}\Psi = E\Psi$. The solution of these equations for a complex system produces great difficulties. The Hamiltonian of Equation 1.204 for many interacting systems can be difficult to determine. The solution of Equation 1.204 is very cumbersome and we often resort to approximate solutions. We have to know beforehand the initial conditions. The wave function is a function of many variables. Further, for a system with a macroscopic number of particles, it may not be possible to exactly specify the wave function because our knowledge about the state of the system is only statistical.

As for classical systems, we can imagine an ensemble of systems. An ensemble here will be understood as a collection of a very large number N of perfectly isolated (conservative) independent systems in various quantum mechanical states $\Psi(q, t)$. We begin the discussion considering that states are given in their coordinate representations; the wave functions depend only on the coordinates, not on the momenta. A wave function is obtained from Dirac's bra-ket notation according to $\Psi(x) = \langle x | \Psi \rangle$.

Let us examine a microscopic body. It is known that the wave function describes the given body with some quantum numbers. Let those numbers be $F, N, L, \ldots$. These are the eigenvalues of a complete set of commuting observables, represented by operators $\hat{F}, \hat{N}, \hat{L}$, and so on. This means that $\hat{F}$ and the other operators commute with $\hat{H}$, that is, $\hat{F}\hat{H} - \hat{H}\hat{F} = 0$. It is a basic postulate of QM that a measurement of any observable for a quantum system can only give one of its eigenvalues. An operator $\hat{F}$ has some set of eigenvalues $F_1, F_2, F_3, \ldots$ where each eigenvalue corresponds to the system being in one of its corresponding eigenstates, $\Psi_1(x), \Psi_2(x), \Psi_3(x), \ldots$, and so on. The eigenstates and eigenvalues are found by solving an eigenvalue problem, $\hat{F}\Psi_i(x) = F_i\Psi_i(x)$, where i labels the different states. When $\hat{F}$ is measured, the different members of the ensemble may be in different eigenstates of $\hat{F}$ with particular probabilities, but we cannot say any more specifically how each measurement will turn out.

Generally, we have what is known as a mixed ensemble. If $N_i(F_i)$ is the number of systems found in a state i with eigenvalue F_i, the probability of that state is

$$W_i = \left. \frac{N_i(F_i)}{N} \right|_{N \to \infty} \geq 0 \tag{1.205}$$

Summing over all possible systems $\left[\sum_i N_i(F_i) = N \right]$ shows that

$$\sum_i W_i = 1 \tag{1.206}$$

Let us apply another quantum mechanical expression for F_i, true in any normalized eigenstate of $\hat{F}$ (it gives a quantum expectation value if the state is not an eigenstate of $\hat{F}$):

$$F_i = \int \Psi_i^*(x) \hat{F}(x) \Psi_i(x) dx \tag{1.207}$$

Here, x represents the aggregate of continuous and discrete variables. The coordinate representation of the operator, $\hat{F}(x)$, is defined from Dirac's bra-ket notation according to

$$\langle x' | \hat{F} | x \rangle \equiv \hat{F}(x) \delta(x - x') \tag{1.208}$$

From the general formula for the statistical expectation value in the coordinate representation we have

$$\bar{F} = \sum_i F_i W_i = \sum_i W_i \int \Psi_i^*(x) \hat{F}(x) \Psi_i(x) dx$$

$$= \sum_i W_i \int \Psi_i^*(x') \delta(x - x') \hat{F}(x) \Psi_i(x) dx dx'$$

$$= \int \hat{F}(x) \sum_i W_i \Psi_i^*(x') \Psi_i(x) \delta(x - x') dx dx' = \int \hat{F}(x) \rho(x, x') \delta(x - x') dx dx'.$$

$$\tag{1.209}$$

This requires the definition

$$\rho(x, x') = \sum_i W_i \Psi_i(x)\Psi_i^*(x') \tag{1.210}$$

$\rho(x, x')$ is the density matrix in the coordinate representation and is positive definite. The relation in Equation 1.210 applies generally to what is called a mixed state. Ψ_i is a member of a set of complex orthonormal wave functions. For a pure state, the wave function is definite, and no probability is needed:

$$\rho(x, x') = \Psi(x)\Psi^*(x') \tag{1.211}$$

Although we used the coordinate representation, that choice is arbitrary. The density matrix (1.210) can be expressed more generally, starting from bra-ket notation for the wave functions ($\Psi(x) = \langle x|\Psi\rangle$, $\Psi^*(x) = \langle x|\Psi\rangle^*$, etc.)

$$\rho(x, x') = \sum_i W_i \langle x|\Psi_i\rangle\langle x'|\Psi_i\rangle^* = \sum_i W_i \langle x|\Psi_i\rangle\langle\Psi_i|x'\rangle$$

$$= \langle x|\left(\sum_i W_i|\Psi_i\rangle\langle\Psi_i|\right)|x'\rangle \tag{1.212}$$

This matrix element between states $\langle x|$ and $|x'\rangle$ allows definition of the density operator between them

$$\hat{\rho} = \sum_i W_i|\Psi_i\rangle\langle\Psi_i| \tag{1.213}$$

The states being used can be any complete set of eigenstates for some operator $\hat{F}$. This then gives the general expression for the statistical expectation value

$$\bar{F} = \int \hat{F}(x)\langle x|\hat{\rho}|x'\rangle\delta(x - x')dxdx' = \int \langle x'|\hat{F}|x\rangle\langle x|\hat{\rho}|x'\rangle dxdx'$$

$$= \int \langle x'|\hat{F}\hat{\rho}|x'\rangle dx' \tag{1.214}$$

or

$$\bar{F} = Tr\{\hat{F}\hat{\rho}\} = \sum_i \langle\Psi_i|\hat{F}\hat{\rho}|\Psi_i\rangle \tag{1.215}$$

The trace operation (Tr) can be evaluated using any complete set of states. There is no need to use the coordinate representation. This result shows the power and beauty of the density operator.

As Equation 1.207 is the expectation value of $\hat{F}$ in the state Ψ_i, then from Equations 1.205, 1.206, and 1.215, we interpret W_i as the probability that the system is in state Ψ_i. If all but one of the W_i are zero, then the given system is in a pure state; otherwise, it is in a mixed state. A necessary and sufficient condition for a pure state is $\hat{\rho}^2 = \hat{\rho}$. We see that

$$\mathrm{Tr}\{\hat{\rho}\} = 1, \quad \mathrm{Tr}\{\hat{\rho}^2\} \leq 1 \tag{1.216}$$

The last equal sign holds only if the state is pure. It should be noted that the states for many-body systems and systems not in equilibrium normally are mixed.

1.14   DENSITY MATRIX IN ENERGY REPRESENTATION

If we want the wave function for the ith system in the ensemble, in the energy representation, we can find it from the time-dependent Schrödinger equation:

$$i\hbar \frac{\partial |\Psi^{(i)}\rangle}{\partial t} = \hat{H}|\Psi^{(i)}\rangle \tag{1.217}$$

This can be accomplished by looking for the energy eigenstates, that have a time dependence

$$|\Psi_n^{(i)}(x,t)\rangle = e^{-iE_n t/\hbar}|\Phi_n(x)\rangle \tag{1.218}$$

leading to the stationary Schrödinger equation for that state:

$$\hat{H}|\Phi_n(x)\rangle = E_n|\Phi_n(x)\rangle \tag{1.219}$$

Here, $|\Phi_n\rangle$ is the nth eigenfunction and E_n the corresponding eigenvalue of the Hamiltonian. The complete set of normalized eigenfunctions acts as an orthonormal basis set. These allow for an expansion of any state for the ensemble, $|\Psi^{(i)}(x,t)\rangle$, as

$$|\Psi^{(i)}(x,t)\rangle = \sum_n C_n^{(i)}(t)|\Phi_n(x)\rangle \tag{1.220}$$

The coefficients $C_n^{(i)}(t)$ describe the wave function in the energy representation.

Expectation values using this representation can be elaborated as described for the coordinate representation:

$$\begin{aligned}\bar{F} &= \sum_i W_i F_i = \sum_i W_i \langle \Psi^{(i)}(t)| \hat{F} |\Psi^{(i)}(t)\rangle \\ &= \sum_i W_i \sum_{m,n} \langle \Psi^{(i)}(t)|\Phi_m\rangle\langle\Phi_m|\hat{F}|\Phi_n\rangle\langle\Phi_n|\Psi^{(i)}(t)\rangle \end{aligned} \tag{1.221}$$

This involves the basic matrix elements

$$\langle \Phi_n | \Psi^{(i)}(t) \rangle = C_n^{(i)}(t), \quad \langle \Psi^{(i)}(t) | \Phi_m \rangle = C_m^{(i)*}(t), \quad \langle \Phi_m | \hat{F} | \Phi_n \rangle = F_{mn} \qquad (1.222)$$

and

$$\rho_{nm} = \sum_i W_i C_n^{(i)}(t) C_m^{(i)*}(t) \qquad (1.223)$$

The latter is the density matrix in the energy representation. Then, as expected, the expectation value is

$$\overline{F} = \sum_{n,m} \rho_{nm} F_{mn} = \sum_n (\hat{\rho}\hat{F})_{nn} = \mathrm{Tr}\{\hat{\rho}\hat{F}\} \qquad (1.224)$$

The expectation value is just the trace, that is, the sum of the diagonal matrix elements of the product of the operators $\hat{\rho}$ and $\hat{F}$.

The density matrix depends on time only via the $C_n^{(i)}(t)$. In a stationary ensemble, it is assumed that the probabilities W_i do not depend on time. In order to find ρ_{nm} it is useful to differentiate Equation 1.223 with respect to time:

$$\frac{\partial \rho_{nm}}{\partial t} = \sum_i W_i \left[\dot{C}_n^{(i)}(t) C_n^{(i)*}(t) + C_n^{(i)}(t) \dot{C}_n^{(i)*}(t) \right] \qquad (1.225)$$

If we substitute Equation 1.220 into Equation 1.217, we have

$$i\hbar \sum_\kappa \dot{C}_\kappa^{(i)}(t) | \Phi_\kappa(x) \rangle = \sum_\kappa C_\kappa^{(i)}(t) \hat{H} | \Phi_\kappa \rangle \qquad (1.226)$$

Taking the overlap with $\langle \Phi_n(t) |$ and using the orthogonality relation

$$\langle \Phi_n | \Phi_\kappa \rangle = \delta_{n\kappa} \qquad (1.227)$$

we obtain

$$i\hbar \, \dot{C}_n^{(i)} = \sum_\kappa C_\kappa^{(i)} \langle \Phi_n | \hat{H} | \Phi_\kappa \rangle \qquad (1.228)$$

But

$$\langle \Phi_n | \hat{H} | \Phi_\kappa \rangle = H_{n\kappa} \qquad (1.229)$$

so that substituting Equation 1.228 into Equation 1.225, we obtain

$$\frac{\partial \rho_{nm}}{\partial t} = \frac{1}{i\hbar} \sum_\kappa \left[H_{n\kappa}\rho_{\kappa m} - \rho_{n\kappa}H^*_{m\kappa} \right] = \frac{1}{i\hbar} \sum_\kappa \left[H_{n\kappa}\rho_{\kappa m} - \rho_{n\kappa}H_{\kappa m} \right] \quad (1.230)$$

The Hamiltonian is Hermitian, so we used $H^*_{m\kappa} = H_{\kappa m}$. Thus, the RHS involves a commutator of the Hamiltonian with the density matrix

$$\frac{\partial \rho_{nm}}{\partial t} = \frac{1}{i\hbar} \left[H\rho - \rho H \right]_{nm} \quad (1.231)$$

This is the quantum Liouville equation. This equation plays the same role for density matrices that the Schrödinger equation plays for wave functions. Equation 1.231 may be written in the operator representation

$$\frac{\partial \hat{\rho}}{\partial t} = \frac{1}{i\hbar} [(\hat{H}\hat{\rho}) - (\hat{\rho}\hat{H})] = \frac{1}{i\hbar}[\hat{H},\hat{\rho}] \quad (1.232)$$

or

$$i\frac{\partial \hat{\rho}}{\partial t} = \frac{1}{\hbar}[\hat{H},\hat{\rho}] \equiv \hat{L}\hat{\rho} \quad (1.233)$$

This is the general equation of motion for the density operator and

$$\hat{L} \equiv \frac{1}{\hbar}[\hat{H},...] \quad (1.234)$$

defines the quantum Liouville operator.

The formal solution of Equation 1.233 yields

$$\hat{\rho}(t) = \exp\{-i\hat{L}t\}\rho(0) = \exp\left\{-\frac{i}{\hbar}\hat{H}t\right\}\rho(0)\exp\left\{\frac{i}{\hbar}\hat{H}t\right\} \quad (1.235)$$

This is the equation of motion in the Schrödinger picture. In the Heisenberg picture

$$\overline{F} = \mathrm{Tr}\{\hat{F}(t)\hat{\rho}(0)\} \quad (1.236)$$

where the observables have time dependence

$$\hat{F}(t) = \exp\left\{\frac{i}{\hbar}\hat{H}t\right\}\hat{F}(0)\exp\left\{-\frac{i}{\hbar}\hat{H}t\right\} \quad (1.237)$$

The equation of motion for $\hat{F}(t)$ is given by the relation

$$i\frac{\partial \hat{F}(t)}{\partial t} = \frac{1}{\hbar}[\hat{H}, \hat{F}(t)] \equiv \hat{L}\hat{F}(t) \tag{1.238}$$

From Equation 1.233 we may draw out the most important property of $\hat{\rho}$. In the case of an equilibrium state, the distribution is independent of time:

$$\frac{\partial \hat{\rho}}{\partial t} = 0, \quad \hat{\rho} \neq \hat{\rho}(t) \tag{1.239}$$

This means that $\hat{\rho}$ commutes with $\hat{H}$, that is

$$[\hat{\rho}, \hat{H}] = [\hat{\rho}\hat{H} - \hat{H}\hat{\rho}] = 0 \tag{1.240}$$

Thus, $\hat{\rho}$ is an integral of motion in the equilibrium state. It follows that $\hat{\rho}$ and $\hat{H}$ have a common basis set. If we multiply from the left on

$$\hat{H}|\Phi_n\rangle = E_n|\Phi_n\rangle \tag{1.241}$$

by $\langle\Phi_m|$ and integrate relative to the coordinates, using Equation 1.235 leads to

$$\hat{\rho}(t) = \sum_n \sum_m \langle\Phi_n|\rho(0)|\Phi_m\rangle \exp\left\{-\frac{i}{\hbar}(E_n - E_m)t\right\}|\Phi_n\rangle\langle\Phi_m| \tag{1.242}$$

This implies that the system is in a stationary state $\hat{\rho}_s \neq \hat{\rho}_s(t)$ when off-diagonal terms vanish. That is to say

$$\hat{H} \to H_{nm} \to \begin{bmatrix} H_{11} & 0 & \cdots & 0 \\ 0 & H_{22} & \cdots & 0 \\ \vdots & 0 & \ddots & \vdots \\ 0 & \cdots & 0 & H_{nn} \end{bmatrix} \tag{1.243}$$

and

$$\hat{\rho} \to \rho_{nm} \to \begin{bmatrix} \rho_{11} & 0 & \cdots & 0 \\ 0 & \rho_{22} & \cdots & 0 \\ \vdots & 0 & \ddots & \vdots \\ 0 & \cdots & 0 & \rho_{nn} \end{bmatrix} \tag{1.244}$$

In this case

$$\rho_{nm} = \rho_{nn}\delta_{nm} \tag{1.245}$$

For the nondegenerate systems (all energy eigenvalues are associated with single eigenstates) it implies that

$$\hat{\rho} = f(\hat{H}) \tag{1.246}$$

and for the degenerate systems (some energy eigenvalues correspond to multiple eigenstates)

$$\hat{\rho} = f(\hat{H}, \hat{I}_1, \hat{I}_2, \ldots) \tag{1.247}$$

Here, $\hat{I}_i$ are integrals of motion that are mutually commutable

$$[\hat{I}_i, \hat{I}_j] = 0, \quad \text{for any } i \text{ and } j \tag{1.248}$$

$$[\hat{H}, \hat{I}_i] = 0 \quad \text{for any } i \tag{1.249}$$

The different states corresponding to a degenerate energy eigenvalue are distinguished by distinct eigenvalues $I_1, I_2, \ldots$.

For the equilibrium case with a diagonal density matrix

$$\overline{F} = \sum_{n,m} \rho_{nm} F_{mn} = \sum_{n,m} \rho_{nn}\delta_{nm} F_{mn} = \sum_{n} \rho_{nn} F_{nn} \tag{1.250}$$

For the equilibrium case, the density matrix is defined by one index (i.e., location along the diagonal, with $\rho_{nn} = W_n$). It follows that

$$\overline{F} = \sum_{n} W_n F_{nn} \tag{1.251}$$

where W_n is the probability of the nth quantum state. It is easily seen that ρ_{nn} is normalized:

$$Tr\hat{\rho} = \sum_{n} \rho_{nn} = \sum_{i,n} W^{(i)} C_n^{(i)} C_n^{(i)*} = \sum_{i} W^{(i)} \sum_{n} C_n^{(i)} C_n^{(i)*}$$

$$= \sum_{i} W^{(i)} = 1 \tag{1.252}$$

and the density matrix is Hermitian. Due to

$$\overline{F} = \overline{F}^{*} \tag{1.253}$$

it follows that

$$\rho_{nm} = \rho_{mn}^{*} \tag{1.254}$$

1.15 ENTROPY

Thermodynamic values are less obvious than mechanical values. Thermodynamic values are obtained as a result of taking expectation values, hence, averaging over the many possible microscopic states that produce an observed macroscopic state. When observing a macroscopic system, we measure only a few state variables, but these are produced by averaging over a much wider phase space. In a very real sense, we always lack full information about the exact state of a system. The fact that many microscopic states work together to realize a given macroscopic state leads to the idea of entropy. Entropy is a measure of the variety of ways to realize (microscopically) the observed macroscopic state. Entropy is also a measure of our lack of complete knowledge about the microscopic state. When there are many microstates that realize a desired macrostate, the entropy is large.

One of the fundamental concepts in statistical physics is the idea that all allowed microstates have equal probabilities. The concept of entropy is intimately related to this idea. Entropy is defined in such a way, mathematically, that this uniform probability among the allowed microstates is realized.

Let us examine the classical distribution function $\rho(q, p)$, expressed in terms of a phase $\eta(q, p)$:

$$\rho(q, p) = \exp\{-\eta(q, p)\} \tag{1.255}$$

Assume that the function $\eta(q, p)$ has the additive property, as follows. Consider an isolated system. Let this system be partitioned into two subsystems, then

$$\rho_{I+II}(q, p) = \exp\{-\eta_{I+II}(q, p)\} \tag{1.256}$$

From the property of multiplicity of ρ, we have

$$\rho_{I+II} = \rho_{I}\rho_{II} = \exp\{-\eta_{I} - \eta_{II}\} \tag{1.257}$$

If we compare Equation 1.256 with Equation 1.257, then we have

$$\eta_{I+II} = \eta_{I} + \eta_{II} \tag{1.258}$$

It may be easily seen that the phase of the distribution function satisfies the Liouville equation. Start from

$$\frac{\partial \rho}{\partial t} + \sum_i \left[\frac{\partial \rho}{\partial q_i} \frac{\partial H}{\partial p_i} - \frac{\partial \rho}{\partial p_i} \frac{\partial H}{\partial q_i} \right] = 0 \qquad (1.259)$$

and from Equation 1.255 we have the stated result:

$$\frac{\partial \eta}{\partial t} + \sum_i \left[\frac{\partial \eta}{\partial q_i} \frac{\partial H}{\partial p_i} - \frac{\partial \eta}{\partial p_i} \frac{\partial H}{\partial q_i} \right] = 0 \qquad (1.260)$$

Let us now introduce entropy S from the definition that it is the expectation value of this phase $\eta(q, p)$. For this we introduce the Naperian logarithm of Equation 1.255 and take the mean of η over phase space (we consider a system with s degrees of freedom). We have

$$\eta(q, p) = -\ln \rho \qquad (1.261)$$

and

$$S = \int \eta(q, p) \rho(q, p) \frac{dq \, dp}{(2\pi \hbar)^s} = \int \eta(q, p) \rho(q, p) d\Gamma = -\int \rho(q, p) \ln \rho(q, p) d\Gamma \quad (1.262)$$

This is called the entropy in the continuum limit. It is understood that Equation 1.262 is not quite explicit. It may be represented in a discrete form for a quantum ensemble through the probability W_i:

$$S = -\sum_i W_i \ln W_i \qquad (1.263)$$

that is, we may define the entropy as the average value of the logarithmic distribution function of a subsystem (taken with opposite sign). These relations express the entropy in units of the Boltzmann constant, k_B. In further chapters, unless otherwise stated, the entropy expressed through the distribution function will be in units of the Boltzmann constant.

The probabilities W_i may define the quantum density operator $\hat{\rho}$ that is diagonal in a set of eigenstates $|\Phi_i\rangle$:

$$S = -\sum_i W_i \ln W_i = -\mathrm{Tr}\{\hat{\rho} \ln \hat{\rho}\} \qquad (1.264)$$

Check the case when all members of the ensemble are in the same quantum state i (pure ensemble) with $W_i = 1$, $W_\kappa = 0$, $\kappa \neq i$. This directly gives $S = 0$. This case of zero entropy reflects the complete order of the ensemble.

If the ensemble's probabilities are different from zero for all possible quantum states, obviously the entropy will be higher. Suppose we consider the probabilities as variables, and ask what values would give the highest entropy. For a total of N quantum states, the probabilities satisfy the normalization

$$\sum_{i=1}^{N} W_i = 1 \tag{1.265}$$

The entropy can be maximized under this constraint, with an unknown Lagrange multiplier λ. This requires for each W_n

$$\frac{\partial}{\partial W_n}\left(S - \lambda \sum_{i=1}^{N} W_i \right) = 0 \tag{1.266}$$

Using definition (1.263) for S gives for any n

$$-\ln W_n - W_n \frac{1}{W_n} - \lambda = 0 \implies W_n = \exp\left\{ -\lambda - 1 \right\} \tag{1.267}$$

Including the constraint determines the probabilities, which are found to be all equal:

$$1 = \sum_{i=1}^{N} W_n = \sum_{i=1}^{N} \exp\{-\lambda - 1\} = N \exp\{-\lambda - 1\} \implies W_n = \frac{1}{N} \tag{1.268}$$

Then the maximum entropy is

$$S = S_{\text{max}} = \ln N \tag{1.269}$$

It is only realized when all states have the same probabilities. This is the macroscopic state with maximum microscopic disorder. Note, however, that usually there will be other constraints that will limit the accessible states, thereby modifying the entropy. Even so, this maximum entropy principle still holds. The macroscopic system attains a configuration of maximum entropy consistent with all applied physical constraints (such as constant energy). It is equivalent to the postulate that all accessible microstates have equal probabilities. From Equation 1.258 it may be seen that the entropy has the property of additivity for statistically independent systems. It is an extensive property of the system, that is, its value is proportional to the size of the system.

1.15.1 Entropy of Microcanonical Distribution

Let us examine the constant energy surface at a selected energy E_0 in Figure 1.11.

For the shaded layer, $\rho = \text{const}$, and outside the shaded layer, $\rho = 0$. From the normalization condition

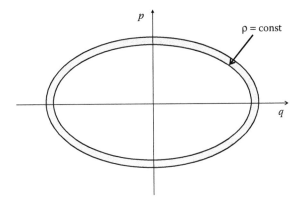

FIGURE 1.11 A constant energy surface at a selected energy E_0. For the shaded layer, $\rho = $ const, and outside the shaded layer, $\rho = 0$.

$$1 = \int \rho(q, p) d\Gamma = \rho \int d\Gamma = \rho \Delta\Gamma \qquad (1.270)$$

where $\Delta\Gamma$ is the number of microstates at chosen energy E_0 in which the system may be found.

It follows that

$$S = -\int \rho \ln\rho \, d\Gamma = -\ln\rho \int \rho d\Gamma = -\ln\rho \qquad (1.271)$$

If we again consider Equation 1.270, then

$$S = -\ln\rho = \ln\Delta\Gamma \qquad (1.272)$$

This relationship between S and $\Delta\Gamma$ was obtained by Ludwig Boltzmann and is therefore called the Boltzmann equation. Expression (1.272) gives the entropy in units of Boltzmann's constant. The relationship allows us to relate the quantities entering in thermodynamic laws to one's microscopic knowledge of the system. It should be noted that if one knows the nature of the particles constituting the system and the interactions between them, then one can use the laws of mechanics to compute the possible quantum states of the system and thus to find $\Delta\Gamma$.

It is interesting to reverse the relation (1.272), and write it as

$$\Delta\Gamma = \exp\{S\} \qquad (1.273)$$

This gives the microscopic interpretation of the entropy as determining the great number of states accessible to the system. It further suggests the connection of entropy with the microscopic disorder. A more disordered system is one spread over a greater region of phase space, hence, corresponding to higher entropy.

The entropy acts as a thermodynamic potential, an idea that will be described in Chapter 2. This implies that *all thermodynamic quantities* may be computed from S or equivalently from $\Delta\Gamma$.

1.15.2 EXACT AND "INEXACT" DIFFERENTIALS

A discussion of differentials will be found useful for further considerations of the properties of entropy and thermodynamics in general.

We consider a purely mathematical problem where $F(\lambda_1, ..., \lambda_n, ..., \lambda_s)$ is some function of s-independent variables $\lambda_1, ..., \lambda_s$. That is to say, the value of F is determined when the values of $\lambda_1, ..., \lambda_s$ are specified. Suppose we move to a neighboring point $\lambda_1 + d\lambda_1, ..., \lambda_n + d\lambda_n, ..., \lambda_s + d\lambda_s$, then the function F changes by the amount

$$dF = F(\lambda_1 + d\lambda_1, ..., \lambda_n + d\lambda_n, ..., \lambda_s + d\lambda_s) - F(\lambda_1, ..., \lambda_n, ..., \lambda_s) \quad (1.274)$$

This may also be written as

$$dF = \sum_n \left(\frac{\partial F}{\partial \lambda_n} \right) d\lambda_n \equiv \sum_n \Lambda_n \, d\lambda_n, \quad \Lambda_n \equiv \frac{\partial F}{\partial \lambda_n} \quad (1.275)$$

This is simply an infinitesimal difference between two adjacent values of the function F. The infinitesimal quantity dF in this case is just an ordinary differential. We call the quantity dF an "exact differential."

Definition

If $F = F(\lambda_1, ..., \lambda_n, ..., \lambda_s)$ and its derivatives are continuous and

$$\frac{\partial \Lambda_n}{\partial \lambda_r} = \frac{\partial \Lambda_r}{\partial \lambda_n} \quad (1.276)$$

then dF is an *exact differential*. An exact differential has the following properties:

1. The path independence of the integral (a and b refer to all variables):

$$F(b) - F(a) \equiv \int_a^b dF \quad (1.277)$$

2. The integral over a closed path is equal to zero:

$$\oint dF = 0 \quad (1.278)$$

3. The total differential is integrable:

$$\int dF = F(\lambda_1, \dots, \lambda_n, \dots, \lambda_s) + \text{const} \tag{1.279}$$

On the other hand, it should be noted that, not every infinitesimal quantity is an exact differential:

$$\sum_n \Lambda_n d\lambda_n \equiv \eth F \neq \text{exact} \tag{1.280}$$

This can hold when

$$\frac{\partial \Lambda_n}{\partial \lambda_r} \neq \frac{\partial \Lambda_r}{\partial \lambda_n} \tag{1.281}$$

It should be noted that $\eth F$ has been introduced merely as an abbreviation for the expression on the left of Equation 1.280. Certainly, although $\eth F$ is an infinitesimal quantity it does not follow necessarily that it is an exact differential. When an infinitesimal quantity $\eth F$ is not an exact differential, it is called an "inexact differential."

How can we get an exact differential from an inexact one? This may be possible if we can find an *integrating factor* $L(\lambda_1, \dots, \lambda_n, \dots, \lambda_s)$ in the neighborhood of $(\lambda_1, \dots, \lambda_n, \dots, \lambda_s)$:

$$df \equiv L \eth F = L \sum_n \Lambda_n d\lambda_n = \text{exact} \tag{1.282}$$

We may see from here that any inexact differential can be given as an exact one through the relation

$$\eth F = \frac{1}{L} df \tag{1.283}$$

1.15.3 PROPERTIES OF ENTROPY

Using the microcanonical ensemble, we show that the entropy S is a maximum when the system is in an equilibrium state, that is, $\rho = \text{const}$ on the constant energy surface. Suppose $\rho'(q, p)$ is a distribution nonhomogeneously distributed in the layer (Figure 1.11). Let us examine the function

$$f(x) = \ln x - 1 + \frac{1}{x}, \quad x > 0 \tag{1.284}$$

It is possible to show that $f(x) \geq 0$, for example

$$x = 1 \quad \rightarrow \quad f(1) = 0 - 1 + \frac{1}{1} = 0$$

$$x = e \quad \rightarrow \quad f(e) = 1 - 1 + \frac{1}{e} > 0 \qquad (1.285)$$

$$x = \frac{1}{e} \quad \rightarrow \quad f\left(\frac{1}{e}\right) = -1 - 1 + e > 0$$

Let x be defined as $x = \rho'/\rho$; then because $\rho > 0$ and $\rho' > 0$

$$f\left(\frac{\rho'}{\rho}\right) = \ln\left(\frac{\rho'}{\rho}\right) - 1 + \frac{\rho}{\rho'} \geq 0 \qquad (1.286)$$

If we multiply Equation 1.286 by ρ' and integrate over $d\Gamma$, the result will be a function greater than or equal to zero:

$$\int f\left(\frac{\rho'}{\rho}\right)\rho'd\Gamma = \int[-\rho'\ln\rho + \rho'\ln\rho' + \rho - \rho']d\Gamma \geq 0 \qquad (1.287)$$

But

$$\int \rho'\ln\rho'\,d\Gamma = -S', \quad \int \rho'\,d\Gamma = \int \rho\,d\Gamma = 1 \qquad (1.288)$$

then

$$\int f\left(\frac{\rho'}{\rho}\right)\rho'd\Gamma = -\int \rho'\ln\rho\,d\Gamma - S' = -\ln\rho\int\rho'd\Gamma - S' = -\ln\rho - S' = S - S' \geq 0$$

$$(1.289)$$

It follows that the entropy of an equilibrium state is greater than the entropy of any nonequilibrium state described by a nonhomogeneous ρ' (see Figure 1.12). Note that the equilibrium state has $\rho = 1/\Delta\Gamma = 1/N$, where the number of quantum micro-states at the selected energy is N.

If initially the system is in a nonequilibrium state, then processes take place in such a way as to move toward the equilibrium state. We know from experience that when external conditions are fixed, on its own the system (spontaneously) arrives at the equilibrium state. As thermodynamics may not explain this phenomenon, we take it as a postulate, that any system left to its own follows the *principle of increase of entropy*: $dS/dt \geq 0$. The entropy increases uniformly (or stays the same) with time. Entropy never decreases.*

---

* This is for an isolated system with no external influences. The entropy of a system acted upon by some external agent that organizes the system or reduces its disorder will decrease. When the total entropy of system plus external agent is analyzed, however, it will be found to increase, just as that for an isolated system.

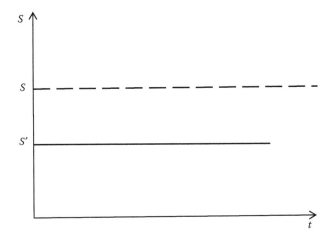

FIGURE 1.12 Comparison of the entropy of the equilibrium (dashed curve) and nonequilibrium (solid curve) state with time.

Let us again examine $dS/dt \geq 0$ from the point of view of reversible mechanical motion. To say a mechanical process is reversible means that it will appear the same if time is reversed. This requires symmetry in time. For example, the flight of a projectile through vacuum traces out a parabola; its reversed flight is the same parabola. The projectile obeys Newton's laws of classical mechanics, which are second order in time, and therefore, reversible. If instead the projectile moves through air, there is friction, and the flight is not reversible. The friction slows the motion of the projectile, leading to asymmetry in time. In the time-reversed path (i.e., a video of the flight, running backwards), it would appear that friction causes the projectile to speed up, a phenomenon that is never observed. Of course, the friction is really caused by the action of the many degrees of freedom in the air molecules that are treated in the mechanics problem in an approximate fashion. In the reversed video, it would look like the air molecules become very organized just before large numbers of them collide in unison with the projectile. Then, after colliding, they would suddenly achieve random velocities.

What about a thermodynamic system, even one following classical mechanics? Consider that at some moment t, a system is not in equilibrium, and has $dS/dt > 0$. Then, in the future, the entropy should continue to increase until equilibrium is attained. If this process is viewed with a video running backwards (i.e., with reversed time $t' = -t$), one will notice that it has decreasing entropy, $dS/dt' < 0$. Of course, for a thermodynamic system, this is impossible. The approach to equilibrium clearly is not reversible. This is because whatever caused the nonequilibrium state has now destroyed the symmetry between the past and the future. In what situation will a thermodynamic process be reversible? The only way that can happen, is when $dS/dt = 0$. When the entropy neither increases forward in time nor backwards in time, the system is in thermodynamic equilibrium. If the entropy is not changing, the probability distribution in phase space stays constant with time. Sometimes it is useful to suppose that a thermodynamic process is carried out very slowly, so that the process is reversible and the probability distribution does not change. This is a useful mental technique when discussing the thermodynamics of heat engines and other machines. These quasi-equilibrium processes are considered reversible.

The equals sign in $(dS/dt) \geq 0$ applies if the process is reversible and the inequality sign holds if the process is irreversible. In other words, the entropy increases during spontaneous processes taking place in a closed, isolated system until it achieves a maximum value at the final equilibrium state.

Consider the example of a gas in equilibrium confined by a partition to one half of a closed isolated container. Suppose the partition is removed suddenly. There is a spontaneous flow of gas into the other half of the container, culminating in a final equilibrium state of maximum entropy. The gas molecules, once spread over the whole container, never find themselves all on one side again. In this example, their available phase space has increased by a factor of 2, due to the removal of the partition, and so their entropy increases by a factor of $\ln(2)$.

Consider now a system and a thermostat (or a thermal reservoir, a very large reservoir at constant temperature) that are thermally isolated from the external medium. Then, for any process, the sum of their entropies may only increase or stay the same. Let us suppose that there is an exchange of energy between the system and the reservoir. To be concrete, assume that the system absorbs an amount of energy dE from the reservoir. Then, from the above, the entropy is equal to the sum of the entropies of the system (S) and the reservoir (S'):

$$\bar{S} = S + S', \quad d\bar{S} = dS + dS' \geq 0 \tag{1.290}$$

A similar relation holds for the energies but the total energy is fixed because the setup is isolated:

$$\bar{E} = E + E', \quad d\bar{E} = dE + dE' = 0 \tag{1.291}$$

It should be noted that all quantities characterizing the reservoir, which is in equilibrium at fixed absolute temperature T, are dependent only on its energy:

$$dS' = \frac{\partial \ln \Delta \Gamma'}{\partial E'} dE' = \frac{1}{T} dE', \quad \frac{1}{T} = \frac{\partial \ln \Delta \Gamma'}{\partial E'} \tag{1.292}$$

$$dS \geq -dS' = -\frac{dE'}{T} = \frac{\delta Q}{T} \tag{1.293}$$

and summarizing as

$$dS \geq \frac{\delta Q}{T} \tag{1.294}$$

The quantity δQ is the energy (or heat) absorbed by the system, but it is *not an exact differential*. Absolute temperature T has been introduced in a fundamental way as a relation between the entropy and energy differentials. The quantity $1/T$ plays the role of an integration factor that changes an infinitesimal quantity δQ into an exact differential. It follows that the entropy of a system is greater than or equal to the quotient of the quantity of heat δQ absorbed divided by the absolute temperature T of the reservoir. This is the *second law of thermodynamics* for nonequilibrium processes.* It should be noted that for the thermally isolated case (i.e., when the process is *adiabatic*), the absorbed heat $\delta Q = 0$ and Equation 1.294 asserts that

$$dS \geq 0 \qquad\qquad (1.295)$$

which is the law of increase of entropy. If we consider the equals sign

$$dS = 0 \qquad\qquad (1.296)$$

we see that it applies when S or $\ln \Delta\Gamma$ do not change, even if the external parameters are varied quasi-statically by a finite amount. This would be the case for a process performed slowly enough that the system always maintains its equilibrium state. The equals sign applies to reversible processes.

It follows that if the external parameters of a *thermally isolated* system are changed quasi-statically by any amount, then $\Delta S = 0$. The execution of a quasi-static work changes the energy of a thermally isolated system, but this does not affect the number of states accessible to it. Such processes are reversible.

It is worth noting that for a nonthermally isolated system, it is possible that the entropy decreases with time. This is true for a system that gives up heat to its surroundings, $\delta Q < 0$.

---

* It may be surprising that the temperature of the reservoir enters here, rather than the temperature of the system. If the system is not in equilibrium, however, its temperature is not well defined. The reservoir has a well-defined temperature. If a quasi-static process is considered, it is supposed that the system has the same temperature as the reservoir, and the latter may slowly change its temperature, always waiting for the system to return to equilibrium as a new temperature is reached.

2 Thermodynamic Functions

In thermodynamics, we are concerned with the macroscopic state of a system, which can be described by specifying its internal energy, entropy, volume, number of particles, temperature, and so on. The distinction from the statistical viewpoint is that these state variables are averages, rather than microstate values. The goal of thermodynamics is to find the relations between the state variables, and especially, find their dependence on external controls such as the temperature.

2.1 TEMPERATURE

In everyday life, there is a common notion of temperature as the degree to which an object is "hot" or "cold." Chemical reactions produce heat and raise the temperature of their surroundings. Refrigeration systems extract heat from an object and reduce its temperature. Temperature scales (i.e., Celsius, Fahrenheit, Kelvin) are based on the temperature points where pure water freezes and vaporizes for a pressure of one atmosphere. At the introductory physics level, the absolute temperature scale (Kelvin) is related to ideal gas laws. Temperature is usually understood as the microscopic internal energy per degree of freedom. In this section, we discuss a more fundamental understanding of the absolute temperature and its connection to entropy.

Using the definition and properties of entropy we expand on the notion of the absolute temperature T. Let us consider an isolated system partitioned into two subsystems as in Figure 2.1. Subsystem I has s_I degrees of freedom and subsystem II has s_{II} degrees of freedom.

If a system is found in an equilibrium state, its entropy is maximized

$$S = S_{max} \tag{2.1}$$

As the phase of the distribution function is an additive (extensive) quantity, then S is also an additive quantity. Especially for the microcanonical distribution we have

$$\Delta \Gamma_{I+II} = \frac{\Delta q_I \Delta q_{II} \Delta p_I \Delta p_{II}}{(2\pi\hbar)^{s_I+s_{II}}} = \Delta \Gamma_I \Delta \Gamma_{II} \tag{2.2}$$

Then it follows from definition that the total entropy is

$$S_{I+II} = \ln \Delta \Gamma_I + \ln \Delta \Gamma_{II} = S_I + S_{II} \tag{2.3}$$

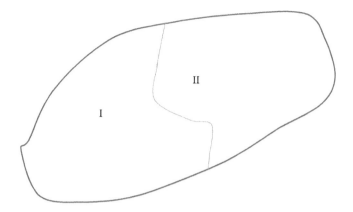

FIGURE 2.1   Two subsystems I and II that make up a larger system.

We assume that both entropy contributions are functions of the energy of the given subsystem. It follows that

$$S_{I+II} = S_I(E_I) + S_{II}(E_{II}) \tag{2.4}$$

and

$$E_{I+II} = E_I + E_{II} = \text{const} \tag{2.5}$$

This shows that the energy differentials are not independent:

$$dE_I = -dE_{II} \tag{2.6}$$

If we consider Equations 2.1 and 2.6, then for the extremum

$$\frac{\partial S}{\partial E_I} = 0 \tag{2.7}$$

and

$$\frac{\partial S_I}{\partial E_I} + \frac{\partial S_{II}}{\partial E_{II}} \frac{\partial E_{II}}{\partial E_I} = 0 \tag{2.8}$$

Using Equation 2.6, one finds a common derivative for the two subsystems in mutual equilibrium

$$\frac{\partial S_I}{\partial E_I} = \frac{\partial S_{II}}{\partial E_{II}} \tag{2.9}$$

The function

$$T \equiv \frac{\partial E}{\partial S} \tag{2.10}$$

is taken to define the *absolute temperature*, which has one and the same value, $T_I = T_{II}$, for the two subsystems. The relation shows a basic physical effect of temperature: when energy is added to a system, causing its temperature to increase, the amount of that increase actually depends on how the system accesses more microstates with the added energy.

Now, let us examine a nonequilibrium state. The system obeys the thermodynamic principle of increase of entropy:

$$\frac{dS}{dt} \geq 0 \tag{2.11}$$

Considering our partitioned system, we have

$$\frac{\partial S_I}{\partial E_I} \frac{\partial E_I}{\partial t} + \frac{\partial S_{II}}{\partial E_{II}} \frac{\partial E_{II}}{\partial E_I} \frac{\partial E_I}{\partial t} = \frac{dE_I}{dt} \left(\frac{\partial S_I}{\partial E_I} - \frac{\partial S_{II}}{\partial E_{II}} \right) \geq 0 \tag{2.12}$$

Let us suppose that the energy in subsystem I is decreasing due to a nonequilibrium process. Assuming $dE_I/dt < 0$ implies that

$$\frac{\partial S_I}{\partial E_I} - \frac{\partial S_{II}}{\partial E_{II}} \leq 0 \tag{2.13}$$

and this implies that

$$\frac{1}{T_I} \leq \frac{1}{T_{II}} \tag{2.14}$$

or simply

$$T_I \geq T_{II} \tag{2.15}$$

In this process where subsystem I lost energy, it gave that energy to subsystem II. When energy is exchanged due to a temperature difference, we say that heat flowed. In this case, we see that the heat must necessarily flow from a hotter region to a cooler region, that is, from a region with higher T to a region with lower T. Heat never flows from a cooler object to a hotter object. If that were to happen, it would violate the principle of increase of entropy.

The definition of absolute temperature

$$\frac{\partial S}{\partial E} = \frac{1}{T} \tag{2.16}$$

is also known as the *thermal equation of state*. It should be noted that for a real macroscopic body, we always have

$$T \geq 0 \tag{2.17}$$

and we prove it as follows. Suppose a system has a negative temperature, with $\partial S/\partial E < 0$, which is identical to $\partial E/\partial S < 0$. Let us have a body with kinetic energy $E_K = p^2/2m$ and the internal energy E_{int} (due to microscopic degrees of freedom), then the total energy is the sum over many degrees of freedom:

$$E = \frac{p^2}{2m} + E_{int} \tag{2.18}$$

It is obvious that $S = S(E_{int})$. Let us examine now $\partial S/\partial E < 0$. Considering the principle of increase of entropy for the transition to the equilibrium state, we have $dS > 0$ and hence from Equation 2.10, $dE_{int} < 0$. It follows that E_K would increase during the relaxation toward equilibrium, but this is not observed anywhere in nature. It follows that we may not have $\partial S/\partial E < 0$ nor $\partial S/\partial E = 0$. This implies that the absolute temperature $T \geq 0$. However, there may be some exceptional cases, for example, say, a spin model with an upper limit on energy, then it is easy to get $\partial S/\partial E < 0$ as the upper energy limit is reached. The spin degrees of freedom by themselves can have a negative temperature.

Q2.1 It is supposed that between the entropy S and probability W of a state of a system there exists some functional dependence (*Boltzmann principle*). Using the general properties of the entropy and the probability establish the Boltzmann relation $S = \text{const} \times \ln W$.

A2.1 According to this Boltzmann principle

$$S = f(W) \tag{2.19}$$

If the system is made up of two parts then

$$S_1 = f(W_1), \quad S_2 = f(W_2) \tag{2.20}$$

Considering the additive properties of the entropy then

$$S = S_1 + S_2 = f(W_1) + f(W_2) = f(W) \tag{2.21}$$

For independent subsystems, the probabilities combine as

$$W = W_1 W_2 \tag{2.22}$$

Thus, from the definition of $f(W)$, we may obtain a functional equation

$$f(W_1) + f(W_2) = f(W_1 W_2) \tag{2.23}$$

If we differentiate Equation 2.23 with respect to W_1 and W_2, then we have

$$f'(W_1) = f'(W_1 W_2)W_2, \quad f'(W_2) = f'(W_1 W_2)W_1 \tag{2.24}$$

where prime indicates differentiation with respect to the indicated argument of f. These can be multiplied by W_1 and W_2, respectively, which gives a result that does not depend on the argument

$$f'(W_1)W_1 = f'(W_2)W_2 = \text{const} \tag{2.25}$$

After integration we have

$$f(W) = \text{const} \times \ln W \tag{2.26}$$

or

$$S_{(\text{state})} = -k_B \ln W_{(\text{state})} \tag{2.27}$$

The constant k_B is the Boltzmann constant; the minus sign is included to give positive entropy, since the probabilities are less than one. We use units where entropies are in units of the Boltzmann constant. Hence, k_B will not be included in the formulas, and temperatures can be considered to have units of energy.* The constant k_B may be evaluated experimentally by applying thermodynamics to a particular case, say, an ideal gas.

The expression (2.27) gives the contribution to total entropy only due to a particular state of the system. As the probability of that state is $W_{(\text{state})}$, the total system entropy is the average over all states, weighted by their respective probabilities

$$S = -k_B \sum_i W_i \ln W_i \tag{2.28}$$

The sum takes place over all possible states in which the system can be found. This recovers the entropy definition (1.263).

---

* If it is desired to replace Boltzmann's constant into the equations, changing T to $k_B T$ while leaving the entropy dimensionless will accomplish this.

2.2   ADIABATIC PROCESSES

We suppose that a body is thermally isolated and that the external fields in which it is found change very slowly. Such a process is called *adiabatic*. We show that for an adiabatic process, the entropy of the body remains unchanged, that is, the process is reversible. An alternative statement is that for an adiabatic process, $\dot{S} = 0$. Let us characterize the external conditions by the parameter λ that is a function of time t. The parameter λ may be set equal to volume or position of surface element or some other physical property. If the parameter λ of the field changes very slowly, then the given process proceeds slowly and as a consequence we have an adiabatic process. Let Δt be the time for which λ changes by $\Delta\lambda = \lambda_2 - \lambda_1$. It must be assumed that $\Delta t \gg t_\kappa$ where t_κ is a "relaxation time" of the system, which is the typical time scale for the system to attain equilibrium if slightly perturbed away from equilibrium. An experimentally realized adiabatic process, however, must be done fast enough so that there is little chance for the system to exchange heat with its surroundings.

The entropy is determined by the change in λ, which depends on $\dot{\lambda}$. From the thermodynamic principle of increase of entropy

$$\dot{S}(\dot{\lambda}) = \dot{S}(0) + \frac{\partial \dot{S}}{\partial \dot{\lambda}}\bigg|_{\dot{\lambda}=0} \dot{\lambda} + \frac{1}{2}\frac{\partial^2 \dot{S}}{\partial \dot{\lambda}^2}\bigg|_{\dot{\lambda}=0} \dot{\lambda}^2 + \cdots \geq 0 \qquad (2.29)$$

We have expanded $\dot{S}(\dot{\lambda})$ in a series of $\dot{\lambda}$, as $\dot{\lambda}$ is very small. The term $\dot{S}(0)$ should be equal to zero since the entropy of a closed system found in thermodynamic equilibrium for constant external conditions should be unchanged. The term $(\partial \dot{S}/\partial \dot{\lambda})\big|_{\dot{\lambda}=0} \dot{\lambda}$ should be equal to zero. That term changes its sign when $\dot{\lambda}$ changes its sign, and considering the law of increase of entropy, the quantity $\dot{S}(\dot{\lambda})$ is always positive. It follows from here that for small $\dot{\lambda}$, we have

$$\frac{dS}{dt} = A_2 \dot{\lambda} \frac{d\lambda}{dt} \qquad (2.30)$$

where

$$A_2 = \frac{1}{2}\frac{\partial^2 \dot{S}}{\partial \dot{\lambda}^2}\bigg|_{\dot{\lambda}=0} \qquad (2.31)$$

This gives

$$\frac{dS}{d\lambda} = A_2 \dot{\lambda} \qquad (2.32)$$

And as $\dot{\lambda} \to 0$ for the adiabatic limit, we have

$$\frac{dS}{d\lambda}\bigg|_{\dot{\lambda}\to0} = A_2 \dot{\lambda}\big|_{\dot{\lambda}\to0} \to 0 \qquad (2.33)$$

Considering finite increments, it follows that

$$\frac{S_2 - S_1}{\lambda_2 - \lambda_1} \to 0 \tag{2.34}$$

and as a consequence

$$S(\lambda) = \text{const} \tag{2.35}$$

This proves the reversibility of an adiabatic process. It should be noted that although the adiabatic process is reversible, not all reversible processes are adiabatic.

2.3 PRESSURE

A fundamental thermodynamic property for gases is the pressure P. Obviously, pressure on the walls of a container is due the average effect of an enormous number of collisions of gas molecules. That is from a statistical viewpoint, but pressure can also be considered simply from thermodynamic and mechanics using energy and force considerations.

Let us introduce the formula that will enable us to evaluate purely thermodynamically a mean value, which can then be applied to find pressure and other quantities. We suppose that a given body executes an adiabatic process. We denote the derivative of its internal energy relative to time as dE/dt. From definition, the thermodynamic energy is $E = \overline{E(q, p, \lambda)}$, where $E(q, p, \lambda)$ is the Hamiltonian of the given body and is dependent on some parameter λ. It is known from mechanics that

$$\frac{dE(q, p, \lambda(t))}{dt} = \frac{\partial E}{\partial q}\dot{q} + \frac{\partial E}{\partial p}\dot{p} + \frac{\partial E(q, p, \lambda(t))}{\partial t} \tag{2.36}$$

and from the equation of motion we have

$$\dot{q} = \frac{\partial E}{\partial p}, \quad \frac{\partial E}{\partial q} = -\dot{p}$$

Substituting this in Equation 2.36, we get

$$\frac{dE}{dt} = \frac{\overline{dE(q, p, \lambda)}}{dt} = \frac{\overline{\partial E(q, p, \lambda)}}{\partial \lambda}\dot{\lambda} \tag{2.37}$$

As the operation of taking the mean value relative to the statistical distribution and the operation of differentiation relative to time may be executed in an arbitrary order, we get Equation 2.37.

It should be noted that

$$\overline{E(q,p,\lambda)} = E(\text{var},\lambda) \tag{2.38}$$

where var stands for variables that may be *intensive* (they do not depend on the size of the system) like T, P, N/V, and so on, or *extensive* (they are proportional to the size of the system) like S, V, N, and so on.

If a system is homogeneous, then it is sufficient to take two parameters to describe a state function, meaning any physical quantity depending on the thermodynamic state. We may write dE/dt in another form and examine the thermodynamic quantity E as a function of the entropy S and an external parameter λ. If we consider the fact that for an adiabatic process the entropy S is a constant, then we have

$$\frac{dE}{dt} = \left(\frac{\partial E}{\partial \lambda}\right)\bigg|_{S} \dot{\lambda} \tag{2.39}$$

If we consider Equations 2.37 and 2.39, then we have

$$\frac{\partial \overline{E(q,p,\lambda)}}{\partial \lambda} = \left(\frac{\partial E(\lambda)}{\partial \lambda}\right)_{S} \tag{2.40}$$

a relation that enables us to evaluate different mean values, using the derivative with respect to the external parameter, taken at constant entropy.

Let us use this to find the force that a body exerts on the boundary of its volume V (as in the pressure force of a gas on its container). We suppose an adiabatic deformation of the boundary. Consider the well-known formula of mechanics for the force exerted on some area element of a surface, $d\vec{A}$, imagining an adiabatic deformation $d\vec{r}$

$$\vec{F}(q,p,\vec{r}) = -\frac{\partial E(q,p,\vec{r})}{\partial \vec{r}} \tag{2.41}$$

$E(q,p,\vec{r})$ is the energy of the body as a function of its coordinates, momenta, and also the radius vector $\vec{r}$ that specifies the location of a boundary element that acts as the parameter λ. If we take the mean of Equation 2.41, then we have

$$\overline{\vec{F}} = -\frac{\partial \overline{E(q,p,\vec{r})}}{\partial \vec{r}} = -\left(\frac{\partial E(q,p,\vec{r})}{\partial \vec{r}}\right)_{S} = -\left(\frac{\partial E}{\partial V}\right)_{S} \frac{\partial V}{\partial \vec{r}} \tag{2.42}$$

where V is the volume, which changes by an increment, $dV = d\vec{r} \cdot d\vec{A}$. Thus

$$\overline{\vec{F}} = -\left(\frac{\partial E}{\partial V}\right)_{S} d\vec{A} \tag{2.43}$$

The mean value of the force acting on the surface element is in the direction of the normal to that element and is proportional to its surface area. This is just *Pascal's law*. The force acting per unit area is the pressure

$$P = -\left(\frac{\partial E}{\partial V}\right)_S \qquad (2.44)$$

The result is fascinating in its simplicity, and only requires knowledge of the energy–volume relationship at constant entropy.

2.3.1 QUESTIONS ON STATIONARY DISTRIBUTIONS FUNCTIONS AND IDEAL GAS STATISTICS

Q2.2 N identical particles of a monatomic ideal gas are found in a volume V described by the microcanonical distribution with energy E_0 and width dE. Evaluate the phase volume $\Delta\Gamma$, entropy S, and temperature T of the gas. Find the equation of state of the gas.

A2.2 As the energy E of the ideal gas is dependent on the generalized coordinates q and momenta p, the phase volume of the system of N particles is the region between the energy contours at E_0 and $E_0 + dE$

$$\Delta\Gamma = \Delta\Gamma(E_0,V) = \int_{E_0 \leq E(q,p) \leq E_0 + dE} dq\, dp \qquad (2.45)$$

where

$$dq\, dp = d\Gamma = dq_1 \ldots dq_{3N} dp_1 \ldots dp_{3N} \qquad (2.46)$$

The number of degrees of freedom of the gas of N particles is $3N$. An energy width dE is assumed. It is expected that the final physical results will not depend on the size of dE.

We first perform the integration over the coordinates of the N particles. It should be noted that the integral over the coordinates of a given particle simply yields the volume of the container, V, since the energy E is independent of the locations of the particles in an ideal gas. There are N such integrals. Thus, integration over the coordinates of the given particles yields the quantity V^N:

$$\Delta\Gamma = V^N \int dp \qquad (2.47)$$

Considering the fact that the gas is ideal, there are no interatomic forces, and the total energy is simply the sum of the individual translational kinetic energies of the gas particles. (If the gas is diatomic or polyatomic, we need to include rotational degrees of freedom.) The domain of integration in the momentum space is defined inside a range of energies of width dE:

$$E_0 \leq \frac{1}{2m}\sum_{i=1}^{3N} p_i^2 \leq E_0 + dE \tag{2.48}$$

The lower limit defines a radius

$$\sum_{i=1}^{3N} p_i^2 \leq 2mE_0 \equiv R^2 \tag{2.49}$$

From the geometrical point of view, it is a sphere of radius $R = \sqrt{2mE_0}$ in the $3N$-dimensional momentum space. The upper limit in Equation 2.48 is another concentric sphere with a slightly larger radius. The phase space volume required corresponds to the change in volume between these two spheres.

If we integrate Equation 2.47 over the momenta then this is the required volume. The volume $\Gamma(E_0, V)$ of the smaller sphere is proportional to R^{3N}, and can be expressed as

$$\Gamma(E_0,V) = A_N V^N E_0^{3N/2} \tag{2.50}$$

Here, A_N is a constant that is independent of the volume and the energy. We evaluate this constant in Q2.3. Then the difference in their volumes is

$$\Delta\Gamma(E_0,V) = \Gamma(E_0 + dE,V) - \Gamma(E_0,V) = \frac{d\Gamma}{dE_0}dE \tag{2.51}$$

Thus, the microcanonical phase volume $\Delta\Gamma(E_0, V)$ is

$$\Delta\Gamma(E_0,V) = A_N V^N E_0^{3N/2-1}\,dE \tag{2.52}$$

For the rest of the calculations, we can drop the 0 subscript on E_0. The entropy S of the system is then

$$S = \ln\Delta\Gamma = \ln A_N + N\ln V + \left(\frac{3N}{2} - 1\right)\ln E + \ln dE \tag{2.53}$$

At this point, we can make the very reasonable approximation that the number of particles is so great that the factor on $\ln E$ can be replaced by $3N/2$. The error in doing so is irrelevant because N is of the order of Avogadro's number for any macroscopic sample. This is the so-called thermodynamic limit. The term $\ln dE$ is a fixed irrelevant constant, considering that dE is fixed. So, to an extremely good approximation, the entropy is

$$S = \ln \Delta\Gamma = \ln A_N + N \ln V + \frac{3N}{2} \ln E \qquad (2.54)$$

The temperature of the ideal gas is defined by

$$T = \left(\frac{\partial E}{\partial S} \right)_V \qquad (2.55)$$

Let us evaluate

$$\left(\frac{\partial S}{\partial E} \right)_V = \frac{3}{2} \frac{N}{E} \qquad (2.56)$$

Then the mean energy E is

$$E = \frac{3}{2} NT \qquad (2.57)$$

It should be noted that the internal energy E is a function of temperature T alone, with no volume dependence. The classical monatomic ideal gas is seen to have an internal energy of $\frac{3}{2}T$ per particle, or $\frac{1}{2}T$ per degree of freedom. For classical systems, this result is known as the *equipartition theorem*. If all terms in the energy are quadratic, then the mean energy is spread equally over all degrees of freedom (hence the name "equipartition").

The pressure p is

$$p = \frac{\partial E}{\partial V} = \frac{\partial E}{\partial \Delta\Gamma} \frac{\partial \Delta\Gamma}{\partial V} \qquad (2.58)$$

But

$$\frac{\partial \Delta\Gamma}{\partial V} = \frac{N}{V} \Delta\Gamma, \qquad \frac{\partial E}{\partial \Delta\Gamma} = \frac{2}{3N} \frac{E}{\Delta\Gamma} \qquad (2.59)$$

Then

$$pV = \frac{2}{3} E = NT \qquad (2.60)$$

This corresponds to the equation of state for an ideal gas. The relationship (2.60) may also be obtained by evaluating p using the thermodynamic identity (this gives a union between the First and Second Laws of Thermodynamics):

$$p = T\left(\frac{\partial S}{\partial V}\right)_E = T\frac{N}{V} \tag{2.61}$$

Q2.3 For N noninteracting linear oscillators with total energy E_0, assume a true microcanonical distribution. Evaluate the phase volume $\Delta\Gamma$, the entropy S, and the temperature T (i.e., find the equation of state involving the temperature).

A2.3 The difference with Equation 2.48 is that the domain of integration is defined now by the condition

$$E_0 \leq \sum_{\kappa=1}^{N}\left[\frac{p_\kappa^2}{2m} + \frac{m\omega^2 q_\kappa^2}{2}\right] \leq E_0 + dE \tag{2.62}$$

where κ labels each oscillator of mass m and identical frequencies ω. For the variables p_κ and $\tilde{q}_\kappa = m\omega q_\kappa$, the lower limit is an equation of the $2N$-dimensional sphere of radius $R = \sqrt{2mE_0}$. Consequently, the volume of that sphere defines a phase volume

$$\Gamma(E_0) = B_N E_0^N \tag{2.63}$$

where B_N is a constant independent of the energy. The desired microcanonical phase volume is the difference of two spheres' volumes that define the constant energy region

$$\Delta\Gamma = \Gamma(E_0 + dE) - \Gamma(E_0) = \frac{d\Gamma}{dE_0}dE = NB_N E_0^{N-1}\, dE \tag{2.64}$$

From this it follows that

$$S = \ln\Delta\Gamma = \ln(NB_N\, dE) + (N-1)\ln E \tag{2.65}$$

and

$$T = \frac{dE}{dS} = \frac{E}{N-1} \tag{2.66}$$

as the volume in the given case is now not the external parameter. From Equation 2.66, the equation of state follows

$$E = (N-1)T \tag{2.67}$$

In the thermodynamic limit, $N - 1$ can be replaced by N. Then, the set of oscillators obeys the classical equipartition theorem, as each oscillator has two degrees of freedom (i.e., a kinetic energy quadratic in p and a potential energy quadratic in q), each contributing an average internal energy of $\frac{1}{2}T$.

Q2.4 Find the normalization factor $\Omega(E)$ of the Gibbs microcanonical distribution for

    a. The ensemble of N particles of an ideal monatomic gas
    b. The ensemble of N independent linear oscillators

A2.4 If we consider a closed system, its energy is a constant:

$$E(q, p, \lambda) = E_0 = \text{const} \tag{2.68}$$

where λ are external parameters that may be the volume and so on. The distribution density in our case is defined by the microcanonical distribution:

$$\rho(q, p; E_0) = C\delta(E(q, p, \lambda) - E_0) \tag{2.69}$$

The normalization factor

$$C = \frac{1}{\Omega(E_0, \lambda)} \tag{2.70}$$

is defined from the condition

$$\Omega(E_0, \lambda) = \int \delta(E(q, p, \lambda) - E_0) \, dq \, dp \tag{2.71}$$

where

$$dq \, dp = dq_1 \ldots dq_{3N} \, dp_1 \ldots dp_{3N} \tag{2.72}$$

It is obvious that

$$\Omega(E_0, \lambda) = \left. \frac{\partial \Gamma(E, \lambda)}{\partial E} \right|_{E=E_0}, \quad \Gamma(E_0, \lambda) = \int_{E < E_0} dq \, dp \tag{2.73}$$

The quantity $\Gamma(E_0, \lambda)$ denotes the volume of phase space enclosed by the constant energy surface given by the energy (2.68). The quantity $\Delta\Gamma = \Omega(E, \lambda) \, dE$ is the phase volume of an infinitesimally thin layer between the constant energy surfaces E and $E + dE$.

a. Ideal monatomic gas. The phase volume $\Gamma(E_0, \lambda)$ as seen above is enclosed by the constant energy surface (2.68). For the given system of N particles this volume is defined by Equation 2.50. How do we find the constant A_N? We evaluate the volume of the sphere of radius $R = \sqrt{2mE}$ in the 3N-dimensional momentum space.

If we examine the $s = 3N$ momentum variables in our space, identified as $Q_i^0 = p_i$, $i = 1,2,3...,s$, then we evaluate the volume $V_s(R)$ bounded by the following sphere:

$$\sum_{i=1}^{3N} p_i^2 = \left(Q_1^0\right)^2 + \left(Q_2^0\right)^2 + \cdots + \left(Q_s^0\right)^2 = R^2 = 2mE \qquad (2.74)$$

and

$$V_s(R) = \int dQ_1^0 dQ_2^0 \ldots dQ_s^0 \qquad (2.75)$$

Consider

$$\left(\frac{Q_1^0}{R}\right)^2 + \left(\frac{Q_2^0}{R}\right)^2 + \cdots + \left(\frac{Q_s^0}{R}\right)^2 = 1 \qquad (2.76)$$

Then, we do the change of variables:

$$Q_i^0 = Q_i R \qquad (2.77)$$

and Equation 2.75 becomes

$$V_s(R) = R^s V_s(1) \qquad (2.78)$$

where

$$V_s(1) = \int dQ_1 dQ_2 \ldots dQ_s \qquad (2.79)$$

Integration in Equation 2.79 is evaluated in the domain bounded by the surface

$$(Q_1)^2 + (Q_2)^2 + \cdots + (Q_s)^2 = 1 \qquad (2.80)$$

Traverse the s-dimensional sphere of radius 1 by a plane at a distance Q_s from the center. We obtain a cross-sectional area of the constant energy surface of radius $\sqrt{1 - Q_s^2}$, that is,

$$V_s(1) = \int dQ_s \int dQ_1 \ldots dQ_2 dQ_{s-1} = \int dQ_s V_{s-1}\left(\sqrt{1-Q_s^2}\right) = V_{s-1}(1)B\left(\frac{1}{2}, \frac{s+1}{2}\right) \tag{2.81}$$

Here, $B(\alpha, \beta)$ is the Euler–Beta function defined by

$$B(\alpha, \beta) = \int_0^1 x^{\alpha-1}(1-x)^{\beta-1}\, dx = \frac{\Gamma(\alpha)\Gamma(\beta)}{\Gamma(\alpha + \beta)} \tag{2.82}$$

and $\Gamma(x)$ is the Gamma function. Thus

$$V_s(1) = V_{s-1}(1)\frac{\Gamma(1/2)\Gamma(s + 1/2)}{\Gamma((s/2) + 1)} \tag{2.83}$$

With the help of this recurrence relationship we have

$$V_s(1) = \frac{\pi^{s/2}}{\Gamma((s/2) + 1)} \tag{2.84}$$

Here, $V_1(1) = 1$. Thus, the general formula for the volume of an s-dimensional sphere is

$$V_s(R) = \frac{\pi^{s/2}}{\Gamma((s/2) + 1)}R^s \tag{2.85}$$

For the case of interest with $s = 3N$, the phase volume of the system of N particles is

$$\Gamma(E,V) = \frac{\pi^{(3N/2)}}{\Gamma((3N/2) + 1)}(2mE)^{(3N/2)}V^N \tag{2.86}$$

Compare this result with Equation 2.50, then we have

$$A_N = \frac{\pi^{(3N/2)}}{\Gamma((3N/2) + 1)}(2m)^{(3N/2)} \tag{2.87}$$

Now, if we use the property

$$\Gamma(x + 1) = x\Gamma(x) \tag{2.88}$$

then

$$\Omega(E_0) = \left.\frac{\partial \Gamma}{\partial E}\right|_{E=E_0} = \frac{(2\pi m)^{(3N/2)}}{\Gamma(3N/2)} V^N E_0^{(3N/2)-1} \tag{2.89}$$

Thus, the normalized microcanonical probability density $\rho(E)$ is

$$\rho(E) = \delta(E - E_0)\frac{\Gamma(3N/2)}{(2\pi m)^{(3N/2)}} V^{-N} E_0^{1-(3N/2)} \tag{2.90}$$

b. For the system of N independent linear oscillators, both the coordinates q_i and the momenta p_i determine the energy. The volume of the $2N$-dimensional sphere of radius $R = \sqrt{2mE_0}$ is needed. If we consider Equation 2.85 with dimension $s = 2N$, the required spherical volume is equal to

$$\frac{\pi^N}{\Gamma(N+1)}(2mE_0)^N \tag{2.91}$$

The phase volume of the system is this quantity divided by $(m\omega)^N$, and is equal to

$$\Gamma(E_0) = \frac{\pi^N}{\Gamma(N+1)}\left(\frac{2E_0}{\omega}\right)^N \tag{2.92}$$

Comparing this expression with Equation 2.63, then we have

$$B_N = \frac{1}{N!}\left(\frac{2\pi}{\omega}\right)^N \tag{2.93}$$

and

$$\Omega(E_0) = \left(\frac{2\pi}{\omega}\right)^N \frac{E_0^{N-1}}{(N-1)!} \tag{2.94}$$

Thus, the distribution density $\rho(E)$ for the oscillators is

$$\rho(E) = \delta(E - E_0)\frac{(N-1)!}{E_0^{N-1}}\left(\frac{\omega}{2\pi}\right)^N \tag{2.95}$$

Q2.5 Develop the Gibbs canonical distribution from the general formula of the microcanonical distribution, by assuming a system of n particles in close contact with a thermostat of N particles (a large thermal reservoir at constant temperature). Assume that $(E/n) \equiv (E'/N) = \text{const}$ as $N \to \infty$ (where E and E' are the energy of the closed system and that of the thermostat, respectively). Consider the two cases:

a. The totality of n particles of an ideal gas
b. The totality of n independent linear oscillators

Show that the results are independent of the selection of the thermostat.

A2.5 a. We consider the ideal gas system with the energy

$$E(q_i, p_i) = E, \quad i = 1, 2, \ldots, n \tag{2.96}$$

that is in thermal constant with the thermostat having the energy

$$E'(q_i', p_i') = E', \quad i = 1, 2, \ldots, N, \quad N \gg n \tag{2.97}$$

These form a closed system with total energy

$$E + E' = E_0 \tag{2.98}$$

In order to find the probability density $\rho(E)$ for the ensemble, we use the micro-canonical distribution for the entire system:

$$\rho(E, E') = C(E_0)\delta(E + E' - E_0) \tag{2.99}$$

We want to find the distribution for only our system of interest. Thus, it is necessary to integrate over the phase space of the thermostat. Considering the rule of summation of probabilities

$$\rho(E) = \int \rho(E, E') \, dq_i' \, dp_i' = \frac{C(E_0)}{C'(E_0 - E)} \tag{2.100}$$

where

$$C'(E_0 - E) = \frac{\partial E'}{\partial \Gamma'} \tag{2.101}$$

Here, Γ' is the volume bounded by the constant energy surface of the thermostat. Using again the result of Q2.4 we have

$$\Omega(E_0 - E) = \frac{d\Gamma'}{dE'} = \frac{(2\pi m)^{(3N/2)}}{\Gamma(3N/2)} V^N E_0^{(3N/2)-1} \left(1 - \frac{E}{E_0}\right)^{(3N/2)-1} \tag{2.102}$$

As $N \to \infty$ and $E_0/N = (3/2)T$ is constant, then we have

$$\rho(E) = \lim_{N\to\infty} \frac{(2\pi m)^{(3N/2)} V^N E_0^{(3N/2)-1}}{\Omega(E_0)\Gamma(3N/2)} \lim_{N\to\infty} \left(1 - \frac{E}{(3NT/2)}\right)^{(3N/2)-1} \qquad (2.103)$$

The first limit in expression (2.103) is independent of the energy E. This first limit is finite as the normalized divisor $\Omega(E_0) \geq \Omega'(E_0)$. If we write it as some constant, then we have

$$\rho(E) = \text{const} \times \lim_{N\to\infty} \left[\left(1 - \frac{E}{(3NT/2)}\right)^{-(3N/2)(T/E)}\right]^{-(E/T)} = \text{const} \times \exp\left\{-\frac{E}{T}\right\} \qquad (2.104)$$

For the canonical ensemble derived here, the system exchanges energy with the thermostat, and hence, there is a wide distribution of possible energies. The most probable states become those with the lowest energy. The width of the distribution increases as the temperature increases.

b. Considering Equation 2.102 for a set of linear oscillators we have

$$\Omega'(E_0 - E) = \left(\frac{2\pi}{\omega}\right)^N \frac{E_0^{N-1}}{(N-1)!}\left(1 - \frac{E}{E_0}\right)^{N-1} \qquad (2.105)$$

We now let $N \to \infty$, then considering equation of state (2.67), where $E_0/N = T$ reaches a finite limit, finally we have

$$\rho(E) = \text{const} \times \lim_{N\to\infty} \left[\left(1 - \frac{E}{NT}\right)^{-(NT/E)}\right]^{-(E/T)} = \text{const} \times \exp\left\{-\frac{E}{T}\right\} \qquad (2.106)$$

It should be noted that formulas (2.104) and (2.106) might be obtained in the general form without any concrete model of the thermostat. This distribution for the canonical ensemble is more physically realizable than the microcanonical ensemble because any real system will be in contact with some surroundings, usually at a specified temperature. Thus, it plays a major role in statistical physics.

2.4   THERMODYNAMIC IDENTITY

Thus far we supposed that the entropy is determined by the states accessible to the system, according to their microscopic energies. In thermodynamics, we do not have knowledge of the microscopic energy. Only its statistical average value (or thermodynamic value) is observed. From the point of view of thermodynamics we can suppose that the expectation value of the energy of a system is related to its entropy, that is, we suppose that $E = E(S, V)$. Then variations in entropy or volume lead to thermodynamic energy changes

$$dE = \left(\frac{\partial E}{\partial V}\right)_S dV + \left(\frac{\partial E}{\partial S}\right)_V dS \tag{2.107}$$

We already know, however, the fundamental identities

$$\left(\frac{\partial E}{\partial V}\right)_S = -p, \quad \left(\frac{\partial E}{\partial S}\right)_V = T \tag{2.108}$$

Then Equation 2.107 becomes a fundamental relation

$$dE = T\,dS - p\,dV \tag{2.109}$$

called the thermodynamic identity.

Let us study the properties of the pressure p. Consider two gases A_1 and A_2 contained in a sealed cylinder and separated by a movable piston (Figure 2.2).

Suppose that the piston is not insulating but clamped in position. Energy (i.e., heat) will in general flow from one gas to the other (although no macroscopic work gets done) and the pressures of the gases will change as a result. This is an example of a purely thermal interaction. Suppose more generally, that the piston in Figure 2.2 conducts heat and is free to move. Then the two systems A_1 and A_2 can interact by exchanging heat and by doing work on each other. Systems A_1 and A_2 are characterized by the energies E_1 and E_2, respectively. Let us suppose that the combined system

$$A = A_1 + A_2 \tag{2.110}$$

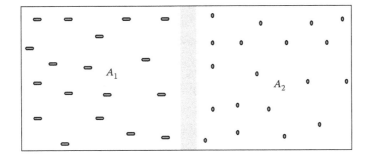

FIGURE 2.2 Two gases A_1 and A_2 contained in a sealed cylinder and separated by a movable piston.

is isolated, then

$$E_1 + E_2 = E = \text{const} \tag{2.111}$$

The gases A_1 and A_2 are described by their volumes V_1 and V_2, respectively, and as the piston moves the total volume remains unchanged, that is

$$V_1 + V_2 = V_0 = \text{const} \tag{2.112}$$

When the system arrives at the equilibrium state then the total entropy is maximum

$$S_{\text{equil}} = S_{\text{max}} \tag{2.113}$$

and

$$S = S_1(V_1) + S_2(V_2) \tag{2.114}$$

If we consider Equation 2.114, then we have

$$\frac{\partial S}{\partial V_1} = \frac{\partial S_1}{\partial V_1} + \frac{\partial S_2}{\partial V_2}\frac{\partial V_2}{\partial V_1} = 0 \tag{2.115}$$

and as

$$\frac{\partial V_2}{\partial V_1} = -1 \tag{2.116}$$

then it follows that

$$\frac{\partial S_1}{\partial V_1} = \frac{\partial S_2}{\partial V_2} \tag{2.117}$$

If we consider Equation 2.109, then we have

$$dS = \frac{dE}{T} + \frac{p}{T}dV \tag{2.118}$$

From this it is obvious that

$$\frac{\partial S}{\partial V} = \frac{p}{T} \tag{2.119}$$

and as a consequence of Equation 2.117

$$\frac{p_1}{T_1} = \frac{p_2}{T_2} \tag{2.120}$$

As the temperatures T_1 and T_2 at equilibrium are the same (if the two sides exchange heat), then it follows that $p_1 = p_2$; the pressures of systems that are in thermal equilibrium are equal.

An Alternate Derivation: When the system arrives at the equilibrium state, then the total entropy is a maximum

$$S_{\text{equil}} = S_{\text{max}} \tag{2.121}$$

and

$$S = S_1(E_1,V_1) + S_2(E_2,V_2) \tag{2.122}$$

Using the thermodynamic identity (2.109) for the entire system, we have $dE = 0$ if the cylinder exchanges no energy with its surroundings, so that

$$dE = T_1\, dS_1 + T_2\, dS_2 - p_1\, dV_1 - p_2\, dV_2 = 0 \tag{2.123}$$

The first terms represent the heat changes of the two sides, and $dV_2 = -dV_1$ because the total volume is fixed, so

$$dE = \delta Q_1 + \delta Q_2 - p_1\, dV_1 + p_2\, dV_1 = 0 \tag{2.124}$$

If the cylinder is thermally insulated from its surroundings, there is no net heat exchanged, $\delta Q_1 + \delta Q_2 = 0$, regardless of whether the piston conducts heat or not! Then, immediately we conclude that $p_1 = p_2$. As long as the cylinder is completely isolated from its surroundings, and the piston is free to move, it will move until the pressures on the two sides are equal. The result does not require that the temperatures on the two sides to be equal. We did not apply a maximum entropy principle; however, look at the entropy variation in equilibrium:

$$dS = dS_1 + dS_2 = \frac{1}{T_1}(dE_1 + p_1\, dV_1) + \frac{1}{T_2}(dE_2 + p_2\, dV_2) \tag{2.125}$$

Using $dE_2 = -dE_1$ and $dV_2 = -dV_1$, with identical pressures $p_1 = p_2$, we get

$$dS = \left(\frac{1}{T_1} - \frac{1}{T_2}\right)(dE_1 + p_1\, dV_1) \tag{2.126}$$

If the temperatures are different, then $dS = 0$ and the entropy is an extremum only if

$$p = -\left(\frac{\partial E}{\partial V}\right)_S \tag{2.127}$$

which recovers the definition of the pressure. Therefore, if the two sides attain the same pressure, the total entropy is a maximum.

It is easily seen that, in any equilibrium state, the pressure of a system should be positive. Based on the thermodynamic identity (2.109) rewritten as

$$dS = \frac{dE}{T} + \frac{p}{T}dV \tag{2.128}$$

an alternative expression for the pressure is

$$p = T\left(\frac{\partial S}{\partial V}\right)_E \tag{2.129}$$

It is obvious that for $p > 0$ we have $(\partial S/\partial V)_E > 0$ and the entropy may increase only for the expansion of the system, which is limited due to the influence of its surroundings. To the contrary, if $p < 0$, then $(\partial S/\partial V)_E < 0$. The system on its own would compress accompanied by an increase in entropy. The system would continue to compress on its own with nothing to prevent its volume from vanishing, an unphysical result.

2.5 LAWS OF THERMODYNAMICS

The thermodynamic properties of a system are seen to obey four very general laws. These laws have distinct fundamental explanations. We briefly mention first, the so-called zeroth law: If two systems are in thermal equilibrium with a third system, they must be in thermal equilibrium with each other. The Zeroth Law allows the introduction of universal scales for temperature, pressure, and so on. It has a fairly straightforward statistical interpretation. This allows us to make contact between the thermodynamic and statistical descriptions.

The First and Second Laws of thermodynamics relate directly to properties of the energy and entropy at any temperatures. The Third Law applies to the limiting value of the entropy at zero absolute temperature. Indeed, it relates to the impossibility of actually cooling any system to zero temperature. We save further discussion of the Third Law for later, and concentrate initially on the First and Second Laws.

2.5.1   FIRST LAW OF THERMODYNAMICS

We consider now the First Law, that deals with the principle of conservation and transformation of energy. Especially, it gives an accounting of how the internal energy of a system can change, either due to the system absorbing heat from its surroundings or due to doing work on it surroundings.

Suppose we have a system of energy E, a thermostat of energy E', and some external bodies. We apply to the system and the thermostat the law of *conservation of energy*: The total change in energy of the system and the thermostat is equal to the work done on them by the external bodies. We let the quantity $-\eth W$ be the work done by the external bodies. Then, $\eth W$ is the macroscopic work done *by* the system as a result of an infinitesimal change in external parameters λ:

$$dE(q,p,\lambda(t)) + dE'(q,p,\lambda(t)) = -\eth W \qquad (2.130)$$

Here, $\lambda = \{\lambda_i\}_{i=1}^{s}$. Interest lies however not on the instantaneous change in energy and work done but on some mean value with time (this is equal to the mean over the ensemble). Thus

$$\overline{dE(q,p,\lambda)} + \overline{\eth W} = -\overline{dE'(q',p',\lambda)} \qquad (2.131)$$

Let

$$\overline{dE(q,p,\lambda)} = dE, \quad -\overline{dE'(q',p',\lambda)} = \eth Q \qquad (2.132)$$

then

$$dE + \overline{\eth W} = \eth Q \qquad (2.133)$$

It should be noted that since the thermostat interacts only with the system and not with the external bodies then the decrease in energy $-dE'$ might come about as a result of transfer of its energy (i.e., heat) to the system. Any motion of any macroscopic body effecting work and so on does not accompany the process.

The difference between the measurable quantities $\eth Q$ and $\eth W$ is found to be the same for any process in which the system evolves between two given states, independently of the path. This is an indication that dE is an exact differential. It follows that the internal energy is a function of state. This is not true of the differentials $\eth Q$ and $\eth W$.

Suppose some quantity of heat

$$\eth Q = -\overline{dE'} \qquad (2.134)$$

is transferred from the thermostat to the system. Then the increase in energy of the system plus the work done by the system is equal to the quantity of heat absorbed by the system, that is

$$\delta Q \equiv dE + \sum_{s'=1}^{s} \overline{\Lambda_{s'} d\lambda_{s'}} \qquad (2.135)$$

For convenience we write $\Lambda_s \equiv \Lambda$, $\lambda_s \equiv \lambda$. The quantity Λ can be thought of as a generalized force and λ as the generalized displacement. It can be seen from Equation 2.135 that

$$\Lambda \neq -\frac{\partial E}{\partial \lambda} \qquad (2.136)$$

The work done is not only due to the decrease in energy but also due to the heat absorbed by the system.

Consider heat exchange between the system and the thermostat to be absent (the system is thermally insulated or is confined in an adiabatic shield, that is, the system is isolated), then $\delta Q = 0$ and

$$\sum \overline{\Lambda} \, d\lambda = -dE \qquad (2.137)$$

or

$$\overline{\Lambda} = -\left(\frac{\partial E}{\partial \lambda}\right)_{\delta Q=0} \qquad (2.138)$$

That is, these generalized forces must be derivable from a potential energy in the absence of heat transfer. We consider the necessity of interaction of the system and the thermostat for the possible heat transfer from the body to the system.

Consider the Hamiltonian of the system $E(q, p, \lambda)$ to be dependent only on its coordinates q, momenta p, and λ. For fixed (constant) λ, the quantity $E(q, p, \lambda)$ is conserved, considering the Hamilton canonical equation. Similarly, as the Hamiltonian $E'(q', p', \lambda)$ of the thermostat is dependent only on the coordinates q', momenta p', then $E'(q', p', \lambda)$ is constant:

$$\frac{dE'}{dt} = \sum_{i=1}^{s'} \left[\frac{\partial E'}{\partial q_i'} \dot{q}_i' + \frac{\partial E'}{\partial p_i'} \dot{p}_i' \right] = \sum_{i=1}^{s'} \left[\frac{\partial E'}{\partial q_i'} \frac{\partial E'}{\partial p_i'} - \frac{\partial E'}{\partial p_i'} \frac{\partial E'}{\partial q_i'} \right] = 0 \quad (2.139)$$

Let now, the Hamiltonian of the interaction between the system and the thermostat be a function

$$E'' = E''(\ldots, q_i, \ldots, p_i, \ldots, q_i', \ldots, p_i', \ldots) \qquad (2.140)$$

This is a function of coordinates and momenta of both the system and the thermostat. Then the conservation property (for constant λ) will be related to the quantity $E + E' + E''$ and then

$$\eth W = \sum \Lambda \, d\lambda = -d(E + E' + E'') \tag{2.141}$$

As E'' is bounded then

$$\frac{\overline{d\tilde{E}''}}{dt} = \frac{1}{T}\left[E''(T) - E''(0)\right] \approx 0 \tag{2.142}$$

The quantity E'' is usually due to what? It is due to the fact that only the interaction of particles at the interface between system and thermostat is important. The energies E and E'' are proportional to the total number of particles in the system. As the range of intermolecular forces is very small (of the order of 10^{-6} to 10^{-7} cm), the number of particles near the interface of thermostat and system is a very small fraction of the total.

What is the difference between the work done and the quantity of heat? The external work done $\eth W$ is dependent on the interaction of the system with a few macroscopic bodies with coordinates λ_s. The quantity of heat $-\eth Q$ is given to the thermostat (a system with many degrees of freedom). This energy is redistributed between the large numbers of particles in a chaotic manner. Thermal energy transferred into the thermostat becomes less organized. On the other hand, we may control the motion of the macroscopic bodies using the work done on them. We may use the energy of macroscopic moving bodies to perform a useful mechanical operation or a technological process. This may not be done typically with the heat transferred to the thermostat. In this sense, we have different notions of the work done $\eth W$ and the quantity of heat transferred $\eth Q$.

2.5.2 SECOND LAW OF THERMODYNAMICS

The First Law expresses the constraint that physical processes must conserve energy. But many processes could be imagined that do not violate the First Law, and yet never happen in the real world. For example, a cup of hot coffee could grab thermal energy from its surroundings (say, only taking energy from the fastest molecules) and get hotter. But that never happens. The Second Law of Thermodynamics expresses an additional constraint, concerning the entropy, which all physical processes must follow.

Recall that an equilibrium macrostate of a system is characterized by the *entropy* S, with the following properties:

- In any process in which a thermally *isolated* system goes from one macrostate to another, the entropy tends to increase or remain the same, that is, *entropy never decreases*

$$\Delta S \geq 0 \tag{2.143}$$

- If the system is not isolated and undergoes a quasi-static infinitesimal process in which it absorbs heat $\eth Q$, then the *amount of its entropy change* is

$$dS = \frac{\eth Q}{T} \tag{2.144}$$

The Second Law of Thermodynamics can be stated in different equivalent ways. The fundamental statement is that in any physical process, the total entropy of a system and its surroundings never decreases. It is a very general result. Although a particular system could experience a decrease in its entropy that must be accompanied by a larger increase in entropy of the surroundings. The total entropy can never decrease. The second property above is a more specific statement that gives a method to calculate the actual amount of entropy change. Typically, one sees that if an object absorbs heat, its entropy increases, whereas, the surroundings that were the source of that heat reduced its entropy. But for heat to flow, the surroundings must be at least slightly hotter than the object. The object's entropy increase will be greater in magnitude then the surroundings' entropy decrease, leading to a net increase overall.

This leads to another statement of the Second Law: heat can only flow from a hotter object to a colder object. There is no physical process where heat flow can transfer energy out of a cold object (surrounded by a warmer environment) and make it colder. By the same principle, no physical process will transfer heat into a hot object when it is surrounded by a colder environment. These events do not take place, even though they would not violate energy conservation.

There is a third well-known statement of the Second Law, which seems only vaguely connected to the first two statements: there are no perfect heat engines. Although it would not violate the law of conservation of energy, a heat engine that burns fuel at some temperature T_H and exhausts heat at a lower temperature T_L cannot convert all of the fuel's energy into mechanical work. In fact, there is a limit given by Carnot concerning the maximum theoretical efficiency in converting thermal energy into work. At first it is surprising that this is connected to entropy. Upon further thought, one sees that a heat engine requires heat flow between objects at different temperatures, leading to entropy increases. These ultimately are connected to production of wasted heat, energy that does not become mechanical work. The only way to make a perfectly efficient engine is to have a device that does not increase entropy. But that would require the high and low temperatures to be the same, a practical impossibility.

Consider further the entropy changes. Let there be external parameters λ that change so slowly so that the system should always be in an equilibrium state. The thermodynamic identity (2.141) is a particular expression of the First Law of Thermodynamics, with the work being the work of expansion, $\eth W = pdV$. The more general expression (2.135) for the First Law can be rearranged as

$$dS = \frac{dE + \sum \overline{\Lambda d\lambda}}{T} = \frac{dE + \overline{\eth W}}{T} = \frac{\eth Q}{T} \tag{2.145}$$

that is, the differential of the entropy is equal to the quantity of heat absorbed δQ divided by the absolute temperature T. Thus, the thermodynamic identity contains the Second Law via the definition of temperature and entropy.

In Section 2.1, we have

$$\frac{\partial S}{\partial E} = \frac{1}{T} \tag{2.146}$$

that is, T^{-1} is the derivative of the entropy with respect to the energy for constant external parameters. Also for

$$dE = T\,dS - \sum \overline{\Lambda}d\lambda, \quad T = \left(\frac{\partial E}{\partial S}\right)_{\lambda=\text{const}} \tag{2.147}$$

we have

$$\overline{\Lambda} = -\left(\frac{\partial E}{\partial \lambda}\right)_S \tag{2.148}$$

as

$$\frac{\partial^2 E}{\partial S \partial \lambda} = \frac{\partial^2 E}{\partial \lambda \partial S} \tag{2.149}$$

Then some general relations can be expressed

$$\left(\frac{\partial T}{\partial \lambda}\right)_S = -\left(\frac{\partial \overline{\Lambda}}{\partial S}\right)_\lambda \tag{2.150}$$

Consider, for example, only one parameter $\lambda_1 = V, \overline{\Lambda}_1 = p$, then one gets an identity

$$\left(\frac{\partial T}{\partial V}\right)_S = -\left(\frac{\partial p}{\partial S}\right)_V \tag{2.151}$$

2.6 THERMODYNAMIC POTENTIALS, MAXWELL RELATIONS

Before moving on to the Third Law of Thermodynamics, we discuss some further aspects of the application of the First and Second Laws.

Thermodynamic energy $E(S,V)$ is an example of a thermodynamic potential. With the use of various thermodynamic identities, it can be used to obtain the desired macroscopic properties of the system. The method of the thermodynamic potentials is based on the use of the First Law of Thermodynamics:

$$dE = T\,dS - p\,dV \tag{2.152}$$

It enable us, for the system in different conditions, to introduce some different functions of state called *thermodynamic potentials*, the change of which when considering a change in state is a total differential. The thermodynamic potentials and their derivatives completely define the thermodynamic behavior of an arbitrary system.

The identity in Equation 2.152 shows that E is dependent on independent variations of the parameters S and V. Thus, we may write E as

$$E = E(S,V) \tag{2.153}$$

The differential of E may be written in a purely mathematical form with the partial derivatives:

$$dE = \left(\frac{\partial E}{\partial S}\right)_V dS + \left(\frac{\partial E}{\partial V}\right)_S dV \tag{2.154}$$

As Equations 2.152 and 2.154 are equal for all possible values of dS and dV, then their corresponding coefficients must be equal:

$$\begin{cases} \left(\dfrac{\partial E}{\partial S}\right)_V = T \\[3mm] \left(\dfrac{\partial E}{\partial V}\right)_S = -p \end{cases} \tag{2.155}$$

In Equation 2.152, the combination of the parameters on the RHS is always equal to the exact differential that in our case is the energy E. Thus, the parameters T, S, p, and V that are found on the RHS of Equation 2.152 cannot be varied arbitrarily. There must exist some link between them guaranteeing that their combination gives the differential dE. To obtain this link, the second derivative of E is independent of the order of differentiation, that is

$$\left(\frac{\partial}{\partial V}\right)_S \left(\frac{\partial E}{\partial S}\right)_V = \left(\frac{\partial}{\partial S}\right)_V \left(\frac{\partial E}{\partial V}\right)_S \tag{2.156}$$

From here, if we consider Equation 2.155, then we have the same result as at the end of the previous section

$$\left(\frac{\partial T}{\partial V}\right)_S = -\left(\frac{\partial p}{\partial S}\right)_V \tag{2.157}$$

This merely reflects the fact that dE is the exact differential of the quantity E, characteristic of the macrostate of the system.

Relationship Equation 2.152 exhibits the effect of the independent variables S and V. We may also exhibit the effect of the independent variables S and p. If we consider $p\,dV$, then we may write it as

$$p\,dV = d(pV) - V\,dp \qquad (2.158)$$

and on substituting this in Equation 2.152, we have an example of a Legendre transformation to a different independent variable

$$d(E + pV) = T\,dS + V\,dp \qquad (2.159)$$

This may be written as

$$dH = T\,dS + V\,dp \qquad (2.160)$$

where

$$H \equiv E + pV \qquad (2.161)$$

is a function called the *enthalpy*. Because its differential involves dS and dp, it is a function of the variables S and p

$$H = H(S, p) \qquad (2.162)$$

The enthalpy is a second example of a thermodynamic potential. It makes sense to use enthalpy as the fundamental potential, rather than energy, if we view a system at fixed entropy and pressure. As an example, a gas might undergo some process while under a constant pressure of one atmosphere. Then enthalpy is the most convenient thermodynamic potential. This is to be contrasted with using energy $E(S, V)$, which is applicable if we want to consider a system at fixed entropy and volume. The transformation from independent variable (S, V) to independent variables (S, p) is an example of a Legendre transformation.

The general differential of the enthalpy is

$$dH = \left(\frac{\partial H}{\partial S}\right)_p dS + \left(\frac{\partial H}{\partial p}\right)_S dp \qquad (2.163)$$

If we compare Equation 2.160 with Equation 2.163, then we have useful identities that give the missing parameters

$$\left(\frac{\partial H}{\partial S}\right)_p = T, \quad \left(\frac{\partial H}{\partial p}\right)_S = V \qquad (2.164)$$

Thus, we can still recover all four desired parameters, (S, p, T, V), but the identification of the independent parameters and dependent parameters has been changed.

In Equation 2.160, the combination of parameters on the RHS is equal to the exact differential dH and the cross derivatives are equal

$$\frac{\partial^2 H}{\partial p \partial S} = \frac{\partial^2 H}{\partial S \partial p}$$

(2.165)

It follows that

$$\left(\frac{\partial T}{\partial p}\right)_S = \left(\frac{\partial V}{\partial S}\right)_p$$

(2.166)

This is analogous to Equation 2.157 and represents the link between the parameters T, S, p, V.

Let us now consider transforming T and V as independent variables. Then, from Equation 2.152 and using $T\,dS = d(TS) - S\,dT$, we have

$$dE = d(TS) - S\,dT - p\,dV$$

(2.167)

which is reexpressed as

$$dF = -S\,dT - p\,dV$$

(2.168)

where

$$F \equiv E - TS$$

(2.169)

F is a thermodynamic potential called the *Helmholtz free energy*, which is perhaps the most important free energy in physics, and in particular, in *phase transitions*. It is especially important because the typical experimental situation under study takes place with temperature under control of the experimentalist. Also, the system is considered to have a fixed volume, meaning, it cannot do a work of expansion. As T and V are the independent variables then

$$F = F(T,V)$$

(2.170)

and

$$dF = \left(\frac{\partial F}{\partial T}\right)_V dT + \left(\frac{\partial F}{\partial V}\right)_T dV$$

(2.171)

Comparing with Equation 2.168, we have identities to give the missing parameters

$$\left(\frac{\partial F}{\partial T}\right)_V = -S, \quad \left(\frac{\partial F}{\partial V}\right)_T = -p$$

(2.172)

From the mixed partial derivatives

$$\frac{\partial^2 F}{\partial V \partial T} = \frac{\partial^2 F}{\partial T \partial V} \tag{2.173}$$

it follows that we get another identity

$$\left(\frac{\partial S}{\partial V} \right)_T = \left(\frac{\partial p}{\partial T} \right)_V \tag{2.174}$$

If we use Equations 2.169 and 2.172, then we have also

$$E = F - T \left(\frac{\partial F}{\partial T} \right)_V = -T^2 \left(\frac{\partial}{\partial T} \frac{F}{T} \right)_V \tag{2.175}$$

Now, let T and p be the independent variables; then from Equation 2.152, we have

$$dG = -S\,dT + V\,dp \tag{2.176}$$

where the function

$$G \equiv E - TS + pV = F + pV \tag{2.177}$$

is called the *Gibbs free energy*. Of course, it applies to a system where temperature and pressure are under the control of the experimentalist, a very common situation.
    If we consider Equation 2.161 or Equation 2.169, then

$$G \equiv E - TS + pV = H - TS = F + pV \tag{2.178}$$

As T and p are its independent variables, then

$$G = G(T, p) \tag{2.179}$$

and

$$dG = \left(\frac{\partial G}{\partial T} \right)_p dT + \left(\frac{\partial G}{\partial p} \right)_T dp \tag{2.180}$$

From here, considering Equation 2.176, we get the remaining parameters from this potential as

$$\left(\frac{\partial G}{\partial T} \right)_p = -S, \quad \left(\frac{\partial G}{\partial p} \right)_T = V \tag{2.181}$$

and from

$$\frac{\partial^2 G}{\partial p \partial T} = \frac{\partial^2 G}{\partial T \partial p} \tag{2.182}$$

it follows that

$$-\left(\frac{\partial S}{\partial p}\right)_T = \left(\frac{\partial V}{\partial T}\right)_p \tag{2.183}$$

From Equations 2.177, 2.161, and 2.181, there follows another identity

$$H = G - T\left(\frac{\partial G}{\partial T}\right)_p = -T^2 \left(\frac{\partial}{\partial T} \frac{G}{T}\right)_p \tag{2.184}$$

A system could also be defined in a fixed volume but have an undetermined number of particles N that can either flow into and out of the system, or be created and destroyed in the system. A discussion of the grand canonical potential $\Omega(T, V, \mu)$ and the chemical potential μ, for a system with a varying number of particles, is given later in Sections 2.14 and 3.5.

We have seen by the above derivations that we are able to establish a large number of relations between different thermodynamic functions. These arise from the mathematical property that the order in which two partial derivatives are made is interchangeable. From the above, these yield relations (2.157), (2.166), (2.174), and (2.183), which are called *Maxwell relations*. These are a direct consequence of the fact that the variables T, S, p, and V are not completely independent but are related through the fundamental identity (2.152). All of Maxwell's relations are basically equivalent as any one of them can be derived from the other by a simple change of independent variables.

The relation (2.152) involves the variables on the RHS in pairs, one pair having T and S and the other pair having p and V. The pair (T, S) involves entropy and temperature, which describe the number of accessible states of the system. The pair (p, V) involves an external parameter and its corresponding generalized force. The essence of the Maxwell relations is the link between the cross derivatives of these two kinds of quantities.

2.7   HEAT CAPACITY AND EQUATION OF STATE

In the previous section, various relations have been described between the thermodynamic variables S, T, p, V. Of course, these are only some of the thermodynamic observables possible to be measured in a complex physical system. Others include the heat capacity (or related specific heat), number density, mass density, magnetization, magnetic susceptibility, and so on. Depending on the complexity of the system, there could be more exotic physical observables. Here, an analysis of the heat capacity can serve as a prototype for how to determine general properties of any thermodynamic observable.

Heat capacity is the ratio of heat absorbed by a sample to the corresponding temperature change that is caused by that heat. We investigate the general relation between the heat capacity C_V at constant volume and the heat capacity C_p at constant pressure. The relation has practical importance as calculations using statistical mechanics are more easily carried out for a fixed volume whereas experimental measurements are usually carried out under conditions of constant (say, atmospheric) pressure. In order to compare the theoretically calculated quantity C_V with the experimentally measured parameter C_p, we must have a notion of the relation between the two quantities.

The equation of state is an equation of the form

$$f(T,p,V) = 0 \tag{2.185}$$

Consider that the volume V and the temperature T define the state of a system. The internal energy E of the system at equilibrium is a function of the temperature T:

$$E = E(T) \tag{2.186}$$

It follows from here that

$$dE = \left(\frac{\partial E}{\partial T}\right)_V dT + \left(\frac{\partial E}{\partial V}\right)_T dV \tag{2.187}$$

The quantity of heat δQ received by the system during a quasi-static process is

$$\delta Q = dE + p\,dV \tag{2.188}$$

From here, considering Equation 2.187, it follows that

$$\delta Q = \left(\frac{\partial E}{\partial T}\right)_V dT + \left[\left(\frac{\partial E}{\partial V}\right)_T + p\right]dV \tag{2.189}$$

Consider that the process takes place at constant volume, then

$$\delta Q = \left(\frac{\partial E}{\partial T}\right)_V dT \tag{2.190}$$

The heat capacity is the ratio of heat transfer to temperature change, so the constant-volume heat capacity is

$$C_V = \left(\frac{\delta Q}{dT}\right)_V = \left(\frac{\partial E}{\partial T}\right)_V \tag{2.191}$$

Similarly, C_p may be evaluated as follows:

$$C_p = \left(\frac{\delta Q}{dT}\right)_p = \left(\frac{\partial E}{\partial T}\right)_V + \left[\left(\frac{\partial E}{\partial V}\right)_T + p\right]\left(\frac{\partial V}{\partial T}\right)_p = C_V + \left[\left(\frac{\partial E}{\partial V}\right)_T + p\right]\left(\frac{\partial V}{\partial T}\right)_p$$

(2.192)

and thus

$$C_p - C_V = \left[\left(\frac{\partial E}{\partial V}\right)_T + p\right]\left(\frac{\partial V}{\partial T}\right)_p$$

(2.193)

Let us apply the given result to an ideal gas that is described by the equation of state, including the units

$$pV = Nk_BT$$

(2.194)

From Joule's law, the energy of an *ideal gas* that is at constant temperature is independent of the volume occupied by the gas. Thus, for an ideal gas

$$\left(\frac{\partial E}{\partial V}\right)_T = 0$$

(2.195)

From the equation of state of an ideal gas

$$\left(\frac{\partial V}{\partial T}\right)_p = \frac{V}{T}$$

(2.196)

and $C_p - C_V = Nk_B = nR$. Here, $n = N/N_A$ is the number of moles of the gas ($N_A = 6.02 \times 10^{23}$/mol is Avogadro's number) and $R = N_Ak_B$ is the gas constant. The heat capacities per mole, $c_p = C_p/n$ and $c_V = C_V/n$ then just differ by the gas constant, $c_p - c_V = R$. The physical reason for this is clear. If the gas absorbs heat at constant volume, it cannot do any work of expansion. Absorbing heat at constant pressure, it also can expand and do work on its surroundings. To get a desired temperature change at constant pressure, more heat must be absorbed, that simply gets converted into the work of expansion.

2.8   JACOBIAN METHOD

In order to establish the relation between two independent state variables, the Jacobian method is often used. Also, Jacobians are useful whenever there is a change of variables in a multidimensional integral. The Jacobian is the determinant of the 2×2 Jacobian matrix,

$$\frac{\partial(u,v)}{\partial(x,y)} = \begin{vmatrix} \dfrac{\partial u}{\partial x} & \dfrac{\partial u}{\partial y} \\ \dfrac{\partial v}{\partial x} & \dfrac{\partial v}{\partial y} \end{vmatrix} \qquad (2.197)$$

where $u(x, y)$ and $v(x, y)$ are functions dependent on x and y. It has the following main properties:

$$\frac{\partial(v,u)}{\partial(x,y)} = -\frac{\partial(u,v)}{\partial(x,y)}, \quad \frac{\partial(u,y)}{\partial(x,y)} = \left(\frac{\partial u}{\partial x} \right)_y \qquad (2.198)$$

$$\frac{\partial(u,v)}{\partial(x,y)} = \frac{\partial(u,v)}{\partial(t,S)} \frac{\partial(t,S)}{\partial(x,y)}$$

$$\frac{d}{dt} \frac{\partial(u,v)}{\partial(x,y)} = \frac{\partial((du/dt),v)}{\partial(x,y)} + \frac{\partial(u,(dv/dt))}{\partial(x,y)} \qquad (2.199)$$

The Jacobian (2.197) can be used to transform between pairs of independent state variables. We evaluate for an adiabatic process the derivatives $(\partial T/\partial V)_S$, $(\partial T/\partial p)_S$, and $(\partial p/\partial V)_S$. We have

$$\left(\frac{\partial T}{\partial V} \right)_S = \frac{\partial(T,S)}{\partial(V,S)} = \frac{\partial(T,S)}{\partial(V,T)} \frac{\partial(V,T)}{\partial(V,S)} = -\frac{\partial(S,T)}{\partial(V,T)} \frac{\partial(V,T)}{\partial(V,S)}$$

$$= -\left(\frac{\partial S}{\partial V} \right)_T \left(\frac{\partial T}{\partial S} \right)_V = -\frac{(\partial S/\partial V)_T}{(\partial S/\partial T)_V} = -\frac{T}{C_V} \left(\frac{\partial S}{\partial V} \right)_T \qquad (2.200)$$

or considering

$$\left(\frac{\partial S}{\partial V} \right)_T = \left(\frac{\partial p}{\partial T} \right)_V \qquad (2.201)$$

it follows that

$$\left(\frac{\partial T}{\partial V} \right)_S = -\frac{T}{C_V} \left(\frac{\partial p}{\partial T} \right)_V \qquad (2.202)$$

Further

$$\left(\frac{\partial T}{\partial p}\right)_S = \frac{\partial(T,S)}{\partial(p,S)} = \frac{\partial(T,S)}{\partial(p,T)}\frac{\partial(p,T)}{\partial(p,S)} = -\frac{\partial(S,T)}{\partial(p,T)}\frac{\partial(p,T)}{\partial(p,S)}$$

$$= -\left(\frac{\partial S}{\partial p}\right)_T \left(\frac{\partial T}{\partial S}\right)_p = -\frac{(\partial S/\partial p)_T}{(\partial S/\partial T)_p} = -\frac{T}{C_p}\left(\frac{\partial S}{\partial p}\right)_T \qquad (2.203)$$

or from

$$\left(\frac{\partial S}{\partial p}\right)_T = -\left(\frac{\partial V}{\partial p}\right)_S \qquad (2.204)$$

We have

$$\left(\frac{\partial T}{\partial p}\right)_S = \frac{T}{C_p}\left(\frac{\partial V}{\partial T}\right)_p \qquad (2.205)$$

Finally

$$\left(\frac{\partial p}{\partial V}\right)_S = \frac{\partial(p,S)}{\partial(V,S)} = \frac{\partial(p,S)}{\partial(p,T)}\frac{\partial(p,T)}{\partial(V,T)}\frac{\partial(V,T)}{\partial(V,S)} = \left(\frac{\partial S}{\partial T}\right)_p \left(\frac{\partial p}{\partial V}\right)_T \left(\frac{\partial T}{\partial S}\right)_V$$

$$= \frac{(\partial S/\partial T)_p (\partial p/\partial V)_T}{(\partial S/\partial T)_V} = \frac{(C_p/T)(\partial p/\partial V)_T}{(C_V/T)} = \frac{C_p}{C_V}\left(\frac{\partial p}{\partial V}\right)_T \qquad (2.206)$$

Let us evaluate the relations

$$\left(\frac{\partial T}{\partial V}\right)_E = \frac{1}{C_V}\left[p - T\left(\frac{\partial p}{\partial T}\right)_V\right] \qquad (2.207)$$

$$\left(\frac{\partial T}{\partial p}\right)_H = \frac{1}{C_p}\left[T\left(\frac{\partial V}{\partial T}\right)_p - V\right] \qquad (2.208)$$

We have

$$\left(\frac{\partial T}{\partial V}\right)_E = \frac{\partial(T,E)}{\partial(V,E)} = \frac{\partial(T,E)}{\partial(V,T)}\frac{\partial(V,T)}{\partial(V,E)} = -\frac{\partial(E,T)}{\partial(V,T)}\frac{\partial(V,T)}{\partial(V,E)}$$

$$= -\left(\frac{\partial E}{\partial V}\right)_T \left(\frac{\partial T}{\partial E}\right)_V = -\frac{(\partial E/\partial V)_T}{(\partial E/\partial T)_V} = -\frac{1}{C_V}\left(\frac{\partial E}{\partial V}\right)_T \qquad (2.209)$$

or considering

$$\left(\frac{\partial E}{\partial V}\right)_T + p = T\left(\frac{\partial p}{\partial T}\right)_V \qquad (2.210)$$

Then we have

$$\left(\frac{\partial T}{\partial V}\right)_E = \frac{1}{C_V}\left[p - T\left(\frac{\partial p}{\partial T}\right)_V\right] \qquad (2.211)$$

Further

$$\left(\frac{\partial T}{\partial p}\right)_H = \frac{\partial(T,H)}{\partial(V,H)} = \frac{\partial(T,H)}{\partial(p,T)}\frac{\partial(p,T)}{\partial(p,H)} = -\frac{\partial(H,T)}{\partial(p,T)}\frac{\partial(p,T)}{\partial(p,H)}$$

$$= -\left(\frac{\partial H}{\partial p}\right)_T\left(\frac{\partial T}{\partial H}\right)_p = -\frac{(\partial H/\partial p)_T}{(\partial H/\partial T)_p} = -\frac{1}{C_p}\left(\frac{\partial H}{\partial p}\right)_T \qquad (2.212)$$

But considering

$$T = \left(\frac{\partial H}{\partial S}\right)_p, \quad V = \left(\frac{\partial H}{\partial p}\right)_S \qquad (2.213)$$

Then

$$\left(\frac{\partial H}{\partial p}\right)_T = T\left(\frac{\partial S}{\partial p}\right)_T + V = V - T\left(\frac{\partial V}{\partial T}\right)_p \qquad (2.214)$$

where we used

$$\left(\frac{\partial S}{\partial p}\right)_T = -\left(\frac{\partial V}{\partial T}\right)_p \qquad (2.215)$$

Thus

$$\left(\frac{\partial T}{\partial p}\right)_H = \frac{1}{C_p}\left[T\left(\frac{\partial V}{\partial T}\right)_p - V\right] \qquad (2.216)$$

Let us now use the Jacobian method to express the difference in heat capacities $C_p - C_V$ through the variables: (a) V, T; and (b) p, T.

a. We have

$$C_p = T\left(\frac{\partial S}{\partial T}\right)_p = T\frac{\partial(S,p)}{\partial(T,p)} = T\frac{\partial(S,p)}{\partial(T,V)}\frac{\partial(T,V)}{\partial(T,p)} = T\frac{(\partial(S,p)/\partial(T,V))}{(\partial(T,p)/\partial(T,V))}$$

$$= T\frac{\left[(\partial S/\partial T)_V(\partial p/\partial V)_T - (\partial S/\partial V)_T(\partial p/\partial T)_V\right]}{(\partial p/\partial V)_T} = C_V - \frac{T(\partial p/\partial T)_V^2}{(\partial p/\partial V)_T} \quad (2.217)$$

where we used the formula

$$\left(\frac{\partial S}{\partial p}\right)_T = \left(\frac{\partial p}{\partial T}\right)_V \quad (2.218)$$

b. Similarly, we transform the formula for C_V through the variables T and p and we have

$$C_V = T\left(\frac{\partial S}{\partial T}\right)_V = T\frac{\partial(S,V)}{\partial(T,V)} = T\frac{\partial(S,V)}{\partial(T,p)}\frac{\partial(T,p)}{\partial(T,V)} = \frac{T(\partial(S,V)/\partial(T,p))}{(\partial(T,V)/\partial(T,p))}$$

$$= T\frac{\left[(\partial S/\partial T)_p(\partial V/\partial p)_T - (\partial S/\partial p)_T(\partial V/\partial T)_p\right]}{(\partial V/\partial p)_T} \quad (2.219)$$

or

$$C_V = T\left[\left(\frac{\partial S}{\partial T}\right)_p - \frac{(\partial S/\partial p)_T(\partial V/\partial T)_p}{(\partial V/\partial p)_T}\right] = C_p + T\frac{(\partial V/\partial T)_p^2}{(\partial V/\partial p)_T} \quad (2.220)$$

or finally

$$C_p - C_V = -T\frac{(\partial V/\partial T)_p^2}{(\partial V/\partial p)_T} \quad (2.221)$$

For the isothermal expansion of a body, the pressure is always decreasing, that is, the derivative $(\partial p/\partial V)_T$ is always negative. From here, it follows that for all bodies

$$C_p > C_V$$

Let us examine the generalized form of Equation 2.210. We consider the general case for arbitrary numbering of the external parameters, $\lambda_1, \ldots, \lambda_s, \ldots, \lambda_m$:

$$\eth Q = dE + \sum_s \overline{\Lambda_s}\, d\lambda_s \quad (2.222)$$

Then

$$dS = \frac{(\partial E/\partial T)_{\lambda_s}\, dT + \sum_s ((\partial E/\partial\lambda_s) + \overline{\Lambda_s})d\lambda_s}{T} \tag{2.223}$$

Thus

$$\frac{\partial}{\partial\lambda_r}\frac{(\partial E/\partial\lambda_s) + \overline{\Lambda_s}}{T} = \frac{\partial}{\partial\lambda_s}\frac{(\partial E/\partial\lambda_r) + \overline{\Lambda_r}}{T} \tag{2.224}$$

or

$$\frac{\partial\overline{\Lambda_s}}{\partial\lambda_r} = \frac{\partial\overline{\Lambda_r}}{\partial\lambda_s} \tag{2.225}$$

Further

$$\frac{\partial}{\partial\lambda_s}\frac{1}{T}\left(\frac{\partial E}{\partial T}\right)_{\lambda_s} = \frac{\partial}{\partial T}\left[\frac{1}{T}\left(\frac{\partial E}{\partial\lambda_s} + \overline{\Lambda_s}\right)\right] \tag{2.226}$$

that is

$$0 = -\frac{1}{T^2}\left(\frac{\partial E}{\partial\lambda_s} + \overline{\Lambda_s}\right) + \frac{1}{T}\frac{\partial\overline{\Lambda_s}}{\partial T} \tag{2.227}$$

or

$$\frac{\partial E}{\partial\lambda_s} + \overline{\Lambda_s} = T\frac{\partial\overline{\Lambda_s}}{\partial T} \tag{2.228}$$

2.9 JOULE–THOMSON PROCESS

We examine the steady-state experimental arrangement first suggested by Joule and Thomson. We consider a pipe with thermally insulated walls and a porous plug providing a constriction to a flow of gas. The situation is sketched in Figure 2.3. The porous plug is darkly shaded in the figure. In the pipe, we have a continuous stream of gas from left to right. The presence of the constriction guarantees a constant pressure difference maintained across it. It follows that the gas pressure p_1 on the LHS is greater than the gas pressure p_2 on the RHS of the constriction.

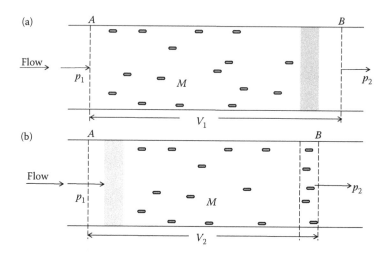

FIGURE 2.3   A gas flowing through a movable porous plug in a pipe (a) at an earlier time and (b) at a later time.

Let M be a mass of gas initially occupying the space between the dashed planes A and B in Figure 2.3a. The planes A and B are chosen far apart, so that the volume occupied by the constriction is negligible compared to the volume contained between A and B. At some initial moment, B coincides with the constriction and virtually the entire mass M of gas lies to the left of the constriction (Figure 2.3b) and occupies a volume V_1 corresponding to pressure p_1. The planes A and B defining its geometrical boundaries move through the pipe as the gas flows. As a consequence, at some later moment, the plane will A coincide with the constriction and virtually the entire mass M of gas will lie to the right of the constriction, occupying some different volume V_2 corresponding to the lower pressure p_2 (Figure 2.3b).

The difference in the internal energy of the mass of gas between the final situation (when it is to the right of the constriction) and the initial situation (when it is to the left of the constriction) is

$$\Delta E = E_2 - E_1 = E(T_2, p_2) - E(T_1, p_1) \tag{2.229}$$

It should be noted that in this process the mass M of gas does work W. The work done in displacing gas to the right of the constriction by the volume V_2 against the constant pressure p_2 is $p_2 V_2$ and that to the left of the constriction is $-p_1 V_1$ due to the displacement of gas of mass M by the volume V_1 with a constant pressure p_1. Consequently, the net work W done by the mass M of gas in the process is

$$W = p_2 V_2 - p_1 V_1 \tag{2.230}$$

It can also be noted that in the processes, there is no heat absorbed by the mass M of gas. This is due to the fact that after the steady-state situation is arrived, there is no temperature difference between the walls and the adjacent gas, so that no heat flows from walls into the gas and consequently, $Q = 0$.

From the First Law of Thermodynamics

$$\Delta E + W = Q = 0 \tag{2.231}$$

And from Equations 2.229 and 2.230, we have

$$E_2 + p_2 V_2 = E_1 + p_1 V_1 \tag{2.232}$$

The quantity

$$H = E + pV \tag{2.233}$$

is the enthalpy introduced earlier. It follows from Equation 2.232 that

$$H(T_2, p_2) = H(T_1, p_1) \tag{2.234}$$

We conclude that in the Joule–Thomson process, the gas passes through the constriction in such a way that the enthalpy H remains constant.

The enthalpy $H = H(T,p)$, then given T_1 and p_1, it is possible to use the conserved enthalpy to determine the unknown final temperature T_2. For the case of an ideal gas

$$H = E + pV = E(T) + nRT \tag{2.235}$$

and we have

$$H = H(T) \tag{2.236}$$

From Equation 2.234, it follows that

$$H(T_2) = H(T_1) \tag{2.237}$$

which further implies $T_2 = T_1$. For an ideal gas, the temperature does not change in the Joule–Thomson process.

The change of the temperature for very small change in pressure as a result of the Joule–Thomson process is defined by the derivative $(\partial T/\partial p)_H = \mu$. The quantity μ is called the Joule–Thomson coefficient (or Joule–Kelvin coefficient). The names Thomson and Kelvin refer to the same person, William Thomson who was elevated to the peerage and thus became Lord Kelvin.

The quantity μ gives the change in temperature produced in the Joule–Thomson process involving an infinitesimal pressure change. For an infinitesimal pressure drop, T decreases if $\mu > 0$ and for $\mu = 0$, there is no temperature change.

Let us rewrite $(\partial T/\partial p)_H = \mu$ by the following algebra:

$$\left(\frac{\partial T}{\partial p}\right)_H = \frac{\partial(T,H)}{\partial(p,H)} = \frac{(\partial(T,H)/\partial(p,T))}{(\partial(p,H)/\partial(p,T))} = -\frac{(\partial H/\partial p)_T}{(\partial H/\partial T)_p} \qquad (2.238)$$

If we consider that

$$\left(\frac{\partial H}{\partial T}\right)_p = C_p \qquad (2.239)$$

and

$$\left(\frac{\partial H}{\partial p}\right)_T = V - T\left(\frac{\partial V}{\partial T}\right)_p \qquad (2.240)$$

then

$$\left(\frac{\partial T}{\partial p}\right)_H = \frac{1}{C_p}\left[T\left(\frac{\partial V}{\partial T}\right)_p - V\right] \qquad (2.241)$$

This formula shows that the positive Joule–Thomson effect, $\mu > 0$, takes place if

$$T\left(\frac{\partial V}{\partial T}\right)_p - V > 0 \qquad (2.242)$$

This condition is necessary so that the gas may cool down after expansion, that is, so that the inequality $dT < 0$ should be satisfied for $dp < 0$.

The temperature for which the square brackets in Equation 2.241 should change sign is called the *inversion temperature* and is denoted by T_i. This temperature is defined by the following formula:

$$\left[T\left(\frac{\partial V}{\partial T}\right)_p - V\right]_{T=T_i} = 0 \qquad (2.243)$$

The change in entropy is related to the derivative $(\partial S/\partial p)_H$.
    If we consider

$$dH = TdS + Vdp \qquad (2.244)$$

it can be rearranged as

$$dS = \frac{1}{T}dH - \frac{V}{T}dp \qquad (2.245)$$

Then

$$\left(\frac{\partial S}{\partial p}\right)_H = -\frac{V}{T} \tag{2.246}$$

This value is always negative. The transformation of a gas to a lower pressure in an irreversible Joule–Thomson process is accompanied by an increase in entropy. If as a result of expansion, the entire volume changes infinitesimally, then the change in temperature is determined by the derivative $(\partial T/\partial V)_E$:

$$\left(\frac{\partial T}{\partial V}\right)_E = \frac{1}{C_V}\left[p - T\left(\frac{\partial p}{\partial T}\right)_V\right] \tag{2.247}$$

and for the change in entropy

$$\left(\frac{\partial S}{\partial V}\right)_E = \frac{p}{T} \tag{2.248}$$

It can be seen that as a result of expansion (increased volume) the entropy increases.

2.10   MAXIMUM WORK

One of the first and the most important problems of thermodynamics is the definition of the conditions for which we may obtain the maximum effect using non-equilibrium to obtain the work done. Let us examine first a system made up of two bodies. Let the system be closed. Suppose that the two bodies are at different temperatures T_1 and T_2 such that $T_1 > T_2$ (Figure 2.4a). If we connect the two bodies (Figure 2.4b) there is transfer of thermal energy $|dE|$ from the hotter one to the colder one.

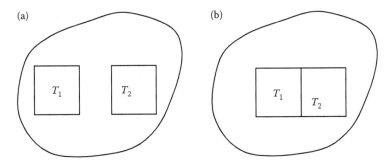

FIGURE 2.4   A system of two bodies. (a) Initially, the bodies are separated and at different temperatures with $T_1 > T_2$. (b) The bodies are placed in thermal contact, allowing heat to be exchanged between them.

The heat lost by the first body is equal to the heat gained by the second body. From definition, we have

$$dS_1 = -\frac{|dE|}{T_1}, \quad dS_2 = \frac{|dE|}{T_2} \tag{2.249}$$

Then the total entropy change in the process is

$$dS = dS_1 + dS_2 = |dE|\left(-\frac{1}{T_1} + \frac{1}{T_2}\right) > 0 \tag{2.250}$$

As $T_1 > T_2$, the process of heat exchange is irreversible.

Now, consider some isolated body as our system. Let the initial energy of the system be E_0 and the final energy $E = E(S)$. As the system is thermally isolated, then the macroscopic work done, W, is simply the change in energy:

$$W(S) = E_0 - E(S), \quad \eth Q = 0 \tag{2.251}$$

Let us differentiate $W(S)$ with respect to the entropy of the final state:

$$\frac{\partial W(S)}{\partial S} = -\left(\frac{\partial E}{\partial S}\right)_V = -T < 0 \tag{2.252}$$

where $T > 0$ is the temperature of the final state. It is seen that

$$\frac{W(S_2) - W(S_1)}{S_2 - S_1} < 0 \tag{2.253}$$

This means that if $S_2 > S_1$, then the work done is negative, and $W_2 < W_1$. Where there is a greater degree of irreversibility, the work is less. Thus, in order to have maximum work, it is necessary that the process should be reversible.

The *Carnot cycle* is an example of such a process. Let us examine such a cycle for a heat engine. Let λ be the external parameter of such an engine. The changes in the given parameter gives rise to the work performed by the engine. Suppose the working substance of the engine is initially in a state where $\lambda = \lambda_a$ and its temperature is that of the cold reservoir $T = T_2$. The Carnot engine then goes through a cycle which consist of four steps, all performed in a quasi-static fashion. Here, quasi-static means that the processes are performed slowly enough, that the entire working substance of the engine has a well-defined thermodynamic state, whose parameters are the same in the whole sample. Let us denote by a, b, c, and d some selected macrostates of the engine (Figure 2.5), connected by the following processes:

1. a → b: Adiabatic compression: The engine's working substance is *thermally insulated*. Its external parameter is changed slowly (adiabatically, without heat exchange) until the engine's temperature rises to T_1. Thus, $\lambda_a \to \lambda_b$ such

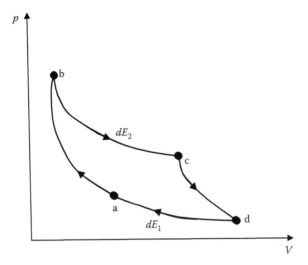

FIGURE 2.5 Carnot cycle of an engine with a, b, c, and d being selected macrostates of the engine.

that $T_2 \rightarrow T_1$. The pressure increases while the volume decreases. There is an increase in the internal energy, associated with the increase in temperature.

2. b → c: Isothermal expansion: The engine is now placed in *thermal contact* with the heat reservoir at the higher temperature T_1. Its external parameter is changed further $\lambda_b \rightarrow \lambda_c$ isothermally at T_1 with the engine absorbing some heat Q_1. As this process is isothermal, the internal energy of the working substance does not change.

3. c → d: Adiabatic expansion: The engine is again *thermally insulated* and its external parameter $\lambda_c \rightarrow \lambda_d$ is changed (again, adiabatically) in such a direction such that its temperature goes back down to T_2: $T_1 \rightarrow T_2$. The internal energy of the working substance returns to the original value it had at point a.

4. d → a: Isothermal compression: The engine is now placed in *thermal contact* with the colder heat reservoir at temperature T_2. Its external parameter is then changed isothermally $\lambda_d \rightarrow \lambda_a$ until it returns to the initial value λ_a, with the engine remaining at T_2 and rejecting some heat Q_2 to this reservoir. There is no change in internal energy.

It is seen that the engine is back to its initial state and the cycle is completed. The ultimate purpose of a mechanical engine is to produce a work output. Further, one wants the maximum work output for a given energy inputted from the higher-temperature reservoir. This suggests the definition of the mechanical efficiency as the ratio of the net work done per cycle to the inputted energy. The inputted energy is the heat Q_1, whereas, the work done must equal the *net heat transfer* to the substance

$$W = |Q_1| - |Q_2| \tag{2.254}$$

so the efficiency is

$$\eta = \frac{W}{Q_1} = \frac{|Q_1| - |Q_2|}{|Q_1|} = 1 - \frac{|Q_2|}{|Q_1|} \tag{2.255}$$

Assuming it was performed quasi-statically, the process is reversible. It follows that $dS = 0$ for the entire cycle. On the paths $c \rightarrow d$ and $a \rightarrow b$, we have $dS = 0$ because those have no heat exchange. On the path $b \rightarrow c$, the engine absorbs heat so we have

$$dS_1 = \frac{|Q_1|}{T_1} \tag{2.256}$$

and on the path $d \rightarrow a$, heat leaves the engine so we have

$$dS_2 = -\frac{|Q_2|}{T_2} \tag{2.257}$$

Then the total entropy change per cycle satisfies

$$dS = \frac{|Q_1|}{T_1} - \frac{|Q_2|}{T_2} = 0 \tag{2.258}$$

and this implies that the ratio of the heats is the same as the absolute temperature ratio

$$\frac{|Q_2|}{|Q_1|} = \frac{T_2}{T_1} \tag{2.259}$$

The result applies to any reversible Carnot machine. Then the efficiency depends only on the temperatures of the heat reservoirs

$$\eta = \eta_{max} = 1 - \frac{T_2}{T_1} \tag{2.260}$$

This is the maximum theoretical efficiency. It applies only to a reversible engine. A real engine is not reversible, it will have $dS > 0$ for the entire cycle, due to friction, turbulence, and so on. Then it is possible to show that a real engine exhausts relatively more heat than the reversible Carnot engine operating between the same heat reservoirs

$$\frac{|Q_2|}{|Q_1|} > \frac{T_2}{T_1} \tag{2.261}$$

This means that its actual operating efficiency is some fraction of the Carnot efficiency

$$\eta = 1 - \frac{|Q_2|}{|Q_1|} < \eta_{max} \tag{2.262}$$

This is the usable efficiency. It is the ratio of the actual work performed to the heat energy obtained from the fuel or other energy source.

These results connect back to the Kelvin–Planck statement of the Second Law of Thermodynamics, which states that it is impossible to build a device or machine whose sole effect is the transformation of heat into work. Even for the Carnot cycle, that transformation is incomplete. All real machines produce an increase in entropy, and as such, the upper limit they can hope to achieve in energy conversion efficiency is the Carnot efficiency.

2.11 CONDITION FOR EQUILIBRIUM AND STABILITY IN AN ISOLATED SYSTEM

We select a small macroscopic portion A of a homogeneous substance (Figure 2.6). In relation to the small macroscopic portion, the rest of the substance A' is very large and can be thought of as an external medium acting as a reservoir at constant temperature T_0 and pressure p_0. The number of particles N_0 in the reservoir is very large compared to that in A. It is assumed that A is in thermal contact with A'. We use the variables E, p, V, and so on for the properties of A, and variables E_0, p_0, V_0, and so on for A'.

The system A can exchange heat with the reservoir A' and the latter is so large that its temperature T_0 remains constant. The system A can change its volume V at

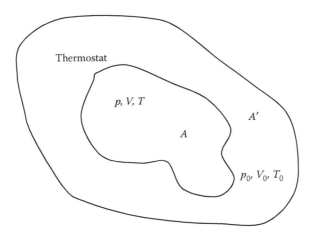

FIGURE 2.6 A small macroscopic portion A of a homogeneous substance in a very large external medium A' acting as a reservoir at constant temperature T_0 and pressure p_0. The variables E, p, V relate to the properties for A, and variables E_0, p_0, V_0 for A'.

the expense of the reservoir A', doing work in that process. It should be noted that the reservoir A' is so large that also its pressure p_0 is unaltered by this change in volume.

From the law of conservation of energy, the small portion's energy can change due to absorbed heat or a work

$$dE = \delta Q + \delta W \tag{2.263}$$

For the equilibrium condition, the entropy of the combined system during any spontaneous process satisfies the condition

$$dS + dS_0 \geq 0 \tag{2.264}$$

The volume of the whole substance is assumed fixed, so

$$dV + dV_0 = 0 \tag{2.265}$$

and the total energy of A and the reservoir is a constant:

$$dE + dE_0 = 0 \tag{2.266}$$

We evaluate now the energy change of the substance A considering the fact that T_0 is a constant. From $dE = -dE_0$ and applying the thermodynamic identity $dE_0 = T_0 \, dS_0 - p_0 \, dV_0$, the energy influx into the body A may be expressed through the parameters of the reservoir and is equal to

$$dE = -T_0 dS_0 + p_0 dV_0 \tag{2.267}$$

Rearranging and using the volume and entropy constraints leads to

$$dE - T_0 dS + p_0 dV \leq 0 \tag{2.268}$$

Now, as p_0 and T_0 are constants, then it follows that

$$d(E - T_0 S + p_0 V) \leq 0 \tag{2.269}$$

Hence, this shows that

$$dG^0 \leq 0, \quad G^0 \equiv E - T_0 S + p_0 V \tag{2.270}$$

where G^0 is the *Gibbs free energy*.

This last result implies that the change in G^0 brings the substance to an equilibrium state. The relation (2.269) ($dG^0 \leq 0$) also implies that the macroscopic substance

A in contact with the reservoir A' can only do work on the reservoir until the stable equilibrium condition is reached, characterized by

$$dG^0 = 0 \qquad (2.271)$$

or

$$G^0 = \text{minimum} \qquad (2.272)$$

Thus

$$G^0_{\text{equil}} = G^0_{\text{min}} \qquad (2.273)$$

One concludes that the Gibbs potential is a minimum in equilibrium. It should be noted that within equilibrium thermodynamics, the macroscopic variables *do not fluctuate* (but microscopically they do because of thermal and quantum fluctuations).

Suppose now the macroscopic substance A has fixed volume V but its temperature T is allowed to vary, due to a deviation away from equilibrium. Let there be a temperature $\tilde{T}$ that minimizes the Gibbs potential

$$G^0_{\text{min}} = G^0_{T=\tilde{T}} \qquad (2.274)$$

then

$$\delta G^0 = \left(\frac{\partial G^0}{\partial T}\right)_V \delta T + \frac{1}{2}\left(\frac{\partial^2 G^0}{\partial T^2}\right)_V (\delta T)^2 + \cdots, \quad \delta T = T - \tilde{T} \qquad (2.275)$$

As $G^0_{\text{min}} = G^0_{T=\tilde{T}}$, it follows that

$$\left(\frac{\partial G^0}{\partial T}\right)_V = 0 \quad \text{for } T = \tilde{T} \qquad (2.276)$$

The fact that

$$G^0_{\text{min}} = G^0_{T=\tilde{T}} \qquad (2.277)$$

requires a positive second derivative to insure a minimum

$$\left(\frac{\partial^2 G^0}{\partial T^2}\right)_V \geq 0 \quad \text{for } T = \tilde{T} \qquad (2.278)$$

In Figure 2.7, δG^0 represents the variation of G^0 around the equilibrium position:

$$\delta G^0 = G^0 - G^0_{\text{min}} \qquad (2.279)$$

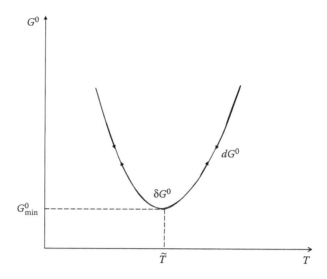

FIGURE 2.7   Variation of G^0 around the equilibrium position.

If we consider

$$G^0_{min} = E - T_0 S + p_0 V \qquad (2.280)$$

and Equation 2.276 then

$$\left(\frac{\partial G^0}{\partial T}\right)_V = \left(\frac{\partial E}{\partial T}\right)_V - T_0\left(\frac{\partial S}{\partial T}\right)_V = C_V - T_0\frac{C_V}{\tilde{T}} = C_V\left(1 - \frac{T_0}{\tilde{T}}\right)_{T_0=\tilde{T}} = 0 \qquad (2.281)$$

This shows that at equilibrium the temperature of the substance A is the same as that of the reservoir A', a result to be expected. If we consider Equation 2.278, then we have

$$\left(\frac{\partial^2 G^0}{\partial T^2}\right)_V = \frac{\partial}{\partial T}\left(\frac{\partial G^0}{\partial T}\right)_V = \frac{\partial}{\partial T}\left[\left(\frac{\partial E}{\partial T}\right)_V - T_0\left(\frac{\partial S}{\partial T}\right)_V\right] = \frac{\partial}{\partial T}\left[\left(\frac{\partial E}{\partial T}\right)_V - \frac{T_0}{T}\left(\frac{\partial E}{\partial T}\right)_V\right]$$

$$= \frac{T_0}{T^2}\left(\frac{\partial E}{\partial T}\right)_V + \left(1 - \frac{T_0}{T}\right)\left(\frac{\partial^2 E}{\partial T^2}\right)_V = \frac{T_0}{T^2}C_V + \left(1 - \frac{T_0}{T}\right)\left(\frac{\partial^2 E}{\partial T^2}\right)_V \qquad (2.282)$$

For $G^0 = G^0_{min}$ it follows that $T_0 = \tilde{T}$ and

$$\left(\frac{\partial^2 G^0}{\partial T^2}\right)_V = \frac{C_V}{T} \geq 0 \qquad (2.283)$$

This is the condition for the intrinsic (thermal) stability of any phase. The condition (2.283) is in accordance with Le Châtelier's principle: For a system in stable equilibrium, any spontaneous change of its parameters will initiate processes that tend to restore the system to its equilibrium state. In our particular case such spontaneous changes raise the Gibbs potential.

Suppose the temperature T of A increases above that of reservoir A' as a result of a spontaneous fluctuation. As a result, heat will be transferred from A to the reservoir A' while there is a decrease in energy E of A, such that $dE < 0$. Le Châtelier's principle requires that a process be induced to return to equilibrium with a temperature decrease, $dT < 0$. Then, with dE and dT having the same sign, their ratio is positive:

$$\frac{\partial E}{\partial T} = C_V > 0 \tag{2.284}$$

Now instead suppose the temperature of the macroscopic system A is fixed at $T = T_0$ and its volume is allowed to vary, then

$$\delta G^0 = \left(\frac{\partial G^0}{\partial V}\right)_T \delta V + \frac{1}{2}\left(\frac{\partial^2 G^0}{\partial V^2}\right)_T (\delta V)^2 + \cdots, \quad \delta V = V - \tilde{V} \tag{2.285}$$

where $\tilde{V}$ is the volume for which G^0 is a minimum. For the stationary value of G^0, it is required that

$$\left(\frac{\partial G^0}{\partial V}\right)_T = 0 \tag{2.286}$$

and from Equation 2.279 we have

$$\left(\frac{\partial G^0}{\partial V}\right)_T = \left(\frac{\partial E}{\partial V}\right)_T - T_0\left(\frac{\partial S}{\partial V}\right)_T + p_0 \tag{2.287}$$

Considering the thermodynamic identity

$$dE = T\,dS - p\,dV \tag{2.288}$$

this yields

$$\left(\frac{\partial G^0}{\partial V}\right)_T = \left(\frac{\partial E}{\partial V}\right)_T + p_0 - \frac{T_0}{T_0}\left[\left(\frac{\partial E}{\partial V}\right)_T + p\right] = p_0 - p \tag{2.289}$$

From Equation 2.286, it follows that

$$p = p_0 \tag{2.290}$$

This shows that at equilibrium the pressure of the subsystem A must be constant and equal to that of the surrounding medium. To see the conditions required so that G^0 is actually a minimum at $V = \tilde{V}$ look at

$$\left(\frac{\partial^2 G^0}{\partial V^2}\right)_T \geq 0 \tag{2.291}$$

and

$$\left(\frac{\partial^2 G^0}{\partial V^2}\right)_T = \frac{\partial}{\partial V}\left[p_0 - p\right] = -\left(\frac{\partial p}{\partial V}\right)_T \geq 0 \tag{2.292}$$

The result involves the isothermal compressibility

$$\chi_T = -\frac{1}{V}\left(\frac{\partial V}{\partial p}\right)_T \geq 0 \tag{2.293}$$

This is the condition for *mechanical stability*. The condition can be expressed by the relation

$$\left(\frac{\partial^2 G^0}{\partial V^2}\right)_T = \frac{1}{\chi_T V} \geq 0 \tag{2.294}$$

The inequality (2.292) is another example of Le Châtelier's principle. If the volume of the system A increases by $dV > 0$ as a result of a fluctuation, then its pressure p must decrease below that of the reservoir, $dp < 0$, in order that the resultant force exerted on A by the reservoir is in such a direction as to reduce the volume back to its initial value. For this to work in a stable fashion, the isothermal compressibility must be positive.

2.12   THERMODYNAMIC INEQUALITIES

We know that

$$0 \leq \delta G^0 = \delta E - T_0\, \delta S + p_0\, \delta V \tag{2.295}$$

and that the energy $E = E\,(S,V)$ is a function of S and V. Considering this let us expand δG^0 up to second order in variations δS and δV:

$$\delta G^0 = \left(\frac{\partial E}{\partial S}\right)_V \delta S + \left(\frac{\partial E}{\partial V}\right)_S \delta V + \frac{1}{2}\left(\frac{\partial^2 E}{\partial S^2}\right)_V (\delta S)^2 + \frac{\partial^2 E}{\partial S \partial V}\delta S\, \delta V$$

$$+ \frac{1}{2}\left(\frac{\partial^2 E}{\partial V^2}\right)_S (\delta V)^2 + \cdots - T_0\, \delta S + p_0\, \delta V \tag{2.296}$$

But in equilibrium

$$\left(\frac{\partial E}{\partial S}\right)_V = T_0, \quad \left(\frac{\partial E}{\partial V}\right)_S = -p_0 \qquad (2.297)$$

Then

$$\delta G^0 = \frac{1}{2}\left(\frac{\partial^2 E}{\partial S^2}\right)_V (\delta S)^2 + \frac{\partial^2 E}{\partial S \partial V}\delta S\,\delta V + \frac{1}{2}\left(\frac{\partial^2 E}{\partial V^2}\right)_S (\delta V)^2 + \cdots \qquad (2.298)$$

Let δS and δV have arbitrary signs, that is

$$\delta S \geq 0, \quad \delta V \lessgtr 0 \qquad (2.299)$$

Consider the fluctuation δS, then from $\delta G^0 \geq 0$, it follows that

$$\left(\frac{\partial^2 E}{\partial S^2}\right)_V \left(\frac{\partial^2 E}{\partial V^2}\right)_S - \left(\frac{\partial^2 E}{\partial S \partial V}\right)^2 > 0 \qquad (2.300)$$

Similarly, we may also consider the fluctuation δV. It follows that

$$\left(\frac{\partial^2 E}{\partial S^2}\right)_V = \frac{\partial}{\partial S}\left[\left(\frac{\partial E}{\partial S}\right)_V\right] = \left(\frac{\partial T}{\partial S}\right)_V = \frac{T}{C_V} > 0, \quad T > 0 \qquad (2.301)$$

We conclude that the heat capacity of a homogeneous system in the equilibrium state is positive:

$$C_V > 0 \qquad (2.302)$$

This is the *first inequality* and is the thermal stability condition that we saw earlier. If we consider Equation 2.300 further, then

$$\begin{vmatrix} \dfrac{\partial^2 E}{\partial S^2} & \dfrac{\partial^2 E}{\partial S \partial V} \\ \dfrac{\partial^2 E}{\partial S \partial V} & \dfrac{\partial^2 E}{\partial V^2} \end{vmatrix} = \begin{vmatrix} \dfrac{\partial T}{\partial S} & \dfrac{\partial T}{\partial V} \\ \dfrac{\partial(-p)}{\partial S} & \dfrac{\partial(-p)}{\partial V} \end{vmatrix} = \frac{\partial(T,-p)}{\partial(S,V)} \qquad (2.303)$$

$$\frac{\partial(T,-p)}{\partial(S,V)} = \frac{\partial(T,-p)}{\partial(T,V)}\frac{\partial(T,V)}{\partial(S,V)} = -\left(\frac{\partial p}{\partial V}\right)_T \left(\frac{\partial T}{\partial S}\right)_V = -\left(\frac{\partial p}{\partial V}\right)_T \frac{T}{C_V} > 0 \qquad (2.304)$$

and as $C_V > 0$ it implies that

$$-\left(\frac{\partial p}{\partial V}\right)_T \geq 0 \tag{2.305}$$

This is the *second inequality*. As we saw earlier, it recovers

$$-\left(\frac{\partial p}{\partial V}\right)_T = \frac{1}{\chi_T V} \geq 0 \tag{2.306}$$

where χ_T is the isothermal compressibility of the substance. The relation (2.305) shows that an increase in volume at constant temperature is always accompanied by a decrease in pressure (this is the mechanical stability condition).

The conditions (2.302) and (2.305) are called *thermodynamic inequalities*. The states in which all these conditions are not satisfied are unstable and may not exist in nature. If we consider Equations 2.305 and 2.176, then

$$C_p - C_V = \frac{VT\alpha_p^2}{\chi_T}, \quad \chi_T - \chi_S = \frac{VT\alpha_p^2}{C_p}, \quad \frac{C_p}{C_V} = \frac{\chi_T}{\chi_S} \tag{2.307}$$

Here

$$\alpha_p = \frac{1}{V}\left(\frac{\partial V}{\partial T}\right)_p \tag{2.308}$$

is the coefficient of thermal expansion and

$$\chi_S = -\frac{1}{V}\left(\frac{\partial V}{\partial p}\right)_S = -\frac{1}{V}\left(\frac{\partial^2 H}{\partial p^2}\right)_S \tag{2.309}$$

is the adiabatic compressibility. We already know that

$$C_p > C_V \tag{2.310}$$

and from Equation 2.302 there results as well

$$C_p > 0 \tag{2.311}$$

The positive values of C_p and C_V imply that the energy is a monotonically increasing function of temperature at constant volume and the Gibbs function is also a monotonically increasing function of temperature but at constant pressure. The entropy is also a monotonically increasing function of temperature at constant volume as well as at constant pressure.

2.13   THIRD LAW OF THERMODYNAMICS

The fact that the heat capacity $C_V = (\partial E/\partial T)_V$ is always positive implies that the energy is monotonically increasing with temperature. As temperature decreases, the energy monotonically decreases and consequently for the least possible temperature, that is, at absolute zero, a body should be found in the state with the least possible energy.

At absolute zero, any part of a body should be found in one defined quantum state. In other words, the statistical weight $\Delta\Gamma$ of the parts of the body should be equal to unity and also their product will be unity, $\Delta\Gamma = 1$. This applies to any system with a nondegenerate* ground state. As the entropy of a system $S = \ln \Delta\Gamma$, it follows that

$$S = \ln \Delta\Gamma\big|_{T\to 0} \to 0 \qquad\qquad (2.312)$$

This thermodynamic state has perfect order, minimum energy and minimum entropy. If the system could actually be found in such a state, there would be no activity of any kind (this is a philosophical question, because the act of measurement would invariably require an energy exchange and cause the temperature to be greater than zero). The Third Law is usually stated: At absolute zero temperature, the entropy of a pure crystalline solid is zero and all motion ceases. That is assuming the perfect order of a crystalline lattice, with all identical molecules. For other types of materials, the entropy should only be expected to be a minimum, although it practically speaking takes a very small value. The Third Law of Thermodynamics is sometimes stated in an alternative form concerning the impossibility of reaching a state of minimum entropy: It is impossible to cause a material to reach the absolute zero of temperature.

To see why this is so, consider how a material can be cooled off. To do so, it must lower its internal energy. Based on conservation of energy and the First Law, $dE = TdS - pdV$, there are only two possibilities. Either the system can be placed in contact with a colder object, and release its heat (i.e., its entropy) to that cooler object, or the system must do a work on its surroundings in order to cool itself. The former process does not help to find a method to cool to any desired temperature because it requires something colder than what is already available. Indeed, it would result in increasing the temperature of the colder object, even if one had an object originally at absolute zero. The latter method, letting the system do work, is the basic principle of refrigeration. For instance, a refrigerant cools itself in an adiabatic expansion, because it does work by expanding. The difficulty is that as it

---

* A quantum ground state could have, for example, an m-fold spin degeneracy. This would lead to non-zero minimum entropy if all the degenerate states are occupied. A stronger statement of the Third Law would be that the entropy is its minimum value at absolute zero.

expands its pressure goes down, and there will be a limit to how much work the system can do, that is of the order of its original temperature times its entropy, at best. When the analysis is done carefully, it is seen that it is impossible for the system to do a work equal to $E - E(0)$, where $E(0)$ is its zero temperature energy, and thus it cannot reach absolute zero.

2.13.1 NERNST THEOREM

The entropy S of a system has the limiting property that

$$\lim_{T \to 0} S \to S_0 \qquad (2.313)$$

Here, S_0 is a constant independent of all parameters of the particular system. The constant S_0 is zero when the ground state is nondegenerate. The theorem is a corollary to quantum statistics in which an appreciable role is played by the notion of discrete quantum states.

The theorem enables us to investigate some thermodynamic values as $T \to 0$. This may be seen easily in the heat capacity definitions

$$\lim_{T \to 0} C_{V,p} \to 0 \qquad (2.314)$$

From the definitions

$$C_{V,p} = T \left(\frac{\partial S}{\partial T} \right)_{V,p} = \left(\frac{\partial S}{\partial \ln T} \right)_{V,p} \qquad (2.315)$$

and as $T \to 0$ then $\ln T \to -\infty$ and also as $T \to 0$ then $S \to S_0 = $ const, which gives $C_{V,p} \to 0$ regardless of the value of S_0.

The limiting behavior $C_{V,p} \to 0$ as $T \to 0$ is not surprising. It shows that as the temperature approaches zero (but not yet equal to zero) the system tends to settle down to its ground state and therefore the mean system energy becomes nearly equal to the ground state energy. Further reduction in temperature does not result in any appreciable reduction of the mean system energy, because there are no lower energy states available.

It should be noted that as $T \to 0$ the limiting entropy value is independent of all external parameters of the system. It is also independent of volume or pressure variations. It follows that

$$S(\delta, T)_{T \to 0} = S_0 \qquad (2.316)$$

where any parameter is denoted by δ. Nernst wrote that

$$S(\delta, T) = S_0 + a(\delta) T^n, \quad n > 0 \qquad (2.317)$$

where $a(\delta)T^n \to 0$ as $T \to 0$. Then the general heat capacity C_δ as $T \to 0$ has the expected behavior:

$$C_\delta = T\left(\frac{\partial S}{\partial T}\right)_\delta = na(\delta)T^n\bigg|_{T \to 0} \to 0 \qquad (2.318)$$

As an example, let us consider a crystal of carbon monoxide (CO) at a very low temperature T (see Figure 2.8). All the spins must be randomly oriented so that $\Delta\Gamma = 2^N$ and $S_0 = N\ln 2 \neq 0$. We can conclude that if a system is found in a degenerate state, then $S \neq 0$ even though S may tend to a minimum.

In a similar spirit, we look at the low temperature thermal properties of solid states. If we consider the Nernst formulation then

$$S(\delta,T) = a(\delta)T^n, \quad n > 0 \qquad (2.319)$$

For $\delta = V$ then

$$\left(\frac{\partial S}{\partial V}\right)_T = \left(\frac{\partial p}{\partial T}\right)_V = T^n\frac{\partial a(V)}{\partial V}\bigg|_{T \to 0} \to 0 \qquad (2.320)$$

and for $\delta = p$ we have

$$\left(\frac{\partial S}{\partial p}\right)_T = -\left(\frac{\partial V}{\partial T}\right)_p = T^n\frac{\partial a(p)}{\partial p}\bigg|_{T \to 0} \to 0 \qquad (2.321)$$

This implies that for low temperatures the volume is independent of temperature. If we consider the difference $C_p - C_V$ it is seen that

$$C_p - C_V = -T\left(\frac{\partial V}{\partial p}\right)_T\left(\frac{\partial p}{\partial T}\right)_V^2 \to T^{2n+1}\bigg|_{T \to 0} \to 0 \qquad (2.322)$$

FIGURE 2.8   A crystal of carbon monoxide CO at a very low temperature T.

Then, $C_p - C_V$ becomes increasingly infinitesimally small compared to C_V as one goes to very low temperatures, that is,

$$\left.\frac{C_p - C_V}{C_V}\right|_{T \to 0} \to 0 \qquad (2.323)$$

It follows that as $T \to 0$ and the system approaches its ground state, quantum mechanical effects become very important and as a consequence the classical equation of state $pV = nRT$ is no longer valid, even if interactions between the particles in a gas are so small that the gas is treated as ideal. However there is no contradiction in the equation $C_p - C_V = nR$ for an ideal gas.

2.14   DEPENDENCE OF THERMODYNAMIC FUNCTIONS ON NUMBER OF PARTICLES

Until now, we have written the internal energy of a system as $E = E(S, V)$. Now we consider systems where the number of particles, N, may change. In principle, the energy must depend on the number of particles, $E = E(S, V, N)$. For different bodies, the dependence of energy E on the number of particles N is universal but approximate. Apart from the energy E and the entropy S the thermodynamic functions F, G, and H also have additive properties. These are also called extensive properties because they depend on the size of the system. This follows immediately from their definition if we consider that the pressure and the temperature are constant at equilibrium. If we have n systems of N particles each, then it follows that $E(nN) = nE(N)$. This shows that the energy E is a homogeneous function. The definition of the homogeneity of a function is based on

$$E(nx, ny) = n^\gamma E(x, y) \qquad (2.324)$$

where γ is the order of homogeneity. The energy is homogeneous with $\gamma \equiv 1$.

Let us take the energy E of the given system as a function of entropy, volume, and the number of particles:

$$E = E(S, V, N) \qquad (2.325)$$

Suppose we define an energy per particle, $\varepsilon = E/N$, and write the total energy as

$$E = N\varepsilon\left(\frac{S}{N}, \frac{V}{N}\right) \qquad (2.326)$$

Here, S and V have additive properties. Relation (2.326) is the most general form of a homogeneous function of the first order relative to N, S, and V. It shows how the energy per particle depends only on the entropy per particle and the volume per particle, all forms of densities. It follows from Equation 2.326 that

$$\left(\frac{\partial E}{\partial N}\right)_{S,V} = \varepsilon + N\left(\frac{\partial \varepsilon}{\partial N}\right)_{S,V} \equiv \tilde{\mu}\left(\frac{S}{N},\frac{V}{N}\right) \tag{2.327}$$

At this point we introduced a function $\tilde{\mu}$ as a shorthand notation for this derivative. Its meaning will become clearer as we look at the other thermodynamic potentials.

If we consider $S = S(E, V, N)$ then it follows that

$$S = N\frac{S(E,V,N)}{N} = N\overline{S}\left(\frac{E}{N},\frac{V}{N}\right) \tag{2.328}$$

But

$$S = \ln \Delta\Gamma \tag{2.329}$$

Then we have

$$\Delta\Gamma = \exp\left\{N\overline{S}\left(\frac{E}{N},\frac{V}{N}\right)\right\} \tag{2.330}$$

If we consider

$$dS = \left(\frac{\partial S}{\partial E}\right)_{V,N} dE + \left(\frac{\partial S}{\partial V}\right)_{E,N} dV + \left(\frac{\partial S}{\partial N}\right)_{E,V} dN \tag{2.331}$$

and

$$dS = \frac{\delta Q}{T} = \frac{1}{T}dE + \frac{p}{T}dV \tag{2.332}$$

then there follows the thermal equation of state:

$$\left(\frac{\partial S}{\partial E}\right)_{V,N} = \frac{1}{T} > 0 \tag{2.333}$$

This implies that S increases relative to an increase in E. It can also be seen that the statistical weight $\Delta\Gamma$ increases with an increase in N as well as with an increase in E. From Equations 2.331 and 2.332, we have

$$\left(\frac{\partial S}{\partial V}\right)_{E,N} = \frac{p}{T} \tag{2.334}$$

If we consider the Helmholtz free energy, $F = F(T, V, N)$, and the fact that the temperature is an intensive quantity of the body while the volume is extensive (or additive), we have free energy per particle, $f(T, V/N)$, defined from

$$F = N \frac{F(T,V,N)}{N} = Nf\left(T, \frac{V}{N}\right), \quad T \neq T(N) \tag{2.335}$$

It follows from here that

$$\left(\frac{\partial F}{\partial N}\right)_{T,V} = f + N\left(\frac{\partial f}{\partial N}\right)_{T,V} \tag{2.336}$$

The free energy per particle just depends on the temperature and the volume per particle.

If we examine instead the Gibbs free energy, $G = G(p, T, N)$, the situation is a little different because both pressure and temperature are intensive and do not scale with particle number. Scaling by particle number gives only an overall factor of N

$$G = N \frac{G(p,T,N)}{N} = N\mu(p,T) \tag{2.337}$$

The function μ does not depend on N, and can be derived by

$$\left(\frac{\partial G}{\partial N}\right)_{p,T} = \mu(p,T) \tag{2.338}$$

μ is called the *chemical potential* per particle and defined so that it has the dimension of energy.

What is the physical significance of the chemical potential? Essentially, it is the energy needed to add one particle to the system, while keeping the pressure and temperature fixed. It determines the direction of flow of particles in a way analogous to temperature determining the flow of heat. Heat flows from regions of higher temperature to regions of lower temperature. Similarly, particles diffuse from regions of higher chemical potential to ones of lower chemical potential. At equilibrium, two systems in thermal contact have the same temperature. Similarly, in equilibrium, two systems that can exchange particles do so until they acquire the same chemical potential. Take the example of ice in contact with water: the ice will tend to melt if its chemical potential is higher than that of water; the water will tend to freeze if its chemical potential is higher than that of ice.

Previously we took the number of particles as a parameter that was constant for a given body. Due to the introduction of the chemical potential, however, the number of particles in an open system can change with time. Therefore, we treat N formally as an independent variable. Then, from $E = E(S, V, N)$, we have

$$dE = \left(\frac{\partial E}{\partial S}\right)_{V,N} dS + \left(\frac{\partial E}{\partial V}\right)_{S,N} dV + \left(\frac{\partial E}{\partial N}\right)_{S,V} dN \qquad (2.339)$$

This contains all the possible thermodynamic information about the system. From Equations 2.332 and 2.326, we have

$$dE = T\,dS - p\,dV + \tilde{\mu}\,dN \qquad (2.340)$$

Alternatively we can write Equation 2.340 in the form

$$dF = d(E - TS) = dE - d(TS) = -S\,dT - p\,dV + \tilde{\mu}\,dN \qquad (2.341)$$

Also in terms of the enthalpy

$$dH = d(E + pV) = dE + p\,dV + V\,dp = T\,dS + V\,dp + \tilde{\mu}\,dN \qquad (2.342)$$

One can also write Equation 2.338 in terms of the Gibbs free energy. Thus

$$dG = d(E - TS + pV) = d(E - TS) + d(pV) = dF + d(pV)$$
$$= -S\,dT + V\,dp + \tilde{\mu}\,dN \qquad (2.343)$$

On the other hand, this differential is also found from

$$dG = d(\mu N) = N\,d\mu + \mu\,dN \qquad (2.344)$$

Comparison shows that the two expressions match only if $\tilde{\mu} = \mu$ and

$$N\,d\mu = -S\,dT + V\,dp \qquad (2.345)$$

This last relation is known as the Gibbs–Duhem relation. It demonstrates how the chemical potential changes in response to temperature and pressure changes. This also shows how the temperature, pressure, and chemical potential, are not independent. The chemical potential is not under the full control of the experimentalist, not as an electric potential might be.

If we consider the formulas (2.338) through (2.342) then it follows that

$$\mu = \tilde{\mu} = \left(\frac{\partial E}{\partial N}\right)_{S,V} = \left(\frac{\partial F}{\partial N}\right)_{T,V} = \left(\frac{\partial G}{\partial N}\right)_{p,T} = \left(\frac{\partial H}{\partial N}\right)_{S,p} \qquad (2.346)$$

All of the thermodynamic potentials then could be applied to systems with varying particle number. This last relation shows directly how the chemical potential is a free energy per particle.

Now let us transform to the generalized thermodynamic potential $\Omega(T, V, \mu)$, also known as the grand potential or grand canonical potential, starting from dF

$$dF = -S \, dT - p \, dV + d(\mu N) - N \, d\mu \tag{2.347}$$

and then rearranging into a differential depending on dT, dV, $d\mu$,

$$d\Omega = d(F - \mu N) = -S \, dT - p \, dV - N \, d\mu \tag{2.348}$$

Therefore, the grand potential for fixed temperature, volume, and chemical potential is

$$\Omega = F - \mu N = E - TS - \mu N \tag{2.349}$$

It is interesting to note that Ω can also be expressed via the two equivalent definitions (2.177) and (2.345) of the Gibbs free energy

$$G = E - TS + pV, \quad G = \mu N \tag{2.350}$$

which combined with Equation 2.349 lead to the intriguing relation

$$\Omega = -pV \tag{2.351}$$

Of course, this results from the fact that the variables T, V, μ, are not all independent. Obviously from the expression for $d\Omega$ we obtain the thermodynamic variables

$$\left(\frac{\partial \Omega}{\partial T} \right)_{V,\mu} = -S, \quad \left(\frac{\partial \Omega}{\partial V} \right)_{T,\mu} = -p, \quad \left(\frac{\partial \Omega}{\partial \mu} \right)_{T,V} = -N \tag{2.352}$$

These results can also be generalized to a system with different species of particles, each with its own chemical potential. The different species typically might be different types of molecules diffusing within a fluid, for example. Suppose μ_i and N_i refer to the ith species. Then the grand potential must include a sum over the species

$$\Omega = F - \sum_i \mu_i N_i = E - TS - \sum_i \mu_i N_i \tag{2.353}$$

and the particle number for the ith species depends only on its respective potential

$$\left(\frac{\partial \Omega}{\partial \mu_i} \right) = -N_i \tag{2.354}$$

The grand thermodynamic potential is convenient to apply to problems having an open system, where temperature and volume are under the control of the experimentalist, while the particle number fluctuates.

2.15   EQUILIBRIUM IN AN EXTERNAL FORCE FIELD

The systems we discussed so far have T = const and p = const because we consider them to have no external force field. If we now introduce an external field that acts to move particles in some preferred direction, the homogeneity of the system will be disturbed. The external field could be gravitational, electric, magnetic, a chemical potential, or any other physical effect that breaks the symmetry and the homogeneity. As the field forces particles in a preferential direction, a gradient in their distribution in space will appear. It is possible that this also causes $p \neq$ const and $T \neq$ const.

Let us examine a system of N particles in which the force field is directed along the z-axis. The system can be partitioned into two subsystems denoted by 1 and 2, separated by a planar boundary at some location on the z-axis. Particles can cross back and forth through the boundary and energy can be transferred across the boundary. The force field is associated with some potential energy that has a gradient across the boundary. In order to define the two subsystems, the boundary must be fixed in place. The entire system is considered isolated, so that its total energy E and its total volume V are both fixed. We denote by N_i the number of particles in the ith subsystem and E_i the energy of that subsystem and V_i its volume, and thus follow the conservation conditions:

$$\begin{cases} E_1 + E_2 = E = \text{const} \\ V_1 + V_2 = V = \text{const} \\ N_1 + N_2 = N = \text{const} \end{cases} \tag{2.355}$$

The entropy of the entire system is a function of the parameters in Equation 2.355. The equilibrium state, corresponding to the most probable situation, is the state with the maximum entropy. That is,

$$S = S(E_1, V_1, N_1; E_2, V_2, N_2) = S_{\text{max}} \tag{2.356}$$

and

$$S = S_1(E_1, V_1, N_1) + S_2(E_2, V_2, N_2) \tag{2.357}$$

Here, S_i is the entropy of the ith subsystem. If we consider Equation 2.356, then

$$dS = dS_1 + dS_2 = 0 \tag{2.358}$$

and from Equation 2.355 we have

$$\begin{cases} dE_1 + dE_2 = 0 \\ dV_1 + dV_2 = 0 \\ dN_1 + dN_2 = 0 \end{cases} \tag{2.359}$$

If we use Equation 2.340 considering the fact that $\tilde{\mu} \equiv \mu$ together with Equations 2.358 and 2.359, then we have

$$dS = dS_1 + dS_2 = \left(\frac{1}{T_1} dE_1 + \frac{p_1}{T_1} dV_1 - \frac{\mu_1}{T_1} dN_1 \right)$$

$$+ \left(\frac{1}{T_2} dE_2 + \frac{p_2}{T_2} dV_2 - \frac{\mu_2}{T_2} dN_2 \right) = 0 \tag{2.360}$$

or

$$dS = \left(\frac{1}{T_1} - \frac{1}{T_2} \right) dE_1 + \left(\frac{p_1}{T_1} - \frac{p_2}{T_2} \right) dV_1 - \left(\frac{\mu_1}{T_1} - \frac{\mu_2}{T_2} \right) dN_1 = 0 \tag{2.361}$$

Now, to define the partition between subsystems 1 and 2 requires a boundary fixed in space. Thus, we must take $dV_1 = dV_2 = 0$, otherwise, we would not know how to distinguish the two regions. As Equation 2.361 is valid for arbitrary variations in dE_1 and dN_1, then it follows that the coefficients of their differentials must separately vanish. This gives the results

$$\frac{1}{T_1} - \frac{1}{T_2} = 0, \quad \frac{\mu_1}{T_1} - \frac{\mu_2}{T_2} = 0 \tag{2.362}$$

or

$$T_1 = T_2, \quad \mu_1 = \mu_2 \tag{2.363}$$

These are the necessary conditions for equilibrium between two subsystems, and reflect (2.355). Note that if the volume also had been allowed to fluctuate, we would additionally obtain a constraint on the pressure, $p_1 = p_2$. But the Gibbs–Duhem relation (2.345) shows that μ, T, p are not all independent; thus, only two of these three variables can be constrained, and the third is then determined implicitly. Considering now that along the z-axis, a system is partitioned into a sequence of subsystems, we conclude that in equilibrium, the chemical potential is the same in all the subsystems, or

$$\mu(p, T, x, y, z) = \text{const} \tag{2.364}$$

Indeed, the result has been written in the most general form, regardless of the direction of any force field. This condition replaces the condition $p = $ const for the case of a homogeneous system with no applied fields. Physically, it means that particles shift around among the subsystems, until finding a state where the entire system is at a uniform chemical potential. If the chemical potential and temperature become constants, then the pressure can vary in space.

Let us see the implications of uniform chemical potential, realizing that it is connected to the definition of the Gibbs free energy, $G = E - TS + pV$. In the presence of an applied field, the potential energy $U(x, y, z)$ associated with that field will be a term in the internal energy. We can write

$$E = E_0 + U(x, y, z) \tag{2.365}$$

Here, E_0 refers to the internal energy in the absence of the applied field. Then the chemical potential is

$$\mu = \frac{G}{N} = \frac{E_0 - TS + pV}{N} + \frac{U(x, y, z)}{N} = \mu_0(p, T) + u(x, y, z) \tag{2.366}$$

$$\mu_0 = \varepsilon_0 - Ts + pv, \quad \varepsilon_0 = \frac{E_0}{N}, \quad s = \frac{S}{N}, \quad v = \frac{V}{N} \tag{2.367}$$

Here, $u(x, y, z)$ is the potential energy per particle due to the force field, and $\mu_0(p, T)$ is the chemical potential in the absence of the field, dependent on the energy density, entropy density, and volume per particle. It is the *total* chemical potential that is uniform

$$\mu = \mu_0(p, T) + u(x, y, z) = \text{const} \tag{2.368}$$

This is the equilibrium condition in any force field. Generally, as a result of $u(x, y, z)$, the pressure as well will depend on position, $p = p(x, y, z)$.

Suppose we return to the special case of a field directed along the z-axis. Imagine the variations in pressure and particle number, and so on for a small displacement along z. It follows from Equation 2.368 that

$$\mu(p + dp, T, z + dz) = \mu(p, T, z) \tag{2.369}$$

which can be expressed as

$$\mu(p + dp, T, z + dz) - \mu(p, T, z) = \frac{\partial \mu}{\partial p} dp + \frac{\partial \mu}{\partial z} dz = 0 \tag{2.370}$$

From the fact that

$$\mu = \varepsilon_0(T) - Ts + pv + u(x, y, z) \tag{2.371}$$

and from Equation 2.370 we have a differential equation

$$v\,dp + \frac{\partial u}{\partial z}dz = 0 \tag{2.372}$$

A simple example is a fluid (of many particles) in a gravitational field. Gravity is a homogeneous force with potential energy per particle $u = mgz$, where m is the mass of a particle, g is the acceleration due to gravity, and z is the vertical coordinate. If we substitute $u = mgz$ in Equation 2.372, then

$$dp + \frac{m}{v}g\,dz = 0 \tag{2.373}$$

$$\frac{dp}{dz} = -\rho g, \quad \rho \equiv \frac{m}{v} = \frac{Nm}{V} \tag{2.374}$$

where ρ is the mass density of the fluid composed from N particles per volume V. If v is considered constant, the fluid is incompressible. Then, we obtain the change in pressure with depth in a fluid

$$p = p_0 - \rho gz \tag{2.375}$$

This is the formula for the hydrostatic pressure in an incompressible liquid (fixed density ρ).

Next, consider instead an ideal gas, where $pV = NT$ or $pv = T$. Writing this as $v = T/p$, which is not constant, leads to the differential equation

$$\frac{dp}{p} = -\frac{mg}{T}dz \tag{2.376}$$

With an equilibrium situation, the temperature is constant, so the pressure decays exponentially with altitude

$$p = p_0 \exp\left\{-\frac{mgz}{T}\right\} \tag{2.377}$$

This is the *barometric formula* describing the change in pressure with respect to height in a gas. It is interesting to make an estimate of the typical height over which the pressure reduces by the factor $e^{-1} \approx 0.37$, known as the *scale height*. As a rough estimate for nitrogen molecules in air, that requires an altitude change on the order of

$$h = \frac{T}{mg} = \frac{(1.38 \times 10^{-23}\,\text{J/K})(300\,\text{K})}{(28 \times 1.66 \times 10^{-27}\,\text{kg})(9.80\,\text{m/s}^2)} \approx 9\,\text{km} \tag{2.378}$$

The actual temperature in the Earth's atmosphere is not constant as assumed here because it is not a true equilibrium system. Also, a better way to estimate is to use the molecular weight averaged over the mixture of gases in air (closer to 29 g/mol). Even so, the results have some interesting implications. The scale height results as a competition between gravitational potential energy mgz and thermal energy k_BT. The thermal energy acts to spread out the molecules while gravity tends to constrain them near the Earth. In a real sense, the presence of a scale height proves the existence of molecules and the molecular theory. As real air is a mixture of gases, it can be seen that the heavier molecules will have smaller scale height; they will be concentrated closer to the Earth. The lighter molecules (such as helium, hydrogen, methane) obviously have the opposite behavior and become more predominate constituents at higher altitudes. By reapplying the ideal gas law in a form to give gas density

$$\rho = \frac{mp}{T}, \quad \rho \equiv \frac{Nm}{V} \tag{2.379}$$

one also sees how the mass density and number density of the molecules both decrease exponentially with altitude.

3 Canonical Distribution

3.1 GIBBS CANONICAL DISTRIBUTION

Let us write once more the Gibbs microcanonical distribution:

$$dW(q, p) = C\,\delta(E(q, p) - E_0)d\Gamma \tag{3.1}$$

where E_0 is the fixed total energy and $E(q, p)$ is the energy of a particular microstate, C is a constant of proportionality independent of the microstate and $d\Gamma$ is the number of states accounted for in $dqdp$. We may use Equation 3.1 as seen earlier to evaluate expectation values.

We consider a small system A (some body of interest) in thermal interaction with a heat reservoir (thermostat) A'. We consider A as having many fewer degrees of freedom than A', that is, $A \ll A'$. A could be any relatively small macroscopic system, for example, a can of drink in a basin of water acting as a heat reservoir. It could sometimes be a microscopic system, say, an atom at some lattice site in a solid that acts as a heat reservoir. As system A is not isolated, the microcanonical distribution does not apply to it by itself. The system is continuously exchanging energy with the reservoir; hence its probability distribution is not microcanonical.

We assume weak interactions between A and A' so that their energies should be additive. The energy

$$E + E' = E_0 \tag{3.2}$$

of the combined system, $A_0 = A + A'$, is assumed to be a constant in some range between E_0 and $E_0 + \delta E$, with E' being the energy of the reservoir (thermostat) A'. We are interested in the microscopic state of the system A, with $dW_A\,(q, p)$ characterizing the probability of coordinates in $q, q + dq$ and $p, p + dp$. The microstate of the thermostat A' is specified by coordinates $q', q' + dq'$ and $p', p' + dp'$. These two microstates give the joint microcanonical distribution $dW(q, p, q', p')$.

Since q and p are continuous quantities, integrating over the reservoir's coordinates gives

$$dW(q, p) = \int_{(q', p')} dW(q, p, q', p') \tag{3.3}$$

It can be seen that

$$d\Gamma_0 = d\Gamma d\Gamma' \tag{3.4}$$

where $d\Gamma$ is the number of states of the system (body) A and $d\Gamma'$ that of the thermostat A'. Considering Equation 3.1, it follows that

$$dW(q,p) = d\Gamma\, C \int \delta(E(q,p) + E'(q',p') - E_0) \frac{d\Gamma'}{dE'} dE' \qquad (3.5)$$

Here, $\Gamma'(E')$ is the total number of quantum states of the thermostat (reservoir) A' with energy less or equal to E'. It should be noted the integrand depends only on E' and that is why we do the change of variable:

$$d\Gamma' = \frac{d\Gamma'}{dE'} dE' \qquad (3.6)$$

Let

$$\frac{d\Gamma'}{dE'} \equiv \frac{\Delta\Gamma'}{\Delta E'} \qquad (3.7)$$

and let us write $\Delta\Gamma'$ according to Boltzmann:

$$\Delta\Gamma' = \exp\{S'(E')\} \qquad (3.8)$$

where S' is the entropy of the thermostat A' as a function of its energy E' (the function E' is finite as well as $\Delta E'$). Thus

$$dW(q,p) = d\Gamma\, C \int \delta(E + E' - E_0) \frac{\exp\{S'(E')\}}{\Delta E'} dE' = d\Gamma\, C \frac{\exp\{S'(E_0 - E)\}}{\Delta E'} \qquad (3.9)$$

Let us consider the fact that A is exceedingly smaller than A'. Then, $E \ll E_0$ and S' can be approximated by an expansion in the Taylor series about the value $E' = E_0$:

$$S'(E_0 - E) = S'(E_0) - \frac{\partial S'(E_0)}{\partial E'} E + \frac{1}{2}\frac{\partial^2 S'}{\partial E'^2} E^2 + \cdots \qquad (3.10)$$

But the reservoir has a temperature given from

$$\frac{\partial S'}{\partial E'} = \frac{1}{T} \qquad (3.11)$$

and

$$\frac{\partial^2 S'}{\partial E'^2} = \frac{\partial}{\partial E'}\frac{1}{T} = -\frac{1}{T^2}\frac{\partial T}{\partial E'} = -\frac{1}{T^2}\frac{1}{C_V} \qquad (3.12)$$

and as $C_V \approx N \to \infty$ it then follows that

$$\frac{\partial^2 S'}{\partial E'^2} \to 0 \tag{3.13}$$

T is the temperature of the system and the thermostat. This is because we suppose that the two are in equilibrium. Thus, from Equations 3.10 and 3.9, we have

$$dW(q,p) = \frac{d\Gamma C}{\Delta E'} \exp\left\{ S'(E_0) - \frac{E}{T} \right\} \tag{3.14}$$

or

$$dW(q,p) = A\exp\left\{ -\frac{E(q,p)}{T} \right\} d\Gamma \tag{3.15}$$

This is called the *Gibbs canonical distribution*, with A being a normalization constant. The canonical distribution is dependent on the energy of the system A, which is not assumed to be constant. The canonical distribution is usually a more convenient tool than the microcanonical distribution, especially because temperature is more easily controlled in experiments than total energy. The microcanonical ensemble deals with states with specified energy $E(S, V, N)$, and T, p, and μ are derived quantities. The system is kept at constant temperature in the canonical ensemble.

The exponential factor in Equation 3.15 is called the *Boltzmann factor*. When including the physical units, it appears with the factor $\beta \equiv (k_B T)^{-1}$ inside the exponential (in further evaluations, we consider the temperature to have units of energy). Then, the statistical weight in phase space varies according to the factor $\exp\{-\beta E\}$. This is one factor that controls how the Gibbs canonical distribution $dW(q, p)$ changes with energy. In principle, $d\Gamma = d\Gamma(E)$ is a rapidly increasing function of the energy of system A (Figure 3.1). We can see from Figure 3.1 that $dW(q, p)$ might be described by a *Gaussian distribution*. The fluctuations of a thermodynamic variable follows a *Gaussian distribution*.

The expression $A\exp\{-E/T\}$ stems from $d\Gamma'$, which is the statistical weight of the thermostat, while $d\Gamma$ is the statistical weight of the system. An increase in E corresponds to a decrease in E', associated with the transfer of energy from the thermostat to the body. An increase in E' results from a decrease in $d\Gamma'$ as E increases. The expression (3.14) is the probability of the microstate of the body inside the interval $q,q + dq$ and $p,p + dp$ in equilibrium with the thermostat at temperature T. If the state is quantized, then the expression in (3.14) becomes

$$W_n = A\exp\left\{ -\frac{E_n}{T} \right\} \tag{3.16}$$

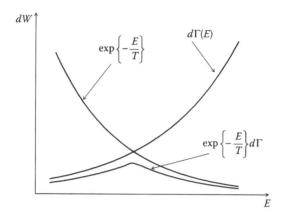

FIGURE 3.1 The variation of the Gibbs canonical distribution, the Boltzmann factor, and the statistical weight in phase space, with the energy.

where n is a quantum number identifying a state. If part of the variables are quantized and the rest are not, then

$$dW_n(q, p) = A \exp\left\{-\frac{E_n(q, p)}{T}\right\} d\Gamma \tag{3.17}$$

We may see from here that the *Boltzmann factors* give the probability for the system to be in a state with energy $E = E_n$. The factor of $d\Gamma = d\Gamma(E)$ is the number of states at the selected energy.

If we apply the normalization condition to Equation 3.16, that the probabilities of all possible states must sum to unity, then

$$\sum_n W_n = A \sum_n \exp\left\{-\frac{E_n}{T}\right\} = AZ = 1 \tag{3.18}$$

and we introduce

$$Z \equiv \sum_n \exp\left\{-\frac{E_n}{T}\right\} = \sum_n \exp\{-\beta E_n\} \tag{3.19}$$

as the *statistical sum or canonical partition function*. Its terms govern the distributions or "partition" of systems among quantum states in phase space. Since the partition function Z extends over all quantum states, it is called *Zustandsumme*—"sum-over-the-states"—in the German language. It will be seen that all thermodynamic information may be derived from the partition function. It follows from above that

$$A = \frac{1}{Z} \tag{3.20}$$

Thus, the probability of the state n is determined by

$$W_n = \frac{1}{Z}\exp(-\beta E_n) \tag{3.21}$$

If there are many states at the same energy, they all have this same probability. All of them would contribute equally to the partition function sum.

We can also apply the normalization condition to Equation 3.17, and then it follows that

$$\sum_n \int dW_n(q,p) = A\sum_n \int \exp\left\{-\frac{E_n(q,p)}{T}\right\}d\Gamma = AZ = 1 \tag{3.22}$$

and

$$Z = \sum_n \int \exp\left\{-\frac{E_n(q,p)}{T}\right\}d\Gamma \tag{3.23}$$

The integrations here are only over the nonquantized coordinates. The quantity (3.23) is again the *statistical integral or canonical partition function* of the system. It is the principal quantity for the evaluation of the thermodynamic properties of the system.

It should be noted that for the evaluation of expectation values, the convenient practical method is the ensemble in which each system is not isolated and is found in thermal equilibrium with another system of large dimensionality (that plays the role of a thermostat). The condition needed to achieve equilibrium is that the energy of interaction between system and thermostat should be sufficiently small, but yet, large enough to enable sufficient exchange of energy between the two to sample a large part of phase space in a short time (ergodic assumption). The probability density for the statistical ensemble of such systems is the Gibbs canonical distribution (3.15). For these reasons, it is probably the most important ensemble in statistical mechanics.

3.2 BASIC FORMULAS OF STATISTICAL PHYSICS

The formula connected with the entropy and the probability of a state is

$$S = -\sum_n W_n \ln W_n = -\sum_n W_n\left(\ln A - \frac{E_n}{T}\right) = -\ln A\sum_n W_n + \frac{1}{T}\sum_n E_n W_n \tag{3.24}$$

From definition

$$\sum_n E_n W_n = \overline{E}, \quad \sum_n W_n = 1 \tag{3.25}$$

(all the degenerate states are included in the sums). Then, it follows that

$$T \ln A = \overline{E} - TS = F \tag{3.26}$$

Considering Equation 3.22, we have found that the Helmholtz free energy can be found from the statistical mechanical definition:

$$F = -T \ln Z \tag{3.27}$$

We need to stress the importance of this relation. As the Helmholtz free energy F is the fundamental thermodynamic potential for a system at fixed T, V, N, the canonical partition function Z represents the same information, but it comes from the statistical description. It is clear then that all desired thermodynamic quantities can be found from Z. For instance, the expectation value of the energy, $E = \overline{E}$, can also be expressed as

$$E = F + TS = F - T\frac{\partial F}{\partial T} = -T^2 \frac{\partial}{\partial T}\frac{F}{T} = \frac{\partial}{\partial(1/T)}\frac{F}{T} = T^2 \frac{\partial}{\partial T}\ln Z \tag{3.28}$$

Thus

$$E = T^2 \frac{\partial}{\partial T}\ln Z \tag{3.29}$$

If we consider Equation 3.26, then

$$S = \frac{\partial}{\partial T}(T \ln Z) \tag{3.30}$$

Equations 3.27 through 3.30 give all the relevant thermodynamics from the partition function Z.

If we consider relations (3.16) and (3.26), then we have

$$W_n = \exp\left\{\frac{F - E_n}{T}\right\} \tag{3.31}$$

This is called *Gibbs canonical distribution*. If we know the Hamilton function E of the system, then we can find Z and then F from Equation 3.26, and as a consequence, all of the other thermodynamic quantities follow.

Q3.1 Given the normalized divisor $\Omega(E)$ of the microscopic distribution, find the integral of states Z. Evaluate the integral of states for an ideal gas made up of N

atoms. Find the probability that the gas has a given energy lying within the interval from E to $E + dE$.

A3.1 We first find the probability $dW(E)$ that the system has energy in the interval from E to $E + dE$. Considering the Gibbs canonical distribution, the probability that the system has the coordinates and the momenta lying in the intervals $q \ldots (q + dq)$ and $p \ldots (p + dp)$ is equal to

$$dW(q, p) = \exp\left\{\frac{F - E(q, p)}{T}\right\} d\Gamma_0 \tag{3.32}$$

Integrating Equation 3.32 with respect to the layer between the surfaces with the constant energies $E(q, p) = E$ and $E(q, p) = E + dE$, we find the probability that the energy of the system lies in the interval, $E \ldots (E + dE)$:

$$dW(E) = \int_{E(q,p)=E} \exp\left\{\frac{F - E(q, p)}{T}\right\} d\Gamma = \exp\left\{\frac{F - E}{T}\right\} \Omega(E) dE \tag{3.33}$$

where $\Omega(E) dE$ is the phase volume between the layers $E(q, p) = E$ and $E(q, p) = E + dE$:

$$\Omega(E) dE = \frac{\partial \Gamma_0(E)}{\partial E} dE \tag{3.34}$$

From here, the probability density of the given value of the energy E in the canonical distribution is

$$W(E) = \exp\left\{\frac{F - E}{T}\right\} \Omega(E) = \frac{1}{Z} \exp\left\{\frac{-E}{T}\right\} \Omega(E) \tag{3.35}$$

Here, we used the definition of the statistical integral from Equation 3.23.
   If we let the minimum value of the energy of the system to be equal to zero, we find from the condition of normalization

$$\int_0^\infty W(E) dE = 1, \quad Z(\beta) = \int_0^\infty \exp\{-\beta E\} \Omega(E) dE, \quad \beta = \frac{1}{T} \tag{3.36}$$

Let us evaluate $\Gamma_0(E)$ for an ideal monatomic gas:

$$\Gamma_0(E) = \int_{K(p) \le E} dpdq = \int_{K(p) \le E} dp \int dq = V^N \int_{K(p) \le E} dp \tag{3.37}$$

The kinetic energy is an additive function of the energy of each particle:

$$K(p) = \sum_{i=1}^{N} \frac{1}{m_i}(p_{ix}^2 + p_{iy}^2 + p_{iz}^2) = \sum_{\alpha=1}^{3N} \frac{1}{m_\alpha} p_\alpha^2, \quad dp = \prod_{i=1}^{N} dp_{ix} dp_{iy} dp_{iz} = \prod_{\alpha=1}^{3N} dp_\alpha$$

(3.38)

Here, we suppose that the particles have different masses. The evaluation is general and simple. Let us introduce new variables:

$$\xi_{3i-2} = \frac{p_{ix}}{\sqrt{2m_i}}, \quad \xi_{3i-1} = \frac{p_{iy}}{\sqrt{2m_i}}, \quad \xi_{3i} = \frac{p_{iz}}{\sqrt{2m_i}}$$

(3.39)

Thus

$$dp = (2)^{\frac{3N}{2}} \prod_{i=1}^{N} m_i^{\frac{3}{2}} \prod_{\alpha=1}^{3N} d\xi_\alpha = \langle 2m \rangle^{\frac{3N}{2}} \prod_{\alpha=1}^{3N} d\xi_\alpha$$

(3.40)

We have introduced the mean geometric mass $\langle m \rangle$ from the formula

$$\langle m \rangle^N = \prod_{i=1}^{N} m_i$$

(3.41)

It follows from above that

$$K(p) = \sum_{\alpha=1}^{3N} \xi_\alpha^2$$

(3.42)

The function $\Gamma(E)$ may be found from the region of momentum space that satisfies the inequality

$$\sum_{\alpha=1}^{3N} \xi_\alpha^2 \leq E$$

(3.43)

In order to evaluate this, let us first consider the volume of a sphere in the s-dimensional space:

$$\Omega_s(R) = A_s R^s = \int \cdots \int \prod_{i=1}^{s} dx_i, \quad \sum_{i=1}^{s} x_i^2 \leq R$$

(3.44)

Here, R is the radius of the sphere. The volume of another sphere with the same radius, but in $(s + 1)$-dimensional space may be expressed through Ω_s (or refer to the equation for the volume of a high-dimensional sphere):

$$\Omega_{s+1} = A_{s+1}R^{s+1} = \int_{-R}^{R} dx_{s+1} \int_{\sum_{i=1}^{s} x_i^2 \le R^2 - x_{s+1}^2} dx_1 \dots dx_s = A_s \int_{-R}^{R} dx_{s+1} \left(R^2 - x_{s+1}^2\right)^{s/2}$$

$$= A_s R^{s+1} \int_0^1 (1 - \zeta)^{s/2} \zeta^{-1/2} d\zeta = A_s R^{s+1} B\left(\frac{1}{2}, \frac{s}{2} + 1\right) = \frac{\sqrt{\pi} A_s R^{s+1} \Gamma(s/2 + 1)}{\Gamma((s + 1/2) + 1)}$$

$$(3.45)$$

From the above this yields the following recurrence formula:

$$A_{s+1} = A_s \frac{\sqrt{\pi} A \Gamma((s/2) + 1)}{\Gamma((s/2) + 1)} \tag{3.46}$$

For the cases $s = 1; 2; 3$, we have, respectively, the numerical factors

$$A_1 = 2; \quad A_2 = \pi; \quad A_3 = \frac{4\pi}{3} \tag{3.47}$$

which give the expected volumes

$$\Omega_1 = 2R, \quad \Omega_2 = \pi R^2, \quad \Omega_3 = \frac{4\pi}{3} R^3 \tag{3.48}$$

If we solve the recurrence formula (3.46), on the other hand, then we get

$$A_s = \pi^{s/2} \left[\Gamma\left(\frac{s}{2} + 1\right)\right]^{-1} \tag{3.49}$$

In our case, $R = \sqrt{E}$, and considering the expression for Ω_s with $s = 3N$, we have

$$\Gamma_0(E) = V^N \Omega_{3N}\left(\sqrt{E}\right) = \frac{(2\pi m E)^{3N/2} V^N}{\Gamma\left(\frac{3N}{2} + 1\right)} \tag{3.50}$$

and

$$Z(\beta) = \frac{(2\pi m)^{3N/2}}{\Gamma\left(\dfrac{3N}{2}\right)} V^N \int_0^\infty \exp\{-\beta E\} E^{3N/2-1} dE \qquad (3.51)$$

The integral with respect to the energy leads to the gamma (special) function and is equal to

$$\int_0^\infty \exp\{-\beta E\} E^{\nu-1} dE = \frac{1}{\beta^\nu} \Gamma(\nu) \qquad (3.52)$$

This gives

$$Z(T) = (2\pi mT)^{3N/2} V^N \qquad (3.53)$$

The probability that the ideal gas has a defined value of the energy is evaluated using the formula (3.33). Substituting here $\Omega(E)$ from Equation 3.50 and considering Equation 3.53, we find the probability density versus energy:

$$dW(E) = \frac{1}{Z} \exp\left\{-\frac{E}{T}\right\} \Omega(E) dE = \frac{1}{\Gamma\left(\dfrac{3N}{2}\right)} \left(\frac{E}{T}\right)^{3N/2} \exp\left\{-\frac{E}{T}\right\} \frac{dE}{E} \qquad (3.54)$$

Q3.2 From the density of the canonical distribution for a monatomic ideal gas made up of N particles, find the expectation value $\overline{E^n}$ $(n \geq 1)$, the dispersion $\sqrt{\overline{\Delta E^2}}$ and the expectation of fluctuations relative to the mean energy, $\delta_E = \left(\sqrt{\overline{\Delta E^2}}/\overline{E}\right)$. Show that the energy of the system with a large number of particles satisfies $\overline{E^m} = \overline{E}^m$ for any whole number m.

A3.2 Using Equations 3.18 and 3.19, we find

$$\overline{E^n} = \int E^n dW(E) = T^n \frac{\Gamma\left(\dfrac{3N}{2} + n\right)}{\Gamma\left(\dfrac{3N}{2}\right)} \qquad (3.55)$$

where Equation 3.52 was used for the evaluation of the integral. From here, using $n = 1$ and $n = 2$, one gets

$$\overline{E} = T\frac{\Gamma\left(\frac{3}{2}N+1\right)}{\Gamma\left(\frac{3}{2}N\right)} = \frac{3}{2}NT, \quad \overline{E^2} = T^2\frac{\Gamma\left(\frac{3}{2}N+2\right)}{\Gamma\left(\frac{3}{2}N\right)} = \left(\frac{3}{2}N+1\right)\frac{3}{2}NT^2 \quad (3.56)$$

$$\sqrt{\overline{\Delta E^2}} = \sqrt{\overline{E^2} - \overline{E}^2} = T\sqrt{\frac{3N}{2}}, \quad \delta_E = \frac{\sqrt{\overline{\Delta E^2}}}{\overline{E}} = \sqrt{\frac{2}{3N}} \quad (3.57)$$

The relative fluctuation is small but finite in a small sample of a gas. As we let $N \to \infty$ (the thermodynamic limit), the relative fluctuation δ_E tends to zero and $\overline{E^2} \cong \overline{E}^2$. Evaluating similarly the relative fluctuations $\delta_{E^{\frac{3}{2}}}, \delta_{E^2}, \ldots$, we have $\overline{E^m} \cong \overline{E}^m$ for $N \to \infty$. For example

$$\lim_{N\to\infty} \delta_{E^2} = \lim_{N\to\infty} \frac{\sqrt{\overline{E^4} - (\overline{E^2})^2}}{\overline{E}^2} \to 0 \quad (3.58)$$

and

$$\overline{E^4} \cong (\overline{E^2})^2 = \overline{E}^4 \quad (3.59)$$

and so on.

Q3.3 Find the statistical sum, the pressure, entropy, internal energy, and heat capacity at constant volume for

a. A monatomic ideal gas of N particles found in a volume V
b. A diatomic ideal gas without considering the vibrations of the atoms in the molecule (suppose each molecule is a rigid rotator)
c. A diatomic ideal gas, including small oscillations of the atoms in the molecule (examine the case at low temperatures)

A3.3 a. For a monatomic gas made of N particles, the Hamiltonian of the system is

$$H = \sum_{i=1}^{N}\left(\frac{\vec{p}_i^2}{2m} + U(\vec{r})\right) = \sum_{i=1}^{N} H_i \quad (3.60)$$

where $U(\vec{r})$ is the potential energy of interaction of the particles with the walls of the containing vessel:

$$U(\vec{r}_i) = \begin{cases} 0, & \vec{r}_i \subset V \\ \infty, & \vec{r}_i \not\subset V \end{cases} \quad (3.61)$$

The particles of an ideal gas only interact occasionally with the walls of the container. On a collision with a wall, a particle can be expected to reverse one component of its momentum. This is a trivial change in the state of that particle that does not change its energy. Thus, this potential can be ignored. Similarly, the particles interact occasionally with each other—these interactions are ignored in this Hamiltonian. For an ideal gas, it is assumed that the interparticle collisions act to provide ergodicity, but any attractive or repulsive forces between the particles are ignored. In an ideal gas, the particles are assumed to move independently. Generally, as long as a real gas is at low-enough density, and far enough above its liquefaction temperature, it should behave approximately as an ideal gas.

The partition function depends only on the kinetic energy terms

$$Z = \int \exp\left\{-\frac{H}{T}\right\} dx_1 dy_1 dz_1 \ldots dx_N dy_N dz_N dp_{1x} \ldots dp_{Nz} = Z_i^N \qquad (3.62)$$

where Z_i is the partition function of the ith molecule:

$$Z_i = \int \exp\left\{-\frac{H_i}{T}\right\} d\Gamma_i = V \int_{-\infty}^{\infty} \exp\left\{-\frac{p_{ix}^2 + p_{iy}^2 + p_{iz}^2}{2mT}\right\} dp_{ix} dp_{iy} dp_{iz} = V(2\pi mT)^{3/2}$$

$$(3.63)$$

and this recovers the result in Equation 3.53

$$Z = V^N (2\pi mT)^{3N/2} \qquad (3.64)$$

From here, we obtain the various thermodynamic functions

$$F = -T \ln Z = -NT \ln V - \frac{3NT}{2} \ln(2\pi mT) \qquad (3.65)$$

$$S = -\left(\frac{\partial F}{\partial T}\right)_V = N \ln V + \frac{3N}{2}\left[\ln(2\pi mT) + 1\right] \qquad (3.66)$$

$$p = -\left(\frac{\partial F}{\partial V}\right)_T = \frac{NT}{V} \qquad (3.67)$$

$$E = F + TS = \frac{3NT}{2}, \quad C_V = \left(\frac{\partial E}{\partial T}\right)_V = \frac{3N}{2} \qquad (3.68)$$

The pressure equation is the usual ideal gas law equation of state. The mean value of the energy recovers the classical equipartition of energy, being $\frac{3}{2}T$ on average for each particle $\frac{1}{2}T$ per active degree of freedom for motion in each of the three coordinate directions).

b. Each rigid rotator molecule has five degrees of freedom: Three of them describe the translational motion of the center of mass with the momentum p. The rotation of a diatomic molecule is described by two spherical coordinate angular variables θ and ϕ, and corresponding canonically conjugate momenta p_θ and p_ϕ, respectively. The angle θ is measured with respect to a chosen z-axis and ϕ is the azimuthal angle measured from the x-axis.

The kinetic energy of a molecule has the form

$$K = \frac{M}{2}\dot{\vec{r}}_c^2 + \frac{\mu}{2}\left(r_0^2\dot{\theta}^2 + r_0^2 \sin^2\theta\,\dot{\phi}^2\right) \tag{3.69}$$

where $\vec{r}_c$ is its center of mass and r_0 is the distance between the atoms, assumed to be constant. The atomic masses m_1 and m_2 lead to total and reduced masses

$$M = m_1 + m_2, \quad \mu = \frac{m_1 m_2}{m_1 + m_2} \tag{3.70}$$

The rotational momenta are

$$p_\theta = \frac{\partial K}{\partial\dot{\theta}} = \mu r_0^2\dot{\theta}, \quad p_\phi = \frac{\partial K}{\partial\dot{\phi}} = \mu r_0^2 \sin^2\theta\,\dot{\phi} \tag{3.71}$$

and then the Hamiltonian per molecule is

$$H_i = \frac{p_i^2}{2M} + \frac{p_{i\theta}^2}{2I} + \frac{p_{i\phi}^2}{2I\sin^2\theta}, \quad I = \mu r_0^2 \tag{3.72}$$

The partition function of each molecule is

$$Z_i = \int\exp\left\{-\frac{H_i}{T}\right\}d\Gamma_i = V(2\pi MT)^{3/2}\int_0^\pi d\theta_i\int_0^{2\pi}d\phi_i\int_{-\infty}^\infty \exp\left\{-\frac{p_{i\theta}^2}{2IT}\right\}dp_{i\theta}$$

$$\times\int_{-\infty}^\infty\exp\left\{-\frac{p_{i\phi}^2}{2IT\sin^2\theta}\right\}dp_{i\phi} = 8\pi^2 IV(2\pi M)^{3/2}T^{5/2} \tag{3.73}$$

There is the same result for every molecule (they move independently), so

$$Z = (8\pi^2 I)^N(2\pi M)^{3N/2}V^N T^{5N/2} = A^N V^N T^{5N/2} \tag{3.74}$$

This results in the free energy and other functions

$$F = -NT \ln A - NT \ln V - \frac{5NT}{2} \ln T \tag{3.75}$$

$$p = \frac{NT}{V} \tag{3.76}$$

$$S = N \ln A + N \ln V + \frac{5N}{2}(\ln T + 1) \tag{3.77}$$

$$E = \frac{5}{2}NT, \quad C_V = \frac{5}{2}N \tag{3.78}$$

The equation of state is the same as for a monatomic ideal gas. The other functions now involve the factor of 5/2 due to the fact that each molecule has five active degrees of freedom, three for translational kinetic energy and two for rotational kinetic energy. Again, this confirms that each active degree of freedom receives an average energy of $1/2\ T$, as expected from the equipartition theorem for classical statistics.

c. When the internal vibrations in the molecule are included, there are additional energy contributions. This is expected to increase the average energy and specific heat per particle.

The small oscillations of the atoms have a potential energy of interaction in the form

$$U(r) = U(r_0) + \frac{\alpha}{2}(r - r_0)^2, \quad \alpha = \frac{\partial^2 U}{\partial r^2}\bigg|_{r=r_0} \tag{3.79}$$

Here, $U(r_0)$ is a constant energy (at equilibrium separation) and α is the effective spring constant for the oscillations. Additionally, the oscillations involve a kinetic energy corresponding to a particle of reduced mass μ in motion. Then, we write the Hamiltonian function of the diatomic molecule:

$$H_i = \frac{p_i^2}{2M} + \frac{p_{ir}^2}{2\mu} + \frac{1}{2\mu r^2}\left(p_{i\phi}^2 + \frac{p_{i\phi}^2}{\sin^2 \theta}\right) + U(r_0) + \frac{\alpha}{2}(r - r_0)^2 \tag{3.80}$$

For simplicity, the potential energy $U(r_0)$ can be set to zero, and we have

$$Z_i = f(r - r_0, T) \times g(r, T)$$

$$= V(2\pi MT)^{3/2}(2\pi \mu T)^{1/2} 4\pi(2\pi \mu T) \int_0^\infty r^2 \exp\left\{-\frac{\alpha(r - r_0)^2}{2T}\right\} dr \tag{3.81}$$

where

$$f(r - r_0, T) = V(2\pi MT)^{3/2} \int_0^\infty \exp\left\{-\frac{\alpha(r - r_0)^2}{2T}\right\} dr \int_{-\infty}^\infty \exp\left\{-\frac{p_{ir}^2}{2\mu T}\right\} dp_r \quad (3.82)$$

$$g(r, T) = \int_0^\pi d\theta_i \int_0^{2\pi} d\phi_i \int_{-\infty}^\infty \exp\left\{-\frac{p_{i\theta}^2}{2\mu r^2 T}\right\} dp_{i\theta} \int_{-\infty}^\infty \exp\left\{-\frac{p_{i\phi}^2}{2\mu r^2 T \sin^2\theta}\right\} dp_{i\phi} \quad (3.83)$$

If we let $r - r_0 = x$, then from Equation 3.81 we have

$$\int_0^\infty r^2 \exp\left\{-\frac{\alpha(r - r_0)^2}{2T}\right\} dr = \int_{-r_0}^\infty (r_0 + x)^2 \exp\left\{-\frac{\alpha x^2}{2T}\right\} dx$$

$$\approx \int_{-\infty}^\infty (r_0^2 + 2r_0 x + x^2)\exp\left\{-\frac{\alpha x^2}{2T}\right\} dx \approx \left(\frac{2\pi T}{\alpha}\right)^{1/2}\left(r_0^2 + \frac{T}{\alpha}\right) \quad (3.84)$$

Considering low temperatures, we have

$$Z_i = 4\pi(2\pi M)^{3/2}(2\pi\mu)^{3/2}\left(\frac{2\pi}{\alpha}\right)^{1/2} r_0^2 VT^{7/2} = BVT^{7/2}, \quad Z = B^N V^N T^{7N/2} \quad (3.85)$$

Then, the free energy and other macroscopic variables are found to be

$$F = -NT \ln B - NT \ln V - \frac{7NT}{2}\ln T, \quad p = \frac{NT}{V},$$

$$S = N \ln B + N \ln V + \frac{7N}{2}(\ln T + 1) \quad (3.86)$$

$$E = \frac{7}{2}NT, \quad C_V = \frac{7}{2}N \quad (3.87)$$

One sees that there is a mean energy of $7/2\ T$ per particle, or again $1/2\ T$ per degree of freedom that contributes with a quadratic term to the Hamiltonian. The oscillatory motion corresponds to two quadratic terms in H_i giving an energy contribution of T per particle. To the contrary, the molecular center of mass coordinates do not appear in the Hamiltonian, giving no average thermal energy nor specific heat contribution. Only the relative atomic coordinate has a potential energy contribution.

Q3.4 Evaluate the partition function and free energy F of an ideal gas that is made up of dipole molecules having fixed electric dipole moments $\vec{d}$, placed in an applied homogeneous electric field $\vec{E}$. Also find the electric polarization $\vec{P}$ of the gas (average dipole moment per unit volume) and its dielectric constant or dielectric permittivity.

A3.4 The applied electric field orientates the dipoles in the direction of the field while thermal motion disorientates them. The equilibrium thermodynamic state balances the ordering effect of the applied field with the disordering effect of temperature or entropy. The potential energy of an electric dipole in the external electric field $\vec{E}$ is equal to

$$U = -\vec{d} \cdot \vec{E} = -d\mathrm{E} \cos\theta \tag{3.88}$$

where θ is the angle between the direction of the field and the dipole moment.

A molecule with a fixed dipole moment can be considered a rigid rotator. Following Equation 3.72, we write the partition function for each molecule in the form

$$Z_1 = V(2\pi mT)^{3/2} \int_0^{2\pi} d\phi \int_0^\pi d\theta \exp\left\{\frac{d\mathrm{E}}{T}\cos\theta\right\} \int_{-\infty}^\infty \exp\left\{-\frac{p_\phi^2}{2IT\sin^2\theta}\right\} dp_\phi$$

$$\times \int_{-\infty}^\infty \exp\left\{-\frac{p_\theta^2}{2IT}\right\} dp_\theta = AVT^{5/2}\left(\frac{2T}{d\mathrm{E}}\right)\sinh\left(\frac{d\mathrm{E}}{T}\right) \tag{3.89}$$

The constant prefactor depends only on inertia factors

$$A = 2\pi^2(2\pi m)^{3/2} I \tag{3.90}$$

The free energy of the entire gas is equal to

$$F = -NT \ln Z_1 = -NT \ln(AV) - \frac{5}{2}NT \ln T - NT \ln\left[\frac{2T}{d\mathrm{E}}\sinh\left(\frac{d\mathrm{E}}{T}\right)\right] \tag{3.91}$$

The mean total electric dipole moment is defined from the relation

$$\overline{\Lambda} = -\left(\frac{\partial F}{\partial \mathrm{E}}\right)_T \tag{3.92}$$

Due to thermal fluctuations, only the component of total dipole moment $\vec{P}$ in the same direction as the applied field will have a nonzero average value. That component is $d_z = d\cos\theta$ and its average is

$$P_z = \overline{d}_z = -\left(\frac{\partial F}{\partial \mathrm{E}}\right)_T = NT\frac{d}{d\gamma}\ln\left[\frac{2\sinh\gamma}{\gamma}\right]\frac{d}{T} = NdL(\gamma), \quad \gamma = \frac{d\mathrm{E}}{T} \tag{3.93}$$

where

$$L(\gamma) = \coth \gamma - \frac{1}{\gamma} \tag{3.94}$$

is the Langevin function.

In the case of weak electric fields and high temperatures, $\gamma = dE/T \ll 1$. Then, the Langevin function can be expanded in a series where the leading term is $L(\gamma) \approx \gamma/3$. This gives the electric polarization from the electric dipole moment per unit volume

$$P = \frac{P_z}{V} \approx \frac{Nd^2E}{3VT} \tag{3.95}$$

From here, the electric susceptibility of the gas is (a scalar rather than a tensor for this simple problem) $\chi = P/E \approx nd^2/3T$, where $n = N/V$ is the number density of dipoles. This gives the dielectric permittivity as

$$\varepsilon = 1 + 4\pi\chi \approx 1 + \frac{4\pi nd^2}{3T} \tag{3.96}$$

Thus with an increase in temperature T, the dipoles become less aligned for a given field, and ε decreases and approaches unity. For the other limit, $\gamma \gg 1$, the Langevin function is $L(\gamma) \cong 1$. At low-enough temperature or large-enough applied field, eventually all dipoles will become aligned, and the total dipole moment becomes saturated at $P_{max} = Nd$. The general dependence of the total electric moment, P_z/P_{max}, on the external field E is represented in Figure 3.2.

Q3.5 Evaluate the partition function, energy, and pressure for a relativistic ideal gas made up of N particles of rest mass m found in a volume V. Examine the limiting cases of nonrelativistic and ultrarelativistic particles.

A3.5 The energy $E(p)$ of a relativistic particle as a function of its momentum magnitude p is defined by the expression

$$E(p) = c\sqrt{p^2 + m^2c^2} \tag{3.97}$$

where c is the speed of light. As the gas is ideal, $Z = Z_i^N$, where

$$Z_i = \int \exp\left\{-\frac{H_i}{T}\right\} d\Gamma_i = V\int \exp\left\{-\frac{E(p)}{T}\right\} p^2 dp \sin\theta\, d\phi$$

$$= 4\pi V \int_0^\infty \exp\left\{-\frac{c}{T}\sqrt{p^2 + m^2c^2}\right\} p^2 dp \tag{3.98}$$

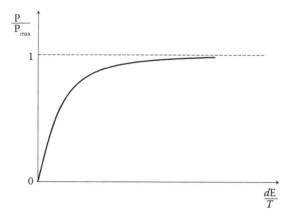

FIGURE 3.2   General dependence of the mean electric dipole P/P_{max} of the gas on $\gamma = d\mathrm{E}/T$, where E is the external field, d the electric dipole moment per molecule, and T the temperature of the gas.

Substituting $p = mc \sinh t$ into this expression, we arrive at

$$Z_i = 4\pi V(mc)^3 \int_0^\infty \exp\{-z\cosh t\}\sinh^2 t \cosh t\, dt, \quad z = \frac{mc^2}{T} \tag{3.99}$$

Using the integral representation of the Bessel function of the second kind (modified Bessel function or Hankel function of imaginary argument) [39]:

$$K_v(z) = \int_0^\infty \exp\{-z\cosh t\}\cosh vt\, dt, \quad \mathrm{Re}\, z > 0 \tag{3.100}$$

we find that its second derivative with respect to z is

$$K_v''(z) = \int_0^\infty \exp\{-z\cosh t\}\cosh^2 t \, \cosh vt\, dt \tag{3.101}$$

Using

$$\cosh^2 t - \sinh^2 t = 1 \tag{3.102}$$

these combine to give what we want

$$Z_i = 4\pi V(mc)^3 [K_1''(z) - K_1(z)] \tag{3.103}$$

The derivative $K_1''(z)$ is evaluated with the help of the recurrence formulas [39] for the modified Hankel function and is equal to

$$K_1''(z) = \left(1 + \frac{2}{z^2}\right)K_1(z) + \frac{1}{z}K_0(z)$$

(3.104)

and finally

$$Z_i = 4\pi V(mc)^3 \frac{T}{mc^2}\left[K_0\left(\frac{mc^2}{T}\right) + \frac{2T}{mc^2}K_1\left(\frac{mc^2}{T}\right)\right]$$

(3.105)

The mean kinetic energy of the particles will be of the order of the temperature T. In the case of nonrelativistic particles, the mean kinetic energy of the particles is small compared with mc^2. If $T \ll mc^2$, the argument of the Bessel function is large. Considering $z \gg 1$, we can use the asymptotic expressions

$$K_0(z) \cong K_1(z) = \sqrt{\frac{\pi}{2z}}\exp\{-z\}\left[1 + O\left(\frac{1}{z}\right)\right]$$

(3.106)

and we have

$$Z_i = (2\pi mT)^{3/2}V\exp\left\{-\frac{mc^2}{T}\right\}, \quad Z = A^N V^N T^{3N/2}\exp\left\{-\frac{Nmc^2}{T}\right\}$$

(3.107)

Thus

$$F = -T\ln Z = -NT\ln A - NT\ln V - \frac{3N}{2}T\ln T + Nmc^2$$

(3.108)

and this leads to

$$p = -\left(\frac{\partial F}{\partial V}\right)_T = \frac{NT}{V}$$

(3.109)

$$S = -\left(\frac{\partial F}{\partial T}\right)_V = N\ln(AV) + \frac{3N}{2}[\ln T + 1]$$

(3.110)

$$E = F + TS = \frac{3}{2}NT + Nmc^2$$

(3.111)

These results recover the expected behavior for a nonrelativistic ideal gas, with the modification that the energy includes the total rest-mass energy. The mean energy above the rest energy follows the classical equipartition result.

For the limiting case of large energies, where the mean energy of the particles is very much greater than mc^2 (ultrarelativistic particles), the energy dispersion is $E \approx cp$. In this case, the argument of the Hankel function $z = (mc^2/T) \ll 1$. For $z \ll 1$, we have

$$K_0(z) \approx -\ln \frac{z}{2}, \quad K_1(z) \approx \frac{1}{z} \tag{3.112}$$

and

$$Z_i = 4\pi V(mc)^3 \frac{T}{mc^2}\left[-1\left(\frac{mc^2}{T}\right) + 2\left(\frac{T}{mc^2}\right)^2\right] \cong \frac{8\pi V}{c^3}T^3 \tag{3.113}$$

$$Z = \left(\frac{8\pi}{c^3}\right)^N V^N T^{3N} \tag{3.114}$$

From here

$$F = -NT \ln\left(\frac{8\pi}{c^3}\right) - NT \ln V - 3NT \ln T, \quad p = \frac{NT}{V} \tag{3.115}$$

$$S = N \ln\left(\frac{8\pi}{c^3}\right) + 3N[\ln T + 1] \tag{3.116}$$

$$E = 3NT \tag{3.117}$$

Thus, for the ultrarelativistic gas, the pressure p is

$$p = \frac{1}{3}\frac{E}{V} \tag{3.118}$$

For the nonrelativistic gas, the pressure p is

$$p = \frac{2}{3}\frac{E}{V} \tag{3.119}$$

Correspondingly, the heat capacity C_V at constant volume V for the ultrarelativistic gas is twice its value in the nonrelativistic case. Also, the mean energy is twice that expected from classical equipartition.

Q3.6 Find the internal energy of an ideal gas found in a cylindrical container of radius R and height h that rotates about its axis with angular velocity ω. Also find the pressure of the gas on the surface of the cylinder. The total number of particles is N and the mass of each particle is m.

A3.6 The energy of the rotating body is equal to

$$E = E' + \frac{\omega L}{2} \qquad (3.120)$$

where E' is the energy of the body in the rotating system of coordinates and L is the angular momentum of the body. The rotation of the gas as a whole together with the vessel having angular velocity ω leads to the appearance of a fictitious external force field (as viewed in a rotating reference frame) acting on the particles equal to

$$F(r) = m\omega^2 r \qquad (3.121)$$

where r is the distance from the axis of rotation. The corresponding potential energy is

$$U(r) = -\int_0^r F(r)dr = -\frac{m\omega^2 r^2}{2} \qquad (3.122)$$

The statistical integral for each molecule, as a result of rotation, has the form

$$Z_i = \int \exp\left\{-\frac{p_i^2}{2mT} - \frac{U}{T}\right\} dp_{ix} dp_{iy} dp_{iz} dx_i dy_i dz_i = (2\pi mT)^{3/2} \int \exp\left\{-\frac{U}{T}\right\} dV \qquad (3.123)$$

where in the cylindrical system of coordinates

$$\int \exp\left\{-\frac{U}{T}\right\} dV = \int_0^R \int_0^{2\pi} \int_0^h \exp\left\{\frac{m\omega^2 r^2}{2T}\right\} r dr d\phi\, dh = 2\pi h \frac{T}{m\omega^2}\left[\exp\left\{\frac{m\omega^2 R^2}{2T}\right\} - 1\right] \qquad (3.124)$$

and

$$Z_i = (2\pi mT)^{3/2} \pi R^2 h \frac{2T}{m\omega^2 R^2}\left[\exp\left\{\frac{m\omega^2 R^2}{2T}\right\} - 1\right] \qquad (3.125)$$

The free energy of the ideal gas is

$$F = -TN \ln Z_i = F_0 - NT \ln\left(\frac{2T}{m\omega^2 R^2}\left[\exp\left\{\frac{m\omega^2 R^2}{2T}\right\} - 1\right]\right) \qquad (3.126)$$

where $\pi R^2 h = V$ is the volume of the cylinder and

$$F_0 = -NT \ln\left[(2\pi mT)^{3/2} V\right] \tag{3.127}$$

is the free energy of the gas at rest. The pressure can be found by imagining a variation in the radius of the container, using the formula

$$p = -\left(\frac{\partial F}{\partial V}\right)_T = -\left(\frac{\partial F}{\partial R}\right)_T \frac{dR}{dV} = -\frac{1}{2\pi Rh}\left(\frac{\partial F}{\partial R}\right)_T \tag{3.128}$$

and

$$p = p_0 + \frac{NT}{V}\left[\frac{m\omega^2 R^2}{2T}\frac{\exp\{(m\omega^2 R^2/2T)\}}{\exp\{(m\omega^2 R^2/2T)\}-1} - 1\right]$$
$$= \frac{NT}{V}\frac{m\omega^2 R^2}{2T}\left[1 - \exp\left\{-\frac{m\omega^2 R^2}{2T}\right\}\right]^{-1} \tag{3.129}$$

where $p_0 = (NT/V)$ is the pressure of the gas at rest. Using Equation 3.122, we have the pressure

$$p = -\frac{NT}{V}\frac{U(R)}{T}\left[1 - \exp\left\{\frac{U(R)}{T}\right\}\right]^{-1} \tag{3.130}$$

As $\omega \to 0$, we have:

$$p = p_0\left(1 + \frac{m\omega^2 R^2}{4T}\right) \tag{3.131}$$

The centripetal pressure is described by the second summand of formula (3.131). The entropy also has a centripetal contribution in addition to the entropy at rest:

$$S = -\left(\frac{\partial F}{\partial T}\right)_V = S_0 - N\left\{\begin{array}{l}1 + \ln\left(\frac{2T}{m\omega^2 R^2}\left[\exp\left\{\frac{m\omega^2 R^2}{2T}\right\}-1\right]\right) \\ -\frac{m\omega^2 R^2}{2T}\left[1 - \exp\left\{-\frac{m\omega^2 R^2}{2T}\right\}\right]^{-1}\end{array}\right\} \tag{3.132}$$

The internal energy of the gas in the rotating system of coordinates is

$$E' = F + TS = E_0 + NT - \frac{Nm\omega^2 R^2}{2[1 - \exp\{-(m\omega^2 R^2/2T)\}]} \tag{3.133}$$

where $E_0 = (3/2)NT$ is the energy of the gas at rest ($\omega \to 0$ limit).
As $\omega \to 0$ (in the sense of $m\omega^2 R^2 \ll 2T$), we have

$$E' = E_0 - \frac{Nm\omega^2 R^2}{4} \tag{3.134}$$

In order to find E (in the laboratory frame), we evaluate first the angular momentum

$$L = -\frac{\partial F}{\partial \omega} = -\frac{2NT}{\omega} + \frac{Nm\omega R^2}{1 - \exp\{-(m\omega^2 R^2/2T)\}} \tag{3.135}$$

Then

$$E = E' + L\omega = E_0 - NT + \frac{Nm\omega^2 R^2}{2[1 - \exp\{-(m\omega^2 R^2/2T)\}]} \tag{3.136}$$

As $\omega \to 0$, we have

$$E = E_0 + \frac{Nm\omega^2 R^2}{4} \tag{3.137}$$

Q3.7 Develop Dalton's law for a mixture of n ideal gases:

$$p_{mix} = \sum_{i=1}^{n} p_i \tag{3.138}$$

where p_i is the partial pressure of the ith gas.

A3.7 As the total Hamilton function of the mixture of ideal gases is additive then

$$H(q,p) = \sum_{i=1}^{n} H_i(q,p) \tag{3.139}$$

Considering the fact that

$$d\Gamma = \prod_{i=1}^{n} d\Gamma_i \tag{3.140}$$

The partition functions Z_i of the ideal gases combine to give the net Z:

$$Z = \int \exp\left\{-\frac{1}{T}\sum_{i=1}^{n} H_i(q,p)\right\} \prod_{i=1}^{n} d\Gamma_i = \prod_{i=1}^{n} Z_i \qquad (3.141)$$

and the free energy F is additive:

$$F = -T \ln Z = -T \sum_{i=1}^{n} \ln Z_i = \sum_{i=1}^{n} F_i \qquad (3.142)$$

Hence the pressure is the sum of the partial pressures due to each gas, according to the number of molecules N_i for each one:

$$p = -\left(\frac{\partial F}{\partial V}\right)_T = -\sum_{i=1}^{n}\left(\frac{\partial F_i}{\partial V}\right)_T = \sum_{i=1}^{n} p_i, \quad p = \frac{T}{V}\sum_{i=1}^{n} N_i \qquad (3.143)$$

Q3.8 Find the mean energy and the constant-volume heat capacity C_V of an ideal gas of N diatomic molecules, considering the anharmonic oscillations of atoms in a molecule. Examine the case of low temperatures.

A3.8 To consider anharmonic oscillations, we represent the potential energy of interaction of the atoms in a molecule in the form

$$U(r - r_0) = \frac{\alpha}{2}(r - r_0)^2 + \beta(r - r_0)^3 + \gamma(r - r_0)^4 + \cdots \qquad (3.144)$$

where r_0 is their equilibrium separation. The partition function of the system is equal to

$$Z = Z_i^N \qquad (3.145)$$

and according to Equation 3.81 we have

$$Z_i = 4\pi V(2\pi MT)^{3/2}(2\pi\mu T)^{3/2}\int_0^\infty r^2 \exp\left\{-\frac{U}{T}\right\} dr \qquad (3.146)$$

The anharmonic correction is considered to be small and can be expanded in a series taking $r - r_0 = x$:

$$Z_i = AT^3 \int_{-\infty}^{\infty} (r_0 + x)^2 \exp\left\{-\frac{\alpha x^2}{2T}\right\}\left(1 - \beta\frac{x^3}{T} - \gamma\frac{x^4}{T} + \frac{1}{2}\beta\frac{x^6}{T^2} + \cdots\right) dx \qquad (3.147)$$

$$A = 4\pi V (4\pi^2 M \mu)^{3/2} \qquad (3.148)$$

For low temperatures, we have

$$Z_i = A T^3 r_0^2 \int_{-\infty}^{\infty} \exp\left\{-\frac{\alpha x^2}{2T}\right\}\left(1 - \gamma \frac{x^4}{T} + \frac{1}{2}\beta \frac{x^6}{T^2} + \cdots\right)dx$$

$$= A r_0^2 \sqrt{\frac{2\pi}{\alpha}} T^{7/2}\left(1 - \frac{3\gamma T}{\alpha^2} + \frac{15}{2}\frac{\beta^2 T}{\alpha^3}\right) \qquad (3.149)$$

The relation defines the average energy of oscillations:

$$E = F - T\left(\frac{\partial F}{\partial T}\right)_V = T^2 \frac{\partial}{\partial T}\ln Z = \frac{7}{2} NT + NT^2\left(\frac{15}{2}\frac{\beta^2}{\alpha^3} - \frac{3\gamma}{\alpha^2}\right) \qquad (3.150)$$

and the heat capacity is

$$C_V = \frac{7}{2} N + 2NT\left(\frac{15}{2}\frac{\beta^2}{\alpha^3} - \frac{3\gamma}{\alpha^2}\right) \qquad (3.151)$$

Thus, the correction to the heat capacity as a result of anharmonicity is proportional to the temperature (at low-enough temperature) and will have a considerable value for high temperatures, when the harmonic approximation will become invalid.

Q3.9 Show that the entropy of a quasi-closed system made up of a large number of particles is proportional to the logarithm of the number of states with energy close to the expectation value.

A3.9 The quasi-closed system made up of a large number of particles $N \to \infty$ will be found with a greater probability in a state with energy E close to the expectation value $\bar{E}$ because the relative fluctuation δ_E will tend to zero in the thermodynamic limit. Thus, the integral of state is found from the contribution only from these states. Then

$$Z(\beta) = \int_0^{\infty} \exp\{-\beta E\}\, \Gamma(E) dE \cong \int_0^{\infty} \exp\{-\beta E\}\, \Gamma(\bar{E})\, \delta(E - \bar{E}) dE$$

$$= \exp\{-\beta\bar{E}\}\Gamma(\bar{E}), \quad \beta = \frac{1}{T} \qquad (3.152)$$

and the entropy of the system is

$$S = \frac{E - F}{T} \cong \frac{\bar{E}}{T} - \frac{1}{T}(-T \ln Z) \cong \ln \Gamma(\bar{E}) \qquad (3.153)$$

3.3 MAXWELL DISTRIBUTION

In an ideal gas, the range of possible energies also determines a range of possible molecular velocities. The width and mean values of this velocity distribution are important because they determine how fast molecular kinetic processes can go. Here, we examine the probability for finding a desired molecular velocity in an ideal gas. The distribution of velocities in an ideal gas is known as the Maxwell distribution because it was discovered by James Clark Maxwell in the middle of the nineteenth century.

It is seen in Section 3.1 that for a quantized state, the distribution function is given using the state's energy as

$$W_n = A \exp\left\{-\frac{E_n(q,p)}{T}\right\} \tag{3.154}$$

This is the Gibbs distribution. From the Gibbs distribution, we may get an expression for the Maxwell distribution using a quasi-classical approximation.

The Maxwell distribution is obtained with the energy represented as the sum of a function of the momentum p and a function of the coordinate q. Thus, in the quasi-classical approximation for N molecules, we have

$$dW_N(q,p) = A \exp\left\{-\frac{E(p) + U(q)}{T}\right\} dpdq \tag{3.155}$$

The kinetic energy $E(p)$ is

$$E(p) = \sum_{l=1}^{N} \frac{p_l^2}{2m_l} \tag{3.156}$$

Then, it follows that

$$dW_N(q,p) = A' \exp\left\{-\frac{E(p)}{T}\right\} A'' \exp\left\{-\frac{U(q)}{T}\right\} dpdq = dW_N(p)dW_N(q) \tag{3.157}$$

$$dW_N(p) = A' \exp\left\{-\sum_{l=1}^{N} \frac{p_l^2}{2m_l T}\right\} \prod_{l=1}^{N} dp_{x_l} dp_{y_l} dp_{z_l}$$

$$= \prod_{l=1}^{N} A' \exp\left\{-\frac{p_l^2}{2m_l T}\right\} dp_{x_l} dp_{y_l} dp_{z_l} = \prod_{l=1}^{N} dW(p_{x_l}, p_{y_l}, p_{z_l}) \tag{3.158}$$

The probability of the momentum results from the product of the probabilities for the individual particles. The function

$$dW(p) = a \exp\left\{-\frac{p^2}{2mT}\right\} dp_x dp_y dp_z \qquad (3.159)$$

is the probability for a single particle, corresponding to the Maxwell distribution for the momentum. From the condition of normalization, we have

$$\int dW(p) = 1 \qquad (3.160)$$

Then

$$1 = a \int_{-\infty}^{\infty} \int_{-\infty}^{\infty} \int_{-\infty}^{\infty} \exp\left\{-\frac{p_x^2 + p_y^2 + p_z^2}{2mT}\right\} dp_x dp_y dp_z = a \left[\int_{-\infty}^{\infty} \exp\left\{-\frac{p_x^2}{2mT}\right\} dp_x\right]^3$$
$$= a(2\pi mT)^{3/2} \qquad (3.161)$$

This implies that

$$a = (2\pi mT)^{-3/2} \qquad (3.162)$$

and thus

$$dW(p) = \frac{1}{(2\pi mT)^{3/2}} \exp\left\{-\frac{p^2}{2mT}\right\} dp_x dp_y dp_z \qquad (3.163)$$

This is the normalized distribution for the momentum. It is the probability that the given value of the momentum is in the volume $dp_x dp_y dp_z$ for a system in equilibrium at a temperature T.

Let us evaluate the probability that the magnitude of the momentum is in the interval from p to $p + dp$. Because the distribution depends only on the magnitude of the momentum, the volume element can be taken as

$$dp_x dp_y dp_z = 4\pi p^2 dp \qquad (3.164)$$

This gives

$$dW(p) = \frac{4\pi}{(2\pi mT)^{3/2}} \exp\left\{-\frac{p^2}{2mT}\right\} p^2 dp \qquad (3.165)$$

Now look at dW as a function of energy E:

$$dW = dW(E) \tag{3.166}$$

We suppose the energy is in the interval from E to $E + dE$, and using

$$p = \sqrt{2mE} \tag{3.167}$$

and

$$dE = \frac{pdp}{m} \tag{3.168}$$

it follows that

$$p^2 dp = \sqrt{2mE}\, mdE = \sqrt{2}m^{3/2}\sqrt{E}dE \tag{3.169}$$

Then, considering Equation 3.165, we obtain

$$dW(E) = \frac{2}{\sqrt{\pi}\, T^{3/2}} \exp\left\{-\frac{E}{T}\right\} E^{1/2} dE \tag{3.170}$$

This gives the distribution function of the energy of the translational motion. Let us transform the distribution function (3.170) to the new variable $E = T\xi$, then

$$dW(\xi) = \frac{2}{\sqrt{\pi}} \exp\{-\xi\}\xi^{1/2} d\xi \tag{3.171}$$

and the expectation value of the energy of the translational motion is found easily

$$\overline{E} = \int_0^\infty E\, dW(E) = \frac{2}{\sqrt{\pi}} T \int_0^\infty \exp\{-\xi\}\xi^{3/2} d\xi = \frac{2}{\sqrt{\pi}} T\Gamma\left(\frac{5}{2}\right) = \frac{3}{2}T \tag{3.172}$$

Obviously, it is the result expected on the basis of the equipartition theorem.
    Next, consider a transformation from the momenta to the velocities

$$\vec{p} = m\vec{v}, \quad \vec{v} = \vec{v}(v_x, v_y, v_z) \tag{3.173}$$

Then, Equation 3.165 becomes

$$dW(v) = \left(\frac{m}{2\pi T}\right)^{3/2} \exp\left\{-\frac{mv^2}{2T}\right\} dv_x dv_y dv_z \qquad (3.174)$$

This is the distribution of velocity; again, it depends only on the speed, not the direction of $\vec{v}$. It is apparent that it has a Gaussian form for each coordinate axis, and a width for each axis given by

$$\Delta v_x = \Delta v_y = \Delta v_z = \sqrt{\frac{T}{m}} \qquad (3.175)$$

We can examine the distribution of speeds looking for speed in the interval from v to $v + dv$ where $v = |\vec{v}|$. That is, we are concerned with all velocity vectors that terminate in velocity space within a spherical shell of inner radius v and outer radius $v + dv$. We see in Figure 3.3 a representation (in two dimensions) of the shell in velocity space containing all particles with velocity $\vec{v}$ such that $v < |\vec{v}| < v + dv$.

Using the spherical symmetry, we can effectively take

$$dv_x dv_y dv_z = 4\pi v^2 dv \qquad (3.176)$$

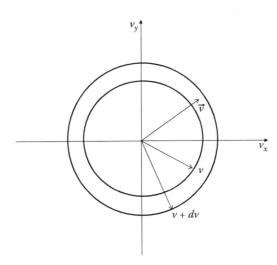

FIGURE 3.3 A representation (in two dimensions) of a shell in velocity space containing all particles with velocity $\vec{v}$ such that $v < |\vec{v}| < v + dv$.

as the volume of this spherical shell. Consequently

$$dW(v) = 4\pi \left(\frac{m}{2\pi T}\right)^{3/2} \exp\left\{-\frac{mv^2}{2T}\right\} v^2 dv \equiv F(v)dv \qquad (3.177)$$

This is the unit-normalized Maxwell distribution of speeds. It has a maximum. If v increases, the exponential factor decreases but the volume of phase space available to the particles is proportional to v^2 and increases. Thus, the net result is a gentle maximum and a long tail at high speeds. We can get an idea of the typical molecular speeds by several different measures, the first of which is the RMS speed.

Let us evaluate the average of one squared velocity component (and all three should be equivalent by symmetry):

$$\overline{v_x^2} = \int v_x^2 dW(v_x) \qquad (3.178)$$

If we factor (3.174) into terms for each Cartesian component, then we have

$$dW(v_x) = \left(\frac{m}{2\pi T}\right)^{1/2} \exp\left\{-\frac{mv_x^2}{2T}\right\} dv_x \qquad (3.179)$$

(and identical results for $dW(v_y)$ and $dW(v_z)$). It follows from here that

$$\overline{v_x^2} = \left(\frac{m}{2\pi T}\right)^{1/2} \int_{-\infty}^{\infty} v_x^2 \exp\left\{-\frac{mv_x^2}{2T}\right\} dv_x = \frac{T}{m} \qquad (3.180)$$

The quantity given by Equation 3.180 is intrinsically positive and is the dispersion of v_x. Symmetry tells us that

$$\overline{v_x^2} = \overline{v_y^2} = \overline{v_z^2} = \frac{T}{m} \qquad (3.181)$$

Then, the average squared speed is the sum of the components

$$\overline{v^2} = \overline{v_x^2} + \overline{v_y^2} + \overline{v_z^2} = 3\overline{v_z^2} = \frac{3T}{m} \qquad (3.182)$$

Hence we recover the expected mean kinetic energy

$$\frac{\overline{mv^2}}{2} = \frac{3T}{2} \qquad (3.183)$$

the RMS speed v_{rms} is thus

$$v_{rms} = \sqrt{\overline{v^2}} = \sqrt{\frac{3T}{m}} \qquad (3.184)$$

The result contains important physics. Obviously, it is directly related to equipartition of the translational kinetic energy (KE). As a result, the uniform sharing of KE results in higher RMS speeds for lighter molecules (at a given temperature), inversely proportional to the square root of their mass. Thus, in a mixture of oxygen (O_2, 32 u) and hydrogen (H_2, 2 u), the hydrogen's RMS speed will be four times faster than that of oxygen. In a very real sense, this makes it much more difficult to keep the hydrogen under control or contained (also because of its small physical size). The RMS speed is probably the most important measure of the velocity distribution; however, it is important to keep in mind that there will be a broad distribution of speeds above and below v_{rms}.

From Equation 3.177, we find the *most probable speed* v_m of a gas particle, that is, the speed for which $F(v)$ is a maximum. We find this from

$$\frac{\partial F}{\partial v} = 0 \quad \text{or} \quad \frac{\partial}{\partial v}\left[\exp\left\{-\frac{mv^2}{2T}\right\}\right]_{v_m} = 0 \qquad (3.185)$$

and this gives

$$v_m^2 = \frac{2T}{m} \qquad (3.186)$$

Thus

$$v_m = \sqrt{\frac{2T}{m}} \qquad (3.187)$$

The fact that the most probable speed is less than the RMS speed gives an indication of the long high-energy tail of the probability distribution.

It may be seen from Equations 3.184 through 3.187 that all the various speeds are proportional to $\sqrt{T/m}$. It follows that molecular speeds increase with temperature, and gas particles with larger mass have smaller speeds. If we consider (3.183), then the microscopic translational kinetic energy of the gas is

$$\overline{E} = \frac{3}{2}NT \qquad (3.188)$$

This is the internal energy of a monatomic ideal gas of N particles. Monatomic gases may include the so-called noble gases (helium, He; neon, Ne; argon, Ar; krypton,

Kr; and xenon, Xe) at normal temperatures and pressures as well as the vapors of the alkali metals (lithium, Li; sodium, Na; potassium, K; rubidium, Ru; and cesium, Cs) at higher temperatures and low pressures. It should be noted that at exceedingly high temperatures, many diatomic gases, such as hydrogen or nitrogen, dissociate and become monatomic. Halides (such as fluorides or chlorides, say, hydrogen fluoride, HF; or hydrogen chloride, HCl) may also dissociate into a mixture of monatomic halogens (fluorine, F; chlorine, Cl; bromine, Br; and iodine, I).

Based on the form of the speed distribution (3.177), it will be useful to evaluate the integral

$$I_n(\alpha) = \int_0^\infty \exp\{-\alpha x^2\} x^n dx = \frac{1}{2}\alpha^{-\left(\frac{n+1}{2}\right)} \Gamma\left(\frac{n+1}{2}\right) \tag{3.189}$$

If

$$n = 2K, \quad K > 0 \tag{3.190}$$

then

$$I_{2k} = \frac{(2K-1)!!}{2^{k+1}} \sqrt{\frac{\pi}{\alpha^{2k+1}}} \tag{3.191}$$

Considering the integral in Equation 3.189, we may evaluate

$$\overline{v^n} = \int F(v)v^n = 4\pi\left(\frac{\alpha}{\pi}\right)^{3/2} \int \exp\{-\alpha v^2\} v^2 v^n dv = 4\pi\left(\frac{\alpha}{\pi}\right)^{3/2} I_{n+2}(\alpha)$$

$$= \frac{2}{\sqrt{\pi}}\left(\frac{2T}{m}\right)^{n/2} \Gamma\left(\frac{n+3}{2}\right) \tag{3.192}$$

If

$$n = 2K, \quad K > 0 \tag{3.193}$$

then

$$\overline{v^{2K}} = \left(\frac{T}{m}\right)^K (2K+1)!! \tag{3.194}$$

If

$$n = 2K + 1 \tag{3.195}$$

then

$$\overline{v^{2K+1}} = \frac{2}{\sqrt{\pi}} \left(\frac{2T}{m}\right)^{\frac{2K+1}{2}} (K + 1)! \tag{3.196}$$

The integral (3.189) is used during the evaluation of expectation values over Maxwell's distribution. If we consider the above, then the average speed is

$$\overline{v} = \sqrt{\frac{8T}{\pi m}} \tag{3.197}$$

From Equations 3.180 (rms speed) and 3.197, the dispersion of the velocity $\overline{\Delta v^2}$ is

$$\overline{(\Delta v)^2} = \overline{v^2} - \overline{v}^2 = \frac{T}{m}\left(3 - \frac{8}{\pi}\right) \tag{3.198}$$

Now consider the number of particles striking a unit surface. Let us recall first that the probability that a particle has velocity component v_z is

$$dW(v_z) = \left(\frac{m}{2\pi T}\right)^{1/2} \exp\left\{-\frac{mv_z^2}{2T}\right\} dv_z \tag{3.199}$$

If the number of particles per unit volume is N/V, then the number per unit volume moving toward a surface perpendicular to z is

$$\frac{dN(v_z)}{V} = \frac{N}{V}\left(\frac{m}{2\pi T}\right)^{1/2} \exp\left\{-\frac{mv_z^2}{2T}\right\} dv_z \tag{3.200}$$

Note, of course, that only the particles within a distance $\Delta z = v_z \Delta t$ of the wall or surface will strike that surface in a time interval Δt. Therefore, the number of particles with velocity component v_z that strike an area ΔA of the wall per unit time is

$$dN(v_z) = \Delta A \Delta t \frac{N}{V}\left(\frac{m}{2\pi T}\right)^{1/2} \exp\left\{-\frac{mv_z^2}{2T}\right\} v_z dv_z \tag{3.201}$$

Summing over particles with different z-components, the total rate at which particles strike a unit area of the wall per unit time is

$$v = \int \frac{dN(v_z)}{\Delta A \Delta t} = \frac{N}{V}\left(\frac{m}{2\pi T}\right)^{1/2} \int_0^\infty \exp\left\{-\frac{mv_z^2}{2T}\right\} v_z dv_z = \frac{NT^{1/2}}{V\sqrt{2\pi m}} \tag{3.202}$$

This is the frequency of molecular collisions with a wall, when multiplied by an area. For an ideal gas, we have $pV = NT$ and thus another equivalent expression is $v = p/\sqrt{2\pi mT}$. This rate might be useful for estimating how fast a gas is likely to escape through a tiny opening in a container.

Q3.10 Find the distribution of probabilities for the kinetic energy of particles. Evaluate

$$\overline{E}, \quad \overline{E^2}, \quad \overline{(\Delta E)^2} \tag{3.203}$$

and the most probable energy E_m of the particles in a gas.

A3.10 We know from Equation 3.170 that

$$dW(E) = \frac{2}{\sqrt{\pi}\,T^{3/2}} \exp\left\{-\frac{E}{T}\right\} E^{1/2} dE \tag{3.204}$$

Then

$$\overline{E} = \frac{2}{\sqrt{\pi}\,T^{3/2}} \int_0^\infty \exp\left\{-\frac{E}{T}\right\} E^{3/2} dE = \frac{3}{2}T \tag{3.205}$$

$$\overline{E^2} = \frac{2}{\sqrt{\pi}\,T^{3/2}} \int_0^\infty \exp\left\{-\frac{E}{T}\right\} E^{5/2} dE = \frac{15}{4}T^2 \tag{3.206}$$

and

$$\overline{\Delta E^2} = \overline{E^2} - \overline{E}^2 = \frac{3}{2}T^2 \tag{3.207}$$

The relative energy fluctuation is

$$\frac{\sqrt{\overline{\Delta E^2}}}{\overline{E}} = \sqrt{\frac{2}{3}} \tag{3.208}$$

which is of the order of unity. The dispersion of the energy E is very wide due to the fact that the energy under consideration is the energy of only one particle, not the total energy of a macroscopic system of particles.

The *most probable energy* for a particle is found from

$$\frac{\partial}{\partial E}\left[E^{1/2}\exp\left\{-\frac{E}{T}\right\}\right]_{E_m} = 0 \tag{3.209}$$

This gives

$$E_m = \frac{T}{2} \tag{3.210}$$

It should be noted that

$$E_m \neq \frac{mv_m^2}{2} = T \tag{3.211}$$

Q3.11 What fraction of the gas molecules has kinetic energy of translational motion greater than the mean kinetic energy $\overline{E} = \frac{3}{2}T$?

A3.11 Using the distribution with respect to the energy (3.170), we write the number N_1 of molecules in the gas for which $E \geq \overline{E}$ as

$$N_1 = \frac{2N}{\sqrt{\pi}\,T^{3/2}}\int_{\overline{E}}^{\infty}\exp\left\{-\frac{E}{T}\right\}E^{1/2}dE \tag{3.212}$$

where N is the total number of molecules in the gas. We evaluate the integral (3.212) with the help of the fractional gamma function

$$\Gamma(\alpha, x) = \int_0^x \exp\{-t\}t^{\alpha-1}dt \tag{3.213}$$

Using the properties of the fractional gamma function

$$\Gamma(\alpha+1, x) = \alpha\Gamma(\alpha, x) - x^{\alpha}\exp\{-x\}, \quad \Gamma\left(\frac{1}{2}, x^2\right) = \sqrt{\pi}\,\mathrm{erf}(x) \tag{3.214}$$

we find

$$\frac{N_1}{N} = \frac{2}{\sqrt{\pi}}\left[\Gamma\left(\frac{3}{2}\right) - \Gamma\left(\frac{3}{2}, \frac{\overline{E}}{T}\right)\right] = 0.39 \tag{3.215}$$

Here, erf (x) is the error integral

$$\text{erf}(x) = \frac{2}{\sqrt{\pi}} \int_0^x \exp\{-\varsigma^2\} d\varsigma \tag{3.216}$$

Q3.12 Find the number of molecules having a velocity with desired components relative to some axis l, where $v_{l,\parallel}$ is the component parallel to the chosen axis and $v_{l,\perp}$ the component perpendicular to the axis. Find the mean velocity of molecules in the given direction.

A3.12 We make the transformation in Equation 3.174 to a cylindrical system of coordinates with polar axis along l. After integration with over $d\phi$, we have

$$dW(v_{l\parallel}, v_{l\perp}) = 2\pi \left(\frac{m}{2\pi T}\right)^{3/2} \exp\left\{-\frac{m(v_{l\parallel}^2 + v_{l\perp}^2)}{2T}\right\} v_{l\perp} dv_{l\perp} dv_{l\parallel} \tag{3.217}$$

In terms of particle numbers, we have

$$dN = 2\pi N \left(\frac{m}{2\pi T}\right)^{3/2} \exp\left\{-\frac{m\left(v_{l\parallel}^2 + v_{l\perp}^2\right)}{2T}\right\} v_{l\perp} dv_{l\perp} dv_{l\parallel} \tag{3.218}$$

In order to find the mean velocity of the molecules in the given direction, we write the distribution of the probability with respect to the component of the velocity in the given direction:

$$dW(v_l) = \left(\frac{m}{2\pi T}\right)^{3/2} \exp\left\{-\frac{mv_l^2}{2T}\right\} dv_l \tag{3.219}$$

Then

$$\overline{|v_l|} = \left(\frac{m}{2\pi T}\right)^{1/2} \int_{-\infty}^{\infty} |v_l| \exp\left\{-\frac{mv_l^2}{2T}\right\} dv_l = \sqrt{\frac{2T}{\pi m}} \tag{3.220}$$

Thus, the mean velocity in any chosen direction is half of the mean speed $\bar{v}$ (Equation 3.197).

Q3.13 Find the portion of the molecules whose velocity is greater than the most probable velocity v_m, and the portion of the molecules whose velocity is found in the interval

$$\frac{v_m}{2} \le v \le 2v_m, \quad v_m = \sqrt{\frac{2T}{m}} \tag{3.221}$$

A3.13 Considering Equation 3.187, the number of molecules whose velocity lies in the interval from v to $v + dv$ is equal to

$$dN_1 = N4\pi \left(\frac{m}{2\pi T}\right)^{3/2} \exp\left\{-\frac{mv^2}{2T}\right\} v^2 dv \tag{3.222}$$

From this, we get

$$\frac{N_1(v \ge v_m)}{N} = 4\pi \left(\frac{m}{2\pi T}\right)^{3/2} \int_{v_m}^{\infty} \exp\left\{-\frac{mv^2}{2T}\right\} v^2 dv \tag{3.223}$$

$$= 1 - \frac{4}{\sqrt{\pi}} \int_0^1 \exp\{-\varsigma^2\}\varsigma^2 d\varsigma$$

Here, the transformation $mv^2/2T = \varsigma^2$ was used. After integration by parts, the last integral leads to the error integral and

$$\frac{N_1(v \ge v_m)}{N} = 1 - 2\,\text{erf}(1) = 0.57 \tag{3.224}$$

Similarly

$$N_1\left(\frac{v_m}{2} \le v \le 2v_m\right) = N4\pi \left(\frac{m}{2\pi T}\right)^{3/2} \int_{\frac{v_m}{2}}^{m} \exp\left\{-\frac{mv^2}{2T}\right\} v^2 dv \tag{3.225}$$

After the same change of variables followed by integration by parts we have

$$\frac{N_1}{N} = \frac{4}{\sqrt{\pi}} \left[-\frac{\varsigma}{2}\exp\{-\varsigma^2\}\bigg|_{1/2}^{2} + \frac{1}{2}\int_{1/2}^{2} \exp\{-\varsigma^2\}\varsigma^2 d\varsigma \right] = 0.87 \tag{3.226}$$

Q3.14 Find the number of gas molecules which collide with a surface ΔA of a wall during a time Δt which have speeds in the interval from v to $v + dv$. Evaluate the total number of collisions of oxygen gas at 300 K and 1.00 atm pressure on a 1 cm² surface during 1s, considering molecules of arbitrary velocities.

A3.14 Let us transform the Maxwell distribution (3.174) into spherical coordinates, and then we have

$$dn = n\left(\frac{m}{2\pi T}\right)^{3/2} \exp\left\{-\frac{mv^2}{2T}\right\} v^2 dv \sin\theta\, d\theta\, d\phi, \quad n = \frac{N}{V} \qquad (3.227)$$

The number of particles arriving on an area ΔA of the wall during Δt, having the given velocity and colliding with it at an angle from θ to $\theta + d\theta$ are those in a volume

$$\Delta V = \Delta z \quad \Delta V = (v\cos\theta \quad \Delta t) \quad \Delta A \qquad (3.228)$$

This gives a total number of collisions as

$$dN_c(\theta) = dn\, \Delta V(\theta) \qquad (3.229)$$

which is equal to

$$dN_c(\theta) = dn\, v\cos\theta\, \Delta A\, \Delta t = n\left(\frac{m}{2\pi T}\right)^{3/2} \exp\left\{-\frac{mv^2}{2T}\right\} v^3 dv \cos\theta \sin\theta\, d\theta\, d\phi\, \Delta A\, \Delta t$$

$$(3.230)$$

The result can be summed over molecules striking the wall at an arbitrary angle to give the collision rate per area

$$\frac{dN_c}{\Delta A \Delta t} = \int_0^{\frac{\pi}{2}} \frac{dN_c(\theta)}{\Delta A \Delta t}\, d\theta = \int_0^{\frac{\pi}{2}} dn\, v\cos\theta\, d\theta = 2\pi n\left(\frac{m}{2\pi T}\right)^{3/2} \exp\left\{-\frac{mv^2}{2T}\right\} v^3 dv$$

$$\times \int_0^{\frac{\pi}{2}} \cos\theta \sin\theta\, d\theta = \pi n\left(\frac{m}{2\pi T}\right)^{3/2} \exp\left\{-\frac{mv^2}{2T}\right\} v^3 dv \qquad (3.231)$$

The integration with respect to θ is from 0 to $\pi/2$ because for $\theta > \pi/2$, the particles are moving away from the wall.

Finally, integrating over possible speeds, the total number of collisions of the molecules per unit area per unit time is

$$\nu \equiv \frac{N_c}{\Delta A\, \Delta t} = \pi n\left(\frac{m}{2\pi T}\right)^{3/2} \int_0^\infty \exp\left\{-\frac{mv^2}{2T}\right\} v^3 dv = \frac{n\bar{v}}{4} \qquad (3.232)$$

where $\bar{v}$ is the mean value of the velocity of the gas molecules. Now consider oxygen gas at 300 K and 1.00 atm pressure. Including Boltzmann's constant, we have

$$\bar{v} = \sqrt{\frac{8T}{\pi m}} = \sqrt{\frac{8(1.38 \times 10^{-23} \text{J/K})(300 \text{K})}{\pi(32)(1.6605 \times 10^{-27} \text{kg})}} = 445 \text{ m/s} \qquad (3.233)$$

and at 1.00 atm pressure, the number density is

$$n = \frac{N}{V} = \frac{p}{T} = \frac{101.3 \text{ kPa}}{(1.38 \times 10^{-23} \text{J/K})(300 \text{K})} = 2.45 \times 10^{25} \text{ m}^{-3} \qquad (3.234)$$

On 1 cm^2 of a wall during a 1 s time interval, these lead to

$$N_c = \frac{n\bar{v}}{4} \Delta A \Delta t = \frac{1}{4}(2.45 \times 10^{25} \text{m}^{-3})(445 \text{m/s})(0.01 \text{m})^2(1.0 \text{s})$$
$$= 2.73 \times 10^{23} \text{ collisions} \qquad (3.235)$$

Q3.15 A molecular beam is emitted from an orifice of a container. Find the mean velocity and mean squared velocity of the particles in the beam.

A3.15 The number of particles escaping through unit surface of an orifice in time Δt and having speed between v and $v + dv$ is described by the formula (3.231). Then, the probability that the speed of a molecule in the beam falls in the given interval is equal to

$$dW(v) = \frac{dN_c}{N_c} = \frac{4\pi}{\bar{v}}\left(\frac{m}{2\pi T}\right)^{3/2} \exp\left\{-\frac{mv^2}{2T}\right\} v^3 dv \qquad (3.236)$$

Using Equation 3.197 for the mean molecular speed $\bar{v}$, and averaging over all molecular speeds, the mean speed of an escaping molecule in the beam is

$$\bar{v}_p = \frac{4\pi}{\bar{v}}\left(\frac{m}{2\pi T}\right)^{3/2} \int_0^\infty \exp\left\{-\frac{mv^2}{2T}\right\} v^4 dv = \frac{3}{2}\sqrt{\frac{\pi T}{2m}} \qquad (3.237)$$

$$\bar{v_p^2} = \frac{4\pi}{\bar{v}}\left(\frac{m}{2\pi T}\right)^{3/2} \int_0^\infty \exp\left\{-\frac{mv^2}{2T}\right\} v^5 dv = \frac{4T}{m} \qquad (3.238)$$

and

$$\sqrt{\bar{v_p^2}} = 2\sqrt{\frac{T}{m}} \qquad (3.239)$$

As shown by the resultant formula, the molecules escaping from the orifice have somewhat larger characteristic velocities than the molecules in the bulk of the gas:

$$\frac{\overline{v_p}}{\overline{v}} = \frac{3\pi}{8}, \quad \frac{\sqrt{\overline{v_p^2}}}{\sqrt{\overline{v^2}}} = \frac{2}{\sqrt{3}} \tag{3.240}$$

One needs to keep in mind, however, that this result only holds in the limit of very low density, where the molecules move independently through the orifice, without macroscopic effects that would take place in a viscous fluid.

Q3.16 Find the probability that the absolute value of the velocity of the relative motion of two particles, $\vec{v}' = \vec{v}_1 - \vec{v}_2$, lies in the interval from v' to $v' + dv'$. Find the mean value $\overline{v'}$.

A3.16 In order to obtain the distribution with respect to the relative velocities, we first find the probability that the first particle has velocity $\vec{v}_1$ and the second has velocity $\vec{v}_2$. This probability is obtained from the Gibbs distribution and is equal to a product

$$dW(\vec{v}_1, \vec{v}_2) = A \exp\left\{-\frac{m_1 v_1^2 + m_2 v_2^2}{2T}\right\} dv_{1x} dv_{1y} dv_{1z} dv_{2x} dv_{2y} dv_{2z} \tag{3.241}$$

The normalization constant is defined with the help of the distribution (3.174) and is found to be

$$A = \frac{(m_1 m_2)^{3/2}}{(2\pi T)^3} \tag{3.242}$$

Now, we introduce new variables—the velocity of the center of mass

$$\vec{v}_c = \frac{m_1 \vec{v}_1 + m_2 \vec{v}_2}{m_1 + m_2} \tag{3.243}$$

and the velocity of the relative motion

$$\vec{v}' = \vec{v}_1 - \vec{v}_2 \tag{3.244}$$

The Jacobian of this transformation is equal to 1. The kinetic energy of the particles is expressed in the new coordinates

$$\frac{m_1 v_1^2}{2} + \frac{m_2 v_2^2}{2} = \frac{M v_c^2}{2} + \frac{\mu v'^2}{2}, \quad M = m_1 + m_2, \quad \mu = \frac{m_1 m_2}{m_1 + m_2} \tag{3.245}$$

The distribution of probabilities for the absolute values of the velocities for this is written in the form

$$dW(v_c, v') = 4\pi \left(\frac{M}{2\pi T}\right)^{3/2} \exp\left\{-\frac{mv_c^2}{2T}\right\} v_c^2 dv_c \, 4\pi \left(\frac{\mu}{2\pi T}\right)^{3/2} \exp\left\{-\frac{mv'^2}{2T}\right\} v'^2 dv'$$

$$(3.246)$$

If we integrate this expression with respect to the velocity of the center of mass v_c, we find the distribution of the particles as a function of the relative velocity:

$$dW(v') = 4\pi \left(\frac{\mu}{2\pi T}\right)^{3/2} \exp\left\{-\frac{mv'^2}{2T}\right\} v'^2 dv' \qquad (3.247)$$

Thus, the distribution of the particles with respect to the relative velocities is described by the usual distribution with respect to the velocities (3.174), with the reduced mass μ in place of the molecular mass. In complete analogy with Equation 3.197, we find the mean relative speed

$$\overline{v'} = \sqrt{\frac{8T}{\pi \mu}} = \sqrt{\frac{8T}{\pi} \frac{m_1 + m_2}{m_1 m_2}} \qquad (3.248)$$

If the gas is made up of one type of particle, then $m_1 = m_2 = m$ and $\overline{v'} = \sqrt{2}\overline{v}$, where $\overline{v}$ is the mean speed of the gas particles.

Q3.17 Examine an ideal two-dimensional gas with particles that move only in two dimensions. Describe the distribution of the particles of the ideal two-dimensional gas with respect to the velocities and find the characteristic velocities of distribution $\overline{v}, \sqrt{\overline{v^2}}, v_m$.

A3.17 The distribution of the particles with respect to the velocities for the two-dimensional ideal gas is easily obtained from Maxwell distribution (3.174) by integrating it with respect to v_z:

$$dW(v_x, v_y) = \frac{m}{2\pi T} \exp\left\{-\frac{m(v_x^2 + v_y^2)}{2T}\right\} dv_x dv_y \qquad (3.249)$$

If we introduce into the velocity space polar coordinates, then we find the probability that a particle of the two-dimensional gas has the speed v found in the interval v to $v + dv$:

$$dW(v) = \frac{m}{T} \exp\left\{-\frac{mv^2}{2T}\right\} v^2 dv \qquad (3.250)$$

From this, the required averages are

$$v_m = \sqrt{\frac{T}{m}}, \quad \bar{v} = \sqrt{\frac{\pi T}{2m}}, \quad \sqrt{\bar{v^2}} = \sqrt{\frac{2T}{m}} \tag{3.251}$$

Q3.18 Find the distribution of probabilities for the angular velocities of rotation of molecules and evaluate the mean squared angular velocity and the mean absolute angular momentum of the molecules.

A3.18 For the same reason as for the translational motion in classical statistics, we may write the distribution of probabilities for the rotation of each molecule separately, as the molecules are considered to move independently between collisions.

The kinetic energy of rotation of a molecule is considered as that of a rigid object, ignoring internal molecular oscillations. It is equal to

$$E_{rot} = \frac{1}{2}(I_1\omega_1^2 + I_2\omega_2^2 + I_3\omega_3^2) \tag{3.252}$$

where I_1, I_2, I_3 are the principal moments of inertia and $\omega_1, \omega_2, \omega_3$ are the projections of the angular velocity on the principal axes of inertia.

The probability that the molecule has components of angular velocity in the intervals

$$[\omega_1, \omega_1 + d\omega_1], \quad [\omega_2, \omega_2 + d\omega_2], \quad [\omega_3, \omega_3 + d\omega_3] \tag{3.253}$$

is

$$dW(\bar{\omega}) = A\exp\left\{-\frac{1}{2T}(I_1\omega_1^2 + I_2\omega_2^2 + I_3\omega_3^2)\right\}d\omega_1 d\omega_2 d\omega_3 \tag{3.254}$$

where A is the normalization constant of the distribution and is equal to

$$A = \frac{\sqrt{I_1 I_2 I_3}}{(2\pi T)^{3/2}} \tag{3.255}$$

The kinetic energy of rotation may also be expressed through the components of the angular momentum L_k, playing the role of the generalized momenta for the angular velocities $\omega_1, \omega_2, \omega_3$:

$$L_k = \frac{\partial E}{\partial \omega_k} = I_k\omega_k \tag{3.256}$$

Then

$$E_{rot} = \frac{1}{2}\left(\frac{L_1^2}{I_1} + \frac{L_2^2}{I_2} + \frac{L_3^2}{I_3}\right) \tag{3.257}$$

The normalized distribution of probabilities for the component of the angular momentum is

$$dW(\vec{L}) = \frac{1}{(2\pi T)^{\frac{3}{2}}(I_1 I_2 I_3)^{\frac{1}{2}}} \exp\left\{-\frac{1}{2T}\left(\frac{L_1^2}{I_1} + \frac{L_2^2}{I_2} + \frac{L_3^2}{I_3}\right)\right\} dL_1 dL_2 dL_3 \tag{3.258}$$

For the evaluation of the mean square, we write the probability that the molecule has the component of the angular velocity ω_1 in the interval $[\omega_1, \omega_1 + d\omega_1]$. From the above, we have

$$dW(\omega_1) = \sqrt{\frac{I_1}{2\pi T}} \exp\left\{-\frac{I_1\omega_1^2}{2T}\right\} d\omega_1 \tag{3.259}$$

This result is a Gaussian distribution with a width $\Delta\omega_1 = \sqrt{T/I_1}$. From here, we have

$$\overline{\omega_1^2} = \sqrt{\frac{I_1}{2\pi T}} \int_{-\infty}^{\infty} \omega_1^2 \exp\left\{-\frac{I_1\omega_1^2}{2T}\right\} d\omega_1 = \frac{T}{I_1} \tag{3.260}$$

and

$$\overline{\omega^2} = \overline{\omega_1^2} + \overline{\omega_2^2} + \overline{\omega_3^2} = T\left(\frac{1}{I_1} + \frac{1}{I_2} + \frac{1}{I_3}\right) \tag{3.261}$$

Similarly, we have

$$\overline{L^2} = \overline{L_1^2} + \overline{L_2^2} + \overline{L_3^2} = T(I_1 + I_2 + I_3) \tag{3.262}$$

Q3.19 Find the current density of thermoelectric emission, assuming that electrons obey classical statistics and the potential energy at the interior of the metal is less than its energy outside the metal by the quantity $w = e\phi$ (the work function). The concentration of electrons in the metal is $n = N/V$ and the mass of an electron is m.

A3.19 Assume that the electrons follow the Maxwell velocity distribution. We obtain the current density in the direction of the ox-axis, perpendicular to the surface of the metal, by averaging the x-velocity component while summing over possible transverse components. Thus, the number of electrons per area per time, multiplied by the charge, gives the current density

$$j_x = en\left(\frac{m}{2\pi T}\right)^{3/2}\int_{v_{x0}}^{\infty} v_x \exp\left\{-\frac{mv_x^2}{2T}\right\}dv_x \int_{-\infty}^{\infty}\int_{-\infty}^{\infty} \exp\left\{-\frac{m(v_y^2 + v_z^2)}{2T}\right\}dv_y dv_z \quad (3.263)$$

This is the magnitude only (the electric current flows in the direction opposite to the electron motion). The lower limit on v_x is not zero because an electron within the metal must have enough kinetic energy to surpass the work function, in order to leave the metal. We assume that only electrons having the normal component of the velocity $v_x \geq v_{x0}$, where

$$\frac{mv_{x0}^2}{2} = w = e\phi \qquad (3.264)$$

have enough energy to emerge from the metal. Further, in thermoelectric emission, that fraction of electrons with this minimum energy increases as the temperature increases. If we do the required integrations, then we find

$$j_x = en\left(\frac{T}{2\pi m}\right)^{1/2}\exp\left\{-\frac{e\phi}{T}\right\} = \frac{en}{4}\bar{v}\exp\left\{-\frac{e\phi}{T}\right\} \qquad (3.265)$$

This is *Richardson's classical formula*. One sees that the current density is limited by the result found in Q3.14 for collisions per area per time on a surface. This result is actually slightly smaller due to the requirement that electron energies must surpass the work function. That barrier clearly becomes more easily crossed with increasing temperature, as the high-energy tail of the distribution grows longer.

Q3.20 An ideal gas made up of N particles with their kinetic energy related to their momentum by $\varepsilon = \alpha\, p^\ell$, is characterized by the equilibrium distribution function with respect to the momentum:

$$dW(p) = 4\pi\, Vf(p)p^2 dp \qquad (3.266)$$

where $f(p)$ is an arbitrary function and V is the volume.
   Find the relationship between the pressure of the gas and the energy of the particles enclosed in a unit volume. It is assumed that the pressure arises as a result of collisions of the molecules with the wall of the container.

A3.20 The mean total energy of the ideal gas is

$$E = N\bar{\varepsilon} = 4\pi\, NV\int_0^{\infty} \varepsilon\, f(p)p^2 dp \qquad (3.267)$$

On the other hand, from the kinetic theory of gases, the pressure P of the gas is

$$P = \frac{1}{3} nm\overline{v^2} = \frac{N}{3V} 4\pi V \int_0^\infty vp\, f(p) p^2 dp \qquad (3.268)$$

If we substitute here a derivative with respect to momentum

$$v = \frac{\partial \varepsilon}{\partial p} = \ell \alpha\, p^{\ell-1} = \frac{\ell \varepsilon}{p} \qquad (3.269)$$

then we have

$$P = \frac{4\pi N}{3} \ell \int_0^\infty \varepsilon f(p) p^2 dp = \ell \frac{E}{3V} \qquad (3.270)$$

In particular, for $\ell = 1$ (ultrarelativistic gas), we have $P = E/3V$ and for $\ell = 2$ (non-relativistic gas), we have $P = 2E/3V$.

3.4 EXPERIMENTAL BASIS OF STATISTICAL MECHANICS

The early days of the development of thermodynamics and statistical physics were greatly motivated to describe properties of dilute systems such as gases, and more condensed systems such as liquids and solids. It is incredible that much of that early work in statistical ideas was carried out before the experimental verification of the existence of molecules, which came via the theoretical description of Brownian motion. In 1827, the botanist Robert Brown noticed that pollen grains of *Clarkia pulchella* suspended in a liquid produced particles that were in continuous movement when viewed in a microscope. Later, he also noticed that inorganic particles have a similar continuous and random movement; hence the phenomenon had nothing to do with any life forces that might be present in the pollen. The statistical theory due to Einstein (1905), Smoluchowski (1906), Wiener, and others could completely describe the motions by the assumption of a multitude of collisions with the molecules of the liquid. To get the correct description of the subsequent random walks, via statistical ideas, is a wonderful confirmation of the power of statistical concepts in helping scientists to understand fundamental physics at the atomic scale. Without the confirmation of the existence of atoms and molecules, many of the ideas of statistical mechanics have no meaning. Statistical mechanics and the microscopic theory of heat essentially require the presence of atomic or molecular constituents in matter.

In addition to the great success of the ideal gas law, the microscopic description of ideal gases has been even more greatly confirmed by direct experimental measurements of the Maxwell velocity distribution. As an example, Miller and Kusch (1955) measured the velocity distributions in potassium and thallium beams from an oven, using a specially designed rotating cylinder with grooves where the atoms pass, that acts as a spiral velocity selector ["Velocity distributions in potassium and thallium atomic beams," R.C. Miller and P. Kusch, *Physical Review* 99, p. 1314 (1955)]. This and other experiments demonstrated beyond any doubt the correctness of the Maxwell velocity distribution, on both sides of the most probable speed. It can be noted that the effusion of the atoms must occur through an orifice whose size is small compared to the mean free path of the atoms, which is aided by using a dilute gas.

An indirect evidence of molecular velocity distributions can be found experimentally in diffusion. An important technological example is the different rates of diffusion of gas molecules according to their mass. The RMS speed being proportional to the inverse square root of the mass, lighter molecules have a faster diffusion. This effect has been used in the separation of the more fissionable uranium-235 isotope from the much more common uranium-238 (used as the gas uranium-hexaflouride UF_6). It was essential for increasing the concentration of U-235 above the natural 0.72%, for the development of the first nuclear fission reactors (both controlled and explosive). This type of uranium enrichment was first carried out at Oak Ridge National Laboratory during the Second World War, although now centrifugal separation has been found to be more practical. By either method, a cascade of reactors is needed, with the slightly enriched output of one reactor being inputted to the next. A more peaceful example of the rapid diffusion rate for low-mass atoms can be found in a child's helium balloon: after a day or so, enough helium escapes through the balloon's rubber via diffusion, that it no longer experiences enough buoyant force to float in the surrounding air. At the same time, heavier oxygen and nitrogen molecules can diffuse into the balloon, increasing its loss of buoyancy.

Another commonly known effect that experimentally supports the kinetic theory is the higher speed of sound in a gas of low-mass atoms. Containers of oxygen (about 32 g/mol) and helium (4 g/mol) both at room temperature and at standard pressure, differ in their densities, but otherwise, obey the ideal gas law almost equally well. The atomic mass ratio of 16:1, however, translates into a 1:4 ratio for the RMS molecular speeds, $v_{rms} = \sqrt{3k_B T/m}$, and hence, the same ratio for the speed of sound in the two different media (the speed of sound cannot be greater than v_{rms}). Without the presence of atoms, as understood through the kinetic theory, it might be difficult to imagine how the speed of sound could be so affected. Another supporting evidence for the validity of the kinetic theory (and really, of atomic theory) in the same context, is that the speed of sound in both gases increases with the square root of the temperature, $v_s = \sqrt{\gamma k_B T/m}$ (where $\gamma = C_p/C_v$ is the adiabatic constant), which again, is connected to the RMS speed.

3.5   GRAND CANONICAL DISTRIBUTION

In some cases, the number of particles in a system may vary. The system might be defined as a region of space surrounded by certain boundaries. Particles can pass into or out of the system, crossing the boundaries, due to interactions with the surroundings. Particles may also be created and destroyed in quantum systems. In these cases, the total number of particles in the system fluctuates and the instantaneous number at any time cannot be known. We can only determine the mean number of particles in the system. This demands a special examination of systems with a variable number of particles. Important examples with variable particle numbers are quantum systems of phonons, photons, and so on, where we may have creation and annihilation of the fundamental quantum excitations, which behave as particles.

If a system exchanges particles and energy with a thermostat, then the probability density for the system will be different from the microcanonical one but the probability density for system plus thermostat is a microcanonical distribution.

We consider, for example, a system A of fixed volume V that is in contact with a large reservoir A' (the thermostat) with which it can exchange not only the energy but also particles. It should be noted that neither the energy E of A nor the number N of particles in A are fixed. Only the total energy E_0 and the total number of particles N_0 of the system plus reservoir

$$A_0 \equiv A + A' \tag{3.271}$$

are fixed, that is

$$E_0 = E + E' = \text{const}, \quad N_0 = N + N' = \text{const} \tag{3.272}$$

where E' and N' are the energy and the number of particles in the thermostat, A'.

The probability in the ensemble of finding system A and thermostat A' in any particular state is given by

$$dW = C\,\delta(E(q,p) + E'(q',p') - E_0)\delta(N + N' - N_0)d\Gamma'_{N'}\,d\Gamma_N dN' \tag{3.273}$$

Let $E_{n,N}$ be the energy of the nth quantum state of the system A for N particles and dW_N be the probability that a body (system A) made up of N particles has energy $E_{n,N}$. If only the system is observed, then we can sum over all states of the reservoir that contribute to the same macroscopic state. Thus

$$dW_N(E_{n,N}) = C\,d\Gamma_N \int \delta(E_N + E'_{N'} - E_0)\delta(N + N' - N_0)d\Gamma'_{N'}dN' \tag{3.274}$$

But

$$d\Gamma'_{N'} = \frac{d\Gamma'_{N'}}{dE'_{N'}}dE'_{N'} \to \frac{\Delta\Gamma'_{N'}}{\Delta E'_{N'}}dE'_{N'} = \frac{\exp\{S'(E'_{N'},N')\}}{\Delta E'_{N'}}dE'_{N'} \tag{3.275}$$

Then

$$dW_N(E_{n,N}) = C\,d\Gamma_N \int \delta(E_N + E'_{N'} - E_0)\delta(N + N' - N_0)dN'$$
$$\times \frac{\exp\{S'(E'_{N'},N')\}}{\Delta E'_{N'}}dE'_{N'} = C'd\Gamma_N \exp\{S'(E_0 - E_N, N_0 - N)\} \tag{3.276}$$

and for the entropy of the reservoir there is the relation

$$S'(E_0 - E_N, N_0 - N) = S'(E_0, N_0) - \frac{\partial S'}{\partial E_N'} E_N - \frac{\partial S'}{\partial N'} N \qquad (3.277)$$

But

$$\left(\frac{\partial S'}{\partial E_N'}\right)_{V,N} = \frac{1}{T}, \quad \left(\frac{\partial S'}{\partial N'}\right)_{V,E} = -\frac{\mu}{T} \qquad (3.278)$$

The second equality is the *chemical equation of state*, where μ is the chemical potential. Here, the derivatives are evaluated for $E' = E_0$ and $N' = N_0$. They are hence constants that characterize the reservoir A'. Thus, the probability for the desired state of the system A is

$$dW_N(E_{n,N}) = A \exp\left\{\frac{\mu N - E_{n,N}}{T}\right\} d\Gamma_N \qquad (3.279)$$

where A is a normalization constant. This is the formula for the *grand canonical distribution*. That is, it is the distribution for a system with a variable number of particles in contact with a thermostat at temperature T in the equilibrium state. The variables N and $E_{n,N}$ are defined by the state of the system under consideration.

The ensemble of systems distributed according to the probability distribution in Equation 3.279 is called the *grand canonical ensemble*. The quantity μ is the *chemical potential* of the reservoir. It is a constant of the system and determined by the number of particles present as well as the distribution of states.

What differentiates the grand canonical ensemble from the canonical ensemble? In distinction with the canonical ensemble, where V, T, and N are fixed, in the grand canonical ensemble, the system is kept at fixed V, T, and μ. The chemical potential μ determines the energy involved in adding a particle to the system. In equilibrium, the chemical potential also determines the mean number of particles in the system, whose instantaneous number is fluctuating.

Let us normalize Equation 3.279: all probabilities must sum to unity, so

$$\sum_{n,N} W_N(E_{n,N}) = \sum_{n,N} A \exp\left\{\frac{\mu N - E_{n,N}}{T}\right\} = A \sum_{n,N} \exp\left\{\frac{\mu N - E_{n,N}}{T}\right\} = A\Xi = 1 \quad (3.280)$$

where

$$\Xi = \sum_{n,N} \exp\left\{\frac{\mu N - E_{n,N}}{T}\right\} = \sum_{N=0}^{\infty} \exp\left\{\frac{\mu N}{T}\right\} \sum_n \exp\left\{-\frac{E_{n,N}}{T}\right\}$$

$$= \sum_{N=0}^{\infty} \exp\left\{\frac{\mu N}{T}\right\} Z(V,T,N) \qquad (3.281)$$

is the *grand canonical partition function*. It corresponds to a summation of the canonical partition function $Z(V, T, N)$ over systems with an arbitrary number of particles, weighted by a power of the activity, defined as

$$\eta = \exp\left\{\frac{\mu}{T}\right\} \qquad (3.282)$$

The activity is essentially another way to parameterize the chemical potential, which can then be expressed as $\mu = T \ln \eta$. From here, it follows that the probability of a certain state n with N particles present is

$$W_N(E_{n,N}) = \frac{1}{\Xi}\exp\left\{\frac{\mu N - E_{n,N}}{T}\right\} \qquad (3.283)$$

If we consider a physical situation where only the *mean* energy $\overline{E}$ and the *mean* number $\overline{N}$ of particles of a system A are known, then the distribution over systems in the ensemble is described by (3.283). But the quantities T and μ no longer characterize any reservoir. They are determined from the conditions that the system A has the specified mean energy $\overline{E}$ and mean number $\overline{N}$ of particles. That is by the equations

$$\overline{E} = \sum_{n,N} W_N E_{n,N}, \quad \overline{N} = \sum_{n,N} W_N N \qquad (3.284)$$

We consider the sum over all possible states of system A irrespective of its number of particles or of its energy.

Consider a macroscopic system A in contact with a reservoir (see Figure 3.4). The relative fluctuations of its energy about its mean energy $\overline{E}$ and of its number of particles about its mean number $\overline{N}$ are very small. Consequently, physical properties of A are not considerably affected if the system is removed from contact with the reservoir so that both its energy and number of particles are rigorously fixed. To evaluate mean values of physical quantities, there is no noticeable difference whether a macroscopic system is isolated or in contact with a reservoir with which it can

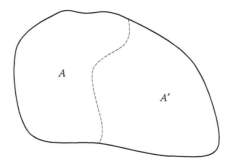

FIGURE 3.4   A macroscopic system A in contact with a reservoir A'.

exchange both energy and particles (once its equilibrium has been established). Its mean values may equally be evaluated considering the system to have equal probability over all states of desired energy $\overline{E}$ and number of particles $\overline{N}$ (microcanonical distribution). It may also be distributed according to the grand canonical distribution (3.283) over all its states irrespective of energy and number.

Let us find the formula of statistical thermodynamics for the entropy, based on the probabilities:

$$S = -\sum_{n,N} W_{n,N} \ln W_{n,N} = -\sum_{n,N} W_{n,N} \left(\ln A + \frac{\mu N - E_{n,N}}{T} \right)$$

$$= -\ln A - \frac{\mu \overline{N}}{T} + \frac{\overline{E}}{T} \qquad (3.285)$$

Thus

$$TS = -T \ln A - \mu \overline{N} + E \qquad (3.286)$$

and rearranging it gives

$$T \ln A = -TS - \mu \overline{N} + E = \Omega \qquad (3.287)$$

Here, $\Omega = \Omega(T, V, \mu)$ is called the *grand thermodynamic potential*. Thus

$$W_N(E_{n,N}) = \exp\left\{ \frac{\Omega + \mu N - E_{n,N}}{T} \right\} \qquad (3.288)$$

This is the *canonical formula of the grand canonical distribution*. Let us set the normalization of Equation 3.288 to unity:

$$\sum_{n,N} W_{n,N} = \exp\left\{ \frac{\Omega}{T} \right\} \sum_{n,N} \exp\left\{ \frac{\mu N - E_{n,N}}{T} \right\} = 1 \qquad (3.289)$$

This becomes a requirement on Ω that is related to the grand canonical partition function Ξ introduced above. Thus, the *grand thermodynamic potential* Ω is determined as

$$\Omega = -T \ln\left\{ \sum_{N=0}^{\infty} \exp\left\{ \frac{\mu N}{T} \right\} \sum_{n} \exp\left\{ -\frac{E_{n,N}}{T} \right\} \right\} = -T \ln \Xi \qquad (3.290)$$

Obviously, the relation is similar in appearance to the relation between the Helmholtz free energy and the canonical partition function, $F = -T \ln Z$. We may use Equation 3.290 together with the thermodynamic identities (Chapter 2) to evaluate thermodynamic quantities of concrete bodies for which models can be developed.

3.6   EXTREMUM OF CANONICAL DISTRIBUTION FUNCTION

The microcanonical distribution as an equilibrium distribution is explained by the maximum of the entropy (in an isolated system at constant energy and particle number). Compared to "nearby" distributions slightly perturbed away from equilibrium, the entropy is

$$S = S_{max} \tag{3.291}$$

Earlier we showed that

$$dS \geq 0 \tag{3.292}$$

This is the second law of thermodynamics for any thermodynamic process occurring in a closed isolated system. The equality sign holds if the process is reversible and the inequality sign holds if the process is irreversible. That is to say, the entropy increases during spontaneous processes occurring in a closed isolated system until it achieves its maximum value at the equilibrium state.

Consider an analysis of a physical system in such a way that mimics what happens in a typical experimental situation. Usually, a system is in contact with a thermal reservoir, rather than being held at strictly fixed energy. The system exchanges heat with the reservoir; therefore, one can imagine that its probability distribution of states can fluctuate. For calculation purposes, we suppose there is a fluctuating probability distribution with probabilities W_n' that is not assumed to be the equilibrium Gibbs distribution. Our goal is to determine what distribution W_n' results under the following conditions: The entropy is maximized, subject to the constraints that there is desired system energy E and that the distribution is unit normalized.

Let us examine along side with the Gibbs distribution W_n the distribution W_n', and we impose the constraint that their mean energies are equal. The entropy of the fluctuating system is

$$S' = -\sum_n W_n' \ln W_n' \tag{3.293}$$

and the energy and normalization constraints are

$$E = \sum_n E_n W_n = \sum_n E_n W_n', \quad \sum_n W_n = \sum_n W_n' = 1 \tag{3.294}$$

This extremum problem subject to constraints can be solved by the *method of Lagrange multipliers*. One introduces an auxiliary function based on the entropy, but including undetermined multipliers α and β on the two constrained quantities:

$$\tilde{S} = -\sum_n W_n' \ln W_n' + \alpha \sum_n W_n' - \beta \sum_n E_n W_n' \tag{3.295}$$

For the extremum, in addition to the above constraints, we require the variation with respect to the distribution to vanish:

$$\frac{\partial \tilde{S}}{\partial W_n'} = -\ln W_n' - 1 + \alpha - \beta E_n = 0 \tag{3.296}$$

The extremum distribution is found to be

$$W_n' \to W_n = \exp\{-1 + \alpha\}\exp\{-\beta E_n\} = \frac{\exp\{-\beta E_n\}}{\sum_m \exp\{-\beta E_m\}} \tag{3.297}$$

The last step follows from applying the normalization, which determines α. At this point, the energy constraint is not yet applied; it would determine the other undetermined multiplier β. From the form of the result, this is the Gibbs canonical distribution, and β must be identified with the inverse temperature.

Now consider comparing an arbitrary distribution W_n' to the Gibbs equilibrium distribution. Let us use an identity

$$\ln x \geq 1 - \frac{1}{x}, \quad x = \frac{W_n'}{W_n} \tag{3.298}$$

Then

$$\ln W_n' - \ln W_n \geq 1 - \frac{W_n}{W_n'} \tag{3.299}$$

and if we multiply both sides by $W_n' > 0$, then we have

$$W_n' \ln W_n' - W_n' \ln W_n \geq W_n' - W_n \tag{3.300}$$

Use the normalization

$$\sum_n (W_n' - W_n) = 1 - 1 = 0 \tag{3.301}$$

then it follows that

$$\sum_n W_n' \ln W_n' \geq \sum_n W_n' \left(\frac{F - E_n}{T} \right) = \frac{F - E}{T} = -S \qquad (3.302)$$

This implies that the arbitrary distribution's entropy satisfies

$$S \geq -\sum_n W_n' \ln W_n' = S' \qquad (3.303)$$

It follows that the Gibbs distribution guarantees the maximum entropy.

4 Ideal Gases

4.1 OCCUPATION NUMBER

Let us examine a generalized Hamiltonian for a set of interacting particles,

$$\hat{H} = \sum_{k=1}^{N} \frac{p_k^2}{2m_k} + \sum_{k=1}^{N} U(q_k) + U(q_1,\ldots,q_N) \tag{4.1}$$

The first term is the kinetic energy, the second, the potential energy due to an external field, and the third, the energy of interaction among the particles. Here, p_k and q_k are the momentum and coordinate of the kth particle, respectively, both with three components, and is the number of particles. If the gas is sufficiently dilute, then interparticle spacing is large, and collisions between particles only take place occasionally (but help to provide ergodicity). Then the interaction between particles is negligible and we take

$$U(q_1,\ldots,q_N) \to 0 \tag{4.2}$$

In this case, we obtain an ideal gas. Thus, we consider a system of N particles with no interaction between them. The energy of the system is the sum of the kinetic energies of the particles, plus the energy of interaction with the external field. If there is no externally applied field, the only energy present is the microscopic kinetic energy. A common situation is a molecular gas; however, the "particles" could also be the fundamental excitations of a quantum system. The Hamiltonian of the system is simply the sum of single particle Hamiltonians

$$\hat{H} = \sum_{k=1}^{N} \hat{H}_k \tag{4.3}$$

Here, each sub-Hamiltonian has two terms:

$$\hat{H}_k = \frac{p_k^2}{2m_k} + U(q_k) \tag{4.4}$$

This is the Hamiltonian of the kth particle. It can be considered as moving independently of the other particles.

Consider a quantum system. Let $\Psi(q)$ be the complete wave function of the system for N particles. Then, ignoring restrictions due to indistinguishable particles

$$\Psi(q) = \prod_{k}^{N} \Psi_k(q_k) \tag{4.5}$$

Here, $\Psi_k(q_k)$ is the state occupied by the kth particle, which must be an eigenstate

$$\hat{H}_k(q_k, p_k)\Psi_k(q_k) = E_k\Psi_k(q_k) \tag{4.6}$$

where E_k is the energy eigenvalue that corresponds to the eigenstate $\Psi_k(q_k)$.

The number of particles in a given level (quantum state) is called the occupation number. Suppose that we solve a problem and we get the energy spectrum. With this we may define the states of the particles. From that information, we want to make a detailed description of the state of the system.

From the quantum perspective, it is not possible to identify individual particles. All electrons look the same, as do all hydrogen molecules. They are said to be indistinguishable. Therefore, it is not possible to identify which particles are in which single-particle states. We can only state the total number of particles occupying a given energy level. This is why we change notation to the occupation number representation.

With that in mind, let n_k be the number of gas particles in the kth quantum state. The number n_k is called the occupation number of the kth quantum state. As E_k is the energy of the kth quantum state, the total energy of all the particles and the total number of particles are given as

$$E_{n,N} = \sum_k n_k E_k \tag{4.7}$$

and

$$N = \sum_k n_k \tag{4.8}$$

respectively. From the perspective of the grand canonical distribution, where particle numbers are allowed to fluctuate, we have the probability distribution

$$W_{n,N} = \exp\left\{\frac{\Omega + \mu N - E_{n,N}}{T}\right\} = \exp\left\{\frac{\Omega + \sum_k n_k(\mu - E_k)}{T}\right\} \tag{4.9}$$

This is the *probability distribution* for different values of n_k. It is clear that the occupations of each state depend on the state energies as well as the system's chemical potential and temperature. Next we consider how to apply this result to different important situations.

4.2 BOLTZMANN DISTRIBUTION

An ideal nondegenerate gas is called a Boltzmann gas. Physically, a system is nondegenerate when the wave functions of different particles have little spatial overlap; the particles have a lot of space in which to move around. This means the particles have negligible interactions, and quantum effects between particles also are negligible. For the available quantum states, it means that the particles are well distributed over all possible levels. There are many levels available for each particle; hence any chosen quantum state has a low probability of being occupied. Thus, a dilute gas at high temperatures, where the particles behave classically rather than quantum mechanically, should satisfy these conditions.

If we consider Equation 4.9 for the kth level, then we have

$$W_k(n_k) = \exp\left\{\frac{\Omega_k + n_k(\mu - E_k)}{T}\right\} \qquad (4.10)$$

The probability that there is no particle in this state is

$$W_k(0) = \exp\left\{\frac{\Omega_k}{T}\right\} \qquad (4.11)$$

The case in which we are interested is when the average occupations are small

$$\overline{n_k} \ll 1 \qquad (4.12)$$

Then the probability of no particles in a state must be close to 1

$$W_k(0) \approx 1 \qquad (4.13)$$

However, the distribution must be normalized, so we also have

$$\sum_{n=0}^{\infty} W_k(n) = \exp\left\{\frac{\Omega_k}{T}\right\} \sum_{n=0}^{\infty} x^n = \exp\left\{\frac{\Omega_k}{T}\right\} \frac{1}{1-x} = 1, \quad x \equiv \exp\left\{\frac{\mu - E_k}{T}\right\}$$

$$(4.14)$$

Then more precisely, we have the probability for no particle in the state to be

$$W_k(0) = 1 - x = 1 - \exp\left\{\frac{\mu - E_k}{T}\right\} \qquad (4.15)$$

In order for this to be close to 1, we need $\{(\mu - E_k)/T\}$ to be large and negative.

If we consider that the kth state has one particle, then

$$W_k(1) = \exp\left\{\frac{\Omega_k + 1(\mu - E_k)}{T}\right\} = \exp\left\{\frac{\Omega_k}{T}\right\}\exp\left\{\frac{\mu - E_k}{T}\right\} \approx \exp\left\{\frac{\mu - E_k}{T}\right\} < 1$$

$$(4.16)$$

For the case of two particles, we have

$$W_k(2) = \exp\left\{\frac{\Omega_k + 2(\mu - E_k)}{T}\right\} = \exp\left\{\frac{\Omega_k}{T}\right\}\exp\left\{\frac{2(\mu - E_k)}{T}\right\}$$

$$\approx \exp\left\{\frac{2(\mu - E_k)}{T}\right\} = \left[\exp\left\{\frac{(\mu - E_k)}{T}\right\}\right]^2 \ll 1 \qquad (4.17)$$

and in general we have

$$W_k(n) = \exp\left\{\frac{\Omega_k}{T}\right\}\exp\left\{\frac{n(\mu - E_k)}{T}\right\} \qquad (4.18)$$

This allows us to evaluate the mean occupation $\overline{n_k}$:

$$\overline{n_k} = 0W_n(0) + 1W_n(1) + 2W_n(2) + \cdots = \sum_{n=0}^{\infty} W_k(0)n\,x^n$$

$$= \frac{W_k(0)x}{(1-x)^2} = \frac{x}{1-x} \approx \exp\left\{\frac{\mu - E_k}{T}\right\}. \qquad (4.19)$$

Thus

$$\overline{n_k} = \exp\left\{\frac{\mu - E_k}{T}\right\} \qquad (4.20)$$

is taken as the *Boltzmann distribution*. It is also the average occupation number for *Maxwell–Boltzmann statistics*. It results from having the mean number of particles occupying a certain quantum state being proportional to the probability that one particle is in that state.

We may also obtain the Maxwell–Boltzmann distribution through the formula

$$\overline{n_k} = -\frac{\partial \Omega_k}{\partial \mu} \qquad (4.21)$$

Let us do the normalization that will help us to obtain the chemical potential.

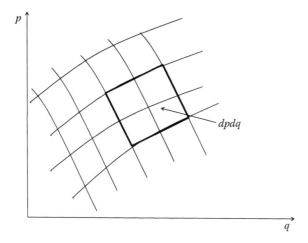

FIGURE 4.1   An element of phase space (q, p) for a single particle.

Examine an element of phase space (q, p) (for a single particle) as in Figure 4.1:

$$d\Gamma = \frac{dqdp}{(2\pi\hbar)^3} \tag{4.22}$$

In this phase space, the distribution density $dN(q,p)$ is

$$dN(q, p) = \bar{n}\, d\Gamma = \exp\left\{\frac{\mu - E(q, p)}{T}\right\}\frac{dqdp}{(2\pi\hbar)^3} \tag{4.23}$$

We assume that there are no external forces and no intermolecular interactions. Then the distribution is homogeneous and the single-particle Hamiltonian is

$$E(p) = \frac{p^2}{2m} \tag{4.24}$$

and as

$$dq \equiv dxdydz = dV \tag{4.25}$$

we get

$$N = \frac{V}{(2\pi\hbar)^3}\exp\left\{\frac{\mu}{T}\right\}\int_{-\infty}^{\infty}\int_{-\infty}^{\infty}\int_{-\infty}^{\infty}\exp\left\{-\frac{p^2}{2mT}\right\}dp_x dp_y dp_z$$

$$= \frac{V}{(2\pi\hbar)^3}\exp\left\{\frac{\mu}{T}\right\}(2\pi mT)^{3/2} \tag{4.26}$$

This shows that the chemical potential depends on the particle number density and on the temperature

$$\mu = T \ln \left[\frac{N}{V} \frac{(2\pi \hbar)^3}{(2\pi mT)^{3/2}} \right] \tag{4.27}$$

If we suppose that occupation numbers are small

$$\overline{n_k} = \exp\left\{ \frac{\mu - E_k}{T} \right\} \ll 1 \tag{4.28}$$

Then it follows that (apply the above result to a very low energy state)

$$\exp\left\{ \frac{\mu}{T} \right\} \ll 1 \tag{4.29}$$

which shows that $|\mu| > T$ and $\mu < 0$. Then it follows that the square brackets in Equation 4.27 is less than unity. This shows that the mathematical requirements for small occupation numbers are

N/V should be small, T should be high, m should be large.

The result (4.27) for the chemical potential can be analyzed from a more physical perspective. The quantity in the square brackets involves the product of particle number density N/V with the cube of the thermal wavelength λ, where

$$\lambda = \frac{h}{\sqrt{2\pi mT}} \tag{4.30}$$

The reciprocal of the number density gives the actual volume occupied per particle of the gas, whereas λ^3 is the volume per particle at which quantum effects become important (where the gas becomes degenerate). If we define a specific volume per particle

$$v_1 = \frac{V}{N} \tag{4.31}$$

then the chemical potential has the simple form

$$\mu = T \ln \left[\frac{\lambda^3}{v_1} \right] \tag{4.32}$$

The chemical potential will be large and negative (leading to small occupation numbers and classical statistics) as long as $v_1 \gg \lambda^3$. Provided there is plenty of room for each particle to move around in, the gas is nondegenerate and the classical Boltzmann distribution applies.

It is instructive to consider the actual sizes of λ and v_1 for a typical ideal gas. Consider (diatomic) oxygen ($m = 32$ u) at 300 K and 1 atm pressure. The thermal wavelength is

$$\lambda = \frac{6.63 \times 10^{-34}\,\text{Js}}{\sqrt{2\pi(32)(1.66 \times 10^{-27}\,\text{kg})(1.38 \times 10^{-23}\,\text{J/K})(300\,\text{K})}} = 17.8\,\text{pm} \quad (4.33)$$

The ideal gas law can be used to obtain the specific volume, written as a cubed length

$$v_1 = \frac{V}{N} = \frac{T}{P} = \frac{(1.38 \times 10^{-23}\,\text{J/K})(300\,\text{K})}{101300\,\text{Pa}} = (3.44\,\text{nm})^3 \quad (4.34)$$

The typical space available per molecule is much greater than λ^3; hence, quantum effects are unimportant. They could become important, however, at lower temperature, where λ becomes larger. The value of the chemical potential at the stated conditions is $\mu = T(3 \ln \{17.8\,\text{pm}/3.44\,\text{nm}\}) = -15.8T$. Clearly, it leads to very small occupation numbers for any state.

4.2.1 DISTRIBUTION WITH RESPECT TO COORDINATES

Let us discuss the Boltzmann distribution when the mechanical energy of the system is a combination of kinetic and potential energies of arbitrary form:

$$E = E(q, p) = E(p) + U(q) \quad (4.35)$$

The number of particles is

$$n(q, p) = \exp\left\{\frac{\mu - E(p) - U(q)}{T}\right\} \quad (4.36)$$

Let us find the number of particles per unit volume in a point of space described by the momentum vector $\vec{p}$ and position $q \to \vec{r}$:

$$dN(\vec{r}, \vec{p}) = \exp\left\{\frac{\mu - E(p) - E(q)}{T}\right\}\frac{d\vec{p}\,dxdydz}{(2\pi\hbar)^3} = \exp\left\{\frac{\mu - E(\vec{p}) - U(\vec{r})}{T}\right\}\frac{d\vec{p}\,dV}{(2\pi\hbar)^3} \quad (4.37)$$

From here, integrating out the momentum coordinates, we have

$$dN(\vec{r}) = \exp\left\{\frac{\mu - U(\vec{r})}{T}\right\}dV\int \exp\left\{-\frac{E(\vec{p})}{T}\right\}\frac{dp_x dp_y dp_z}{(2\pi\hbar)^3} = dV\,C\exp\left\{-\frac{U(\vec{r})}{T}\right\} \quad (4.38)$$

We can take the zero of the potential arbitrarily at the origin

$$U(\vec{r} = 0) = 0 \tag{4.39}$$

Then

$$\frac{dN}{dV}\bigg|_{r=0} = C = n_0 \tag{4.40}$$

from where we have

$$n(\vec{r}) = n_0 \exp\left\{-\frac{U(\vec{r})}{T}\right\} \tag{4.41}$$

This is the *Boltzmann distribution with respect to the coordinate*. In a force field like a gravitational field

$$U = mgz \tag{4.42}$$

we have

$$n(\vec{r}) = n_0 \exp\left\{-\frac{mgz}{T}\right\} \tag{4.43}$$

This is the same as the *barometric formula* discussed earlier in Section 2.15.

4.3   ENTROPY OF A NONEQUILIBRIUM BOLTZMANN GAS

We consider again the definition in Equation 1.263 of the entropy S as the average value of the logarithmic distribution function of a subsystem (taken with opposite sign):

$$S = -\sum_n W_n \ln W_n \tag{4.44}$$

where W_n is an arbitrary distribution. For an arbitrary state of a macroscopic system, the entropy may be defined as in Equation 4.44:

$$S = \ln \Delta\Gamma \tag{4.45}$$

where $\Delta\Gamma$ is the statistical weight.

Let us examine the Gibbs $2s$ phase space where s is the number of degrees of freedom of the body. In particular, for the Boltzmann phase space, the state of the

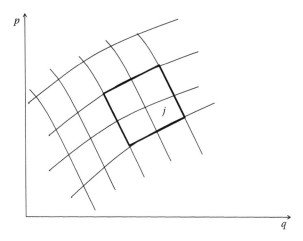

FIGURE 4.2 Different groups of quantum states $j = 1,2,\ldots$

particles is six dimensional (x, y, z, p_x, p_y, p_z). Let us examine different groups of quantum states $j = 1,2,\ldots$ and let G_j be the number of quantum states (cells in Figure 4.2) in the jth group and N_j the number of particles in the jth group (see Figure 4.3). Due to symmetries, any real system has energetic degeneracies: groups of states at the same energy, only differing from each other by some symmetry operations. Suppose that each group of degenerate states is labeled by index $j = 1,2,\ldots$, with energy E_j. The degeneracy of group j (the number of states at that same energy E_j) is an integer G_j, and we suppose there are N_j particles in the jth group or energy level. In a quantum system, the degeneracy of an energy level is usually a small integer (due to discrete symmetries), whereas, in classical systems, it could be approaching infinity (degeneracy due to a continuous symmetry).

For Boltzmann statistics, assume that the particles are distinguishable and there is no limit on the number of particles per quantum state. The state of the system is determined by its total energy and by how the particles are distributed among the energy levels E_j. For a system of N particles, we have

$$N = \sum_j N_j \tag{4.46}$$

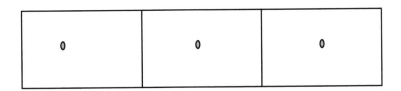

FIGURE 4.3 An occupation of quantum states by particles.

and the choice of particular N_j describes the state of a gas. We can think of the jth energy level as having G_j cells into which particles are placed. The number of cells (the partitions in Figure 4.3) in the jth domain is

$$d\Gamma_j = \frac{dqdp}{(2\pi\hbar)^s} = G_j \tag{4.47}$$

We suppose that we have a nondegenerate Boltzmann gas, then each level is much less than fully occupied and

$$G_j \gg N_j \gg 1 \tag{4.48}$$

Let us introduce the mean occupation in an energy level, n_j:

$$n_j = \frac{N_j}{G_j} \tag{4.49}$$

Each separate occupation defines a quantum state. For example, a quantum level with degeneracy $G_j = 3$ is depicted in Figure 4.3 as a set of three cells into which particles can be placed. Different arrangements of N_j particles into the cells will still have the same total energy. The set of n_j values defines the macroscopic state. If we do not change n_j but change the microscopic state (i.e., how the cells are occupied), we get a set of microstates for one and the same macrostate. It follows that the number of different microstates is defined by all possible ways of arrangement. Thus, we need to count the different ways in which N_j particles can be placed into G_j cells. This number is the statistical weight of the desired macrostate. First, we find how many ways to occupy just one energy level E_j. Initially, each of the N_j particles is placed into one of the G_j cells. There are G_j choices for each particle; thus, the number of ways to do that is simply

$$M = G_j \times G_j \times \cdots = G_j^{N_j} \tag{4.50}$$
$$(N_j \text{ terms})$$

$$\Delta\Gamma_j = \begin{cases} 1 \to G_j \\ 2 \to G_j^2 \\ \vdots \\ N_j \to G_j^{N_j} \end{cases} \tag{4.51}$$

Although the particles are considered as distinguishable, it does not matter in what order they were placed into the cells. Suppose they all went into the first cell. It is the same microstate regardless of which particle went in there first, or second, or third, and so on. So the result needs to be divided by the number of permutations in selecting which particle was placed first, then second, and so on. That number of permutations is $N_j!$. Thus, the statistical weight for N_j particles placed into G_j distinct states is

$$\Delta\Gamma_j = \frac{G_j^{N_j}}{N_j!} \tag{4.52}$$

Using Stirling's formula for large numbers, we have

$$\ln(N!) \cong N \ln\frac{N}{e} \tag{4.53}$$

For many particles occupying a collection of different energy levels, we need to combine their statistical weights as a product. Then the total entropy of the whole system is found to be a sum over entropy terms for each level:

$$
\begin{aligned}
S = \ln\Delta\Gamma &= \ln\left(\prod_j \Delta\Gamma_j\right) = \sum_j \ln\Delta\Gamma_j = \sum_j\left(N_j \ln G_j - N_j \ln\frac{N_j}{e}\right) \\
&= \left(\sum_j N_j \ln\left(e\frac{G_j}{N_j}\right)\right) = \sum_j G_j\left[\frac{N_j}{G_j}\ln G_j - \frac{N_j}{G_j}\ln\left(\frac{N_j}{G_j}\frac{G_j}{e}\right)\right] \\
&= \sum_j G_j\left[n_j \ln G_j - n_j \ln n_j - n_j \ln G_j + n_j\right] \\
&= \sum_j G_j n_j\left[1 - \ln n_j\right] = -\sum_j G_j n_j \ln\frac{n_j}{e} \tag{4.54}
\end{aligned}
$$

Thus, the total entropy involves the degeneracies and the occupation numbers for each level:

$$S_{\text{Boltzmann}} = -\sum_j G_j n_j \ln\frac{n_j}{e} \tag{4.55}$$

At the equilibrium state, the entropy should have a maximum value. Let us find the distribution function at the equilibrium state. Our problem is to find the n_j for which the sum (4.55) has a maximum value, subject to the auxiliary conditions of a desired energy

$$E = \sum_j E_j N_j = \sum_j G_j E_j n_j \tag{4.56}$$

and a total number of particles

$$N = \sum_j G_j n_j \tag{4.57}$$

If we use Lagrange's multipliers then we have

$$\frac{\partial}{\partial n_j}(S + \alpha N - \beta E) = 0 \tag{4.58}$$

Here, α and β are the undetermined multipliers. From Equation 4.58, we have

$$G_j(-\ln n_j + \alpha - \beta E_j) = 0 \tag{4.59}$$

which gives

$$\ln n_j = \alpha - \beta E_j \tag{4.60}$$

or

$$n_j = \exp\{\alpha - \beta E_j\} \tag{4.61}$$

This is the Boltzmann distribution, and based on earlier results for the occupation numbers, one needs to identify the multipliers as

$$\alpha = \frac{\mu}{T}, \quad \beta = \frac{1}{T} \tag{4.62}$$

4.4   FREE ENERGY OF THE IDEAL BOLTZMANN GAS

We already know that the free energy can be expressed as

$$F = -T \ln Z \tag{4.63}$$

where the canonical partition function is

$$Z = \sum_n \exp\left\{-\frac{E_n}{T}\right\} \tag{4.64}$$

This is the sum over states, where n is the quantum number identifying the state of the entire body, and E_n is the energy of the body.

Let us write E_n in the form of the sum of energies E_k, where k refers to the state of one particle of the gas. We may start from the summation of the energies of separate particles. Thus, for N gas particles

$$E = \sum_{k=1}^{N} E_k \tag{4.65}$$

Here, E_k is the energy of the kth particle, taken itself as a sum

$$E_k = \frac{p^2}{2m_k} + E_k', \quad E_k' = E_{rot} + E_{pot} + E_{el} \tag{4.66}$$

where E_{rot} is the rotational energy, E_{pot} the potential energy, and E_{el} the electronic energy (there are no interparticle terms). We may be tempted to say that the sum over states, Z, with respect to all quantum states, is

$$Z = \sum_{k}\prod_{k=1}^{N} \exp\left\{-\frac{E_k}{T}\right\} \stackrel{?}{=} \prod_{k=1}^{N}\sum_{k} \exp\left\{-\frac{E_k}{T}\right\} = \prod_{k=1}^{N} Z_k = (Z_1)^N \tag{4.67}$$

where Z_1 is the sum over states for one particle.

Let us evaluate the sum of states for a system made up of two states and two distinguishable (or classical) particles. The two states are separated by energy E; for simplicity, the lower state's energy is set to zero in Figure 4.4.

Based on the four configurations of Figure 4.4 for distinguishable particles, it follows that

$$Z = 1 + 2\exp\left\{-\frac{E}{T}\right\} + \exp\left\{-\frac{2E}{T}\right\} = \left(1 + \exp\left\{-\frac{E}{T}\right\}\right)^2 \tag{4.68}$$

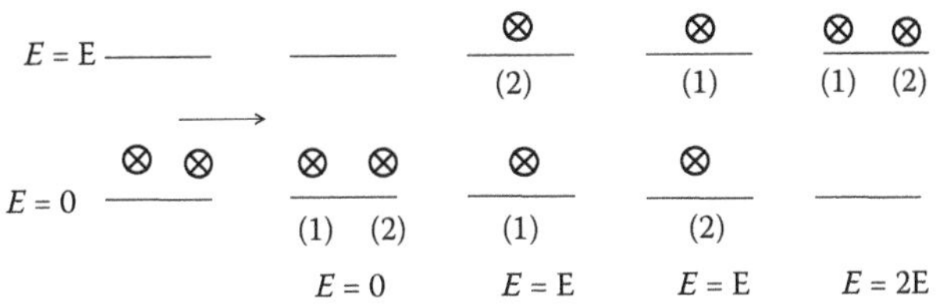

FIGURE 4.4 A system made up of two states and two distinguishable (or classical) particles. The two states are separated by energy E; the lower state's energy is set to zero.

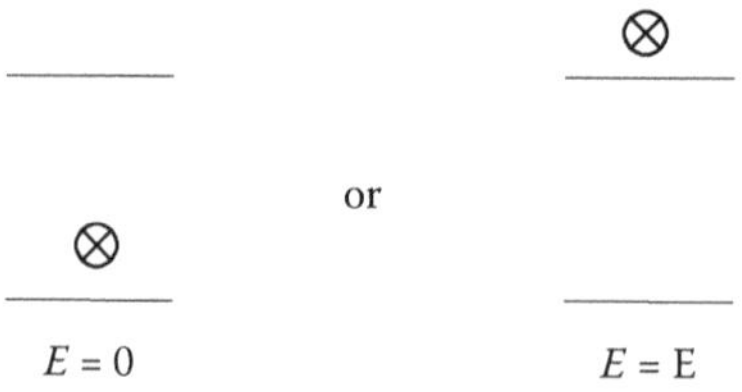

FIGURE 4.5   A system made up of two states and one particle. The two states are separated by energy E; the lower state's energy is set to zero.

For the same system but with one particle (Figure 4.5), the partition function is

$$Z = 1 + \exp\left\{-\frac{E}{T}\right\} \tag{4.69}$$

From Equations 4.68 and 4.69, it follows that we may remove the question mark "?" in Equation 4.67 for *distinguishable particles*

$$Z = (Z_1)^N \tag{4.70}$$

Now let us examine the sum over states for two identical and indistinguishable (quantum or classical) particles. From the configurations in Figure 4.6, one gets

$$Z = 1 + \exp\left\{-\frac{E}{T}\right\} + \exp\left\{-\frac{2E}{T}\right\} \tag{4.71}$$

In the case of a nondegenerate state for nonidentical particles, we do not have configurations (1) and (3) in Figure 4.6 and thus there exists only configuration (2) as shown in Figure 4.7, and as a consequence

$$Z \approx \exp\left\{-\frac{E}{T}\right\} \tag{4.72}$$

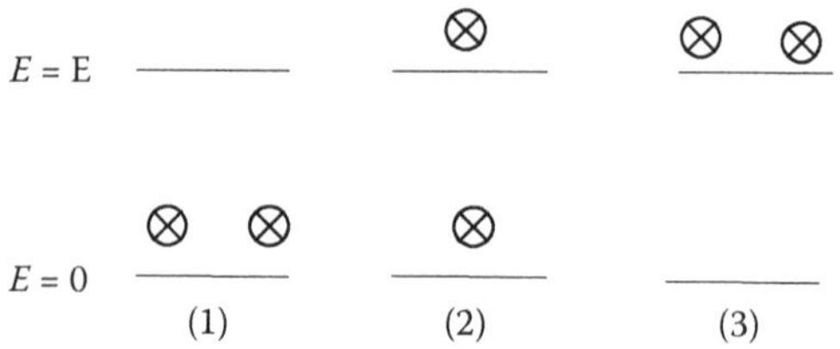

FIGURE 4.6   A system made up two identical and indistinguishable (classical or quantum) particles. The two states are separated by energy E; the lower state's energy is set to zero.

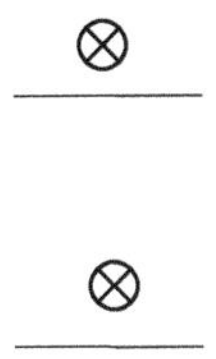

(2)

FIGURE 4.7 A system made up of two nonidentical (quantum) particles. The two states are separated by energy E; the lower state's energy is set to zero.

On the other hand, for nonidentical particles, there are two ways to realize configuration (2) in Figure 4.7, which gives

$$Z \approx 2\exp\left\{-\frac{E}{T}\right\} \qquad (4.73)$$

It follows that for identical particles, (indistinguishable, whether classical or quantum)[*]

$$Z = \frac{(Z_1)^N}{N!} \qquad (4.74)$$

We may use the above argument to evaluate the thermodynamic quantities for an ideal nondegenerate gas. For such a gas, we have the single-particle partition function

$$Z_1 = \frac{V}{(2\pi\hbar)^3}\int\exp\left\{-\frac{E(p)}{T}\right\}d\vec{p}\sum_k\exp\left\{-\frac{E'_k}{T}\right\} = \frac{V}{(2\pi\hbar)^3}(2\pi mT)^{3/2}\sum_k\exp\left\{-\frac{E'_k}{T}\right\} \qquad (4.75)$$

Here, $E(p)$ is the kinetic energy of translational motion and E'_k is the energy level corresponding to the rotation of the particle and its internal state. The quantity E'_k is independent of velocity and the coordinate of the center of mass of the particle. Thus, the free energy for the whole gas is

$$F = -T\ln\frac{(Z_1)^N}{N!} = -T\ln\left[\left(\frac{e}{N}\right)^N Z_1^N\right] = -NT\ln\left[\frac{eV}{N}\frac{(2\pi mT)^{3/2}}{(2\pi\hbar)^3}\sum_k\exp\left\{-\frac{E'_k}{T}\right\}\right] \qquad (4.76)$$

---

[*] One can also argue that even in a classical ideal gas at low density, far from degeneracy, the atoms or molecules are indistinguishable. One cannot label individual oxygen molecules, for example.

This is the *free energy of an ideal nondegenerate gas*, that is, a Boltzmann gas. We may write Equation 4.76 in the form

$$F = -NT \ln \frac{eV}{N} + Nf(T) \tag{4.77}$$

The function $f(T)$ depends on the details of the Hamiltonian.

Q4.1 Find the position of the center of mass of a cylinder of a classical ideal gas in a homogeneous gravitational field if the temperature of the gas is T, the molecular mass is m, and the acceleration due to gravity is g.

A4.1 From definition, the vertical center of mass is

$$z_c = \frac{\int z \, dm(z)}{\int dm(z)} \tag{4.78}$$

where $dm(z)$ is the mass of all gas molecules in the layer $[z, z + dz]$, which can be expressed using the barometric formula introduced earlier

$$dm(z) = n_0 \exp\left\{-\frac{mgz}{T}\right\} dz \tag{4.79}$$

Then, the mean value is

$$z_c = \frac{\int_0^\infty z \exp\left\{-(mgz/T)\right\} dz}{\int_0^\infty \exp\left\{-(mgz/T)\right\} dz} = -\frac{d}{d\alpha} \ln \frac{1}{\alpha} = \frac{T}{mg} \tag{4.80}$$

where $\alpha = (mg/T)$. The result shows that the average potential energy (mgz_c) is proportional to the average kinetic energy $\left(\frac{3}{2}T\right)$.

Q4.2 A mixture of ℓ ideal gases made up of the same quantities of particles but different atoms with masses $m_1, m_2, \ldots, m_\ell$ is contained in a cylinder of height h and radius R. The entire system is placed in the gravitational field of the Earth. Find the center of mass of the mixture.

A4.2 For one kind of particle the center of mass z_k is

$$z_k = \frac{\int_0^h z \exp\{-(m_k gz/T)\}dz}{\int_0^h \exp\{-(m_k gz/T)\}dz} = -\frac{d}{d\alpha_k} \ln\frac{1}{\alpha_k}(1 - \exp\{-\alpha_k h\})$$

$$= \frac{1}{\alpha_k} - \frac{h}{\exp\{\alpha_k h\} - 1}, \quad \alpha_k = \frac{m_k g}{T} \tag{4.81}$$

The common center of mass z_c is

$$z_c = \frac{\sum_{k=1}^{\ell} m_k z_k}{\sum_{k=1}^{\ell} m_k} = \frac{\ell T}{Mg} - \frac{h}{M}\sum_{k=1}^{\ell}\frac{m_k}{\exp\{(m_k gh/T)\} - 1}, \quad M = \sum_{k=1}^{\ell} m_k \tag{4.82}$$

As $h \to \infty$, the second summand in Equation 4.82 tends to zero and the net result depends on only the average molecular mass M/ℓ.

Q4.3 Find the mean gravitational potential energy of the molecules of an ideal gas found in the vertical cylinder of height h at a temperature T.

A4.3 The potential energy of a molecule in the homogeneous gravitational field is

$$U(z) = mgz \tag{4.83}$$

Thus

$$\bar{U} = mg\bar{z} = \frac{mg\int_0^h z\exp\{-(mgz/T)\}dz}{\int_0^h \exp\{-(mgz/T)\}dz} = -mg\frac{d}{d\alpha}\ln\frac{1}{\alpha}(1 - \exp\{-\alpha h\}) \tag{4.84}$$

This can be expressed as

$$\bar{U} = T\left(1 - \frac{\alpha h}{\exp\{\alpha h\} - 1}\right) = T\left(1 - \frac{mg\,h/T}{\exp\{(mg\,h/T)\} - 1}\right) \tag{4.85}$$

As $h \to 0$, then

$$\alpha h \ll 1, \quad \bar{U} = \frac{mgh}{2} \tag{4.86}$$

on the other hand, as $h \to \infty$, we have

$$\bar{U} \approx T \tag{4.87}$$

That is, the mean potential energy increases linearly with temperature, the same as the mean altitude of the particles. However, this result ignores changes in the gravitational field with altitude that would occur for any real system.

Q4.4 One mole of a monatomic ideal gas at a temperature T is placed in a vertical cylindrical container of height h. Find the energy and heat capacity of the gas, considering a homogeneous gravitational field. Examine the limiting cases of small and large h.

A4.4 If we consider the following relation:

$$C_p = \left(\frac{\partial E}{\partial T}\right)_p, \quad C_V = C_p + \left[\left(\frac{\partial E}{\partial p}\right)_T - V\right]\left(\frac{\partial p}{\partial T}\right)_V \tag{4.88}$$

and then the energy of the gas, using Equation 4.85 to get the potential energy, is

$$E = \frac{3}{2}N_A T + N_A T\left(1 - \frac{(mg\,h/T)}{\exp\{(mg\,h/T)\} - 1}\right) = \frac{5}{2}RT - \frac{M_A\,gh}{\exp\{(M_A\,g\,h/RT)\} - 1} \tag{4.89}$$

where $M_A = mN_A$ is the molar mass and $R = k_B N_A$ is the gas constant. This gives

$$C_V = \frac{5}{2}R - R\left(\frac{M_A\,gh}{RT}\right)^2 \frac{\exp\{(M_A\,g\,h/RT)\}}{\left(\exp\{(M_A\,g\,h/RT)\} - 1\right)^2} \tag{4.90}$$

For a very light gas or a short cylinder

$$M_A g h \ll RT, \quad C_V = \frac{3}{2}R \tag{4.91}$$

In this case, the gas density at all heights of the cylinder is practically the same. The mean potential energy is independent of the temperature, and the gravitational field practically does not influence the heat capacity of the ideal gas.

When $M_A\,g\,h \gg RT$ and as $h \to \infty$, then $C_V = \frac{5}{2}R$. The gravitational field in this case leads to an increase in the heat capacity as the mean potential energy per molecule appears to be proportional to the temperature.

Q4.5 Find the total number of molecules found above the surface of the Earth in a column of air with cross-sectional area 1 cm² and height h. Find the pressure produced by that column of air, assuming the entire column is at a uniform temperature.

A4.5 The number of molecules for the column of air with base area A (= 1 cm²) and height h is equal to

$$v(h) = \int_0^h n(z)Adz = An_0 \int_0^h \exp\left\{-\frac{mgz}{T}\right\}dz = An_0 \frac{T}{mg}\left(1 - \exp\left\{-\frac{mgh}{T}\right\}\right) \quad (4.92)$$

where n_0 is the number density of molecules at Earth's surface. Presumably, this is determined by the thermodynamic conditions of pressure and temperature there, and assuming 273 K and 1.00 atm, we have

$$n_0 = \frac{N}{V} = \frac{p_0}{T} = \frac{101.3\,\text{kPa}}{(1.38 \times 10^{-23}\,\text{J/K})(273\,\text{K})} = 2.69 \times 10^{25}\,\text{m}^{-3} \quad (4.93)$$

This can be expressed in molar units:

$$n_0 = \frac{p_0 N_A}{RT} = (44.6\,\text{mol/m}^3)N_A \quad (4.94)$$

Then, multiplication by the column width gives a number per length

$$An_0 = (0.01\,\text{m})^2(44.6\,\text{mol/m}^3)N_A = 4.46\,N_A/\text{km} \quad (4.95)$$

The other factor needed is the scale height, and using an average molecular mass of 28.8 u, we have

$$\frac{T}{mg} = \frac{(1.38 \times 10^{-23}\,\text{J/K})(273\,\text{K})}{(28.8 \times 1.66 \times 10^{-27}\,\text{kg})(9.81\,\text{m/s}^2)} = 8.03\,\text{km} \quad (4.96)$$

The product of these gives the number of molecules in an infinitely high column of air:

$$v_\infty = An_0 \times \frac{T}{mg} = \left(4.46 \frac{N_A}{\text{km}}\right)(8.03\,\text{km}) = 35.8\,N_A \quad (4.97)$$

This implies that an infinite column of air with base area 1 cm² contains 35.8 mol.

Note that the result really does not depend directly on the temperature, if the problem is specified by the pressure p_0 at the Earth's surface:

$$V_\infty = A\left(\frac{p_0}{T}\right)\left(\frac{T}{mg}\right) = \frac{Ap_0}{mg} = \frac{Ap_0}{M_A g}N_A \tag{4.98}$$

On the other hand, thinking of temperature as the control variable, this result is expressed equivalently as

$$V_\infty = \left(\frac{A}{mg}\right)(n_0 T) \tag{4.99}$$

The first factor in this expression is fixed for a given air column confined within the cross section A. The second factor does not increase with temperature, but rather, it is constant. If the column is heated, the air expands and its density decreases inversely with the temperature, so that $n_0 T$ remains constant. The pressure at the bottom of the column, ultimately, depends only on the total weight of air above the cross section, which does not change with temperature. (We are assuming a perfectly still atmosphere without weather, etc.)

Q4.6 Find the height in an atmosphere at a uniform temperature of 273 K where the pressure is reduced by a factor of $1/\chi$ compared to that at ground level.

A4.6 The pressure at any altitude is the weight of air above that height divided by the area of the column. The relation

$$p(h) = \frac{mg}{A}\left[V_\infty - V(h)\right] \tag{4.100}$$

defines the pressure with respect to the height in the air h. The quantities $v(h)$ and v_∞ are defined correspondingly by the formulas (4.92) and (4.98). These give

$$p(h) = n_0 T \exp\left\{-\frac{mgh}{T}\right\} = p_0 \exp\left\{-\frac{mgh}{T}\right\} \tag{4.101}$$

As we require

$$\frac{p(h)}{p_0} = \frac{1}{\chi} \tag{4.102}$$

then

$$h = \frac{T}{mg}\ln\chi = \frac{RT}{M_A g}\ln\chi \tag{4.103}$$

or

$$h = (8.03 \text{ km})\ln\chi \tag{4.104}$$

TABLE 4.1
Various Values of h for Different Values of χ

χ	1.5	2	3	5	10
h/km	3.26	5.57	8.82	12.9	18.5

We display various values of h for different values of χ in Table 4.1.

As the barometric formula is only approximately true for Earth's atmosphere, the resulting values must be viewed with caution. At high altitudes, a deviation from the barometric formula is present, connected with the nonhomogeneity of the atmosphere, different temperatures at different heights, and the fact that the atmosphere is not in a state of equilibrium.

Q4.7 A planet has radius r_0 and mass M. Determine the portion of molecules of its atmosphere that are capable of overcoming its gravitational field and escaping from the planet. The mass of a molecule is m and the temperature T of the atmosphere is assumed to be constant with respect to the height.

A4.7 In the previous problems, we assumed the gravitational field to be homogeneous, that is, we did not consider its change with distance from the center of the planet. However, for distances from a planet large compared to the planet's radius, its gravitational field decays as $1/r^2$, and the gravitational potential should be described by Newton's expression:

$$U(r) = -G\frac{mM}{r} + C \tag{4.105}$$

G is the gravitational constant and C is selected to set the potential to zero at the surface of the planet, $U(r_0) = 0$. If we introduce the acceleration due to gravity on the planet's surface, g_0:

$$g_0 = \frac{GM}{r_0^2} \tag{4.106}$$

we find

$$U(r) = mg_0r_0^2\left(\frac{1}{r_0} - \frac{1}{r}\right) \tag{4.107}$$

The distribution of the particles with respect to the height now has the form

$$n(r) = n_0 \exp\left\{-\frac{mg_0r_0^2}{T}\left(\frac{1}{r_0} - \frac{1}{r}\right)\right\} \tag{4.108}$$

At infinity, the density of the particles of the gas is different from zero and has the finite value

$$n_\infty = n_0 \exp\left\{-\frac{mg_0 r_0}{T}\right\}$$ (4.109)

Thus, the molecules may be moving infinitely away from the planet, that is, they escape from its atmosphere. For this to occur, it is necessary that their kinetic energy should be more than $U(\infty) = mg_0 r_0$, that is, the velocity is

$$v \geq v_c = \sqrt{2g_0 r_0}$$ (4.110)

Assuming that the distribution of the molecules with respect to the velocities is described by the Maxwell distribution (3.177), we find

$$N_1(v \geq v_c) = N4\pi\left(\frac{m}{2\pi T}\right)^{3/2} \int_{v_c}^{\infty} \exp\left\{-\frac{mv^2}{2T}\right\} v^2 dv$$ (4.111)

Let us do the change of variables

$$\frac{mv^2}{2T} = \varsigma^2, \quad \varsigma_c^2 = \frac{mv_c^2}{2T} = \frac{mg_0 r_0}{k_B T}$$ (4.112)

Then we have

$$\frac{N_1(v \geq v_c)}{N} = \frac{4}{\sqrt{\pi}} \int_{\varsigma_c}^{\infty} \exp\left\{-\varsigma^2\right\} \varsigma^2 d\varsigma = 1 - \frac{4}{\sqrt{\pi}} \int_{0}^{\varsigma_c} \exp\left\{-\varsigma^2\right\} \varsigma^2 d\varsigma$$ (4.113)

After integration by parts, the integral (4.113) is transformed to the error integral:

$$\frac{N_1(v \geq v_c)}{N} = 1 - \frac{2}{\sqrt{\pi}} \varsigma_c \exp\left\{-\varsigma_c^2\right\} - \mathrm{erf}(\varsigma_c)$$ (4.114)

Using further the case of large $\varsigma_c \gg 1$, we have the asymptotic representation of the error integral:

$$1 - \mathrm{erf}(z) = \frac{1}{\pi} \exp\left\{-z^2\right\} \sum_{n=0}^{\infty} (-1)^n \frac{\Gamma\left(n + \frac{1}{2}\right)}{z^{2n+1}}$$ (4.115)

We find finally

$$\frac{N_1(v \geq v_c)}{N} \cong \frac{2}{\sqrt{\pi}} \sqrt{\frac{mg_0 r_0}{k_B T}} \exp\left\{-\frac{mg_0 r_0}{k_B T}\right\} \tag{4.116}$$

For the Earth (assuming 273 K)

$$r_0 = 6380 \text{ km}, \quad \frac{mg_0 r_0}{k_B T} = \frac{r_0}{8.03 \text{ km}} = 795 \tag{4.117}$$

and

$$\frac{N_1}{N} = 31.8 \exp\{-795\} = 3 \times 10^{-344} \tag{4.118}$$

That is, the portion of the molecules capable of escaping from the Earth's atmosphere is infinitely small. The examination of the escape velocity compared to the velocity distribution shows that the danger of the Earth losing its atmosphere is completely negligible.

On the other hand, planets with low values of $mg_0 r_0/T$ are constantly losing their atmosphere. On mercury and the moon, this process is completely finished and there is no discernible atmosphere.

Q4.8 The main reason that the temperature of the atmosphere decreases with altitude is the following: The processes in the atmosphere are close to adiabatic because the atmosphere exchanges very little heat with its surroundings. (There is heat exchanged with the sun and with empty space, but these processes are slow.)

Using the adiabatic equation of an ideal gas, evaluate the temperature gradient with respect to height.

A4.8 Let us use the ideal gas equation:

$$pV = nRT, \quad n = \frac{M}{M_A} \tag{4.119}$$

We find the adiabatic equation of an ideal gas in the differential form (in the variables p and T):

$$dS = \frac{1}{T}(nC_V dT + pdV) = n\left(C_p \frac{dT}{T} - R\frac{dp}{p}\right) = 0 \tag{4.120}$$

or

$$\frac{dT}{T} = \frac{\gamma - 1}{\gamma} \frac{dp}{p}, \quad \gamma = \frac{C_p}{C_V} \tag{4.121}$$

The following relation defines the change in pressure of the ideal gas with height if the atmosphere is found in a stable mechanical equilibrium (i.e., pressure forces balance the weight of gas in thickness dh, for constant acceleration due to gravity):

$$dp = -\rho \, g \, dh = -\frac{M_A g}{RT} p \, dh \tag{4.122}$$

or from Equation 4.101 we have

$$p = p_0 \exp\left\{ -\frac{M_A g h}{RT} \right\} \tag{4.123}$$

Within the interval $[h, h + dh]$ we assume that the temperature is constant.
From Equations 4.121 and 4.122, we find that

$$\frac{dT}{dh} = -\frac{\gamma - 1}{\gamma} \frac{M_A g}{R} \tag{4.124}$$

For air, primarily diatomic gases, with $\ell = 5$ quadratic coordinates per molecule, and $C_v = \frac{5}{2} R$, $C_p = \frac{7}{2} R$,

$$\gamma = \frac{\ell + 2}{\ell} = \frac{7}{5} = 1.4 \tag{4.125}$$

or

$$\frac{dT}{dh} = -\frac{2}{7} \frac{M_A g}{R} = -9.7 \text{ K/km} \tag{4.126}$$

Actually the adiabatic temperature gradient in absolute value is a little bit less (the mean temperature gradient is approximately -8 K/km). It should be noted that if expansion releases heat, it goes back into the same atmosphere. So $\eth Q = 0$ and hence $dS = 0$.

Q4.9 We saw in the previous problem that the mechanical stability of the atmosphere together with an adiabatic constraint results in the atmospheric temperature gradient. The convection fluxes of air, however, mix the atmosphere's layers and distort its mechanical stability. Find the maximum temperature gradient (with respect to the height) for which the atmosphere will still be found in stable equilibrium, with respect to the process of equilibrating the temperature during convection.

A4.9 For stable equilibrium, the temperature T, density ρ, and pressure p are dependent only on the height h above the Earth's surface, and related by the ideal gas law, $p = (RT/M_A)\rho$. For a change in h we have

$$\frac{dp}{p} = \frac{d\rho}{\rho} + \frac{dT}{T} \tag{4.127}$$

From here the equilibrium density at the height $h + dh$ is equal to

$$\rho(h + dh) = \rho(h) + d\rho = \rho(h) + \rho\left(\frac{dp}{p} - \frac{dT}{T}\right) \tag{4.128}$$

Now consider some accidental perturbation that may involve a small mass of air being transferred from height h to height $h + dh$. The pressure at the interior of the transferred mass of air is equal to the pressure of the surrounding air and thus its density changes and becomes $\rho^*(h + dh)$. The star refers to this moving air.

Assuming as before that the process of transfer is adiabatic, we have, considering Equation 4.121 through 4.123 for the change of the density of the transferred mass

$$\frac{d\rho^*}{\rho} = \frac{dp}{p} - \frac{dT}{T} = \frac{1}{\gamma}\frac{dp}{p} \tag{4.129}$$

and

$$\rho^*(h + dh) = \rho(h) + d\rho^* = \rho(h) + \frac{\rho}{\gamma}\frac{dp}{p} \tag{4.130}$$

If for $dh > 0$ we have

$$\rho^*(h + dh) > \rho(h + dh) \tag{4.131}$$

then the convecting air is heavier than the surrounding air and it tends to return to its initial position and we have stable equilibrium. Conversely, if the convecting air is less dense than the air it moves into, it will tend to rise even more, becoming even less dense, and we have unstable equilibrium. From Equations 4.128 and 4.130, we find the condition of stability:

$$\frac{1}{\gamma}\frac{\rho}{p}dp \geq \rho\left(\frac{dp}{p} - \frac{dT}{T}\right) \tag{4.132}$$

or

$$\frac{dT}{T} \geq -\frac{1-\gamma}{\gamma}\frac{dp}{p} = \frac{\gamma-1}{\gamma}\left(-\frac{M_A g}{RT}\right)dh \qquad (4.133)$$

Here, we use for the mechanical equilibrium the formula (4.122). Finally

$$\frac{dT}{dh} \geq = -\frac{\gamma-1}{\gamma}\frac{M_A g}{R} = -\frac{M_A g}{C_p} \qquad (4.134)$$

The limiting value is the same as the equilibrium temperature gradient found in the previous question. The equilibrium atmosphere undergoes perturbations until it reaches the limiting temperature gradient for stability.

Q4.10 Find the mean magnetic moment and the magnetic susceptibility of the gas made up of molecules with constant magnetic moments $\vec{\mu}_0$ and located in a constant homogeneous magnetic field $\vec{H}$ (a paramagnetic gas).

A4.10 The magnetic field $\vec{H}$ orients the magnetic moments $\vec{\mu}_0$ in the direction of the field and thermal motion disorients them. From the classical point of view, the moment $\vec{\mu}_0$ may be arbitrary directed relative to the field $\vec{H}$. The potential energy U in the field $\vec{H}$ is equal to

$$U = -\vec{\mu}_0 \cdot \vec{H} = -\mu_0 H \cos\theta \qquad (4.135)$$

where θ is the angle between $\vec{\mu}_0$ and $\vec{H}$. The probability that $\vec{\mu}_0$ may point at an angle between θ and $\theta + d\theta$ with respect to $\vec{H}$ is defined by the Boltzmann distribution (for simplicity, take the z-axis to be directed along $\vec{H}$)

$$dW = A \exp\left\{\frac{\mu_0 H}{T}\cos\theta\right\}d\Omega \qquad (4.136)$$

where

$$d\Omega = 2\pi\sin\theta\, d\theta \qquad (4.137)$$

is the solid angle corresponding to the interval of the angle $d\theta$ and A is a constant defined from the condition of normalization:

$$\int dW = 1 \qquad (4.138)$$

Thus

$$dW(\theta) = \frac{\exp\{(\mu_0 H/T)\cos\theta\}\sin\theta\,d\theta}{\displaystyle\int_0^\pi \exp\{(\mu_0 H/T)\cos\theta\}\sin\theta\,d\theta} \tag{4.139}$$

The mean value of the projection of the moment $\vec{\mu}_0$ on the direction of the magnetic field is the mean of its z-component:

$$\overline{\mu_0} = \int_0^\pi \mu_0\cos\theta\,dW(\theta) = \frac{\displaystyle\int_0^\pi \mu_0\cos\theta\exp\left\{\frac{\mu_0 H}{T}\cos\theta\right\}\sin\theta\,d\theta}{\displaystyle\int_0^\pi \exp\left\{\frac{\mu_0 H}{T}\cos\theta\right\}\sin\theta\,d\theta} = \mu_0\frac{d}{d\alpha}\ln J(\alpha) \tag{4.140}$$

where

$$J(\alpha) = \int_0^\pi \exp\{\alpha\cos\theta\}\sin\theta\,d\theta = \frac{\exp\{\alpha\} - \exp\{-\alpha\}}{\alpha}, \quad \frac{\mu_0 H}{T} = \alpha \tag{4.141}$$

From the above, we have

$$\overline{\mu_0} = \mu_0\left(\coth\alpha - \frac{1}{\alpha}\right) = \mu_0 L(\alpha) = \mu_0 L\left(\frac{\mu_0 H}{T}\right) \tag{4.142}$$

where $L(\alpha)$ is the Langevin function. The quantity $\mu_0 H/T$ measures the ratio of a typical magnetic energy to a typical thermal energy. In the case of a weak field when $\mu_0 H \ll T$ or $\alpha \ll 1$, we find that the function $L(\alpha)$ has the linear approximation

$$L(\alpha) \cong \frac{\alpha}{3} \tag{4.143}$$

and

$$\overline{\mu_0} = \frac{\mu_0^2 H}{3T} \tag{4.144}$$

Suppose the substance contains n atoms (or molecules) per unit volume. Then the mean magnetic moment per unit volume of gas (the magnetization) is

$$M = n\overline{\mu} = n\frac{\mu_0^2 H}{3T} \tag{4.145}$$

This is proportional to the magnetic field (paramagnetism). From here the initial *magnetic susceptibility* of the gas is

$$\chi = \frac{dM}{dH} = \frac{n\mu_0^2}{3T} \tag{4.146}$$

This shows the *Pierre Curie law*, where we have

$$\chi \propto \frac{1}{T} \tag{4.147}$$

In the opposite limit, for

$$\alpha \gg 1 \tag{4.148}$$

we have

$$L(\alpha) \approx 1, \quad \bar{\mu} = \mu_0 \tag{4.149}$$

Here, the magnetic moment of the gas attains *saturation*. This means that all the dipoles tend to become aligned with the field for adequately strong applied field.

In the quantum theory of *paramagnetism* of atoms, it is necessary to consider an important modification: the discreteness of the space quantization of the spin angular momentum of the electrons. In a common situation, each atom has spin $\frac{1}{2}$ (in units of $\hbar$) corresponding to one unpaired electron; however, higher spin values are also possible. From the quantum mechanical point of view, the magnetic moment of each spin-$\frac{1}{2}$ atom is pointing either parallel or antiparallel to the external field. (Really, all that matters is the quantization of the spin component along the applied field.)

We assume each atom interacts only weakly with other atoms and with the other degrees of freedom of the gas. From this argument, we partition our system into two, that is, a single atom (*small system*) and all other atoms and other degrees of freedom (*heat reservoir*). Each atom can possibly have either spin up with spin component $S_z = +\hbar/2$ or spin down with spin component $S_z = -\hbar/2$. The state with spin $+\hbar/2$ is parallel to $\vec{H}$ and that with spin $-\hbar/2$ is antiparallel to $\vec{H}$. It is convenient to label these states by the parameter $\sigma = +1$ for spin up and $\sigma = -1$ for spin down. To get the corresponding energies, one needs the atomic magnetic moment, related to the spin angular momentum via the gyromagnetic ratio γ, where $\vec{\mu} = \gamma \vec{S}$.

The quantum gyromagnetic ratio is $\gamma = -(2e/m_e)$, where m_e is the electron mass, and the value is negative because the electron charge is negative. Then, the spin magnetic energy is

$$U = -\vec{\mu} \cdot \vec{H} = \frac{2e}{m_e} \frac{\hbar}{2} \sigma H = \sigma g \mu_B H \tag{4.150}$$

where the Landé g-factor is $g = 2$ and the Bohr magneton is the fundamental atomic unit of magnetic moment

$$\mu_B = \frac{e\hbar}{2m} = 9.27 \times 10^{-24} \text{ J/T} \tag{4.151}$$

Here, the atomic moment is $\mu_z = -\sigma g \mu_B$, $\sigma = \pm 1$. Then, it is easy to work out the average component of magnetic moment along $\vec{H}$:

$$\bar{\mu} = \frac{\sum_{\sigma=\pm 1} -\sigma g \mu_B \exp\left\{-\left(\sigma g \mu_B H/T\right)\right\}}{\sum_{\sigma=\pm 1} \exp\left\{-(\sigma g \mu_B H/T)\right\}} = g \mu_B \frac{\sinh\alpha}{\cosh\alpha} = g \mu_B \tanh\alpha, \quad \alpha = \frac{g \mu_B H}{T} \tag{4.152}$$

Again there is linear paramagnetic behavior at low field, with saturation at high field. The rest of the quantum solution can be worked out by the student.

Q4.11 A cylinder with a vertical axis and end faces located at heights h_1 and h_2 is filled with an ideal gas and located in a uniform gravitational field. Find the gas pressure on the end faces and show that their difference is determined by the weight of the gas inside the cylinder of length $h = h_2 - h_1$.

A4.11 For the evaluation of pressure P it is sufficient to know the configurational part of the free energy, due to:

$$P = -\left(\frac{\partial F}{\partial V}\right)_T \tag{4.153}$$

As the gas is ideal, the canonical partition function is

$$Z = \tilde{z}^N \tag{4.154}$$

where $\tilde{z}$ is the statistical integral of one molecule:

$$\tilde{z} = \int \exp\left\{-\frac{p^2}{2mT}\right\} dp_x dp_y dp_z \int_{h_1}^{h_2} \exp\left\{-\frac{mgz}{T}\right\} dz \int dx dy$$

$$= (2\pi mT)^{3/2} A \frac{T}{mg}\left[\exp\left\{-\frac{mgh_1}{T}\right\} - \exp\left\{-\frac{mgh_2}{T}\right\}\right] \tag{4.155}$$

where A is the area of the base of the cylinder. The free energy is

$$F = -NT \ln \left[\exp\left\{-\frac{mgh_1}{T}\right\} - \exp\left\{-\frac{mgh_2}{T}\right\} \right] + f(T) \qquad (4.156)$$

where the second function does not depend on heights.

   Using now the relation

$$V = A\left(h_2 - h_1\right) \qquad (4.157)$$

we find

$$P_2 = -\frac{1}{A}\frac{\partial F}{\partial h_2} = \frac{Nmg}{A}\left[\exp\left\{\frac{mgh}{T}\right\} - 1\right]^{-1}, \quad P_1 = \frac{1}{A}\frac{\partial F}{\partial h_1} = \frac{Nmg}{A}\frac{\exp\{mgh/T\}}{\exp\{mgh/T\} - 1}$$

$$(4.158)$$

and

$$P_1 - P_2 = \frac{Nmg}{A} \qquad (4.159)$$

As $h \rightarrow 0$, then

$$\frac{mgh}{T} \ll 1, \quad P_2 \approx P_1 = \frac{NT}{V} = \frac{2E}{3V} \qquad (4.160)$$

As $h \rightarrow \infty$, then

$$P_2 = 0, \quad P_1 = \frac{Nmg}{A} \qquad (4.161)$$

Q4.12 A mixture of two ideal gases made of N_1 and N_2 particles of masses m_1 and m_2, respectively, is contained in a cylinder of height h with base area A. The mixture is found in a gravitational field. Find the pressure on the upper surface of the container and also the position of the center of mass.

A4.12 The configurational part of the statistical sum, Z', in the given case, is equal to

$$Z' = (Z'_1)^{N_1}(Z'_2)^{N_2} \qquad (4.162)$$

where

$$Z'_k = A\int_0^h \exp\left\{-\frac{m_k g z}{T}\right\} dz = A\frac{T}{m_k g}\left[1 - \exp\left\{-\frac{m_k g h}{T}\right\}\right] \tag{4.163}$$

This is the configuration part of the statistical sum of one particle of the kth type ($k = 1, 2$). From here, the configuration part of the free energy of the mixture is

$$F = -T\sum_{k=1,2} N_k \ln\left(\frac{AT}{m_k g}\right) - T\sum_{k=1,2} N_k \ln\left[1 - \exp\left\{-\frac{m_k g h}{T}\right\}\right] \tag{4.164}$$

The pressure at height h is

$$p(h) = -\left(\frac{\partial F}{\partial V}\right)_T = -\frac{1}{A}\frac{\partial F}{\partial h} = \sum_{k=1,2}\frac{N_k m_k g}{A}\left[\exp\left\{\frac{m_k g h}{T}\right\} - 1\right]^{-1} \tag{4.165}$$

The center of mass of the gas mixture is

$$z_c = \frac{\displaystyle\sum_{k=1,2} N_k m_k \overline{z_k}}{\displaystyle\sum_{k=1,2} N_k m_k} = \frac{\overline{U}}{Mg} \tag{4.166}$$

where

$$M = \sum_{k=1,2} N_k m_k \tag{4.167}$$

is the mass of the gas mixture and $\overline{U}$ is its mean energy

$$\overline{U} = F + TS = F - T\left(\frac{\partial F}{\partial T}\right)_V = \sum_{k=1,2} N_k\left[T - \frac{m_k g h}{\exp\{(m_k g h / T)\} - 1}\right] \tag{4.168}$$

Considering the fact that the gas is ideal, U and P are additive from the respective values for each component of the mixture. As $h \to 0$, then

$$P = \frac{T}{Ah}(N_1 + N_2) = \frac{2}{3}\frac{E}{V}, \quad \overline{U} = \frac{Mgh}{2} \tag{4.169}$$

And as $h \to \infty$, then

$$P \to 0, \quad \overline{U} = (N_1 + N_2)T, \quad z_c = (N_1 + N_2)\frac{T}{Mg} \qquad (4.170)$$

That is, the center of the gas mixture is proportional to the absolute temperature.

Q4.13 Consider a centrifuge of radius R rotating with a constant angular velocity ω. The centrifuge separates a mixture of two gas molecules. The masses of the gas molecules are m_1 and m_2. Find the partition coefficient q, defined by

$$q = \left(\frac{n_1}{n_2}\right)_{r=R} \qquad (4.171)$$

where n_1/n_2 is the ratio of the concentration of the molecules with masses m_1 and m_2, respectively, measured at the outer radius of the centrifuge.

A4.13 The rotation of the centrifuge is equivalent to the appearance of a centripetal external field with the potential energy

$$U = -\frac{m\omega^2 r^2}{2} \qquad (4.172)$$

The Boltzmann distribution of the probabilities for a molecule in the element of the volume

$$dV = r\,dr\,d\varphi\,dz \qquad (4.173)$$

is equal to

$$dW(r,\varphi,z) = B\exp\left\{\frac{m\omega^2 r^2}{2T}\right\}r\,dr\,d\varphi\,dz \qquad (4.174)$$

where the normalization constant is obtained from

$$\int dW = \int_0^R r\,dr \int_0^{2\pi} d\varphi \int_0^h dz\, B\exp\left\{\frac{m\omega^2 r^2}{2T}\right\} = 2\pi h B\frac{T}{m\omega^2}\left[\exp\left\{\frac{m\omega^2 R^2}{2T}\right\} - 1\right] = 1,$$

$$B = \frac{m\omega^2}{2\pi hT}\left[\exp\left\{\frac{m\omega^2 R^2}{2T}\right\} - 1\right]^{-1} \qquad (4.175)$$

From here we obtain the distribution of particles with respect to r by integrating out the height and azimuthal angle:

$$dW(r) = \int_0^{2\pi} d\varphi \int_0^h dz\, B\exp\left\{\frac{m\omega^2 r^2}{2T}\right\} = 2\pi h B\exp\left\{\frac{m\omega^2 r^2}{2T}\right\} r\,dr$$

$$= \frac{m\omega^2}{T}\frac{\exp\{(m\omega^2 r^2/2T)\}}{\exp\{(m\omega^2 R^2/2T)\} - 1} r\,dr \equiv w(r)dr \qquad (4.176)$$

where $w(r)$ is the desired probability density. It should be noted that the probability density $w(r)$ is a monotonic function of r.

As the concentration of the particles of a defined type at some distance r from the axis of the centrifuge is proportional to the probability density $w(r)$, the ratio of the concentration of the particles with masses m_1 and m_2 equal to

$$\frac{n_1(r)}{n_2(r)} = \frac{w_1(r)}{w_2(r)} = \frac{m_1}{m_2}\frac{\exp\{(m_2\omega^2 R^2/2T)\} - 1}{\exp\{(m_1\omega^2 R^2/2T)\} - 1}\exp\left\{\frac{(m_1 - m_2)\omega^2 r^2}{2T}\right\} \qquad (4.177)$$

From here, the partition coefficient can be approximated, supposing $m\omega^2 R^2 \ll 2T$ for both molecular types:

$$q \approx \exp\left\{\frac{(m_1 - m_2)\omega^2 R^2}{2T}\right\} \qquad (4.178)$$

We can check that the approximation is reasonable, say, for oxygen gas at 300 K. In a centrifuge with a 5.0 cm radius producing a maximum centripetal acceleration of $\omega^2 R = 10{,}000\,g$, the angular speed needs to be $\omega = 1400$ rad/s. This gives $m\omega^2 R^2/2T = 0.031$, so the approximation is good.

The partition coefficient increases as the temperature is reduced. This is because the thermal motion of the molecules tends to mix the different types, rather than separate them.

Q4.14 Find the mean potential energy of the molecules of an ideal gas found in a centrifuge of radius R rotating with a constant angular velocity ω.

A4.14 Let us consider Equations 4.172 and 4.176, then

$$\overline{U} = -\frac{m\omega^2 \overline{r^2}}{2} = -\frac{m^2\omega^4}{2T}\frac{\int_0^R \exp\{(m\omega^2 r^2/2T)\}r^3\,dr}{\exp\{(m\omega^2 R^2/2T)\} - 1} \qquad (4.179)$$

If we evaluate this integral then we have

$$\overline{U} = T - \frac{m\omega^2 R^2}{2} \frac{\exp\{(m\omega^2 R^2/2T)\}}{\exp\{(m\omega^2 R^2/2T)\} - 1} \tag{4.180}$$

As $\omega \to 0$, then

$$\overline{U} \approx -\frac{m\omega^2 R^2}{4} \tag{4.181}$$

and as $\omega \to \infty$, then

$$\overline{U} \approx T - \frac{m\omega^2 R^2}{2} \tag{4.182}$$

Q4.15 In a diatomic molecule, the atoms interact according to the law

$$U(r) = \frac{A}{r^{12}} - \frac{B}{r^6}, \quad A, B > 0 \tag{4.183}$$

where r is the interatomic separation. Find the coefficient of linear expansion of such a molecule.

A4.15 The atoms in a molecule are found at some mean distance r_0 (when at exceedingly low temperature). If $\overline{x} = \overline{r} - r_0$ is the mean displacement of the atoms from their equilibrium position, then the coefficient of linear expansion may be defined as

$$\alpha = \frac{\overline{r} - r_0}{r_0 T} = \frac{\overline{x}}{r_0 T} \tag{4.184}$$

The quantity $\overline{x}$ is

$$\overline{x} = \frac{\displaystyle\int_{-\infty}^{\infty} x \exp\{-U(x)/T\} dx}{\displaystyle\int_{-\infty}^{\infty} \exp\{-U(x)/T\} dx} \tag{4.185}$$

where

$$U(x) = U(r_0 + x) = U(r_0) + \left.\frac{\partial U}{\partial r}\right|_{r_0} x + \left.\frac{1}{2}\frac{\partial^2 U}{\partial r^2}\right|_{r_0} x^2 + \left.\frac{1}{6}\frac{\partial^3 U}{\partial r^3}\right|_{r_0} x^3 + \cdots \tag{4.186}$$

The zero-temperature equilibrium distance between the atoms as defined from the condition $(\partial U/\partial r)\big|_{r_0} = 0$ and is equal to $r_0 = (2A/B)^{1/6}$. From this

$$U(x) \cong \frac{A}{r_0^{12}} - \frac{B}{r_0^6} + \frac{3}{r_0^2}\left(\frac{26A}{r_0^{12}} - \frac{7B}{r_0^6}\right)x^2 - \frac{28}{r_0^3}\left(\frac{13A}{r_0^{12}} - \frac{2B}{r_0^6}\right)x^3$$

$$\equiv \frac{A}{r_0^{12}} - \frac{B}{r_0^6} + \gamma x^2 - \delta x^3 \tag{4.187}$$

where

$$\gamma = \frac{3}{r_0^2}\left(\frac{26A}{r_0^{12}} - \frac{7B}{r_0^6}\right), \quad \delta = \frac{28}{r_0^3}\left(\frac{13A}{r_0^{12}} - \frac{2B}{r_0^6}\right) \tag{4.188}$$

This gives

$$\overline{x} = \frac{\displaystyle\int_{-\infty}^{\infty} x\exp\{-(\gamma x^2/T)\}\left(1 + (\delta x^3/T)\right)dx}{\displaystyle\int_{-\infty}^{\infty} \exp\{-(\gamma x^2/T)\}\left(1 + (\delta x^3/T)\right)dx} \tag{4.189}$$

Using the value of the integral

$$\int_{-\infty}^{\infty} x^4 \exp\{-\alpha x^2\}\,dx = \frac{3\sqrt{\pi}}{4\alpha^{5/2}} \tag{4.190}$$

we have

$$\overline{x} = \frac{3}{4}T\frac{\delta}{\gamma^2} \tag{4.191}$$

from which we obtain

$$\alpha = \frac{3}{4}\frac{\delta}{\gamma^2}\frac{1}{r_0} \tag{4.192}$$

That is, the coefficient of thermal expansion in the given approximation is independent of temperature.

4.5 EQUIPARTITION THEOREM

This classical theorem represents some results that may be used for the evaluation of certain quantities. Earlier we mentioned some applications of the equipartition theorem in simple cases. It pertains to models in which there are quadratic degrees of freedom; ones whose coordinates or momenta appear as squares in the Hamiltonian. As an example, the classical energy E of a molecule includes the translational, rotational, and vibrational energy:

$$E = \frac{p_x^2 + p_y^2 + p_z^2}{2m} + \frac{J_x^2}{2I_{xx}} + \frac{J_y^2}{2I_{yy}} + \frac{J_z^2}{2I_{zz}} + \sum_{i,j}(a_{ij}\dot{q}_i\dot{q}_j + b_{ij}q_iq_j) \quad (4.193)$$

where I and J are, respectively, the rotational inertia and the angular momentum.

If the energy is sufficiently quantized, then we have, respectively, the following rotational and vibrational quantized energies:

$$E_{\text{rot}} = \frac{\hbar^2 J(J+1)}{2I}, \quad E_{\text{vib}} = \hbar\omega\left(v + \frac{1}{2}\right) \quad (4.194)$$

where $J = 0, 1, 2, \ldots$ and $v = 0, 1, 2, \ldots$ are, respectively, the rotational and vibrational quantum numbers. At higher temperatures, however, classical statistics can be applied, and the momenta, angular momenta, and vibrational coordinates all appear quadratically in the energy in Equation (4.193). We consider any system like this where all coordinates and momenta in $E(q,p)$ appear as squares or as cross terms like the vibrational terms shown above.

Let us write the classical partition function for the internal energy:

$$Z = C\int \exp\left\{-\frac{E(q,p)}{T}\right\}dqdp \quad (4.195)$$

Transform all quadratic coordinates and momenta to some scaled variables

$$q' = \frac{q}{\sqrt{T}}, \quad p' = \frac{p}{\sqrt{T}} \quad (4.196)$$

and let ℓ be the number of quadratic variables in the energy, then

$$E(q,p) = TE(q',p') \quad (4.197)$$

and

$$dqdp \approx T^{\ell/2}dq'dp' \quad (4.198)$$

and thus

$$Z = T^{\ell/2}C'\int \exp\left\{-E(q',p')\right\}dq'dp' = AT^{\ell/2} \quad (4.199)$$

Now, let us evaluate the free energy F:

$$F = -NT \ln Z = -NT \ln A - NT \frac{\ell}{2} \ln T \qquad (4.200)$$

This leads to

$$S = -\frac{\partial F}{\partial T} = N \ln A + N \frac{\ell}{2} \ln T + N \frac{\ell}{2} \qquad (4.201)$$

and mean energy

$$E = F + TS = N \frac{\ell}{2} T \qquad (4.202)$$

Then the average energy per particle is proportional to ℓ and the temperature:

$$\varepsilon = \frac{E}{N} = \frac{\ell}{2} T \qquad (4.203)$$

The heat capacity per particle (in units of k_B) follows:

$$c_V = \frac{\partial \varepsilon}{\partial T} = \frac{\ell}{2} \qquad (4.204)$$

Relation (4.203) represents the *equipartition theorem* of classical statistical mechanics. It implies that in the equilibrium state, the mean energy of each independent quadratic term in the energy is equal to $T/2$.

It should be noted that the equipartition theorem is valid only in classical statistical mechanics. If we consider a correct quantum mechanical description, a system has a set of possible energy levels. For exceedingly low temperatures, there is a dominance of quantum effects and there arise discrete energy levels with well-defined energy spacing. The classical description of the system in this case does not hold. When the temperature is exceedingly high, the spacing between levels around the mean energy is small compared to the thermal energy. Considering this, the discrete energy levels are not particularly important. Thus, the classical description, that is, the equipartition theorem can be a good approximation at sufficiently high temperature.

Let us study the following examples:

1. Particles with translational kinetic energy only (monatomic ideal gas)

$$E = \frac{p_x^2 + p_y^2 + p_z^2}{2m} \qquad (4.205)$$

It follows that $\ell = 3$ and for such a gas $E = \frac{3}{2} NT$ and thus $C_V = \frac{3}{2} N$.

2. Linear oscillators that apply to gas molecules whose axis coincides with the oz-axis.

$$E = \frac{p_x^2 + p_y^2 + p_z^2}{2m} + \frac{J_x^2}{2I_{xx}} + \frac{J_y^2}{2I_{yy}} + \frac{m\dot{q}^2}{2} + \frac{m\omega^2 q^2}{2} \qquad (4.206)$$

In this case

$$\ell = 7, \quad \varepsilon = \frac{7}{2}T, \quad c_V = \frac{7}{2} \qquad (4.207)$$

However, for a real gas at room temperature, the vibrations are hardly excited, and

$$E_{\text{vib}} \approx 0, \quad \varepsilon = \frac{5}{2}T, \quad c_V = \frac{5}{2} \qquad (4.208)$$

We use the above knowledge to evaluate the distribution function. Let us show that the statistical weight $\Delta\Gamma$ is a rapidly increasing function of energy. We examine a hyper energy surface and consider a system of s degrees of freedom. The phase volume will be $(2\pi\hbar)^s$ and

$$\Gamma(E) = \frac{1}{(2\pi\hbar)^s} \int_0^E dqdp = \frac{E^s}{(2\pi\hbar)^s} \int_0^1 dq'dp' = CE^s \qquad (4.209)$$

where

$$q' = \frac{q}{\sqrt{E}}, \quad p' = \frac{p}{\sqrt{E}} \qquad (4.210)$$

and

$$\Delta\Gamma = \frac{\partial\Gamma}{\partial E} dE = CsE^{s-1}dE \qquad (4.211)$$

The Gibbs distribution with respect to energy is

$$dW(E) = \frac{1}{(2\pi\hbar)^s} \exp\left\{ -\frac{E(q,p)}{T} \right\} \frac{dqdp}{dE} dE \qquad (4.212)$$

Our problem is to find $d\Gamma$ in terms of dE. From Equation 4.211, we have

$$\Gamma(E) = CE^{\ell/2} \qquad (4.213)$$

where $\ell = 2s$ is the number of quadratic variables in the energy. Thus

$$d\Gamma = C\frac{\ell}{2}E^{\frac{\ell}{2}-1}dE \tag{4.214}$$

and this leads to

$$dW(E) = A\exp\left\{-\frac{E}{T}\right\}E^{\frac{\ell}{2}-1}dE \tag{4.215}$$

which is the distribution in the interval $[E, E+dE]$ for any number of degrees of freedom for the system. We may find A from the normalization condition, then

$$dW(E) = \frac{1}{T^{\ell/2}\Gamma(\ell/2)}\exp\left\{-\frac{E}{T}\right\}E^{\frac{\ell}{2}-1}dE \tag{4.216}$$

4.6 MONATOMIC GAS

We have so far considered only the translational part of the molecular motion. Though this aspect of motion is invariably present in gaseous systems, other aspects that are essentially concerned with the *internal motion* of the gas particles also exist. It is important that in the evaluation of the physical properties of such systems, those contributions due to the internal motions also be taken into account.

First, consider a general gas of N identical particles of mass m enclosed in a container of volume V. Let us write the general formula

$$F = -NT\ln\frac{eV}{N} - NT\ln\left[\frac{(2\pi mT)^{3/2}}{(2\pi\hbar)^3}Z'\right] \tag{4.217}$$

where Z' is as follows:

$$Z' = \sum_k \exp\left\{-\frac{E'_k}{T}\right\} \tag{4.218}$$

This is the sum of states of the system where

$$E'_k = E_{el} + E_{vib} + E_{rot} + E_{nuc} \tag{4.219}$$

is the energy of the particle associated with the internal state of motion that is characterized by the quantum numbers k. We denote by el—electronic, vib—vibrational, rot—rotational, and nuc—nuclear. The contributions made by the internal motions of the particles to the various thermodynamic quantities of the system follow from the partition function:

$$Z' = \sum \exp\left\{-\frac{E_{el}}{T}\right\} \sum \exp\left\{-\frac{E_{vib}}{T}\right\} \sum \exp\left\{-\frac{E_{rot}}{T}\right\}$$

$$\times \sum \exp\left\{-\frac{E_{nuc}}{T}\right\} = Z_{el} Z_{vib} Z_{rot} Z_{nuc} \qquad (4.220)$$

It follows that

$$F = -NT \ln \frac{eV}{N} - NT \left(\ln\left[\frac{(2\pi mT)^{3/2}}{(2\pi \hbar)^3}\right] + \ln Z_{el} + \ln Z_{vib} + \ln Z_{rot} + \ln Z_{nuc} \right)$$

$$(4.221)$$

Now consider monatomic gases. As a single particle has no vibrational or rotational energy, then those terms are absent for a monatomic gas and there exists only

$$Z' = \sum_k \exp\left\{-\frac{E_k}{T}\right\} \qquad (4.222)$$

where the energy levels are due to electronic and nuclear states. The gas is becoming ionized for $T \approx I$ and is nearly completely ionized for $T \gg I$, where I is the ionization energy of the atoms. Consequently, we intend to study the gas for temperatures satisfying the condition $T \ll I$, where practically no atoms are ionized, and very few atoms are in excited states. Thus, we may assume that all the atoms are found in their normal states (ground or very low electronic states).

We know that in Equation 4.222, E_k represents the quantum levels of an atom. When evaluating the sum with respect to different energy levels, we must consider the fact that an energy level may be degenerate. Thus, the corresponding term should appear in the sum as many times as its degeneracy. Let us denote the multiplicity of this degeneracy by g_k. The multiplicity of degeneracy is at times called its *statistical weight*. This means we can write Equation 4.222 in the form

$$Z' = \sum_k g_k \exp\left\{-\frac{E_k}{T}\right\} \qquad (4.223)$$

where index k refers to energy levels.

We may examine a simple case of atoms that in their normal state do not have either an orbital moment L or spin S. This may be seen in the noble gases (column VIII of the periodic table, with filled electronic shells), which possess neither orbital angular momentum nor spin ($L = S = 0$) in their ground state. Their (electronic) ground state is clearly a singlet and $g_e = 1$. In this case, we have a nondegenerate state and the statistical sum has only one term

$$Z' = \exp\left\{-\frac{E_0}{T}\right\} \tag{4.224}$$

For a monatomic gas, often we take $E_0 = 0$, for the energy of the normal state of the atom. Thus, $Z' = 1$.

The nucleus, however, possesses a degeneracy that arises from the possibility of different orientations of the nuclear spin. The presence of the nuclear spin as we know may lead to the so-called *hyperfine structure in the electronic state.* If at the ground state, either $L = 0$ ($S \neq 0$) or $S = 0$ ($L \neq 0$), then there is no fine structure.

It may happen that the ground state of the atom may possess both orbital angular momentum and spin ($L \neq 0$, $S \neq 0$). Examples may be found in alkali atoms. In this case, the ground state would then possess a definite fine structure characterized by Δ, the interlevel distances of the fine structure. Consequently, in the statistical sum, we have to consider all the components of the normal terms of the fine structure. The different levels of the fine structure are distinguished by total angular momentum:

$$J = |L + S|, \ldots, |L - S| \tag{4.225}$$

Each level with total angular momentum J is $(2J + 1)$-fold degenerate and is denoted by E_J. Thus, the electronic statistical sum is

$$Z' = \sum_J (2J + 1)\exp\left\{-\frac{E_J}{T}\right\} \tag{4.226}$$

The summation is with respect to all possible values of J for the given L and S. The significant quantity in Equation 4.226 is the argument of the exponential function. Consider the following limiting cases:

1. The temperature T is reasonably large such that

$$T \gg E_J \sim \Delta \tag{4.227}$$

Here, Δ is the interval of the fine structure, and then the states are strongly occupied

$$\exp\left\{-\frac{E_J}{T}\right\} \approx 1 \qquad (4.228)$$

and from Equation 4.226 considering summation for all possible values of J for the given L and S, we have

$$Z' = (2S + 1)(2L + 1) = g_0 \qquad (4.229)$$

The *internal motion* in this case makes no contribution to the specific heat but contribute only the quantity

$$\varsigma_{SL} = \ln\left[(2S + 1)(2L + 1)\right] \qquad (4.230)$$

to the *chemical constant (potential) and the entropy of the gas.* There are no contributions to the internal energy and the specific heat.

2. Let the temperature T be exceedingly small such that

$$T \ll E_J \sim \Delta \qquad (4.231)$$

We may neglect in this case all the terms in Equation 4.226 except $E_J = 0$ (the ground state of the atom). This results in

$$Z' = (2J_0 + 1) = g_0 \qquad (4.232)$$

where J_0 is the total angular momentum of the atom in the ground state. Thus, the quantity

$$\varsigma_J = \ln(2J_0 + 1) \qquad (4.233)$$

is added to the chemical constant. Also in this case, the electronic motion makes no contribution to the specific heat of the gas. Thus, the free energy is

$$F = -NT \ln\left[\frac{eV}{N}\frac{(2\pi mT)^{3/2}}{(2\pi\hbar)^3}g_0\right] = -NT\left(\ln\frac{eV}{N} + \frac{3}{2}\ln T + \frac{3}{2}\ln\frac{m}{2\pi\hbar^2} + \ln g_0\right)$$

$$(4.234)$$

Let us make the following denotations:

$$c_V = \frac{3}{2}, \quad \varsigma_{trans} = \frac{3}{2}\ln\frac{m}{2\pi\hbar^2} \qquad (4.235)$$

Then

$$F = -NT \ln\left[\frac{eV}{N}\frac{(2\pi\, mT)^{3/2}}{(2\pi\hbar)^3} g_0\right] = -NT\left(\ln\frac{eV}{N} + c_V \ln T + \varsigma_{trans} + \ln g_0\right) \quad (4.236)$$

Here, trans stands for translational and ς_{trans} is the *translational chemical constant*. The specific heat is primarily due to the translational degrees of freedom of the atom. Hence, when the ground state of the atom has fine structure, the specific heat of the gas has the constant value $c_V = 3/2$ at exceedingly low and high temperatures. At intermediate temperatures $T \sim \Delta$, it is dependent on temperature and passes through a maximum.

Consider now the effect of the nuclear spin i that causes the hyperfine splitting of atomic levels. The degeneracy due to this is $(2i + 1)$-fold. The differences of the energies between the hyperfine components are so small that we neglect them in the evaluation of the partition function. The nuclear partition function is

$$Z' = 2i + 1 \quad (4.237)$$

and the additional term to the free energy is

$$-NT \ln(2i + 1) = F_{nuc} \quad (4.238)$$

This term does not change the heat capacity of the gas. The corresponding nuclear energy is $E_{nuc} = 0$. The free energy leads to a change in the entropy by

$$S_{nuc} = N \ln(2i + 1) \quad (4.239)$$

This implies that the free energy term changes the chemical constant by

$$\varsigma_{nuc} = \ln(2i + 1) \quad (4.240)$$

Thus, for most situations, the effects of electronic and nuclear energy levels on the specific heat will not be noticeable, except around temperatures similar to the quantum energy splitting.

4.7 VIBRATIONS OF DIATOMIC MOLECULES

In Section 4.6, we studied monatomic gases under the condition $T \ll I$. We consider now the case $T \ll D$ for diatomic gases, where D is the dissociation energy. At temperatures $T \ll D$, the number of dissociated molecules in the gas is negligible. Further, all molecules of the gas are considered to be in their lowest electronic state.

Diatomic molecules have contributions from the vibrational and rotational degrees of freedom to the partition function. This is in addition to the terms due to the electronic angular momentum and nuclear spin. The effect of the nuclear spin is more pronounced only for diatomic molecules that have like atoms, due to quantum restrictions. Examples are isotopes of hydrogen and deuterium. For diatomic molecules with unlike atoms, these quantum effects are negligible. An example may be observed in hydrogen chloride.

In this section, we discuss the vibrational motions of a molecule. Consider a one-dimensional harmonic oscillator on the ox-axis. The oscillator has a limiting frequency ω (for small motions of the atoms). If the gas temperature is not exceedingly high, oscillations of gas atoms are small in amplitude, equivalent to harmonic oscillators. The mechanical vibrational energy E_v of one oscillator is

$$E_v = \hbar\omega\left(v + \frac{1}{2}\right), \quad v = 0,1,2,\ldots \tag{4.241}$$

Here, $\hbar\omega$ is the vibrational quantum and v is the vibrational quantum number. The *zero point energy* is $E_0 = \hbar\omega/2$. Vibrational energy levels are nondegenerate, so $g_v = 1$. Then we write the vibrational partition function of a diatomic molecule (or any quantum oscillator) as follows:

$$Z = \sum_{v=0}^{\infty} \exp\left\{-\frac{\hbar\omega\left(v + \frac{1}{2}\right)}{T}\right\} \tag{4.242}$$

We take the sum over all quantum states, then

$$Z = \frac{\exp\{-\hbar\omega/2T\}}{1 - \exp\{-\hbar\omega/T\}} \tag{4.243}$$

which is an exact solution. From the statistical sum of the vibrational motion of a diatomic molecule (4.243), we get the statistical sum of all N molecules as the product of the statistical sum of all molecules, that is, the quantity in Equation 4.243 raised to the N th power:

$$Z_{vib} = \left[\frac{\exp\{-\hbar\omega/2T\}}{1 - \exp\{-\hbar\omega/T\}}\right]^N \tag{4.244}$$

The free energy of the vibrational motion is

$$F_{vib} = -T\ln Z = \frac{N\hbar\omega}{2} + NT\ln\left(1 - \exp\left\{-\frac{\hbar\omega}{T}\right\}\right) \tag{4.245}$$

This is an exact formula. The entropy is

$$S = -\frac{\partial F}{\partial T} = -N\ln\left(1 - \exp\left\{-\frac{\hbar\omega}{T}\right\}\right) + NT\frac{(\hbar\omega/T^2)\exp\{-(\hbar\omega/T)\}}{1 - \exp\{-(\hbar\omega/T)\}} \qquad (4.246)$$

The internal energy is

$$E = F + TS = \frac{N\hbar\omega}{2} + \frac{N\hbar\omega}{\exp\{(\hbar\omega/T)\} - 1} \qquad (4.247)$$

The heat capacity C_V is

$$C_V = \frac{\partial E}{\partial T} = N\left(\frac{\hbar\omega}{T}\right)^2 \frac{\exp\{(\hbar\omega/T)\}}{(\exp\{(\hbar\omega/T)\} - 1)^2} \qquad (4.248)$$

As $T \to \infty$, then $C_V \to N$. The mean energy per oscillator is (see Figure 4.8)

$$\varepsilon = \frac{E}{N} = \frac{\hbar\omega}{2} + \frac{\hbar\omega}{\exp\{(\hbar\omega/T)\} - 1} \qquad (4.249)$$

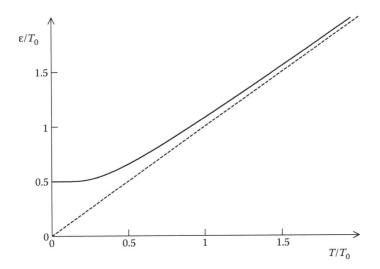

FIGURE 4.8   Variation of the mean energy per oscillator with temperature. The energy and temperature are scaled by $T_0 = \hbar\omega$.

For $T \gg \omega$ (thermal energy high compared to the separation ω between energy levels)

$$\exp\left\{\frac{\hbar\omega}{T}\right\} \approx 1 + \frac{\hbar\omega}{T} \qquad (4.250)$$

and

$$\varepsilon = \frac{\hbar\omega}{2} + \frac{\hbar\omega}{1 + (\hbar\omega/T) - 1} = \frac{\hbar\omega}{2} + T \approx T \qquad (4.251)$$

This is in agreement with the classical result $\varepsilon = T$, which is the equipartition result.

The value $\hbar\omega = T_0$ is called the characteristic temperature. It is the boundary between low and high temperatures. The typical values are rather high, for instance

$$T_{0\,H_2} \cong 6000\,\text{K}, \quad T_{0\,O_2} \cong 2000\,\text{K}, \quad T_{0\,N_2} \cong 3000\,\text{K} \qquad (4.252)$$

This means that room temperature is found in the domain of low temperatures, $\omega \gg T$, and then

$$\varepsilon = \frac{\hbar\omega}{2} + \hbar\omega\exp\left\{-\frac{\hbar\omega}{T}\right\} \qquad (4.253)$$

This approaches properly the zero point energy $\hbar\omega/2$ or the ground state as $T \to 0$. Also, the specific heat per particle is (see Figure 4.9)

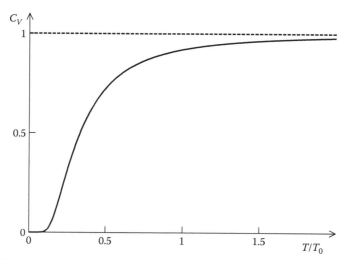

FIGURE 4.9 Variation of the heat capacity of an oscillator with temperature. The temperature is scaled by $T_0 = \hbar\omega$.

$$c_V = \frac{\partial \varepsilon}{\partial T} = \left(\frac{\hbar\omega}{T}\right)^2 \exp\left\{-\frac{\hbar\omega}{T}\right\} \tag{4.254}$$

As $T \to 0$, then $c_V \to 0$. Indeed, c_V is very small for any $T < 0.1T_0$, as seen in Figure 4.9. The vibrational degrees of freedom at very low temperature are said to be *frozen*. The contribution due to the vibrational degree of freedom to the heat capacity of a diatomic gas has the limits: $C_V \to 0$ as $T \to 0$, and $C_V \to N$ as $T \to \infty$. For the intermediate temperature range, the heat capacity ranges from 0 to N.

4.8 ROTATION OF DIATOMIC MOLECULES

The quantum mechanical formula for the rotational energy of a diatomic molecule is

$$E_{\text{rot}} = \frac{\hbar^2 J(J+1)}{2I}, \quad J = 0,1,2,..., \quad I = \mu r_0^2, \quad \mu = \frac{m_1 m_2}{m_1 + m_2} \tag{4.255}$$

where J is the rotational quantum number, I is the rotational inertia of the molecule, μ is the reduced mass of the two atoms, and r_0 is the equilibrium distance between them (see Figure 4.10).

The molecule may be considered as a rigid rotator (Figure 4.10a). Parameters defining the position of the rigid rotator $m_1 m_2$ whose center of mass is at the origin 0 of the reference frame are as follows: The vectors r_1 and r_2 are fixed with respect to

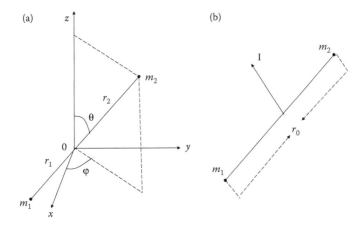

FIGURE 4.10 Rigid rotor model for a diatomic molecule. (a) The positions r_1 and r_2 of the masses m_1 and m_2 are shown relative to a fixed xyz coordinate system with angular positions θ and φ. (b) The masses are shown with the equilibrium separation r_0 and rotational inertia I.

each other; the polar angles θ and φ vary. Let $\hbar \vec{J}$ denote the angular momentum of the system; then its mechanical energy is

$$E_{rot} = \frac{\hbar^2 \vec{J}^2}{2I}, \quad I = 0,1,2,\dots \tag{4.256}$$

From quantum mechanics, $\vec{J}^2$ can assume the possible values $J(J + 1)$. This gives the rotational energy levels in Equation 4.255. It should be noted that a small rotational inertia implies a large spacing of the rotational energy levels. Vector $\vec{J}$ may have several discrete spatial orientations that we denote by quantum number M, that determine its projection along some axis, usually taken as the z-axis:

$$M = -J, -J + 1, \dots, (J - 1), J \tag{4.257}$$

Each value of J corresponds to $2J + 1$ possible quantum states of the same energy (4.255). The energy levels of a rigid rotator with angular momentum $\hbar \vec{J}$ are $(2J + 1)$-fold degenerate.

Let us evaluate the rotational partition function, Z_{rot}:

$$Z_{rot} = \sum_{J,M} \exp\left\{ -\frac{\hbar^2 J(J + 1)}{2IT} \right\} = \sum_{J} (2J + 1)\exp\left\{ -\frac{\hbar^2 J(J + 1)}{2IT} \right\} \tag{4.258}$$

The significant parameter in Equation 4.258 is the argument of the exponential. That is the ratio of the rotational to thermal energy. Then it makes sense to define a characteristic rotational temperature by $T_0 = \hbar^2/2I$. Some characteristic temperatures are

$$T_{0\,H_2} \cong 85\,K, \quad T_{0\,D_2} \cong 43\,K, \quad T_{0\,O_2} \cong 2.1\,K, \quad T_{0\,N_2} \cong 2.9\,K \tag{4.259}$$

These temperatures are considerably less than the characteristic vibrational temperatures. This implies that typically, rotational degrees of freedom are indeed excited for room temperature gases.

1. Let us examine some limiting cases for the temperature T:

$$T \gg T_0 \tag{4.260}$$

We consider an exceedingly high temperature T with the rotational inertia not exceedingly small, then

$$E_{rot} \ll T \tag{4.261}$$

This is the case for many diatomic molecules for which the spacing between the rotational energy levels (4.255) is of the order of 10^{-4} eV, while room temperature of 300 K is around 0.026 eV. Exceptions are molecules like hydrogen molecules well below room temperature. This is because they have small mass and hence small rotational inertia.

From Equation 4.261 the spacing of the rotational energy levels is small compared with T. Thus, the rotational levels may be treated by classical statistics. In the sum (4.258), the main role is played by the terms with larger J. But for larger J the rotation of the molecule is even more classical. Then the rotational energy may be approximated by a continuum and the summation in Equation 4.258 is replaced by the integral:

$$Z = \int_0^\infty (2J+1)\exp\left\{-\frac{\hbar^2 J(J+1)}{2IT}\right\} dJ = \frac{2IT}{\hbar^2} \tag{4.262}$$

Let two nuclei of a molecule be identical. Then we consider their essential indistinguishability. Let us turn the molecule end-for-end. This is the same as interchanging the two identical nuclei. But when they are identical, the exchange does not lead to a new state. Therefore, a turning over by 180°, we count as a distinct state only for unlike nuclei, and Equation 4.262 applies only to heteronuclear molecules. The turning over does not give a distinguishable state for like nuclei. If the nuclei are alike, the result in Equation 4.262 must be divided by 2. Consequently, Equation 4.262 should be replaced by

$$Z = \frac{2IT}{\hbar^2 \sigma} \tag{4.263}$$

where

$$\sigma = \begin{cases} 1, & \text{if the nuclei are unlike (heteronuclear molecules)} \\ 2, & \text{if the nuclei are identical (homonuclear molecules)} \end{cases} \tag{4.264}$$

Using the modification (4.263), the free energy of the rotational motion is

$$F_{rot} = -NT\ln Z = -NT\ln\frac{2IT}{\hbar^2\sigma} = -NT\ln T - NT\ln\frac{2I}{\hbar^2\sigma} \tag{4.265}$$

Here

$$c_V = 1 \tag{4.266}$$

and

$$\varsigma_{\text{rot}} = \ln \frac{2I}{\hbar^2 \sigma} \qquad (4.267)$$

is the chemical constant for the rotational motion.

2. We examine the case when $T \ll T_0$, that is, for low temperature T or small rotational inertia. Then, practically all molecules are in the very lowest rotational state and all terms in the sum (4.258) (with $\sigma = 1$) beyond the first few are negligible. Using the first two terms

$$Z \approx 1 + 3\exp\left\{-\frac{\hbar^2}{IT}\right\} + \cdots \qquad (4.268)$$

From here

$$\ln\left(1 + 3\exp\left\{-\frac{\hbar^2}{IT}\right\}\right) \approx 3\exp\left\{-\frac{\hbar^2}{IT}\right\} \qquad (4.269)$$

as

$$\ln(1 + x) \approx x, \quad |x| \ll 1 \qquad (4.270)$$

The free energy F_{rot} in this case is approximately

$$F_{\text{rot}} = -3NT\exp\left\{-\frac{\hbar^2}{IT}\right\} \qquad (4.271)$$

and the energy E is

$$E_{\text{rot}} = F_{\text{rot}} + TS_{\text{rot}} = F_{\text{rot}} - T\frac{\partial F_{\text{rot}}}{\partial T} = -T^2\frac{\partial}{\partial T}\left(\frac{F_{\text{rot}}}{T}\right) = 3N\frac{\hbar^2}{I}\exp\left\{-\frac{\hbar^2}{IT}\right\} \qquad (4.272)$$

From this, the energy per molecule is

$$\varepsilon_{\text{rot}} = \frac{E_{\text{rot}}}{N} = 3\frac{\hbar^2}{I}\exp\left\{-\frac{\hbar^2}{IT}\right\} \qquad (4.273)$$

The entropy S_{rot} is

$$S_{\text{rot}} \cong 3\frac{N\hbar^2}{IT}\left(1 + \frac{IT}{\hbar^2}\right)\exp\left\{-\frac{\hbar^2}{IT}\right\} \qquad (4.274)$$

and the heat capacity C_V is

$$C_V \cong 3N\left(\frac{\hbar^2}{IT}\right)^2 \exp\left\{-\frac{\hbar^2}{IT}\right\} \qquad (4.275)$$

Thus as $T \to 0$, the specific heat and entropy tend toward zero:

$$C_V \to 0, \quad S \to 0 \qquad (4.276)$$

This is in agreement with the Nernst theorem. It means that at low temperatures a diatomic gas behaves as a monatomic gas. At low temperatures, the rotational and vibrational degrees of freedom are frozen.

In Equation 4.258, we do not consider the angular momentum component parallel to the axis of the molecule. This is because the rotational inertia about that axis is very small. Any state with such an angular momentum that is different from zero in analogy to Equation 4.258 has a very high energy compared to T and can then be neglected.

4.9 NUCLEAR SPIN EFFECTS

If we consider the case where the quasi-classical treatment of rotation is not applicable (e.g., the two isotopes of hydrogen, H_2 and deuterium D_2) at low temperatures, then we arrive at a complicated situation. The molecular quantum states depend on whether the nuclei are bosons or fermions. This leads to the involvement of nuclear spins. This nuclear spin effect is important for diatomic molecules with like atoms (homonuclear) as for the case of H_2 and D_2.

Consider a diatomic molecule with like atoms each having nuclear spin i. The nuclear spin causes hyperfine splitting of atomic levels. If this splitting is not observed, then the atomic levels have a degeneracy of $2i + 1$. The diatomic molecule's resultant total nuclear spin I may take values $2i, 2i - 1, \ldots, 1, 0$. These values are always integers.

Even more importantly, the nuclear spin is coupled to the symmetry of the wave function. For *integral* nuclear spins, the two nuclei are bosons, and the wave function must be *symmetric* with respect to their interchange. For *half-integral* nuclear spins, the two nuclei are fermions, and the wave function must be *antisymmetric* with respect to their interchange. Further, the total molecular wave function contains a product of a nuclear part and a rotational part, as well as an electronic part and a vibrational part. When considering the interchange of *only* the nuclei, the electronic part is unaffected, and the vibrational part is an even function of the nuclear separation, so it is also symmetric. The rotational part, however, is *symmetric* when the rotational quantum number J (total angular momentum) is *even*, and it is *antisymmetric* when J is *odd*. These properties come about because the spherical harmonics, which represent orbital angular momentum, have the property $Y_J^M(\theta, \varphi) = (-1)^J Y_J^M(\theta + \pi, \varphi)$, which is their symmetry under inversion through the origin. We need to couple the nuclear and rotational wave functions together with the correct symmetry, to get the right symmetry for the whole molecule.

Now suppose the nuclei are fermions and have half-integral spin i. This would be the case for an H_2 molecule ($i = 1/2$). The overall states of the molecule must be *antisymmetric* under nuclear interchange (fermions). Each nucleus has $2i + 1$ nuclear spin states; hence, there are a total of $(2i + 1)^2$ nuclear spin states for the entire molecule. But some of these states are symmetric and some are antisymmetric. The symmetric nuclear spin states must be combined with antisymmetric rotational states, and, the antisymmetric nuclear spin states must be combined with symmetric rotational states. For addition of half-integral spins, the nuclear spin states are symmetric when they have odd total spin, $I = 1,3,5,\ldots$, and antisymmetric when they have even total spin, $I = 0,2,4,\ldots$. Take the spin-1/2 case as an example. The total nuclear spin states can be denoted $|I, M\rangle$. There is an antisymmetric singlet state, with $I = 0$ and $M = 0$, usually written like

$$|0,0\rangle = \frac{1}{\sqrt{2}}\left(|\uparrow\rangle_1 |\downarrow\rangle_2 - |\downarrow\rangle_1 |\uparrow\rangle_2\right) \tag{4.277}$$

where the subscripts label the up/down states in which each particle is found. This state changes sign if 1 and 2 are interchanged, hence it is antisymmetric. There are also the three members of the spin triplet, with $I = 1$ and $M = 1, 0, -1$,

$$|1,1\rangle = |\uparrow\rangle_1 |\uparrow\rangle_2, \quad |1,0\rangle = \frac{1}{\sqrt{2}}\left(|\uparrow\rangle_1 |\downarrow\rangle_2 + |\downarrow\rangle_1 |\uparrow\rangle_2\right), \quad |1,-1\rangle = |\downarrow\rangle_1 |\downarrow\rangle_2 \tag{4.278}$$

Obviously, these are symmetric under interchange of the pair. This set of states is something like "parallel" nuclear spins, but with different orientations in space. The singlet is sometimes considered as "antiparallel," although these designations really cannot be taken seriously in the quantum world due to the uncertainty principle. For any general half-integral spins, the states of odd total spin are symmetric and those of even total spin are antisymmetric.

Then for half-integral spins, the even (odd) total nuclear spin values I are paired with the even (odd) values of J. We can count the total number of even (antisymmetric) nuclear spin states as

$$\text{number of antisymmetric nuclear states} = 1 + 5 + 9 + \cdots + \left[2(2i) - 1\right]$$
$$\text{for even total nuclear spin } I = 0, 2, 4, \ldots, 2i - 1 \tag{4.279}$$

The total number of odd (symmetric) nuclear spin states can be counted instead as

$$\text{number of symmetric nuclear states} = 3 + 7 + 11 + \cdots + \left[2(2i) + 1\right]$$
$$\text{for odd total nuclear spin } I = 1, 3, 5, \ldots, 2i \tag{4.280}$$

It can be left as an exercise for the reader to show that these sum of the degeneracies for even (antisymmetric) and odd (symmetric) nuclear spin states is

$$g_{\text{nuc}}^{\text{even}} = i(2i + 1) \quad \text{(antisymmetric, half-integral spin } i) \tag{4.281}$$

$$g_{\text{nuc}}^{\text{odd}} = (i + 1)(2i + 1) \quad \text{(symmetric, half-integral spin } i) \tag{4.282}$$

Thus, the relative degeneracies for the even and odd total nuclear spins are found by dividing by the total number of nuclear spin states, giving

$$g_g = \frac{i}{2i + 1}, \quad g_u = \frac{i + 1}{2i + 1} \quad \text{(half-integral spin } i) \tag{4.283}$$

Here, g and u denote even and odd, respectively, taken from the German *gerade* and *ungerade*. This is for half-integral spin i.

On the other hand, for integral spin i, the states of total nuclear spin $I = 2i, 2i - 2$, $2i - 4, \ldots$, which are all even numbers, are symmetric, and the states with total nuclear spin $I = 2i - 1, 2i - 3, 2i - 5, \ldots$, which are odd numbers, are antisymmetric. These nuclei are now bosons and must have total wave functions that are symmetric. So symmetric nuclear spin states must be paired with symmetric rotational states, and antisymmetric nuclear spin states must be paired with antisymmetric rotational states. Or, even (odd) nuclear spin states must be paired with the even (odd) rotational states to make symmetric total wave functions for boson nuclei. Now the total number of even (symmetric) nuclear spin states is

$$\text{number of symmetric nuclear states} = 1 + 5 + 9 + \cdots + [2(2i) + 1]$$
$$\text{for even total nuclear spin } I = 0, 2, 4, \ldots, 2i \tag{4.284}$$

The total number of odd (antisymmetric) nuclear spin states are counted as

$$\text{number of antisymmetric nuclear states} = 3 + 7 + 11 + \cdots + [2(2i) - 1]$$
$$\text{for odd total nuclear spin } I = 1, 3, 5, \ldots, 2i - 1 \tag{4.285}$$

These are found to sum to the results

$$g_{\text{nuc}}^{\text{even}} = (i + 1)(2i + 1) \quad \text{(symmetric, integral spin } i) \tag{4.286}$$

$$g_{\text{nuc}}^{\text{odd}} = i(2i + 1) \quad \text{(antisymmetric, integral spin } i) \tag{4.287}$$

The symmetric spin states are the so-called ortho states and the antisymmetric spin states the so-called para states. The ortho states are $g_{\text{nuc}}^{\text{even}}$-fold degenerate and the para states $g_{\text{nuc}}^{\text{odd}}$-fold degenerate. Thus, the relative degeneracy for the even and odd total nuclear spins are

$$g_g = \frac{i}{2i + 1}, \quad g_u = \frac{i + 1}{2i + 1} \tag{4.288}$$

This is for the half-integral spin i. For integral spin i we have

$$g_g = \frac{i + 1}{2i + 1}, \quad g_u = \frac{i}{2i + 1} \tag{4.289}$$

The even/odd formulas have switched for the integral and half-integral spins. However, the formulas are the same for integral and half-integral spins when identified with either the symmetric or antisymmetric nuclear spin states. Thus, we might define the relative degeneracies of symmetric and antisymmetric nuclear spin states:

$$g_{\text{nuc}}^{\text{sym}} = \frac{i + 1}{2i + 1}, \quad g_{\text{nuc}}^{\text{anti}} = \frac{i}{2i + 1} \quad \text{(all spins } i\text{)} \tag{4.290}$$

Consider hydrogen atom isotopes. The molecules with greater statistical weight are called ortho-hydrogen molecules. Molecules with lower nuclear statistical weight are called para-hydrogen molecules. For H_2 and D_2, the nuclear statistical sums are

$$H_2\left(i = \frac{1}{2}\right) \begin{cases} g_u = \dfrac{3}{4}, & \text{ortho} \\[2mm] g_g = \dfrac{1}{4}, & \text{para} \end{cases}, \quad D_2\left(i = 1\right) \begin{cases} g_g = \dfrac{2}{3}, & \text{ortho} \\[2mm] g_u = \dfrac{1}{3}, & \text{para} \end{cases} \tag{4.291}$$

We see that the molecules with higher statistical weight have symmetric nuclear spins, that is, the ortho form for both isotopes.

The rotational partition function for hydrogen molecules in equilibrium may take the form

$$Z_{\text{rot}} = g_g Z_g + g_u Z_u \tag{4.292}$$

where

$$Z_g = \sum_{J=0,2,\cdots} (2J + 1)\exp\left\{-\frac{\hbar^2}{2IT}J(J + 1)\right\},$$

$$Z_u = \sum_{J=1,3,\cdots} (2J + 1)\exp\left\{-\frac{\hbar^2}{2IT}J(J + 1)\right\} \tag{4.293}$$

The rotational free energy F_{rot} follows:

$$F_{rot} = -NT \ln \left(g_g Z_g + g_u Z_u \right) \tag{4.294}$$

For high temperatures

$$Z_g \cong Z_u \cong \frac{1}{2} Z_{rot} = \frac{TI}{\hbar^2} \tag{4.295}$$

This is the classical approximation for the free energy. Thus, at high temperatures, the specific heat of a diatomic molecule is unity and is independent of the nature of the atoms in the molecule. As $T \to 0$, the statistical sum Z_g tends to unity and Z_u tends exponentially to zero. Thus, as $T \to 0$, the gas behaves as monatomic $(C_V = \frac{3}{2}T)$ and to the chemical constant the term $\varsigma_{nuc} = \ln g_g$ is added.

The relative concentrations of the ortho and para in equilibrium are defined as follows:

$$n_{H_2} = \frac{N_{ortho\,H_2}}{N_{para\,H_2}} = \frac{g_u Z_u}{g_g Z_g} = \frac{3Z_u}{Z_g}, \quad n_{D_2} = \frac{N_{ortho\,D_2}}{N_{para\,D_2}} = \frac{g_g Z_g}{g_u Z_u} = \frac{2Z_g}{Z_u} \tag{4.296}$$

As $T \to 0$, all the molecules are found in the lowest rotational state with $J = 0$, and the H_2 gas is pure para while D_2 gas is pure ortho. For high temperatures when Equation 4.295 is satisfied, the relative concentration value is three for H_2 and two for D_2. For intermediate temperatures, we use Equation 4.296 to evaluate the thermodynamic properties of the gas. However, for equilibrium to hold, there must be transitions between the ortho and para states. D. M. Dennison [*Proc. Roy. Soc.* A115, 483, 1927] realized that these transitions take place at a very slow rate unless a catalyst is present. Thus, a gas prepared at room temperature and then cooled to low temperature may be in a metastable state with the relative ortho/para ratio closer to the room temperature value rather than the equilibrium value for the temperature being studied.

Consider varying the temperature from 0 to ∞. Then, in thermodynamic equilibrium, n_{H_2} would vary from 0 to 3 and n_{D_2} from 0 to 2. However, for practical purposes, the gases prepared at room temperature have $n_{H_2} = 3$ and $n_{D_2} = 2$ when brought to low temperatures. Real hydrogen and deuterium gases are a mixture of the ortho and para in the ratios 3:1 and 2:1, respectively. This implies that

$$N_{H_2} = \frac{3}{4} N_{ortho} + \frac{1}{4} N_{para}, \quad N_{D_2} = \frac{2}{3} N_{ortho} + \frac{1}{3} N_{para} \tag{4.297}$$

and correspondingly the heat capacity can be estimated as

$$C_{rotH_2} = \frac{3}{4} C_{ortho} + \frac{1}{4} C_{para}, \quad C_{rotD_2} = \frac{2}{3} C_{ortho} + \frac{1}{3} C_{para} \tag{4.298}$$

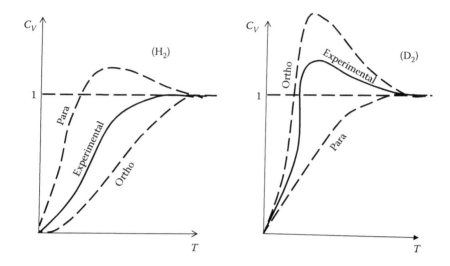

FIGURE 4.11   The variations of C_{ortho}, C_{para}, and C_{rot} with temperature for H_2 and D_2.

Figure 4.11a and b shows the evaluated values of C_{ortho}, C_{para}, and C_{rot} for H_2 and D_2.

The free energy of the mixture of para and ortho components is the sum of the energies of both components. For pure ortho-H_2 or para-D_2, we have

$$F_{rot} = -NT \ln g_u Z_u \qquad (4.299)$$

For low temperatures $\left(\hbar^2/2IT \gg 1\right)$, we need only the first term in the partition function:

$$Z_u = 3\exp\left\{-\frac{\hbar^2}{IT}\right\} \qquad (4.300)$$

The free energy becomes

$$F_{rot} = \frac{N\hbar^2}{I} - NT\ln 3g_u \qquad (4.301)$$

This implies that the gas behaves as a monatomic type ($C_{rot} = 0$). In the chemical constant there appears an additional term $\ln 3g_u$ and in the energy the term $N\hbar^2/I$. This corresponds to the rotational energy with all molecules having $J = 1$.

4.10   ELECTRONIC ANGULAR MOMENTUM EFFECT

Earlier we described diatomic molecules in their lowest electronic state. It was assumed that $L = S = 0$, where S is the spin angular momentum and L is the orbital angular momentum. Notwithstanding this, certain molecules in their lowest electronic state have

1. Nonzero orbital angular momentum $L \neq 0$
2. Nonzero spin $S \neq 0$
3. Both (1) and (2), that is, $L \neq 0$, $S \neq 0$

For case (3), there is a twofold degeneracy that corresponds to two possible directions of the orbital angular momentum relative to the molecular axis. This results in $g_e = 2$ and leads to the addition of the quantity $\varsigma_L = \ln 2$ to the chemical constant. Consider the condition $L = 0$, $S \neq 0$. This results in the splitting of $(2S + 1)$-levels. The interval of this fine structure is so small that in evaluation of thermodynamic quantities it may be neglected. For the condition of $L = 0$, $S \neq 0$ the levels are $g_S = (2S + 1)$-fold degenerate. The chemical constant is increased by $\varsigma_S = \ln(2S + 1)$.

Fine structure also occurs for $L \neq 0$, $S \neq 0$. The interval of the fine structure is generally of the order of T. This should be considered in the evaluation of thermodynamic quantities. We develop formulas for the case of a doublet fine structure. The electronic partition function is approximated as

$$Z_e = g_0 + g_1 \exp\left\{-\frac{\Delta}{T}\right\} \tag{4.302}$$

where g_0 and g_1 are the degeneracies of the components of the doublet and Δ is their separation. The electronic free energy F_e is

$$F_e = -NT \ln\left[g_0 + g_1 \exp\left\{-\frac{\Delta}{T}\right\}\right] \tag{4.303}$$

The electronic heat capacity c_e becomes

$$c_e = \frac{C_e}{N} = \frac{(\Delta/T)^2}{\left[1 + (g_0/g_1)\exp\{(\Delta/T)\}\right]\left[1 + (g_1/g_0)\exp\{-(\Delta/T)\}\right]} \tag{4.304}$$

This must be added to other parts of the heat capacity. One sees that as $T \to 0$ or $T \to \infty$, this electronic contribution $c_e \to 0$. When the temperature is the same order as the level spacing, $T \approx \Delta$, then $c_e \approx 1$, which should be the maximum value of the electronic heat capacity.

4.11   EXPERIMENT AND STATISTICAL IDEAS

The various gas laws, including Charles's law ($V \propto T$ for fixed pressure), Boyle's law ($PV = $ const for fixed temperature), and eventually the ideal gas law ($PV = Nk_BT$ for a given number of molecules), come out as consequences of the kinetic theory for gases with noninteracting molecules. The most logical and simple theory (i.e., kinetic theory) to get these laws requires the assumption of atoms or molecules in thermal motion bombarding the walls of the container. There is not really an acceptable theory of gas simply as a continuous fluid that would make the same connection between pressure, volume, and temperature as that contained in the ideal gas law. The gas constant and Boltzmann's constant are experimentally determined; however, we know that in terms of kinetic theory, they really only give a conversion from temperature in kelvin to the equivalent internal energy or thermal energy in joules.

4.11.1 Specific Heats

Measurements of specific heats of diverse materials have long been a standard for testing our statistical and atomic theories of matter. The reason is simple. If the theory cannot describe the heat capacity or specific heat per particle as observed in real experiments, then it means that something is missing in the understanding of the active degrees of freedom and their interactions in that system. Thus, in the example of oxygen gas (as diatomic O_2) versus monatomic helium gas, oxygen's specific heat is considerably higher (about $C_V \approx \frac{5}{2}R$) due primarily to the rotational degrees of freedom it has, that are absent for helium ($C_V \approx \frac{3}{2}R$). In addition, the vibrational modes in O_2 make an additional modification. Thus, the specific heat reflects indirectly what states of the system are physically accessible for the given temperature. For example, at low temperature, a quantum system does not have sufficient energy to access all but its lowest energy levels. We saw that vibrational modes of a diatomic molecule only get excited significantly when the temperature is well above $0.1\,\hbar\omega$.

The specific heat can also inform us about the interactions among the particles of the system. One can compare the real experimental C_p with that expected for a noninteracting model system. The interactions could have a number of effects on C_p, especially, leading to phase transitions exhibited as discontinuous changes in C_p (first-order transitions) or smoother but significant peaks in C_p (second-order transitions). At the fundamental level, these features in C_p are related significant changes in the physical structure of the system, whether that might be in locations of atom/molecules, or a magnetic structure, or some kind of angular arrangement, and so on. For example, a magnetic system at low temperature is likely to possess some kind of magnetic order in its accessible states, which could vanish either suddenly or smoothly as the temperature is increased through a critical point. Depending on the details of its internal interactions, there will likely be some kind of peak in the specific heat versus temperature.

We talk about C_p or C_V here as fundamental measures of the accessibility of the energy states. But it is good to keep in mind that there will always be other quantities, referred to as order parameters, whose thermal averages and thermal fluctuations will be important quantities to measure in experiments, as basic tests of the theory. Some simple examples of order parameters are the magnetization of magnetic materials (an example was given in Q4.10) and the electric polarization of dielectric media, both discussed in Chapter 7.

5 Quantum Statistics of Ideal Gases

5.1 MAXWELL–BOLTZMANN, BOSE–EINSTEIN, AND FERMI–DIRAC STATISTICS

In this section, we are concerned with systems consisting of particles with negligible mutual interaction, that is, *ideal gases*. Let us first consider a *classical case* where the particles are considered distinguishable. For this classical description, no symmetry requirements are imposed on the wave function when two particles are interchanged and any number of particles can be in the same single particle state, say, k. In this case the particles are said to obey *Maxwell–Boltzmann statistics* (abbreviated *MB statistics*). In the real world, however, identical particles are indistinguishable, so a quantum treatment is more correct. Usually, however, MB statistics is applied to systems that are nondegenerate, meaning the density of particles is so low that their quantum wave functions have insignificant overlap. Then it is unlikely that any two particles can even be in the same single-particle quantum state, and quantum symmetry constraints are not necessary.

For gases at high-enough density, or low-enough temperature, quantum degeneracy effects need to be accounted for. Here, the "gas" may be composed of atoms, molecules, photons, or perhaps nuclei or electrons in the core of a star. Or it could be composed from other elementary excitations that might be considered as particles. At high-enough densities, the wave functions from individual identical particles can be thought to have considerable overlap. This also becomes more likely as the temperature is lowered because the particles then have less kinetic energy and access to a lesser set of available states. Then the symmetry of the states needs to be accounted for, depending on whether the particles of the gas are bosons (integral spins) or fermions (half-integral spins). Recall that if the particles are bosons, then the system wave function is symmetrical with respect to any pair interchange, and any number of bosons may occupy a single-particle state. On the other hand, for fermions, the system wave function is antisymmetric with respect to a pair interchange, and no more than one identical fermion can occupy a single-particle state.

For counting the possible states of the gas and calculating the statistics and thermodynamic quantities, it does not matter which particle is in which single-particle state, but only on how many particles there are in each single-particle state, say, k. Then we will see that counting allowed boson states is very different from counting allowed fermion states.

The statistics for a boson gas is called *Bose–Einstein statistics* (abbreviated *BE statistics*). Because there is no restriction on the number of particles in an elementary state, BE statistics can lead to very unusual low-temperature macroscopic quantum effects, to be discussed below, grouped under the name of "Bose–Einstein condensation" that includes superconductivity and superfluidity. It is an impressive success of the theory that the symmetry requirements make these effects appear only when the particle spins are integral, such as for photons or helium-4 nuclei or electron pairs (Cooper pairs) in superconductors.

For a fermion gas, the statistics are called *Fermi–Dirac statistics* (abbreviated *FD statistics*). FD statistics also leads to a different kind of unusual low-temperature macroscopic quantum behavior that typically produces a quantum pressure to degenerate fermion gases. It is a direct consequence of the Pauli exclusion principle, where fermions effectively must repel each other to avoid occupying the same quantum states. This effect will be discussed below; it is further impressive that such a microscopic quantum pressure is responsible for determining the sizes of very large stellar objects such as white dwarf and neutron stars.

Generally, quantum statistics effects are expected at "low temperature," but we will see that the definition of low temperature is relative. From the physical point of view, the temperature is in the low (quantum) regime when the typical distance between particles is less than their typical de Broglie wavelength.

5.2 GENERALIZED THERMODYNAMIC POTENTIAL FOR A QUANTUM IDEAL GAS

Let us consider a quantum gas of identical particles in a volume V in equilibrium at a temperature T. Let k label the quantum states, E_k is the energy of a single particle in the state k, and n_k is the number of particles in the state k. The assumption of small interaction between particles enables us to write the total energy of the gas as

$$E = \sum_k E_k n_k \tag{5.1}$$

The fluctuating quantum state of the system is described by the collection of *occupation numbers* n_k, which we denote for simplicity as $n \equiv \{n_1, n_2, n_3, \ldots\}$. These sum up to the total number of particles, $N = \sum_k n_k$.

If we know the mean number of particles $\overline{n_k}$ for each quantum state k, then we may evaluate the mean energy $\overline{E}$:

$$\overline{E} = \sum_k E_k \overline{n_k} \tag{5.2}$$

We take the sum with respect to all quantum states k. Statistically, the system is completely described with the help of the mean number of particles in each state.

Because particle numbers are not conserved, the statistics is described with the help of the *grand canonical distribution*:

$$W_{N,E_n} = \exp\left\{\frac{\Omega + \sum_k n_k(\mu - E_k)}{T}\right\} \tag{5.3}$$

This depends on the chemical potential $\mu = \mu(p, T)$. From the normalization condition, we have

$$1 = \sum_{n,N} W_{N,E_n} = \exp\left\{\frac{\Omega}{T}\right\} \sum_{n,N} \exp\left\{\frac{\sum_k n_k(\mu - E_k)}{T}\right\} \tag{5.4}$$

From here, we have the thermodynamic potential

$$\Omega = -T \ln \sum_{n,N} \exp\left\{\frac{\sum_k n_k(\mu - E_k)}{T}\right\} \tag{5.5}$$

or

$$\Omega = -T \ln \sum_{n,N} \prod_k \exp\left\{\frac{n_k(\mu - E_k)}{T}\right\} = -T \ln \prod_k \sum_{n_k} \exp\left\{\frac{n_k(\mu - E_k)}{T}\right\} \tag{5.6}$$

This is the formula for the grand thermodynamic potential of a quantum ideal gas.

It follows from Equation 5.6 that the total potential is a sum over terms defined for each state

$$\Omega_k = -T \ln \sum_{n_k} \left[\exp\left\{\frac{\mu - E_k}{T}\right\}\right]^{n_k} \tag{5.7}$$

We can also define the grand canonical partition function Ξ for the entire system as

$$\Omega = -T \ln \Xi = -T \sum_k \ln \Xi_k \tag{5.8}$$

where the partition function for each state is a factor that is easy to evaluate

$$\Xi_k = \sum_{n_k} \left[\exp\left\{\frac{\mu - E_k}{T}\right\}\right]^{n_k} \tag{5.9}$$

Note that the sum over n_k is only over the allowed values for the type of quantum particles being considered.

5.3 FERMI–DIRAC AND BOSE–EINSTEIN DISTRIBUTIONS

We know already that the average occupation number is given by the following relation:

$$\bar{n}_k = -\frac{\partial \Omega_k}{\partial \mu} \tag{5.10}$$

Dirac showed that particles with half-integral spin angular momentum, $S = \frac{1}{2}, \frac{3}{2}, \frac{5}{2}, \ldots$, obey the Pauli exclusion principle, where one quantum state cannot be occupied by more than one particle. In a state with a given momentum $\vec{p}$ and spin $\vec{S}$, we may have only one or no particle. It follows that $n_k = 0, 1$. The particles are considered indistinguishable so that just specifying the numbers $n_k = 0, 1$ for all allowed k is enough to specify the state of the gas. Some examples are some elementary particles with masses different from zero such as electrons, protons, and so on. Fermi did the concrete evaluation of $\overline{n_k}$ and as a consequence, we call the statistics Fermi–Dirac statistics. From the Pauli exclusion principle and $n_k = 0, 1$, and Equation 5.7, we have

$$\Omega_k = -T \ln\left(1 + \exp\left\{\frac{\mu - E_k}{T}\right\}\right) \tag{5.11}$$

Using Equation 5.10 gives

$$\overline{n_k} = -\frac{\partial \Omega_k}{\partial \mu} = \frac{1}{\exp\{(E_k - \mu)/T\} + 1} \tag{5.12}$$

This is the Fermi–Dirac distribution function. It could be derived alternatively by evaluating Ξ_k and finding the expectation value of n_k, summing over only the two possibilities $n_k = 0, 1$. If $E_k \gg \mu$, then $\overline{n_k} \approx \exp\{-(E_k/T)\}$, which can approach zero for high-energy states. In this limit, the FD distribution carries over into Maxwell–Boltzmann statistics. Also, in Equation 5.12, the denominator can never be less than unity. Then, for fermions, we have the limits

$$0 \leq \overline{n_k} \leq 1 \tag{5.13}$$

This relation properly reflects the Pauli exclusion principle.

Equation 5.12 gives the mean occupation number for fermions in the kth quantum state for an equilibrium system at temperature T. The Fermi–Dirac distribution is normalized using

$$N = \sum_k \overline{n_k} = \sum_k \frac{1}{\exp\{(E_k - \mu)/T\} + 1} = N\left(\frac{\mu}{T}\right) \tag{5.14}$$

This equality implicitly defines the chemical potential $\mu(p, T)$ as a function of temperature T and total number of particles N in the gas. Given the set of values E_k and fixed temperature, the RHS of Equation 5.14 depends only upon μ, which therefore must be selected to yield the appropriate number of particles. A connection between μ, N, and the pressure $p = f((N/V), T)$ is implied through the Gibbs–Duhem relation (2.345). The thermodynamic potential Ω of the entire gas is then found as

$$\Omega = -T\sum_k \ln\left(1 + \exp\left\{\frac{\mu - E_k}{T}\right\}\right) \tag{5.15}$$

Consider instead the Bose–Einstein distribution. The state of particles in this case is described by symmetric wave functions. This distribution applies to particles with integral spin angular momentum, $S = 0, 1, 2, \ldots$. The occupation number n_k of the kth quantum state for symmetric wave functions is not bounded and may have arbitrary values $n_k = 0, 1, 2, \ldots$. Here, the particles are also considered as indistinguishable such that just specifying the numbers $n_k = 0, 1, 2, \ldots$ is enough to specify the state of the gas. Examples are photons, phonons, ^{4}He atoms, and so on.

If we consider Equations 5.7 and 5.9, for bosons

$$\Omega_k = -T\ln\sum_{n_k}\left[\exp\left\{\frac{\mu - E_k}{T}\right\}\right]^{n_k} = -T\ln\Xi_k \tag{5.16}$$

then there results a geometric series for the terms in the grand canonical partition function

$$\Xi_k = \sum_{n_k=0}^{\infty}\left[\exp\left\{\frac{\mu - E_k}{T}\right\}\right]^{n_k} = \frac{1}{1 - \exp\left\{(\mu - E_k)/T\right\}} \tag{5.17}$$

For the geometric series to be convergent, we require $\mu \leq 0$. It follows that

$$\Omega_k = T\ln\left[1 - \exp\left\{\frac{\mu - E_k}{T}\right\}\right] \tag{5.18}$$

which gives

$$\overline{n_k} = -\frac{\partial\Omega_k}{\partial\mu} = \frac{1}{\exp\left\{(E_k - \mu)/T\right\} - 1} \tag{5.19}$$

This is the distribution function satisfying Bose–Einstein statistics or more simply a Bose gas. Note that in this case, n_k can become very large. Including the multiplicity of degeneracy g, both the FD and BE statistics can be combined into one formula

$$n_{k\,\text{BE(FD)}} = \frac{g}{\exp\left\{(E_k - \mu)/T\right\} \pm 1} \tag{5.20}$$

where the plus sign is FD and the minus sign is BE statistics.

The total number of particles N in the Bose gas may be obtained similar to Equation 5.14:

$$N = \sum_k \frac{1}{\exp\{(E_k - \mu)/T\} - 1} \tag{5.21}$$

For practical application, however, this implicitly determines the chemical potential if the number of particles is known (or if the number of particles per unit volume N/V is given). When the volume density of particles is low, this causes the chemical potential to be large and negative. This is the classical limit of the distribution.

From Equation 5.12, it is obvious that $\bar{n}_{kFD} \leq 1$ and from Equation 5.19, we may have $\bar{n}_{kBE} \gg 1$. The thermodynamic potential for bosons is

$$\Omega = T \sum_k \ln\left[1 - \exp\left\{\frac{\mu - E_k}{T}\right\}\right] \tag{5.22}$$

For states at high energy, $E_k \gg \mu$, or low-enough chemical potential, the occupation number for either FD or BE gases is approximated as

$$\bar{n}_k \approx \exp\left\{-\frac{(E_k - \mu)}{T}\right\} \ll 1 \tag{5.23}$$

which again recovers the Maxwell–Boltzmann distribution. This is the classical limit of both quantum distributions. It applies at high temperature, where the chemical potential is large and negative. This causes the occupation of the states to be very low; then, the quantum symmetry restrictions are relaxed because it is very unlikely for more than one particle to occupy any state k. When using either type of quantum distribution, it is important first to check whether quantum statistics is truly needed, or whether the Boltzmann approach will be valid.

5.4 ENTROPY OF NONEQUILIBRIUM FERMI AND BOSE GASES

We take a look at the statistical properties of quantum gases from a different viewpoint, using ideas from the classical phase space. This might be considered as a quasi-classical approach, in contrast to the assignment of an abstract index k to label the states. In essential aspects, it leads to the same results we found earlier. It is based on analyzing the occupation of states within the available phase space. An arbitrary macroscopic state of the system is assumed, that could be out of equilibrium.

5.4.1 FERMI GAS

The Boltzmann formula connects the entropy S and the statistical weight $\Delta\Gamma$:

$$S = \ln \Delta\Gamma \tag{5.24}$$

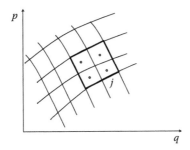

FIGURE 5.1 Different groups of quantum states $j = 1, 2, \ldots$ in a selected domain of phase space.

The mean occupation number in a nonequilibrium state defines the macrostate:

$$n_j = \frac{N_j}{G_j} \tag{5.25}$$

where $j = 1, 2, \ldots$ are different groups of quantum states. G_j is the number of quantum states in a selected domain of phase space in Figure 5.1 and N_j is the number systems of the ensemble in the jth group. N points define the state of the entire ensemble for the gas. A cell is meant to indicate the volume of phase space that corresponds to one available single-particle quantum state.

The domain of phase space is selected in such a way that the energy should remain constant. An individual quantum state has a cell of volume $(2\pi\hbar)^s$, for a phase space in s dimensions. Then, a small domain j has the number of quantum states or cells equal to

$$G_j = \frac{\Delta q \Delta p}{(2\pi\hbar)^s} \tag{5.26}$$

Let us find the number of microstates in order to evaluate $\Delta\Gamma$. If we consider Fermi particles, then the inequality $N_j \leq G_j$ must be satisfied. We need to find the number of possible ways for the distribution of N_j indistinguishable particles over G_j states (where no quantum state or cell can accommodate more than one particle). The number of ways to place N_j particles in G_j states with no more than one particle per state involves taking the combinations of G_j elements N_j at a time (see Figure 5.2). The number of ways of placing N_j indistinguishable particles in G_j distinguishable cells with a restriction of not more than one particle per cell ($N_j \leq G_j$) is

$$\Delta\Gamma_j = \frac{G_j!}{N_j!(G_j - N_j)!} \tag{5.27}$$

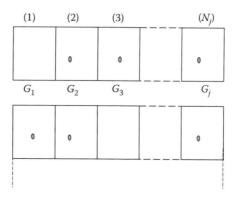

FIGURE 5.2 Some different ways of placing N_j particles in G_j states with no more than one particle per state.

Here, the $N_j!$ term in the denominator accounts for particle indistinguishability. The total number of ways of obtaining an arbitrary particle distribution then becomes

$$\Delta\Gamma = \prod_j \Delta\Gamma_j \tag{5.28}$$

and denotes the generic number of microstates per macrostate for Fermi–Dirac statistics. Hence, the entropy of the (nonequilibrium) system is

$$S = \ln\Delta\Gamma = \sum_j \ln\Delta\Gamma_j = \sum_j \ln\left[\frac{G_j!}{N_j!(G_i - N_j)!}\right] \tag{5.29}$$

This may be evaluated using the values of $\Delta\Gamma_j$ for different systems. We use Stirling's formula for these evaluations when N_j is very large

$$N_j! = \sqrt{2\pi N_j}\,N_j^{N_j}\exp\{-N_j\},\ N_j \gg 1 \tag{5.30}$$

Then, considering that $n_j = N_j/G_j$, we have

$$S = \sum_j \left[G_j \ln\frac{G_j}{e} - N_j \ln\frac{N_j}{e} - (G_j - N_j)\ln\frac{G_j - N_j}{e}\right]$$

$$= \sum_j G_j \left[\ln\frac{G_j}{e} - \frac{N_j}{G_j}\ln\left(\frac{N_j}{G_j}\frac{G_j}{e}\right) - \left(1 - \frac{N_j}{G_j}\right)\ln\left(\frac{G_j - N_j}{G_j}\frac{G_j}{e}\right)\right]$$

$$= \sum_j G_j \left[\ln\frac{G_j}{e} - n_j \ln\left(n_j \frac{G_j}{e}\right) - (1 - n_j)\ln\left((1 - n_j)\frac{G_j}{e}\right)\right]$$

$$= \sum_j G_j \left[\ln \frac{G_j}{e} - n_j \ln n_j - n_j \ln \frac{G_j}{e} - (1 - n_j) \ln(1 - n_j) - \ln \frac{G_j}{e} + n_j \ln \frac{G_j}{e} \right]$$

$$= \sum_j G_j \left[-n_j \ln n_j - (1 - n_j) \ln(1 - n_j) \right]$$

$$= -\sum_j G_j \left[n_j \ln n_j + (1 - n_j) \ln(1 - n_j) \right] \tag{5.31}$$

Thus

$$S = -\sum_j G_j \left[n_j \ln n_j + \left(1 - n_j \right) \ln \left(1 - n_j \right) \right] \tag{5.32}$$

This is the entropy for a nonequilibrium system.

If j is a continuous label for states in phase space, then instead, the general expression for the entropy is

$$S = -\frac{1}{(2\pi\hbar)^s} \int dq dp [n \ln n + (1 - n) \ln(1 - n)] \tag{5.33}$$

Finally, it is usual that the number of particles and total energy are to be constrained in the system. We want to know the entropy when N and E have been assigned desired thermodynamic values. This corresponds to finding the extremum of S under these constraints. If we use the method of Lagrange's undetermined multipliers, we apply

$$\frac{\partial}{\partial n_j} (S + \alpha N - \beta E) = 0 \tag{5.34}$$

where α and β are the multipliers and the constrained quantities are expressed

$$N = \sum_j N_j = \sum_j G_j n_j, \quad E = \sum_j E_j N_j = \sum_j E_j G_j n_j \tag{5.35}$$

Then we find that occupation numbers

$$n_j = \frac{1}{\exp\{\beta E_j - \alpha\} + 1} \tag{5.36}$$

will maximize the entropy. This defines the equilibrium distribution. It coincides with the Fermi–Dirac distribution, once α and β are properly identified and related to the chemical potential and the temperature. For comparison, see the derivations for classical statistics following Equations (3.295) and (4.58).

5.4.2 BOSE GAS

We now discuss the case of Bose statistics, in which any number of particles is allowed to occupy any one-particle state (Figure 5.3).

Thus, the statistical weight $\Delta\Gamma_j$ is the number of all possible ways to arrange N_j particles in G_j states or cells as in Figure 5.3. In this case, $G_j \geq N_j$. We introduce a direct but slightly abstract way of evaluating the contribution from the jth group of the distribution as follows. In the jth group, there are G_j states containing N_j identical points (that represent particles) with no restriction on occupation numbers. We represent a typical microstate as in Figure 5.3 by $G_j + 1$ partitions and N_j crosses. The partitions represent divisions between G_j states, and the crosses represent an occupying particle. We may obtain new microstates representing the same distribution, that is, the same value of N_j for the group by shuffling the partitions and crosses in Figure 5.3. There are G_j cells (states) and $G_j + 1$ partitions, but only $G_j - 1 = G_j + 1 - 2$ partitions contained inside the system.

We want to examine how the particles should be distributed in the cells such that

$$n_j = \frac{N_j}{G_j} \tag{5.37}$$

The statistical weight $\Delta\Gamma_j$ is the number of distinct ways in which N_j particles may be distributed in G_j states such that each state can contain any number of particles: 0, 1, 2, 3, ... In our evaluation, we view N_j particles to be arranged in positions and separated by $G_j - 1$ movable partitions. The total number of positions that are occupied by the particles and partitions is $G_j + N_j - 1$. Out of this number, N_j positions are selected for the particles. The permutations of the particles do not change a state because the particles are indistinguishable. In addition, the permutations of the partitions among themselves do not change the states. Then, the statistical weight $\Delta\Gamma_j$ is the number of ways of placing N_j indistinguishable particles in G_j distinguishable cells with no restriction on the number of particles per cell, which is

$$\Delta\Gamma_j = \frac{(G_j + N_j - 1)!}{(G_j - 1)!N_j!} \tag{5.38}$$

Because each energy level represents an independent event, the total number of ways of obtaining an arbitrary particle distribution becomes

$$\Delta\Gamma = \prod_j \Delta\Gamma_j \tag{5.39}$$

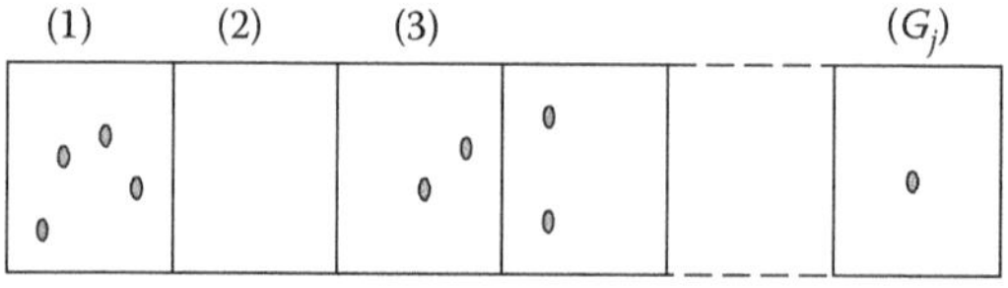

FIGURE 5.3 One of many possible ways of arranging N_j particles in G_j states.

This identifies the generic number of microstates per macrostate for Bose–Einstein statistics.

If we assume that the numbers $G_j + G_j$ and G_j are exceedingly large, then

$$\Delta\Gamma_j = \frac{(G_j + N_j)!}{(G_j - 1)!N_j!} \tag{5.40}$$

Then, applying Stirling's formula, the entropy is

$$S = \sum_j \ln\Delta\Gamma_j = \sum_j \ln\left[\frac{(G_j + N_j)!}{(G_j - 1)!N_j!}\right]$$

$$= \sum_j \left[(G_j + N_j)\ln\frac{G_j + N_j}{e} - G_j \ln\frac{G_j}{e} - N_j \ln\frac{N_j}{e}\right] \tag{5.41}$$

or

$$S = \sum_j\left[(G_j + N_j)\ln(G_j + N_j) - (G_j + N_j) - G_j \ln G_j + G_j - N_j \ln N_j + N_j\right]$$

$$= \sum_j\left[(G_j + N_j)\ln(G_j + N_j) - G_j \ln G_j - N_j \ln N_j\right] \tag{5.42}$$

and using Equation 5.37

$$S = \sum_j G_j\left[(1 + n_j)\ln(1 + n_j) - n_j \ln n_j\right] \tag{5.43}$$

This is the entropy of an arbitrary nonequilibrium state of a Bose gas. We may use Lagrange's multipliers to show that the condition of the maximum of the entropy (5.43) brings us to the Bose distribution:

$$\frac{\partial}{\partial n_j}(S + \alpha N - \beta E) = 0 \tag{5.44}$$

The constrained quantities are

$$N = \sum_j N_j = \sum_j G_j n_j, \quad E = \sum_j E_j N_j = \sum_j E_j G_j n_j \tag{5.45}$$

Then, there results

$$n_j = \frac{1}{\exp\{\beta E_j - \alpha\} - 1} \tag{5.46}$$

for all values of j. When the occupation numbers obey this rule, the entropy is maximized, and the system is in equilibrium. Equation 5.46 is called the Bose–Einstein distribution law.

If in Equations 5.32 and 5.43 we have $N_j \ll G_j$, then we arrive at the Boltzmann case, corresponding to very small occupation of any quantum state:

$$S = \sum_j G_j \ln \frac{eN_j}{G_j} \tag{5.47}$$

Q5.1 Five independent particles may have access to two energy levels in a thermo-dynamic assembly. A particular particle distribution for the given system and the associated degeneracies for each energy level are as follows: (1) $N_1 = 2$, $G_1 = 4$; (2) $N_2 = 3$, $G_2 = 6$. Determine the number of microstates for the given macrostate if the particles are (a) fermions and (b) bosons.

A5.1 a. The number of ways that a single energy level can be occupied is

$$\Delta \Gamma_j = \frac{G_j!}{N_j!(G_j - N_j)!} \tag{5.48}$$

Thus, for each energy level, $\Delta\Gamma_1 = \frac{4!}{2!2!} = 6$ and $\Delta\Gamma_2 = \frac{6!}{3!3!} = 20$. Consequently, the number of microstates for the given macrostate is $\Delta\Gamma = \Delta\Gamma_1 \times \Delta\Gamma_2 = 120$.

b. The number of ways that a single energy level can be occupied is

$$\Delta\Gamma_j = \frac{(G_j + N_j - 1)!}{(G_j - 1)!N_j!} \tag{5.49}$$

Thus, for each energy level $\Delta\Gamma_1 = \frac{5!}{2!3!} = 10$ and $\Delta\Gamma_2 = \frac{8!}{3!5!} = 56$. As a result, the number of microstates for the given macrostate is $\Delta\Gamma = \Delta\Gamma_1 \times \Delta\Gamma_2 = 560$.

5.5 THERMODYNAMIC FUNCTIONS FOR QUANTUM GASES

Among other quantities, an important property for quantum gases is their density, $\rho = N/V$. To that end, let us describe how to count the number of particles in states, depending on the energies of the states. From the general structure of quantum gases, the mean state occupation numbers are

$$n_k = \frac{1}{\exp\{(E_k - \mu)/T\} + \delta} \tag{5.50}$$

where

$$\delta = \begin{cases} 1, & \text{Fermi} \\ -1, & \text{Bose} \\ 0, & \text{Boltzmann} \end{cases} \tag{5.51}$$

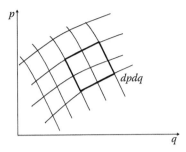

FIGURE 5.4   An element of volume in phase space.

Consider a phase space (q, p) as in Figure 5.4. We select a small volume $dqdp$ in phase space. Let dN be the number of particles in the volume $dqdp$:

$$dN(q, p) = \frac{dq d\vec{p}}{(2\pi\,\hbar)^3}\, g\bar{n}(q, p) \tag{5.52}$$

where g is the multiplicity of degeneracy. But $dq = dxdydz = dV$ and then

$$dN(p) = \frac{gVd\vec{p}}{(2\pi\,\hbar)^3}\, \frac{1}{\exp\{(E - \mu)/T\} + \delta} \tag{5.53}$$

We count the number of particles found within the momentum interval

$$\vec{p} \to \vec{p} + d\vec{p} : d\vec{p} = dp_x dp_y dp_z = p^2 dp\, d\varphi\, d(\cos\theta) \tag{5.54}$$

Suppose the particles have a nonrelativistic kinetic energy or dispersion relation. For any direction of the momentum p if the energy, $E = p^2/2m$, is independent of the angles, then in the spherical coordinates the relation in Equation 5.53 becomes

$$dN(p) = \frac{4\pi\, gV}{(2\pi\,\hbar)^3}\, \frac{p^2 dp}{\exp\{(E - \mu)/T\} + \delta} \tag{5.55}$$

If we consider the interval of the energy $E \to E + dE$ and $p = \sqrt{2mE}$, then relation (5.55) yields

$$dN(E) = \frac{gVm^{3/2}}{\sqrt{2}\pi^2\hbar^3}\, \frac{E^{1/2}dE}{\exp\{(E - \mu)/T\} + \delta} \tag{5.56}$$

This is the number of particles in the state with a defined energy in the interval $E \rightarrow E + dE$. If we integrate Equation 5.56, then we obtain the total number of particles in the gas:

$$N = \sum_k n_k = \frac{gVm^{3/2}}{\sqrt{2}\pi^2\hbar^3} \int_0^\infty \frac{E^{1/2}dE}{\exp\{(E-\mu)/T\} + \delta} \tag{5.57}$$

Note, however, that if the number of particles per unit volume is already known, then this gives an implicit relation for determining the chemical potential, μ.

The *Fermi function* is defined as follows:

$$F_m\left(\frac{\mu}{T}\right) \equiv A_m \int_0^\infty \frac{z^m dz}{\exp\{z-(\mu/T)\} + 1} \tag{5.58}$$

It is a tabular function. A_m is a normalization constant. We apply the Fermi function in Equation 5.58 as follows:

$$N = \frac{gVm^{3/2}}{\sqrt{2}\pi^2\hbar^3} \int_0^\infty \frac{E^{1/2}dE}{\exp\{(E-\mu)/T\} + 1} = \frac{gVm^{3/2}T^{3/2}}{\sqrt{2}\pi^2\hbar^3} \int_0^\infty \frac{z^{1/2}dz}{\exp\{z-(\mu/T)\} + 1}$$

$$= \frac{gVm^{3/2}T^{3/2}\sqrt{\pi}}{2^{3/2}\pi^2\hbar^3} F_{1/2}\left(\frac{\mu}{T}\right) \tag{5.59}$$

where $z = E/T$ and

$$F_{1/2}\left(\frac{\mu}{T}\right) = \frac{2}{\sqrt{\pi}} \int_0^\infty \frac{z^{1/2}dz}{\exp\{z-(\mu/T)\} + 1} \tag{5.60}$$

Then the total number of particles in the interval $E \rightarrow E + dE$ is

$$N = \frac{gVm^{3/2}T^{3/2}\sqrt{\pi}}{2^{3/2}\pi^2\hbar^3} F_{1/2}\left(\frac{\mu}{T}\right) \tag{5.61}$$

Suppose the gas is nondegenerate (a nondegenerate gas obeys classical statistics), then, $\mu < 0$ and

$$\exp\left\{-\frac{\mu}{T}\right\} \gg 1 \tag{5.62}$$

From this condition, the Fermi function in Equation 5.60 is approximated as

$$F_{1/2}\left(\frac{\mu}{T}\right) = \frac{2}{\sqrt{\pi}}\exp\left\{\frac{\mu}{T}\right\}\int_0^\infty \exp\{-z\}z^{1/2}dz = \exp\left\{\frac{\mu}{T}\right\} \tag{5.63}$$

Then, the total number of particles in Equation 5.61, for a classical gas, now becomes

$$N = \frac{gVm^{3/2}T^{3/2}\sqrt{\pi}}{2^{3/2}\pi^2\hbar^3}\exp\left\{\frac{\mu}{T}\right\} \tag{5.64}$$

Consider finding the formula for the total energy E of the quantum gas, based on

$$\bar{E} = \sum_k E_k \bar{n}_k \tag{5.65}$$

If we change from summation to integration

$$\bar{E} = \frac{gVm^{3/2}}{\sqrt{2}\pi^2\hbar^3}\int_0^\infty \frac{E^{3/2}dE}{\exp\{(E-\mu)/T\} + \delta} \tag{5.66}$$

This expression is true for $E = p^2/2m$, but it is not true for a photon gas, whose energy dispersion is relativistic, $E = cp$.

If we consider the procedures that yield the relations in Equations 5.57 and 5.66, then we obtain the *generalized thermodynamic potential* $\Omega_{FD,BE}$:

$$\Omega_{FD,BE} = \mp T\sum_k \ln\left(1 \pm \exp\left\{\frac{\mu - E}{T}\right\}\right)$$

$$= \mp T\frac{gVm^{3/2}}{\sqrt{2}\pi^2\hbar^3}\int_0^\infty \ln\left(1 \pm \exp\left\{\frac{\mu - E}{T}\right\}\right)E^{1/2}dE \tag{5.67}$$

Integrating by parts yields

$$\Omega_{FD,BE} = -\frac{2}{3}\frac{gVm^{3/2}}{\sqrt{2}\pi^2\hbar^3}\int_0^\infty \frac{E^{3/2}dE}{\exp\{(E-\mu)/T\} \pm 1} = -\frac{2}{3}\bar{E} \tag{5.68}$$

as

$$\Omega = -pV = -\frac{2}{3}\bar{E} \tag{5.69}$$

then

$$pV = \frac{2}{3}\bar{E} \qquad (5.70)$$

The expression (5.70) for the equation of state is true for any gas in any state of degeneracy. For example, the energy of the classical ideal Boltzmann gas is

$$\bar{E} = \frac{3}{2}NT \qquad (5.71)$$

If we substitute this in Equation 5.70, then we have the classical ideal gas law

$$pV = NT \qquad (5.72)$$

Now let us develop a formula for adiabatic processes. We can represent N and Ω, considering Equations 5.61 and 5.66, as follows:

$$N = \frac{gVm^{3/2}T^{3/2}\sqrt{\pi}}{2^{3/2}\pi^2\hbar^3}F_{1/2}\left(\frac{\mu}{T}\right) = VT^{3/2}f_1\left(\frac{\mu}{T}\right) \qquad (5.73)$$

$$\Omega = -\frac{2}{3}\frac{gVm^{3/2}}{\sqrt{2}\pi^2\hbar^3}T^{5/2}\int_0^\infty \frac{z^{3/2}dz}{\exp\{z-(\mu/T)\}\pm 1} = VT^{5/2}f_2\left(\frac{\mu}{T}\right) \qquad (5.74)$$

Then the entropy is as follows:

$$S = -\frac{\partial\Omega}{\partial T} = \frac{5}{2}VT^{3/2}f_2\left(\frac{\mu}{T}\right) + VT^{5/2}\left(-\frac{\mu}{T^2}\right)f_2'\left(\frac{\mu}{T}\right) = VT^{3/2}\phi\left(\frac{\mu}{T}\right) \qquad (5.75)$$

The entropy per particle is

$$\frac{S}{N} = \frac{\phi(\mu/T)}{f_1(\mu/T)} \qquad (5.76)$$

For an adiabatic process

$$\frac{S}{N} = \text{const} \qquad (5.77)$$

Then it follows that

$$\frac{\mu}{T} = \text{const} \tag{5.78}$$

Then, from Equation 5.75, we have

$$S = CVT^{3/2} = \text{const}, \quad C = \text{const} \tag{5.79}$$

or

$$VT^{3/2} = \text{const} \tag{5.80}$$

Using the thermodynamic potential

$$\Omega = -pV = VT^{5/2} f_2\left(\frac{\mu}{T}\right) \tag{5.81}$$

it follows that

$$pV^{5/3} = \text{const} \tag{5.82}$$

and also

$$\frac{T^{5/2}}{p} = \text{const} \tag{5.83}$$

These equalities coincide with the Poisson adiabatic equation for an ordinary (i.e., nonrelativistic) monatomic gas, with $\gamma = C_p/C_v = \frac{5}{3}$:

$$TV^{\gamma-1} = \text{const}, \quad pV^{\gamma} = \text{const} \tag{5.84}$$

5.6 PROPERTIES OF WEAKLY DEGENERATE QUANTUM GASES

5.6.1 FERMI ENERGY

For a nonrelativistic degenerate gas, we have the following thermodynamic potential:

$$\Omega_{\text{FD,BE}} = -\frac{2}{3}\frac{gVm^{3/2}}{\sqrt{2\pi^2\hbar^3}} \int_0^\infty \frac{E^{3/2}dE}{\exp\{(E-\mu)/T\} \pm 1} \tag{5.85}$$

For a weakly degenerate gas, we have the condition:

$$\exp\left\{-\frac{\mu}{T}\right\} > 1 \tag{5.86}$$

It follows that

$$\Omega_{\text{FD,BE}} \cong -\frac{2}{3}\frac{gVm^{3/2}}{\sqrt{2\pi^2\hbar^3}}\int_0^\infty \exp\left\{\frac{\mu - E}{T}\right\}\left(1 \mp \exp\left\{\frac{\mu - E}{T}\right\}\right)E^{3/2}dE \tag{5.87}$$

The number of quantum states of the translational motion of the particle with the absolute value of the momentum in the interval $p \to p + dp$ is equal to

$$\frac{4\pi p^2 dp V}{(2\pi\hbar)^3} \tag{5.88}$$

and the total number of quantum states is

$$\frac{4\pi p^2 dp V g}{(2\pi\hbar)^3} \tag{5.89}$$

Now consider a Fermi gas. The Fermi distribution with respect to the quantum states is

$$n_F = \frac{1}{\exp\{(E - \mu)/T\} + 1} \tag{5.90}$$

We may easily determine the value of μ in the limit of low temperatures. This is sketched out as follows. As $T \to 0$, there are *no thermal fluctuations (excitations)* and the function n_F approaches a step function, being equal to 1 for all $E < \mu$ and equal to 0 for all $E > \mu$ (see Figure 5.5). It is seen that for $T = 0$, we have an abrupt cut off when the energy passes the value $E = \mu$, which we call the *Fermi energy*, $E_F = \mu$. The Fermi–Dirac distribution is thus considered a *Fermi sea*. From Figure 5.5, at $T = 0$, all single particle states with $E < \mu$ are completely occupied with one particle per state, satisfying the Pauli exclusion principle. All single particle states with $E > \mu$ are empty. It can be seen that the Fermi energy E_F is the energy of the topmost occupied level when $T = 0$. Applied exactly at the Fermi energy, the formula gives $n_F(E_F) = \frac{1}{2}$.

When the energy states are viewed in phase space, the highest energy occupied levels map out a surface in the phase space, called the Fermi surface. Let us assume the Fermi surface has a spherical geometry in the momentum space of the gas particles. This is the case when the kinetic energy obeys the nonrelativistic

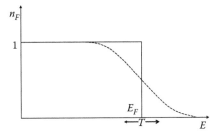

FIGURE 5.5 The Fermi distribution over the energy.

form, $E = p^2/2m$. We denote the radius of the highest energy occupied states by p_F, called the *Fermi momentum*. From the energy relation, the Fermi energy and Fermi momentum are simply related by $p_F = \sqrt{2mE_F}$. Then it follows that the number of particles filling all states with momentum from $0 \to p_F$, that is, with energy below the Fermi energy, $E < E_F = \mu$, based on Equation 5.89, is

$$N = \frac{gV}{2\pi^2\hbar^3} \int_0^{p_F} p^2 dp = \frac{gVp_F^3}{6\pi^2\hbar^3} \tag{5.91}$$

From here, the Fermi momentum p_F is

$$p_F = \left(\frac{6\pi^2}{g}\right)^{1/3} \left(\frac{N}{V}\right)^{1/3} \hbar \tag{5.92}$$

and the bounding Fermi energy E_F is

$$E_F = \frac{p_F^2}{2m} = \left(\frac{6\pi^2}{g}\right)^{2/3} \left(\frac{N}{V}\right)^{2/3} \frac{\hbar^2}{2m} \tag{5.93}$$

The Fermi energy E_F has a simple thermodynamic sense, and is directly determined by the number density of the quantum particles, and their mass. Commonly, it is applied to spin-1/2 particles (electrons, neutrons, etc.), in which case the multiplicity is $g = 2$.

The total energy of the Fermi gas, E, can be obtained using

$$dE = \frac{4\pi p^2 dp Vg}{(2\pi\hbar)^3} E = \frac{4\pi p^2 dp Vg}{(2\pi\hbar)^3} \frac{p^2}{2m} \tag{5.94}$$

Momentum integration (from 0 to the Fermi momentum p_F) leads to

$$E = \frac{gV}{4m\pi^2\hbar^3} \int_0^{p_F} p^4 dp = \frac{gVp_F^5}{20m\pi^2\hbar^3} = \frac{3}{10}\left(\frac{6\pi^2}{g}\right)^{2/3}\frac{\hbar^2}{m}\left(\frac{N}{V}\right)^{2/3} N = \frac{3}{5}\mu N \tag{5.95}$$

The energy per particle is proportional to the Fermi energy

$$\bar{E} = \frac{E}{N} = \frac{3}{5}\mu = \frac{3}{5}E_F \tag{5.96}$$

This is the mean energy per particle for the case of a degenerate Fermi gas. From Equation 5.70

$$pV = \frac{2}{3}E = \frac{2}{3}\frac{3}{5}\mu N = \frac{2}{3}\frac{3}{10}\left(\frac{6\pi^2}{g}\right)^{2/3}\frac{\hbar^2}{m}\left(\frac{N}{V}\right)^{2/3}N \tag{5.97}$$

Then, we have the pressure due to quantum degeneracy

$$p = \frac{1}{5}\frac{\hbar^2}{m}\left(\frac{6\pi^2}{g}\right)^{2/3}\left(\frac{N}{V}\right)^{5/3} \tag{5.98}$$

Thus, the pressure of the Fermi gas at $T = 0$ is proportional to its number density $n_0 = N/V$ to the $\frac{5}{3}$ power. If more particles are added to the system at constant volume, it reacts with an increasing quantum pressure. For strong degeneracy, the temperature is very low compared to the relevant energy scale, given by

$$T \ll \frac{\hbar^2}{2m}\left(\frac{N}{V}\right)^{2/3} = T_F \equiv E_F \tag{5.99}$$

The temperature $T_F \cong E_F$ is called the *degeneracy temperature*. For temperatures well above this scale, quantum effects are negligible and the gas has classical properties.

5.7 DEGENERATE ELECTRONIC GAS AT TEMPERATURE DIFFERENT FROM ZERO

A primary example of a degenerate Fermi gas is the conduction electrons in metals. We consider the conduction electrons in metals as nearly free and thus they behave like ideal gas particles. Electrons have half-integral spins and of course obey Fermi–Dirac statistics. Furthermore, spin-1/2 usually leads to double degeneracy of the electronic states, so $g = 2$.

For temperatures that are low compared to the degeneracy temperature T_F, the Fermi distribution function n_F has the form of the broken curve in Figure 5.5. The occupation falls slightly below 1 for energies below μ, and slightly above 0 for energies above μ. One can think that some of the states below μ transferred their electrons into states at energies above μ, as a result of thermal energy. Summerfield studied the properties of such a gas. He used the Fermi distribution and was able to find μ and its effects on the thermodynamics. Our problem here is the evaluation of the thermodynamic potential:

$$\Omega = -\frac{2}{3}\frac{gVm^{3/2}}{\sqrt{2\pi^2\hbar^3}}\int_0^{\infty}\frac{E^{3/2}dE}{\exp\{(E-\mu)/T\}+1} \tag{5.100}$$

Consider the following integral:

$$I\left(\frac{\mu}{T}\right) = \int_0^{\infty}\frac{f(E)dE}{\exp\{(E-\mu)/T\}+1} \tag{5.101}$$

Here, $f(E)$ is some function for which the integral $I(\mu/T)$ is convergent. Let us consider the following approximation:

$$\frac{\mu}{T} \gg 1 \tag{5.102}$$

The function

$$f(E)\exp\left\{-\frac{E}{T}\right\} \tag{5.103}$$

increases very slowly such that

$$f(E)\exp\left\{-\frac{E}{T}\right\}_{E\to\infty} \to 0 \tag{5.104}$$

Let

$$z = \frac{E-\mu}{T} \tag{5.105}$$

then

$$E = \mu + zT, \quad dE = Tdz \tag{5.106}$$

From Equation 5.101, we have

$$I\left(\frac{\mu}{T}\right) = T\int_{-\frac{\mu}{T}}^{\infty}\frac{f(\mu+Tz)dz}{\exp\{z\}+1} = T\int_{-\frac{\mu}{T}}^{0}\frac{f(\mu+Tz)dz}{\exp\{z\}+1} + T\int_0^{\infty}\frac{f(\mu+Tz)dz}{\exp\{z\}+1}$$

$$= T\int_0^{\infty}\frac{f(\mu-Tz)dz}{\exp\{-z\}+1} + T\int_0^{\infty}\frac{f(\mu+Tz)dz}{\exp\{z\}+1} \tag{5.107}$$

If we consider the following function:

$$\frac{1}{\exp\{-z\} + 1} \equiv 1 - \frac{1}{\exp\{z\} + 1} \tag{5.108}$$

then the integral $I(\mu/T)$ now becomes:

$$I\left(\frac{\mu}{T}\right) = T\int_0^{\frac{\mu}{T}} f(\mu - Tz)dz - T\int_0^{\frac{\mu}{T}} \frac{f(\mu - Tz)dz}{\exp\{z\} + 1} + T\int_0^{\infty} \frac{f(\mu + Tz)dz}{\exp\{z\} + 1} \tag{5.109}$$

If in the second integral we change the upper limit of integration to infinity, considering $\mu/T \gg 1$, then the integral converges faster. Thus

$$I\left(\frac{\mu}{T}\right) = \int_0^{\mu} f(E)dE + T\int_0^{\infty} \frac{f(\mu + Tz) - f(\mu - Tz)}{\exp\{z\} + 1} dz \tag{5.110}$$

We expand the second integrand with respect to z in a Taylor series, which gives

$$I\left(\frac{\mu}{T}\right) = \int_0^{\mu} f(E)dE + 2T^2 f'(\mu)\int_0^{\infty} \frac{zdz}{\exp\{z\} + 1} + \frac{1}{3}T^4 f'''(\mu)\int_0^{\infty} \frac{z^3dz}{\exp\{z\} + 1} + \cdots$$

$$= \int_0^{\mu} f(E)dE + \frac{\pi^2}{6}T^2 f'(\mu) + \frac{7\pi^4}{360}T^4 f'''(\mu) + \cdots \tag{5.111}$$

The third summand in Equation 5.111 is just a control. It is not needed. Now putting $f(E) = E^{3/2}$, then

$$\int_0^{\mu} f(E)dE = \int_0^{\mu} E^{3/2}dE = \frac{2}{5}\mu^{5/2} \tag{5.112}$$

and

$$f'(\mu) = \frac{3}{2}\mu^{1/2} \tag{5.113}$$

From here, considering Equations 5.111 and 5.100, we have

$$I\left(\frac{\mu}{T}\right) \cong \frac{2}{5}\mu^{5/2} + \frac{\pi^2}{6}\frac{3}{2}T^2\mu^{1/2} = \frac{2}{5}\mu^{5/2}\left[1 + \frac{5\pi^2}{8}\left(\frac{T}{\mu}\right)^2\right] \tag{5.114}$$

and thus

$$\Omega = -\frac{2}{3}\frac{gVm^{3/2}}{\sqrt{2\pi^2\hbar^3}}\frac{2}{5}\mu^{5/2}\left[1+\frac{5\pi^2}{8}\left(\frac{T}{\mu}\right)^2\right] \tag{5.115}$$

or

$$\Omega = \Omega(\mu,T) \tag{5.116}$$

It follows that μ and T are independent variables. If $T=0$, then

$$\Omega = \Omega_0 = -\frac{2^{3/2}}{15}\frac{gVm^{3/2}}{\pi^2\hbar^3}\mu^{5/2} \tag{5.117}$$

Thus, for low temperature, a slight variation is present

$$\Omega = \Omega_0\left[1+\frac{5\pi^2}{8}\left(\frac{T}{\mu}\right)^2\right] \tag{5.118}$$

This leads to the temperature-dependent thermodynamic functions

$$S = -\frac{\partial\Omega}{\partial T} = -\Omega_0\frac{5\pi^2}{4}\frac{T}{\mu^2} \tag{5.119}$$

$$C_V = T\left(\frac{\partial S}{\partial T}\right) = -\Omega_0\frac{5\pi^2}{4}\frac{T}{\mu^2} \tag{5.120}$$

We see that as $T\rightarrow 0$, then $C_V\rightarrow 0$ (Nernst theorem). The physical meaning of Equation 5.120 is that only electrons with the energy on the order of T from the Fermi energy can be excited.

As

$$(\delta\Omega)_{T,V,N} = (\delta E)_{S,V,N} = (\delta F)_{T,V,N} \tag{5.121}$$

Then, from Equation 5.118, we get

$$E = E_0 + |\Omega_0|\frac{5\pi^2}{8}\left(\frac{T}{\mu}\right)^2 \tag{5.122}$$

where E_0 is the average energy at $T=0$ and

$$F = E - TS = E_0 - |\Omega_0|\frac{5\pi^2}{8}\left(\frac{T}{\mu}\right)^2 \tag{5.123}$$

or

$$F = F_0 - |\Omega_0| \frac{5\pi^2}{8} \left(\frac{T}{\mu} \right)^2 \qquad (5.124)$$

The quantum equation of state, however, holds here

$$pV = \frac{2}{3} E \qquad (5.125)$$

Then, that the temperature dependence of the pressure is

$$p = \frac{2}{3} \frac{E_0}{V} + \frac{5\pi^2}{12} \frac{|\Omega_0|}{V} \left(\frac{T}{\mu} \right)^2 \qquad (5.126)$$

In a metal, the electron gas is in a degenerate state and this degeneracy is removed as the temperature increases. These formulas show how the mean energy and the pressure of a strongly degenerate gas vary slowly as $T \to 0$ (Figure 5.6). This is consistent with Equation 5.120, which shows that $C_V \to 0$ as $T \to 0$. From Equation 5.119, there results $S \to 0$ as $T \to 0$, consistent with the Planck formulation of the third law. The boundedness of the energy and the pressure and the vanishing entropy as $T \to 0$ are connected to the properties of Fermi particles.

As every state may be occupied at most by only one particle, for $T = 0$, all the states with energies from 0 to E_F are completely occupied and all higher states are free. Thus, there is only one microstate for this macrostate. For such a distribution, the entropy $S = 0$.

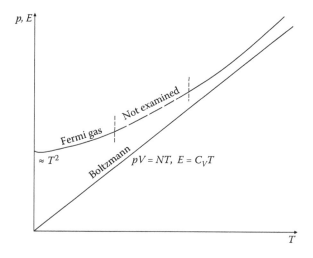

FIGURE 5.6 Variation of pressure and energy with temperature for Fermi and Boltzmann gases.

Electrons in the conduction band of semiconductors are an interesting example of changing degeneracy. The electron concentration $N = N(T)$ is dependent on the temperature. For a low concentration (say, intrinsically generated by thermal effects), there is no degeneracy. When the temperature is increased there is a transition of the electrons to the conduction band and we have a degenerate state. When temperatures are further increased the degeneracy is removed. The properties of electrons in semiconductors are further discussed in Sections 5.9 and 5.10.

5.8   EXPERIMENTAL BASIS OF STATISTICAL MECHANICS

Metals and their electronic contributions to the thermodynamics provide an interesting example of the importance of quantum effects on specific heat. This example also lends strong support to both the quantum theory for electrons in conductors and the application of statistical mechanics to a fermion-dominated system. The electrons, being fermions, can be highly degenerate even at room temperature, and not described by Maxwell–Boltzmann statistics. For example, take the case of copper, a very important technological metal, commonly used in household wiring and many other applications. There is expected to be one free conduction electron per atom; the core electrons remain home with their nuclei. The mass density of copper at room temperature is $\rho \approx 8.94$ g/cm^3 and its molar mass is 63.546 g/mol. Thus, the number density of conduction electrons is $n = N_A(\rho/M_A) \approx 8.47 \times 10^{28}$ m^{-3}. Thus, the Fermi wave vector is $k_F = (3\pi^2 n)^{1/3} \approx 1.36 \times 10^{10}$ m^{-1} and the Fermi energy using the bare electron mass is then a rather high value: $E_F = (\hbar^2 k_F^2/2m_e) \approx 1.13 \times 10^{-18}$ J ≈ 7.05 eV, greatly above room temperature (295 K, $T \approx 0.025$ eV). It would be a similar order of magnitude in other monovalent metals such as gold, silver, sodium, magnesium, and so on. This is so high, in fact, that most electrons are in their ground levels (i.e., filling states up to E_F), and they cannot contribute to the specific heat at room temperature. Only the small fraction within about T of E_F are "active" to contribute to C. This significantly reduces C below what would be its classical value, by a factor of about T/E_F.

At low temperatures, where phonons are not excited, the electronic degrees of freedom dominate in a metal. Consider the theoretical prediction for the low-temperature electronic-specific heat of a fermion gas found above. The chemical potential μ is the same as E_F at zero temperature when the temperature is very low. The volume in the system can be expressed as $V = N/n$, for N electrons at a volume density of n. Then the specific heat per electron is linear with temperature and can be written in the form

$$\frac{C_v}{N} = \gamma T \tag{5.127}$$

where the electronic specific heat coefficient is

$$\gamma = -\frac{5\pi^2}{4\mu^2}\frac{\Omega_0}{N} = \frac{g}{12}\left(\frac{2m}{\hbar^2}\right)^{3/2}\frac{\mu^{1/2}}{n} \tag{5.128}$$

Once the definition of the Fermi energy is substituted for $\mu = E_F$, some simple algebra leads to the famous result

$$\gamma_{th} = \frac{\pi^2 k_B^2}{2E_F} \qquad (5.129)$$

Here, Boltzmann's constant has been included to produce an expression with the correct units, so one can compare with experiment. The subscript "th" refers to "theory." For copper, using $E_F = 7.05$ eV, the expression gives $\gamma_{th} = 8.32 \times 10^{-28}$ J/K^2, or, converted to a molar quantity with Avogadro's number, $N_A \gamma_{th} = 0.501$ mJ $\cdot$ K^{-2}mol^{-1}. This same constant was measured in careful experiments below 30 K by Martin (1973) [40] for copper, silver, and gold. For copper, Martin found a good linear fit for C_v at low temperatures, but instead with $N_A \gamma_{expt} = 0.691$ mJ $\cdot$ K^{-2}mol^{-1}. The fact that the behavior is linear with temperature already is a great confirmation of the Fermi–Dirac statistical theory. The discrepancy is accounted for very simply, by assuming that there is an effective mass m^* for the conduction electrons, which actually move in bands rather than freely. One can then define the Fermi energy using the effective mass

$$E_F^* = \frac{\hbar^2 k_F^2}{2m^*} = \frac{m}{m^*} E_F \qquad (5.130)$$

Now when E_F^* is used to define the experimental values, γ_{expt}, we have

$$\gamma_{expt} = \frac{\pi^2 k_B^2}{2E_F^*} = \frac{m^*}{m} \frac{\pi^2 k_B^2}{2E_F} = \frac{m^*}{m} \gamma_{th} \qquad (5.131)$$

The experimental value can be made to "match" the theory by using the effective mass ratio as

$$\frac{m^*}{m} = \frac{\gamma_{expt}}{\gamma_{th}} \qquad (5.132)$$

For copper, the needed ratio is $m^*/m = 1.38$, while silver has $m^*/m = 1.00$ and gold has $m^*/m = 1.08$. The specific heats of all these metals fit very well to the Fermi–Dirac theory. This is a wonderful confirmation of the fermion theory, the Pauli principle, and of course, the statistical theory, in what can be considered fairly complex physical systems.

5.9 APPLICATION OF STATISTICS TO AN INTRINSIC SEMICONDUCTOR

Semiconductors are composed of atoms such as Si, Ge, Ga, As, and so on, on some three-dimensional lattice. Some sites of that lattice could be occupied by atoms of other undesired impurities, or the regular atomic arrangement in the crystal could have what are called defects. A defect is a region or sites where

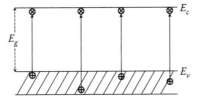

FIGURE 5.7   Transition of electrons from the valence band to the conduction band in a semiconductor.

the periodicity of the lattice is interrupted. A semiconductor that contains no impurities or defects is called an *intrinsic semiconductor*, otherwise it is called an *extrinsic semiconductor*. Calculation of the electronic structure of semiconductors shows that the electronic states form bands (Figure 5.7). A simplified model is that there is a set of states, called the valence band, and above that, separated by a gap, are other states, called the conduction band. The separation of the two bands is called the band gap, E_g. There are no energy states within the gap, so it is a forbidden region. For an intrinsic semiconductor at $T = 0$ K, the valence band states are fully occupied, and the conduction band states are unoccupied. At $T = 0$ K the electrical conductivity of an intrinsic semiconductor is zero as there are no free charge carriers. Usually this is explained by saying that, because the valence band is completely filled, there are no available states into which electrons can move, that would correspond to charge motion. The conduction band states are not available because of the energy gap. At $T > 0$ K, there appears the probability of transition of electrons from the valence band to the conduction band (Figure 5.7). A consequence is that in the valence band there appear "holes," which are states that are missing electrons that can be treated as positively charged carriers, of positive mass. It is obvious that the concentration N_e of electrons is equal to the concentration of holes N_h: $N_e = N_h$. This concentration is determined by the temperature.

Simultaneously, alongside the process of free charge carrier generation, there is also the process of recombination. A certain number of electrons from the conduction band make a transition back to the valence band and combine with the holes there. At a given temperature, thanks to the competition between generation and recombination, an equilibrium concentration of charge carriers will be achieved. At ordinary temperatures, the concentration of free electrons and holes in silicon (Si) is about 10^{10} cm^{-3} and in germanium (Ge) about 10^{13} cm^{-3}. This is a statistical process for which the equilibrium can be analyzed.

5.9.1   CONCENTRATION OF CARRIERS

The concentration of electrons at equilibrium in the conduction band may be determined by the expression (5.61), if the chemical potential is known. To do so, the energy of conduction band electrons is measured relative to the bottom of the conduction band, which is considered the energy level E_c. The mean electron occupation number

$$n_e = \frac{1}{\exp\left\{((E - E_c) - (\mu - E_c))/T\right\} + 1} \tag{5.133}$$

This is used in Equations 5.60 and 5.61, with the integration variable as $\varepsilon_e = E - E_c$, and taking into account a dispersion relation for the electrons as $\varepsilon_e = p^2/2m_e$, where m_e is the electron effective mass. This leads to electron concentration

$$N_e = \frac{gVm_e^{3/2}T^{3/2}\sqrt{\pi}}{2^{3/2}\pi^2\hbar^3} F_{1/2}\left(\frac{\mu - E_c}{T}\right) \tag{5.134}$$

This can be written as

$$N_e = N_c F_{1/2}\left(\frac{\mu - E_c}{T}\right) \tag{5.135}$$

The value

$$N_c = \frac{gVm_e^{3/2}T^{3/2}\sqrt{\pi}}{2^{3/2}\pi^2\hbar^3} \tag{5.136}$$

is the effective density of states in the conduction band. The function $F_{1/2}(\mu - E_c/T)$ is dependent on the chemical potential μ relative to the conduction band edge, and on the temperature T.

We evaluate the occupation number of holes at equilibrium in the valence band, using an identity

$$n_h = 1 - n_e \tag{5.137}$$

as an electron may either occupy a state or not. If an electron does not occupy a state, then the state is occupied by a hole. The hole occupation number is

$$n_h = 1 - \frac{1}{\exp\left\{(E - \mu)/T\right\} + 1} = \frac{1}{\exp\left\{(\mu - E)/T\right\} + 1} \tag{5.138}$$

The hole energy is measured with respect to the top of the valence band energy, E_v. The difference of the two band edges is the band gap energy, $E_g = E_c - E_v$. The energy dispersion for holes can be written as $\varepsilon_h = p^2/2m_h$, using as shifted variable for integration, $\varepsilon_h = E_v - E$. The mean hole occupation number

$$n_h = \frac{1}{\exp\left\{((E_v - E) - (E_v - \mu))/T\right\} + 1} \tag{5.139}$$

Then the hole concentration is found to be

$$N_h = N_v F_{1/2} \left(\frac{E_v - \mu}{T} \right)$$

(5.140)

where

$$N_v = \frac{g V m_h^{3/2} T^{3/2} \sqrt{\pi}}{2^{3/2} \pi^2 \hbar^3}$$

(5.141)

is the effective density of the states in the valence band and $F_{1/2}(E_v - \mu/T)$ is the Fermi–Dirac integral in the valence band. It depends on the valence band edge relative to the Fermi level.

It is seen that in order to evaluate the concentration of holes and electrons, it is necessary to evaluate the Fermi–Dirac integrals. These integrals may not have exact evaluations but there exist approximate expressions in some given domain of its arguments:

$$F_{1/2}(\varsigma) = \frac{2}{\sqrt{\pi}} \begin{cases} \dfrac{\sqrt{\pi}}{2} \exp\{\varsigma\}, & -\infty < \varsigma < -1 \\[2mm] \dfrac{\sqrt{\pi}}{2} \dfrac{1}{0.27 + \exp\{\varsigma\}}, & -1 < \varsigma < 5 \\[2mm] \dfrac{2}{3} \varsigma^{3/2}, & 5 < \varsigma < \infty \end{cases}$$

(5.142)

For the nondegenerate case, $\varsigma < -1$. This corresponds to Boltzmann's statistics. The condition for classical statistics for the electrons N_e is

$$\varsigma < -1 \quad \text{or} \quad \frac{\mu - E_c}{T} < -1$$

(5.143)

This should be when the Fermi level μ is far below the bottom of the conduction band E_c by a value greater than T. In this case, for

$$F_{1/2} \left(\frac{\mu - E_c}{T} \right) \approx \exp \left\{ \frac{\mu - E_c}{T} \right\}$$

(5.144)

then

$$N_e \approx N_c \exp \left\{ \frac{\mu - E_c}{T} \right\}$$

(5.145)

Here, the semiconductor is described by classical statistics, that is, it is nondegenerate.

If the Fermi level μ is far above the conduction band edge E_c by a value greater than $5T$, then the semiconductor is completely degenerate. For

$$E_c - T < \mu < E_c + 5T \tag{5.146}$$

the semiconductor has intermediate properties, that is, between those of the nondegenerate and completely degenerate. In this case

$$F_{1/2}\left(\frac{\mu}{T}\right) \approx \frac{1}{0.27 + \exp\{\mu/T\}} \tag{5.147}$$

It should be noted that the condition of degeneracy is determined by the temperature and the Fermi level relative to the conduction band. If we consider a completely degenerate case, then $5T < \mu < \infty$ and

$$F_{1/2}\left(\frac{\mu}{T}\right) \approx \left(\frac{\mu}{T}\right)^{3/2} \tag{5.148}$$

and

$$N_e \approx N_c \left(\frac{\mu - E_c}{T}\right)^{3/2} \approx (\mu - E_c)^{3/2} \tag{5.149}$$

Here, it is seen that the conduction of the electron gas is independent of the temperature. The Fermi level in this case is inside the conduction band at a distance of $5T$ above the bottom of the band. We have seen that for $\varsigma < -1$ then for any $x > 0$, $\exp\{-\varsigma\} > 1$ and the Fermi–Dirac function may be conveniently replaced by the Boltzmann's function:

$$\frac{1}{\exp\{x - \varsigma\} + 1} \approx \exp\{-x + \varsigma\} \tag{5.150}$$

From this

$$F_{1/2}(\varsigma) \cong \exp\{\varsigma\} \tag{5.151}$$

We may also get an approximate expression for $F_{1/2}(E_v - \mu/T)$ to apply the same analysis for the holes.

If the position of the Fermi level μ is known, then the expressions N_e and N_h give the concentration of electrons and holes. The position of the Fermi level is determined from the condition of electrical neutrality of an intrinsic semiconductor, which may be found by solving the equation $N_e = N_h$, or

$$N_c F_{1/2}\left(\frac{\mu - E_c}{T}\right) = N_v F_{1/2}\left(\frac{E_v - \mu}{T}\right) \tag{5.152}$$

Suppose $N_c = N_v$, then Equation 5.152 becomes

$$F_{1/2}\left(\frac{\mu - E_c}{T}\right) = F_{1/2}\left(\frac{E_v - \mu}{T}\right) \tag{5.153}$$

and this requires the arguments to match

$$\frac{\mu - E_c}{T} = \frac{E_v - \mu}{T} \tag{5.154}$$

or

$$\mu \equiv E_F = \frac{E_v + E_c}{2} \tag{5.155}$$

Hence, the Fermi level is independent of the temperature and is located exactly at the middle of the gap. In this way, the tail of the Fermi–Dirac distribution has the same strength at the conduction band edge as it does at the valence band edge, which of course gives equal concentrations. If the band gap is greater than $2T$, then we have

$$\frac{\mu - E_c}{T} < -1 \quad \text{and} \quad \frac{E_v - \mu}{T} < -1 \tag{5.156}$$

For a nondegenerate semiconductor the concentration of the electrons is then approximated by

$$N_e = N_c \exp\left\{\frac{\mu - E_c}{T}\right\} = N_c \exp\left\{-\frac{E_c - E_F}{T}\right\} = N_c \exp\left\{-\frac{\Delta E_a}{T}\right\} \tag{5.157}$$

and that of the holes is

$$N_h = N_v \exp\left\{\frac{E_v - \mu}{T}\right\} = N_v \exp\left\{-\frac{E_F - E_v}{T}\right\} = N_V \exp\left\{-\frac{\Delta E_a}{T}\right\} \tag{5.158}$$

Here, $\Delta E_a = E_g/2$ is the activation energy equal to half of the size of the band gap for intrinsic semiconductors. In finding Equations 5.157 and 5.158, we assumed

$N_v = N_c$. Instead, let us discuss the case where $N_v \neq N_c$. Consider Equation 5.152, then

$$N_c \exp\left\{\frac{\mu - E_c}{T}\right\} = N_v \exp\left\{\frac{E_v - \mu}{T}\right\} \tag{5.159}$$

can be solved in the form

$$\frac{N_v}{N_c} = \exp\left\{\frac{2\mu - E_c - E_v}{T}\right\} \tag{5.160}$$

or

$$E_F = \frac{E_c + E_v}{2} + T\ln\left(\frac{N_v}{N_c}\right)^{1/2} \tag{5.161}$$

If we consider Equation 5.141, then this gives

$$E_F = \frac{E_c + E_v}{2} + T\ln\left(\frac{m_h}{m_e}\right)^{3/4} \tag{5.162}$$

If $m_h = m_e$, then the Fermi level in an intrinsic semiconductor is independent of the temperature and is located at the middle of the forbidden gap. If $m_h \neq m_e$ then E_F is located at the middle of the forbidden band only when $T = 0$ K. For an increase in the temperature, the Fermi level is displaced linearly toward the band where the effective mass of the charge carriers is smaller. If we consider that in an intrinsic semiconductor, $N_e = N_h = N_i$, where N_i refers to the intrinsic concentration, then it can be seen that the product of the concentration of electrons and holes for a nondegenerate intrinsic semiconductor is independent of the Fermi level:

$$(N_e N_h)^{1/2} = N_i = (N_c N_v)^{1/2} \exp\left\{\frac{-E_g}{2T}\right\} \tag{5.163}$$

With N_c and N_v given in Equations 5.136 and 5.141, respectively, then the intrinsic concentrations are

$$N_i = \frac{gVT^{3/2}\sqrt{\pi}}{2^{3/2}\pi^2\hbar^3}(m_e m_h)^{3/4}\exp\left\{\frac{-E_g}{2T}\right\} \tag{5.164}$$

With spin degeneracy, $g = 2$, this is widely used to determine the width E_g of the forbidden band, given experimental results for the concentration N_i with temperature T or $\ln N_i$ versus $1/T$.

From Equation 5.164, we have

$$\ln N_i = \text{const} - \frac{3}{2}\ln\frac{1}{T} - \frac{E_g}{2}\frac{1}{T} \tag{5.165}$$

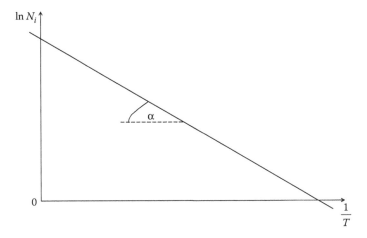

FIGURE 5.8 Variation of the logarithm of the concentration $\ln N_i$ with inverse temperature.

We may neglect $\ln(1/T)$ compared to $1/T$ and thus the result from Equation 5.165 is nearly linear (Figure 5.8), with negative slope determined by the gap

$$\tan \alpha = -\frac{E_g}{2} \tag{5.166}$$

From Equations 5.158 and 5.162, an increase in temperature makes the Fermi level approach the band with the lighter carriers and the semiconductor may be transformed from nondegenerate to degenerate. This degeneracy appears when the distance between E_F and the nearest band edge is of the order of T. For instance, if the degeneracy takes place in the conduction band, then in the valence band it is absent, as the Fermi level moves further away from it as the temperature increases. In such a case with degeneracy in the conduction band, the concentration of carriers in an intrinsic degenerate semiconductor results from the relation

$$N_i = N_c F_{1/2}\left(\frac{\mu - E_c}{T}\right) = N_v \exp\left\{\frac{E_v - \mu}{T}\right\} \tag{5.167}$$

The Fermi integral may not be replaced by an exponential function here. In an intrinsic semiconductor, degeneracy appears when the effective masses of electrons and holes differ considerably. This can be seen strongly in the example of indium antimony (InSb), where $m_h \approx 10\, m_e$.

5.10 APPLICATION OF STATISTICS TO EXTRINSIC SEMICONDUCTOR

It is necessary to find the position of the Fermi level in an extrinsic semiconductor. Typically, an extrinsic semiconductor is one doped with some kind of impurities to

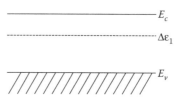

FIGURE 5.9 Energy level of a donor impurity relative to the conduction and valence bands of a semiconductor.

make it more conductive. The impurity atoms substitute for some of the original atoms of the host semiconductor. Impurity atoms have different energy levels than the host atoms. If an impurity has an electronic state close to and slightly below the conduction band edge of the semiconductor, it is called a *donor* impurity. It will tend to "donate" or give one of its electrons to the conduction band that can contribute to the conductivity. If an impurity has an electronic state close to and slightly above the valence band edge, it is called an *acceptor*. It will tend to "accept" an electron from the valence band, creating a hole in the semiconductor that can contribute to its conductivity.

Consider a semiconductor with donors and let the depth of the donor level below the conduction band edge, E_c, be $\Delta\varepsilon_1 \equiv \Delta\varepsilon_d = E_c - E_d$ as in Figure 5.9. Suppose $\Delta\varepsilon_d \ll E_g$ where the band gap is E_g. Suppose that the temperature is exceedingly low. The intrinsic conductivity of the semiconductor will also be extremely low (without any doping), because of the exponential behavior in (5.164), leading to very few free carriers. Let N_d be the concentration of the donors. Thermal fluctuations will lift electrons from the donor atoms into the conduction band. We find the position of the chemical potential under the conditions of total neutrality.

We construct the Fermi statistics for the impurity electron localized below the bottom of the conduction band E_c. One electron may be localized on each donor level E_d, with an arbitrary orientation of the spin. Let m_j label the number of quantum states, and ε_j and n_j are, respectively, the number and energy of an electron in the given state such that $n_j < m_j$. Here, $j = 1, 2, \ldots$ are different groups of quantum states. The number of ways of placing n_j electrons in m_j quantum states with a restriction of not more than one electron per state is

$$C_{n_j}^{m_j} = \frac{m_j!}{n_j!(m_j - n_j)!} \tag{5.168}$$

The number of electrons with spin up is n_j' and that with spin down is $n_j - n_j'$. The number of ways of selecting n_j' electrons from n_j is

$$C_{n_j'}^{n_j} = \frac{n_j!}{n_j'!(n_j - n_j')!}, \quad 0 \le n_j' \le n_j \tag{5.169}$$

The number of possible arrangements of n_j electrons into m_j quantum states, considering the spin, is

$$W_j = \frac{m_j!}{n_j!(m_j - n_j)!} \sum_{n_j'} C_{n_j'}^{n_j} = \frac{m_j!2^{n_j}}{n_j!(m_j - n_j)!} \tag{5.170}$$

Then from all possible groups of quantum states, we have

$$W = \prod_j W_j = \prod_j \frac{m_j!2^{n_j}}{n_j!(m_j - n_j)!} \tag{5.171}$$

From Boltzmann's definition, the entropy is

$$S = \ln W \tag{5.172}$$

If we use the method of Lagrange's undetermined multipliers, we apply

$$\frac{\partial}{\partial n_j}(S + \alpha n + \beta \varepsilon) = 0 \tag{5.173}$$

where the constrained quantities are expressed

$$n = \sum_j n_j, \quad \varepsilon = \sum_j \varepsilon_j n_j \tag{5.174}$$

that are, respectively, the total number of electrons and total energy. Then we see that occupation numbers

$$\frac{n_j}{m_j} = \frac{1}{1/2 \exp\{\alpha + \beta \varepsilon_j\} + 1} \tag{5.175}$$

will maximize the entropy. This defines the equilibrium distribution. It coincides with the Fermi–Dirac distribution with

$$\alpha = -\frac{\mu}{T}, \quad \beta = \frac{1}{T} \tag{5.176}$$

Thus

$$\frac{n_j}{m_j} = \frac{1}{1/2 \exp\{(\varepsilon - \mu)/T\} + 1} \tag{5.177}$$

It should be noted that each donor level has two possible spin orientations for the donor electron. Hence, each donor level has two quantum states. However the insertion of an electron into one quantum state precludes putting an electron into the second quantum state. The vacancy requirement of the atom is satisfied by adding one electron. The addition of a second electron to the donor level is not feasible. This confirms the fact why the distribution function in Equation 5.177 of donor electrons in the donor energy states is then somewhat different from the Fermi–Dirac function seen earlier. The factor $\frac{1}{2}$ in the distribution function in Equation 5.177 is a direct consequence of the spin factor as mentioned above.

Considering relation (5.177) and from the condition of neutrality in the semiconductor, there follows the relation

$$N = N_d \frac{1}{1/2 \exp\left\{-(\Delta\varepsilon_d + \mu)/T\right\} + 1} + 2\left(\frac{2\pi m_e^* T}{(2\pi\hbar)^2}\right)^{3/2} \exp\left\{\frac{\mu}{T}\right\} \qquad (5.178)$$

is the concentration of the electrons in the conduction band with energy E_c. We consider in Equation 5.178 $N \equiv N_d$, the donor concentration and N is not large. It follows that for $\mu < 0$, we suppose that

$$|\Delta\varepsilon_d| > |\mu| \gg T \qquad (5.179)$$

from where

$$\left[1 + \frac{1}{2}\exp\left\{-\frac{\Delta\varepsilon_d + \mu}{T}\right\}\right]^{-1} \cong 1 - \frac{1}{2}\exp\left\{-\frac{\Delta\varepsilon_d + \mu}{T}\right\} \qquad (5.180)$$

under the condition $|\mu| \gg T$, the function $\mu = E_F(T)$ approximates to its value at absolute zero. The distribution is degenerate when $T \ll E_F$ and nondegenerate when $T \gg E_F$ (the classical case). If we substitute Equation 5.180 in Equation 5.178, then we have

$$N \exp\left\{-\frac{\Delta\varepsilon_d + \mu}{T}\right\} = 4\left(\frac{2\pi m_e^* T}{(2\pi\hbar)^2}\right)^{3/2} \exp\left\{\frac{\mu}{T}\right\} \qquad (5.181)$$

from where taking the natural logarithm we have

$$\mu = -\frac{\Delta\varepsilon_d}{T} + \frac{T}{2}\ln\frac{4}{N}\left(\frac{2\pi m_e^* T}{(2\pi\hbar)^2}\right)^{3/2} \exp\left\{\frac{\mu}{T}\right\} \qquad (5.182)$$

and considering Equation 5.158 we have

$$\mu = E_c - \frac{\Delta\varepsilon_d}{2} - \frac{T}{2}\ln\left(2\frac{N_c}{N_d}\right) \qquad (5.183)$$

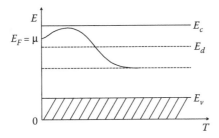

FIGURE 5.10   Relation between the Fermi level and the temperature for a semiconductor with a donor impurity.

Equation 5.161 gives the position of the Fermi level for an extrinsic semiconductor. Assuming low temperatures, we have neglected the transition of electrons from the valence band to the conduction band. In Equation 5.183, the quantity N_c is the effective density of states in the conduction band determined by the relation $N_c = \left(gVm_e^{3/2}T^{3/2}\sqrt{\pi}/2^{3/2}\pi^2\hbar^3\right)$. What should be the physical interpretation of the factor 2 in the logarithm? The factor 2 is the *spin degeneracy* of the impurity level. The total number of electronic states in the forbidden band is equal to the number of impurity atoms. This is N_d per unit volume of the crystal. Each impurity atom can donate only one electron to the conduction band. That electron, however, can be promoted to the conduction band in two different ways, depending on the direction of the spin. As a result, the impurity level is twofold degenerate. This implies that the neutral state of a donor impurity has twice the statistical weight of the ionized state. A similar result holds for acceptors.

From Equation 5.183 for $T = 0$, in a semiconductor with a donor impurity, we have $\mu = E_c - (\Delta\varepsilon_d/2)$, which implies that the Fermi level is situated halfway between the bottom of the conduction band and the impurity level (see Figure 5.10). As the temperature T increases (but still for low temperatures) and assuming $2N_c < N_d$, the Fermi level first approaches the conduction band and then begins to move back downward.

For $2N_c = N_d$ we also have $\mu = E_c - (\Delta\varepsilon_d/2)$. If we substitute Equation 5.183 in the expression of the concentration, then we have

$$n = N_c \exp\left\{-\frac{\Delta\varepsilon_d}{2T} - \frac{1}{2}\ln\left(\frac{2N_c}{N_d}\right)\right\} = \sqrt{\frac{N_c N_d}{2}}\exp\left\{-\frac{\Delta\varepsilon_d}{2T}\right\} \tag{5.184}$$

The progressive concentration of electrons in the conduction band becomes comparable to N_d. In this case, the expression for E_F becomes inapplicable. A more careful detailed analysis shows that

$$E_F = E_c - T\ln\left(\frac{N_c}{N_d}\right) \tag{5.185}$$

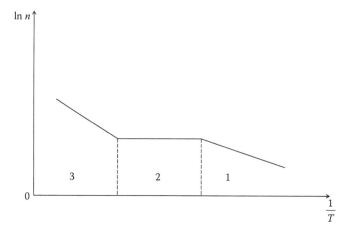

FIGURE 5.11 Variation of the logarithm of the concentration with inverse temperature of a donor semiconductor; 1—ionization of impurities; 2—exhaustive region; 3—transition from the valence band into the conduction band.

and the concentration of electrons becomes

$$n = N_d \qquad (5.186)$$

which implies that all the donor impurities are ionized. The temperature for which the relation (5.186) is true is called the *exhaustive region of the impurity*. When E_F is less than E_d, there is total ionization which takes place above some temperature (see Figure 5.11).

5.11 DEGENERATE BOSE GAS

5.11.1 CONDENSATION OF BOSE GASES

The quantum symmetry restrictions on a Bose gas have consequences completely different from those for a Fermi gas. At low temperature, a Bose gas has nothing in common with a Fermi gas. At $T = 0$ for a Bose gas, all particles can be in the ground state, giving total energy $E = 0$, while the Pauli exclusion principle leads to a nonzero macroscopic ground state energy for the Fermi gas, $E\big|_{T=0} \neq 0$. Here, we consider the details of how a Bose gas behaves at low temperature in its degenerate state. The behavior depends crucially on how the chemical potential changes with temperature, and how it affects the distribution of the particles among the allowed energy states.

Let us write the number N of Bose particles from the usual normalization condition, assuming the single-particle kinetic energy $E = p^2/2m$

$$N = \frac{gVm^{3/2}}{\sqrt{2}\pi^2\hbar^3} \int\limits_0^\infty \frac{E^{1/2}\,dE}{\exp\{(E-\mu)/T\}-1} = \frac{gVm^{3/2}T^{3/2}}{\sqrt{2}\pi^2\hbar^3} \int\limits_0^\infty \frac{z^{1/2}\,dz}{\exp\{z-(\mu/T)\}-1} \qquad (5.187)$$

In the above relation, we let $z = E/T$. The lowest energy single-particle state lies at $E = 0$, and for nonzero temperature, the chemical potential is negative. This is necessary so that the Bose–Einstein occupation number remains finite for all $E > 0$. As the temperature is lowered, the factor $T^{3/2}$ decreases, and the factor $|\mu|/T$ must then tend to decrease so that the normalization integral remains invariant for a fixed number of particles. For a given fixed number N, eventually the chemical potential must attain the value $\mu = 0$ for a special temperature T_0, defined by the equality

$$N = \frac{gVm^{3/2}T_0^{3/2}}{\sqrt{2}\pi^2\hbar^3} \int_0^\infty \frac{z^{1/2}dz}{\exp\{z\} - 1} \qquad (5.188)$$

Evaluation of the integral leads to

$$T_0 = \frac{3.3}{g^{2/3}} \frac{\hbar^2}{m} \left(\frac{N}{V}\right)^{2/3} \qquad (5.189)$$

the so-called condensation temperature. If we continue to reduce the temperature below T_0, while assuming that μ now remains at 0, Equation 5.187 suggests that there would be a further reduction in N. Of course, this cannot take place; the number of particles is considered fixed. There is a difficulty in using Equation 5.187, considering the fact that the integral is not exact. Recall that this integral replaced (approximately) a discrete sum over single-particle states in phase space. We see that it cannot be applied with confidence below the condensation temperature. We need a different analysis for low temperatures, $T < T_0$.

In general, the total number of particles is given by a discrete sum

$$N = \sum_k \bar{n}_k \qquad (5.190)$$

but the density of states used in a continuum sense was

$$d\Gamma = \frac{gVm^{3/2}E^{1/2}dE}{\sqrt{2}\pi^2\hbar^3} \qquad (5.191)$$

Expression (5.191) gives $d\Gamma = 0$ for $E = 0$. However, that is incorrect; experiment shows that $d\Gamma(E = 0) = 1$. The level $E = E_0 = 0$ is the ground state level for a single particle. For a Bose gas, there is nothing to prevent this lowest energy state from being occupied by a very large number of particles. This is in strong contrast to a Fermi gas. By replacing the discrete sum by the integral, a grievous error was made in counting the contribution of the ground state level. This needs to be corrected to get the proper description of a Bose gas for $T < T_0$. The physical effect at low T is that the ground state can acquire a macroscopically large occupation of particles. The mathematics needs to be able to reflect this fact.

The correction is to count the number of particles N_0 in the ground level, $E_0 = 0$, separately from the particles in excited states, $N_{E>0}$. We consider that the integral (5.187) with $\mu = 0$ is only the number of particles in the excited states. The expression (5.189) relates the condensation temperature to the total number of particles, N. When combined with Equation 5.187, the number of particles in excited states can be written as

$$N_{E>0} = N \left(\frac{T}{T_0} \right)^{3/2}, \quad T < T_0 \tag{5.192}$$

The total number of particles in the system can be expressed as the sum of those in the ground state and those in excited states, $N = N_0 + N_{E>0}$. Then, the number of particles in the ground state for temperatures below the condensation temperature is

$$N_0 = N - N_{E>0} = N \left[1 - \left(\frac{T}{T_0} \right)^{3/2} \right], \quad T < T_0 \tag{5.193}$$

This describes the process of accumulation of particles as a function of temperature into the level with $E = 0$. From Equation 5.193 it is obvious that $N_0 = N$ for $T = 0$; all particles fall into the ground state at absolute zero. At higher temperatures, only a fraction of the particles occupy the ground state. Nevertheless, it is still a *macroscopic* fraction of the particles. The accumulation of the particles in the ground state is called *Bose–Einstein condensation*. Here, we are dealing with condensation at the center of phase space ($\vec{p} = 0$), not in real space. At finite but exceedingly low temperatures $T < T_0$, the condensation exists. For any temperature $T > T_0$, we have no condensate. Bose–Einstein condensation differs from the ordinary condensation of vapor into liquid that occurs in ordinary physical space when the temperature is lowered. There, even in the condensed phase, the particles still occupy different energy states. Bose–Einstein condensation is a purely quantum mechanical effect.

At temperatures $T \leq T_0$, an ideal Bose gas is composed from two phases that are completely intermixed (unlike the usual phases of matter, solid, liquid, and gas):

1. A gaseous phase: it has $N_{E>0} = N(T/T_0)^{3/2}$ particles distributed over the exited states.
2. A condensed phase: it has N_0 particles that occupy the ground state ($E = 0$).

Let us evaluate some thermodynamic functions for the condensation of the Bose gas. We use the notation $z = E/T$, and take $\mu = 0$ for $T < T_0$. Then the total energy E is only due to the gaseous phase

$$E = \frac{gVm^{3/2}}{\sqrt{2}\pi^2 \hbar^3} \int_0^\infty \frac{E^{3/2} dE}{\exp\{E/T\} - 1} = \frac{gVm^{3/2}T^{5/2}}{\sqrt{2}\pi^2 \hbar^3} \int_0^\infty \frac{z^{3/2} dz}{\exp\{z\} - 1} = AVT^{5/2} \tag{5.194}$$

where A is a constant. The heat capacity C_V is

$$C_V = \left(\frac{\partial E}{\partial T}\right)_V = \frac{5}{2} AVT^{3/2} \tag{5.195}$$

and the entropy S is

$$S = \int \frac{C_V}{T} dT = \frac{5E}{3T} \tag{5.196}$$

and the free energy F is

$$F = E - TS = -\frac{2}{3}E \tag{5.197}$$

This leads to the equation of state

$$p = -\frac{\partial F}{\partial V} = \frac{2}{3} AT^{5/2} = \frac{2}{3}\frac{E}{V} \tag{5.198}$$

The pressure depends only on temperature; it is independent of volume, but the density is a constant. The dependence in Equation 5.198 is expected because the particles in the ground state have no momentum and make no contribution to the pressure. Once the temperature reaches $T = 0$, all particles are in the ground state, and the pressure vanishes. Figure 5.12 shows the relation between the pressure, energy, and temperature for the different particle types.

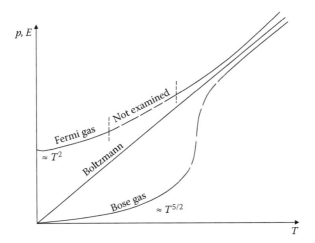

FIGURE 5.12 Relation between pressure, energy, and temperature for the different particle types.

For $T < T_0$, the system can be considered to consist of two intermixed fluids: The condensate that is a superfluid and the rest that is a normal (gaseous) fluid. The superfluid carries no energy or entropy. It forms a single quantum state. The superfluid does not have a heat capacity or viscosity. We examine helium-4 further in the chapters because it has these unusual and intriguing properties and can be approximated by an ideal Bose gas.

5.12   EQUILIBRIUM OR BLACK BODY RADIATION

5.12.1   Electromagnetic Eigenmodes of a Cavity

A photon is a Bose particle, considered as the quantum of electromagnetic radiation energy. A photon has two possible polarizations associated with its spin angular momentum, which takes the values $\pm\hbar$ (spin-1 object); hence, it is a boson. Here, we consider how to describe the collection of photons, or electromagnetic radiation, within a hollow cavity, whose walls are held at some temperature T. In order to have an equilibrium radiation distribution, we suppose that photons are continuously absorbed and reemitted by the walls of the enclosure. It is by virtue of this mechanism that the radiation inside the container is dependent on the temperature of the walls. It is not necessary for us to investigate the exact mechanisms that bring about the thermal equilibrium. This is because the general probability arguments of statistical mechanics are sufficient to describe the equilibrium situation.

We consider the radiation as an ensemble of photons (indistinguishable particles). Each photon has energy of $\varepsilon = \hbar\omega = \hbar cq$, where $\vec{q}$ is the wave vector and c is the speed of light. The momentum of a photon is $\vec{p} = \hbar\vec{q}$. A photon has the relativistic energy dispersion, $\varepsilon = pc$. The total number of photons in the enclosure is a variable and depends on the temperature T of the walls of the enclosure. The radiation field that is in thermal equilibrium inside the enclosure is completely described with the use of the photon mean number $\bar{n}(\omega)$, where we must also keep track of the number of photons at different wavelengths or frequencies. It is important to be able to describe the *spectrum* of the radiation.

The term *black body* refers to any object that completely absorbs any incident light energy. A black body, when heated, also easily emits radiation. While a real physical object has an emissivity $e(\omega) < 1$ that determines the efficiency of absorption and emission, a black body has emissivity $e(\omega) = 1$. Equilibrium black body radiation has a universal property, that is, the spectral characteristics are independent of the type of body radiating in the equilibrium state. The radiation density emitted by all black bodies in equilibrium at the same temperature is the same.

Using the occupation number $\bar{n} = \bar{n}(\omega)$, let us find the number of photons in a given state described by some occupation of photon momenta:

$$dN(\vec{p}) = 2\bar{n}\, d\Gamma = 2\bar{n}\,\frac{V dp_x dp_y dp_z}{(2\pi\hbar)^3} \tag{5.199}$$

We included a factor of 2 for both directions of polarization. We let $\hbar\omega = pc$ and change to spherical coordinates, using $\int dp_x dp_y dp_z \rightarrow \int 4\pi p^2 dp$, then we find the number of photons with the frequency ω in the range $\omega \rightarrow \omega + d\omega$:

$$dN(\omega) = \frac{\bar{n}(\omega)V\omega^2 d\omega}{\pi^2 c^3} \tag{5.200}$$

From Equation 5.200, we obtain the expression for the increment of energy in the photon gas

$$dE(\omega) = \hbar\omega dN(\omega) = \frac{V\hbar\omega^3 d\omega}{\pi^2 c^3} \frac{1}{\exp\{\hbar\omega/T\} - 1} \tag{5.201}$$

Here, the chemical potential for photons is $\mu = 0$. The formula (5.201) for the spectral distribution of energy of a black body radiation is called the Planck radiation law (1900). The exponential factor eliminates the ultraviolet catastrophe (divergence at short wavelengths) that had plagued the classical electromagnetic theory. This was a great success at the time, but still many more years passed before the quantum theory was fully accepted. In Planck's hypothesis, the radiation corresponds to a set of harmonic oscillators where only the energies $E = n\hbar\omega$ are allowed for each oscillator, n being an integer. The oscillators are the normal modes of the electromagnetic (EM) field in the cavity. Because this is in contradiction with the intuitive classical notions of an oscillation, acceptance of the theory was slow, although its ability to describe real data was exceptional.

Let us find the free energy $F = F(T,V,N) = -T \ln Z$, using Planck's hypothesis to evaluate the partition function. One oscillator at frequency ω can be populated by any number of photons. Thus the contribution from this one oscillator is

$$Z_1 = \sum_{n=0}^{\infty} \exp(-n\hbar\omega/T) = \frac{1}{1 - \exp(-\hbar\omega/T)} \tag{5.202}$$

The total partition function involves a similar term for each available mode at wave vector $\vec{q}$,

$$Z = \prod_{\vec{q}} \frac{1}{1 - \exp(-\hbar\omega_{\vec{q}}/T)}, \quad -\ln Z = \sum_{\vec{q}} \ln(1 - \exp(-\hbar\omega_{\vec{q}}/T)) \tag{5.203}$$

The sum over wave vectors can be done with the help of the density of states as in Equation 5.199, and changing to continuum integration over the photon energy. The number of states in a desired energy interval $d\varepsilon = \hbar d\omega$ is

$$d\tilde{N} = 2V\frac{4\pi p^2 dp}{(2\pi\hbar)^3} = \frac{V\varepsilon^2 d\varepsilon}{\pi^2 c^3 \hbar^3} \tag{5.204}$$

then we have

$$-\ln Z = \frac{V}{\pi^2 c^3 \hbar^3} \int_0^\infty \ln\left[1 - \exp(-\varepsilon/T)\right]\varepsilon^2 d\varepsilon \qquad (5.205)$$

Now it is convenient to change to the variable, $z = \varepsilon/T$, which leads to

$$-\ln Z = \frac{VT^3}{\pi^2 c^3 \hbar^3} \int_0^\infty \ln\left[1 - \exp(-z)\right]z^2 dz \qquad (5.206)$$

In the latter integral, the logarithm can be expanded in its power series, and then each term is easy to integrate separately. The result is related to the Riemann zeta function, $\zeta(s)$, for $s = 4$

$$-\int_0^\infty \ln\left[1 - \exp(-z)\right]z^2 dz = \int_0^\infty \sum_{m=1}^\infty \frac{e^{-mz}}{m}z^2 dz = \sum_{m=1}^\infty \frac{2}{m^4} = 2\zeta(4) = \frac{\pi^4}{45} \qquad (5.207)$$

Thus, we have the total free energy for the gas, which depends only on the volume of the cavity and the temperature of the walls

$$F = -T \ln Z = -\frac{V\pi^2 T^4}{45(c\hbar)^3} \qquad (5.208)$$

The free energy has no dependence on the total number of photons because the photon number is not conserved, it cannot be specified as an independent variable. This explains why the chemical potential for photons is zero:

$$\left(\frac{\partial F}{\partial N}\right)_{T,V} = \mu = 0 \qquad (5.209)$$

Working similarly, the spectral energy density $u(\omega, T)$ can be defined:

$$u(\omega, T) = \frac{1}{V}\frac{dE(\omega)}{d\omega} = \frac{\hbar\omega^3}{\pi^2 c^3}\frac{1}{\exp\{(\hbar\omega/T)\} - 1} \qquad (5.210)$$

Planck developed the expression (5.210) in 1900. Planck's hypothesis for each oscillator, $E = n\hbar\omega$, has a very nonclassical interpretation. It means that an EM field may emit (or absorb) only finite packets of energy, which are the quanta we are now familiar with. We can find the mean energy $\bar{E}_1$ of EM radiation in one mode of the cavity, in thermal equilibrium:

$$\bar{E}_1 = \frac{\sum_n \hbar\omega n \exp\{-(\hbar\omega n/T)\}}{\sum_n \exp\{-(\hbar\omega n/T)\}} = \frac{\hbar\omega}{\exp\{(\hbar\omega/T)\} - 1} \qquad (5.211)$$

The factor $(\exp\{\hbar\omega/T\} - 1)^{-1}$ can be interpreted as the number of photons in the state characterized by the photon energy $\hbar\omega$. Of course, this is the occupation number for that mode. From Planck's hypothesis, for a harmonic oscillator, only the quantized energies $E = n\hbar\omega$ are legitimate, contrary to the intuitive notion of a classical oscillator that would have a continuum of available energy states.

Before Planck's theory and the introduction of quantized oscillator energy levels, the theory for classical electromagnetic radiation in a cavity was well developed (Rayleigh–Jeans). As mentioned earlier, the classical theory leads to a divergence in the energy density at short wavelengths, the ultraviolet catastrophe. The evaluation of Rayleigh–Jeans formula (1880) is presented below. The density of states gives the number of standing waves in the frequency interval $d\omega$, which is equal to the number of oscillators

$$d\Gamma = \frac{V\omega^2 d\omega}{\pi^2 c^3} \tag{5.212}$$

For the classical limit, the average energy in any oscillator mode is $\overline{E} = T$, from which it follows that the averaged cavity energy in the frequency interval $d\omega$ is

$$dE(\omega, T) = \frac{V\omega^2 d\omega}{\pi^2 c^3} T \tag{5.213}$$

This is the Rayleigh–Jeans law. We may also represent it with the spectral energy density,

$$u(\omega, T) = \frac{1}{V}\frac{dE}{d\omega} = \frac{\omega^2 T}{\pi^2 c^3} \tag{5.214}$$

If we also consider Equation 5.211, then we can compare with the quantum result

$$dE(\omega) = \frac{V\hbar\omega^3 d\omega}{\pi^2 c^3}\frac{1}{\exp\{(\hbar\omega/T)\} - 1} \tag{5.215}$$

that is, the *Planck radiation law*. Classical mechanics should be the limit of quantum mechanics at high temperature, or equivalently, for large quantum numbers (the correspondence principle). Applying the condition, $\hbar\omega \ll T$, then the exponential term is approximated as

$$\exp\left\{\frac{\hbar\omega}{T}\right\} \approx 1 + \frac{\hbar\omega}{T} \tag{5.216}$$

Then, the classical limit of Planck's theory gives

$$dE(\omega) = \frac{V\hbar\omega^3 d\omega}{\pi^2 c^3}\frac{1}{\exp\{(\hbar\omega/T)\} - 1} \approx \frac{V\omega^2 d\omega}{\pi^2 c^3} T \tag{5.217}$$

Thus, the Rayleigh–Jeans formula (5.213) is the classical limit of Planck's formula. If we consider instead the condition for low temperatures or for short wavelengths:

$$\hbar\omega \gg T \qquad (5.218)$$

then we have

$$\exp\left\{\frac{\hbar\omega}{T}\right\} - 1 \approx \exp\left\{\frac{\hbar\omega}{T}\right\} \qquad (5.219)$$

and

$$dE(\omega) = \frac{V\hbar\omega^3 d\omega}{\pi^2 c^3}\exp\left\{-\frac{\hbar\omega}{T}\right\} \qquad (5.220)$$

This is called the *Wien's radiation formula*. Thus, in the domain of high frequencies, Planck's formula behaves as a Boltzmann distribution.

Let us find the frequency or wavelength where the maximum radiation density is present. For that we let $\hbar\omega/T = z$, and express the radiation spectrum as

$$dE(z,T) = \frac{Vz^3 dz T^4}{\pi^2 c^3 \hbar^3}\frac{1}{\exp\{z\} - 1} \qquad (5.221)$$

From evaluations it may be seen that the maximum of $z^3/(e^z - 1)$ occurs at the value

$$z_{max} = 2.822 \qquad (5.222)$$

This corresponds to a famous frequency/temperature relationship

$$\hbar\omega_{max} = 2.822\, T \qquad (5.223)$$

This is known as the *Wien displacement law*. It can be expressed alternatively using the wavelength

$$\omega_{max} = \frac{2\pi c}{\lambda_{max}} \qquad (5.224)$$

then

$$\lambda_{max} T = \frac{\hbar 2\pi c}{2.822} \cong 0.29\,\mathrm{cm}\cdot\mathrm{K} \qquad (5.225)$$

This means that the wavelength emitted most intensely by a black body is inversely proportional to the temperature of the body. λ_{max} shifts (or "displaces") with the temperature. This is why it is called a "displacement law." This law may be used to determine the temperature of bodies (such as stars) by looking for their maximum spectral density. To see the typical numbers, consider the sun. Inserting the solar surface temperature of 6000 K into Wien's displacement law we find

$$\lambda_{max} = \frac{0.29}{6000} \text{ cm} \approx 500 \text{ nm} \tag{5.226}$$

This is approximately the wavelength of yellow light. We may also obtain Equation 5.223 from Wien's formula:

$$dE(\omega) = \frac{V\hbar\omega^3 d\omega}{\pi^2 c^3} \exp\left\{-\frac{\hbar\omega}{T}\right\}\Bigg|_{z=\frac{\hbar\omega}{T}} = \frac{VT^4 z^3 dz}{\pi^2 c^3 \hbar^3} \exp\{-z\} = dE(z,T) \tag{5.227}$$

and if we let

$$\varepsilon = \frac{dE(z,T)}{dz} = B(T)z^3 \exp\{-z\} \tag{5.228}$$

then for z_{max} we have

$$\frac{d\varepsilon}{dz} = B(T)(3z^2 - z^3)\exp\{-z\} = 0 \tag{5.229}$$

It follows that

$$\hbar\omega_{max} \cong 3T \tag{5.230}$$

This recovers, approximately, the Wien displacement law.

The total energy of the radiation E is found from Equation 5.201 by letting $z = \hbar\omega/T$:

$$E = \frac{VT^4}{\pi^2 c^3 \hbar^3} \int_0^\infty \frac{z^3 dz}{\exp\{z\} - 1} = \frac{4\sigma VT^4}{c} \tag{5.231}$$

This is the *Stefan–Boltzmann law* and $\sigma = \pi^2/60c^2\hbar^3$ is the *Stefan–Boltzmann constant*. The required integral can be evaluated as we did earlier for a similar expression. The heat capacity C_V of the gas of photons in equilibrium with the container is

$$C_V = \left(\frac{\partial E}{\partial T}\right)_V = \frac{16\sigma VT^3}{c} \tag{5.232}$$

The entropy S of the gas is

$$S = \int \frac{C_V}{T} dT = \frac{16\sigma V T^3}{3c} \tag{5.233}$$

and as $T \to 0$, we have $S \to 0$, which satisfies the Nernst law (third law of thermodynamics). The free energy F of the gas of photons can also be expressed as

$$F = E - TS = -\frac{4\sigma V T^4}{3c} \tag{5.234}$$

which gives the pressure

$$p = -\left(\frac{\partial F}{\partial V}\right)_T = \frac{4\sigma T^4}{3c} = \frac{1}{3}\frac{E}{V} \tag{5.235}$$

This is known as the *Lebedeev formula*. The coefficient $\frac{1}{3}$ in Equation 5.235 is the result of averaging with respect to angles.

Finally, we also consider the total rate at which a black body emits radiant energy. Suppose a small hole of cross-sectional area A is made in a box containing black body radiation at a temperature T. The energy radiated through the hole will have the Planck spectrum (5.201). The flux of energy that passes through the hole per unit time can be expressed counting only the photons moving perpendicular to the area. We need to account for only half of the photons in the box (those moving with a velocity component toward the hole), and need to include the average component of their velocity perpendicular to the hole. The photons in the box have the energy density as above in the Stefan–Boltzmann law. This leads to the energy flux through the hole, which is also the same as the intensity through it

$$I = \frac{\Delta E/t}{A} = \frac{E}{V}\frac{c}{2}\langle\cos\theta\rangle = \frac{Ec}{4V} = \sigma T^4 \tag{5.236}$$

This is the usual form of the Stefan–Boltzmann law. It is reinterpreted as the intensity of black body radiation emitted from the surface of any hot body. It shows how the total flux of radiant energy from the surface of a hot body depends only on fundamental constants (in the Stefan–Boltzmann constant) and the absolute temperature to the fourth power.

5.13 APPLICATION OF STATISTICAL THERMODYNAMICS TO ELECTROMAGNETIC EIGENMODES

We have just established plane monochromatic waves (as a gas of photons) as being possible types of oscillations (or, a basis) of the EM field in a vacuum. This field represents the oscillation of a system that has an infinite number of degrees of freedom. Further, we can introduce normal coordinates of this oscillating system, by writing the equations of motion in Hamiltonian form. This type of approach appears to be convenient when solving problems involving the interaction of a particle with the EM field (on the basis of classical mechanics). In addition, the Hamiltonian method is useful in the quantum mechanical description of the EM field. By having the Hamiltonian description of electromagnetic modes in vacuum, the statistical mechanical averages can be found with the procedures already outlined as for any Hamiltonian system. At the basic level, if we know the fundamental modes and their energies, then we can find all desired statistical properties. In this section, the formulas are presented in CGS units, for greater simplicity (where electric and magnetic fields have the same dimensions). Further, we consider initially the application of classical statistical mechanics to the problem, as was initially done in the late 1800s by Max Planck and others before the advent of the quantum theory.

The vector potential of a single mode with wave vector $\vec{K}$ and angular frequency ω

$$\vec{A} = \vec{A}_o \exp\left\{i\left[\vec{K}\cdot\vec{r} - \omega t\right]\right\} \tag{5.237}$$

satisfies the following equations (using the *Coulomb gauge* with electric potential $\Phi = 0$)

$$\operatorname{div} \vec{A} = 0$$

$$\Delta \vec{A} - \frac{1}{c^2}\frac{\partial^2 \vec{A}}{\partial t^2} = 0 \tag{5.238}$$

and describes the EM field. The vector potential gives the magnetic induction according to $\vec{B} = \nabla \times \vec{A}$ and the electric field according to $\vec{E} = -(1/c)\partial\vec{A}/\partial t$. It is convenient to study the fields within the interior of a parallelepiped with sides A, B, C (Figure 5.13), and then suppose that the system is repeated in the sense of periodic boundary conditions. This means that translations across the system, relative to each coordinate, leave the field invariant:

$$\vec{A}(x, y, z, t) = \vec{A}(x + A, y, z, t) = \vec{A}(x, y + B, z, t) = \vec{A}(x, y, z + C, t) \tag{5.239}$$

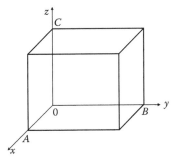

FIGURE 5.13   A parallelepiped with sides A, B, C on which is affixed a coordinate system.

This implies that we partition the entire space into such parallelepipeds and assume that in all such domains, the field behavior is the same. From the physical point of view, for exceedingly large A, B, C, the behavior of the field at the interior of such a domain will depend only weakly on the boundary conditions. Also, all points within the parallelepiped are equivalent. Thus, no physical result should depend on the volume V of the main domain, when the domain is sufficiently large compared to the important wavelengths in any problem of interest. We should note that for a system in a thermostat all the possible states are realized and an arbitrary EM field is given by appropriate values of the vector potential $\vec{A}(\vec{r},t)$ at any point in space. In addition, the periodic boundary conditions (5.239) make the eigen oscillations discrete, although their number is infinite. Thus, we may now expand all quantities characterizing the field in a Fourier series. We write this expansion of the vector potential in the form:

$$\vec{A}(\vec{r},t) = \sum_{\vec{K}} \vec{A}_{\vec{K}} \exp\left\{i\vec{K} \cdot \vec{r}\right\} \qquad (5.240)$$

where the $\vec{A}_{\vec{K}}$ are expansion coefficients and

$$\sum_{\vec{K}} = \sum_{K_x} \sum_{K_y} \sum_{K_z} \qquad (5.241)$$

The sum is relative to all possible values of the wave vector $\vec{K}$ that are consistent with the periodic boundary conditions being applied.

From the periodic condition in Equation 5.239 applied to Equation 5.240, we get these constraints:

$$\exp\left\{iK_x A\right\} = \exp\left\{iK_y B\right\} = \exp\left\{iK_z C\right\} = 1 \qquad (5.242)$$

Then, the allowed wave vectors are

$$K_x = \frac{2\pi}{A} n_x, \quad K_y = \frac{2\pi}{B} n_y, \quad K_z = \frac{2\pi}{C} n_z \qquad (5.243)$$

where n_x, n_y, n_z are positive and negative whole numbers.

As $\vec{A}$ is a real quantity, then the coefficients of expansion $A_{\vec{K}}$ in Equation 5.240 satisfy the relation:

$$\vec{A}_{-\vec{K}} = \vec{A}_{\vec{K}}^* \qquad (5.244)$$

From the equation

$$\text{div } \vec{A} = 0 \qquad (5.245)$$

it follows that for each $\vec{K}$ we have

$$\vec{K} \cdot \vec{A}_{\vec{K}} = 0 \qquad (5.246)$$

Condition (5.246) shows that vector $\vec{A}_{\vec{K}}$ is perpendicular to the corresponding wave vector $\vec{K}$. This shows that the waves are transverse. Also, because space is three dimensional, this allows for two independent directions for $\vec{A}_{\vec{K}}$ for a chosen wave vector. The two directions correspond to two independent polarizations at any wave vector. Thus, $\vec{A}_{\vec{K}}$ could also carry an index to label the polarization, but we suppress that in most formulas, for simplicity. The two polarizations may be chosen as either linear polarizations or as circular polarizations (say, right circular and left circular), this detail does not affect the arguments about the statistics. $\vec{A}_{\vec{K}}$ is also a function of time, due to the wave Equation 5.238, and then we have

$$\ddot{\vec{A}}_{\vec{K}} + c^2 K^2 \vec{A}_{\vec{K}} = 0 \qquad (5.247)$$

The solutions are exponentials in time

$$\vec{A}_{\vec{K}}(t) = \vec{A}_{\vec{K}}(0) e^{-i\omega_K t} \qquad (5.248)$$

where the frequencies follow the linear dispersion relation for light

$$\omega_K = cK \qquad (5.249)$$

In various situations, we want to know the number of independent EM modes for a selected frequency (or $|\vec{K}|$) or a small range of frequency. More specifically, given some small range ΔK in the magnitude of the wave vector (or $\Delta \omega = c \Delta K$), how many EM modes are allowed within that range? Suppose A, B, C are exceedingly large. Then neighboring values of wave numbers K_x, K_y, K_z are very close to each other. We look for possible values in the intervals ΔK_x, ΔK_y, ΔK_z respectively. Neighboring values of K_x correspond to values of n_x that differ only by unity. Hence, we can apply Equation 5.243 in differential form

$$\Delta n_x = \frac{A}{2\pi} \Delta K_x, \quad \Delta n_y = \frac{B}{2\pi} \Delta K_y, \quad \Delta n_z = \frac{C}{2\pi} \Delta K_z \qquad (5.250)$$

These relations give a translation from differential elements in wave vector space to the counting of wave vectors. The total number Δn of values of vector $\vec{K}$ in some 3D differential wave vector element is

$$\Delta n = \Delta n_x \Delta n_y \Delta n_z = \frac{ABC}{(2\pi)^3} \Delta K_x \Delta K_y \Delta K_z = \frac{V}{(2\pi)^3} \Delta K_x \Delta K_y \Delta K_z \qquad (5.251)$$

Hence, $V = ABC$ is the volume of the box in which the field is found.

The above counting was done arbitrarily in a rectangular box. This can be considered instead in an isotropic system of very large volume. In the spherical coordinate system for $\vec{K}$-space, with $\Delta K_x \Delta K_y \Delta K_z \rightarrow K^2 \Delta \Omega \Delta K$, this becomes

$$\Delta n = \frac{V}{(2\pi)^3} K^2 \Delta K\, \Delta \Omega \tag{5.252}$$

where $\Delta \Omega$ is the solid angle. Integration over all directions for the wave vector effectively gives the factor

$$\Delta \Omega = 4\pi \tag{5.253}$$

then the number of wave vectors with magnitude from K to K + ΔK is

$$\Delta n = \frac{V K^2 \Delta K}{2\pi^2} \tag{5.254}$$

This includes wave vectors in any direction. Note that if the number of EM modes is to be counted in this same interval, an additional factor of 2 should be applied, corresponding to the two independent polarization states.

Next, consider the energy density w of the EM field

$$w = \frac{1}{8\pi}(\vec{E}^2 + \vec{B}^2) \tag{5.255}$$

The total energy stored in a given volume due to the EM field is its integral

$$W = \frac{1}{8\pi} \int (\vec{E}^2 + \vec{B}^2)dV \tag{5.256}$$

We consider that inside the volume being considered, the EM radiation is in equilibrium with the surrounding walls at some fixed temperature that play the role of the thermostat. The walls are continuously absorbing and emitting EM radiation, at a rate determined by the temperature, and thus they exchange energy with the radiation field. In order to apply statistical physics, it is necessary that we introduce the generalized coordinates and momenta for the radiation.

The following is one approach to get these coordinates. We know the electric and magnetic fields strength are given at any point of space and time from the vector potential. The electric field strength is determined by

$$\vec{E} = -\frac{1}{c}\frac{\partial \vec{A}}{\partial t} \tag{5.257}$$

For a monochromatic plane wave (5.237), this reduces to

$$\vec{E} = iK\vec{A} \tag{5.258}$$

The magnetic induction is determined by

$$\vec{B} = \operatorname{curl}\vec{A} = \nabla \times \vec{A} = \frac{\partial}{\partial \vec{r}} \times \vec{A} \tag{5.259}$$

And again in the monochromatic wave we have

$$\vec{B} = i\vec{K} \times \vec{A} \tag{5.260}$$

Then, from Equations 5.258 and 5.260

$$\left|\begin{array}{l} \vec{E} = -\dfrac{1}{c}\displaystyle\sum_{\vec{K}} \vec{A}_{\vec{K}}\,\exp\left\{i\vec{K}\cdot\vec{r}\right\} \\[2mm] \vec{B} = i\displaystyle\sum_{\vec{K}} \vec{K}\times\vec{A}_{\vec{K}}\,\exp\left\{i\vec{K}\cdot\vec{r}\right\} \end{array}\right. \tag{5.261}$$

Substituting Equation 5.261 in Equation 5.256, the energy becomes

$$W = \frac{V}{8\pi}\sum_{\vec{K}}\left\{\frac{1}{c^2}\dot{\vec{A}}_{\vec{K}}\cdot\dot{\vec{A}}_{\vec{K}}^{*} + \left(\vec{K}\times\vec{A}_{\vec{K}}\right)\cdot\left(\vec{K}\times\vec{A}_{\vec{K}}^{*}\right)\right\} \tag{5.262}$$

From the transverse condition (5.246), there results

$$\left(\vec{K}\times\vec{A}_{\vec{K}}\right)\cdot\left(\vec{K}\times\vec{A}_{\vec{K}}^{*}\right) = \vec{K}^2\vec{A}_{\vec{K}}\cdot\vec{A}_{\vec{K}}^{*} \tag{5.263}$$

Thus

$$W = \frac{V}{8\pi c^2}\sum_{\vec{K}}\left\{\dot{\vec{A}}_{\vec{K}}\cdot\dot{\vec{A}}_{\vec{K}}^{*} + \vec{K}^2 c^2 \vec{A}_{\vec{K}}\cdot\vec{A}_{\vec{K}}^{*}\right\} \tag{5.264}$$

Considering the dynamics in Equation 5.247, the vectors $\vec{A}_{\vec{K}}$ are harmonic functions of time with frequency $\omega_{\vec{K}} = c K$. The frequency depends only on the absolute value of the wave vector. The manner of selection of these functions is very important in the expansion in Equation 5.243. This has to give us either standing waves or moving plane waves.

Consider Equation 5.240 for the moving waves. $\vec{A}_{\vec{K}}$ can be forced to be real by using a linear combination, based on dependencies

$$\vec{a}_{\vec{K}} \approx \exp\{-i\omega_{\vec{K}}t\}, \quad \vec{a}_{\vec{K}}^* \approx \exp\{i\omega_{\vec{K}}t\} \tag{5.265}$$

Then combining so that Equation 5.244 is automatically satisfied, we have

$$\vec{A}_{\vec{K}} = \vec{a}_{\vec{K}} + \vec{a}_{-\vec{K}}^*, \quad \dot{\vec{A}}_{\vec{K}} = -icK(\vec{a}_{\vec{K}} - \vec{a}_{-\vec{K}}^*) \tag{5.266}$$

Substituting this in Equation 5.264, one finds

$$W = \sum_{\vec{K}} W_{\vec{K}}, \quad W_{\vec{K}} = \frac{K^2 V}{2\pi} \vec{a}_{\vec{K}} \cdot \vec{a}_{\vec{K}}^* \tag{5.267}$$

Here, $W_{\vec{K}}$ is the energy of a plane monochromatic wave. Each wave essentially vibrates as an independent simple harmonic oscillator. The Hamiltonian form for a set of linear oscillators is

$$W = H = \sum_{\vec{K}} \frac{1}{2}\left(P_{\vec{K}}^2 + \omega_{\vec{K}}^2 Q_{\vec{K}}^2\right) = \sum_{\vec{K}} H_{\vec{K}} \tag{5.268}$$

where $\vec{P}_{\vec{K}}$ and $\vec{Q}_{\vec{K}}$ are the generalized momentum and coordinate for each oscillator. Each oscillator corresponds to one mode of the EM field, for a selected wave vector and polarization. Then to make Equations 5.267 and 5.268 match, the coefficients of expansion $\vec{a}_{\vec{K}}$ need to be

$$\vec{a}_{\vec{K}} = \frac{i}{K}\sqrt{\frac{\pi}{V}}\left(\vec{P}_{\vec{K}} - i\omega_{\vec{K}}\vec{Q}_{\vec{K}}\right) \tag{5.269}$$

Thus, Equation 5.261 is the expansion of the fields with respect to oscillators, and Equation 5.269 defines the amplitudes needed to actually calculate the vector potential. The equation of motion of one of the oscillators is

$$\ddot{\vec{Q}}_{\vec{K}} + \omega_{\vec{K}}^2 \vec{Q}_{\vec{K}} = 0 \tag{5.270}$$

Each of the vectors $\vec{Q}_{\vec{K}}$ and $\vec{P}_{\vec{K}}$ is perpendicular to the wave vector $\vec{K}$. That is, they each have two independent components. Directions of these vectors are defined by the direction of the polarization of the corresponding moving waves. Consider the plane perpendicular to vector $\vec{K}$. Denote the two components of vectors $\vec{Q}_{\vec{K}}$ and $\vec{P}_{\vec{K}}$ in that plane by $\vec{Q}_{\vec{K}j}$ and $\vec{P}_{\vec{K}j}$, $j = 1, 2$, respectively. Then the Hamiltonian actually involves sums also over the polarization index j:

$$H = \sum_{\vec{K}j} H_{\vec{K}j} = \sum_{\vec{K}j} \frac{1}{2}\left(P_{\vec{K}j}^2 + \omega_{\vec{K}j}^2 Q_{\vec{K}j}^2\right) \tag{5.271}$$

From Equations 5.266 and 5.269, the vector potential is

$$\vec{A} = \sqrt{\frac{4\pi}{V}} \sum_{\vec{K}j} \frac{1}{K} \left[cK\vec{Q}_{\vec{K}j} \cos(\vec{K} \cdot \vec{r}) - \vec{P}_{\vec{K}j} \sin(\vec{K} \cdot \vec{r}) \right] \qquad (5.272)$$

Consider again Equations 5.258 and 5.260, then the EM fields are

$$\vec{E} = -\sqrt{\frac{4\pi}{V}} \sum_{\vec{K}} \left[cK\vec{Q}_{\vec{K}} \sin(\vec{K} \cdot \vec{r}) + \vec{P}_{\vec{K}} \cos(\vec{K} \cdot \vec{r}) \right]$$

$$\vec{B} = -\sqrt{\frac{4\pi}{V}} \sum_{\vec{K}} \frac{1}{K} \left[cK\vec{K} \times \vec{Q}_{\vec{K}} \sin(\vec{K} \cdot \vec{r}) + \vec{K} \times \vec{P}_{\vec{K}} \cos(\vec{K} \cdot \vec{r}) \right]$$

$$(5.273)$$

From here, we have the usual transverse EM wave relation

$$\vec{B} = \frac{\vec{K}}{K} \times \vec{E} = \hat{K} \times \vec{E} \qquad (5.274)$$

We use relation (5.271) to evaluate the classical canonical partition function, which involves only simple Gaussian integrals:

$$Z = \prod_{\vec{K}} \prod_{j=1}^{2} \iint \exp \left\{ -\frac{P_{\vec{K}j}^2 + \omega_{\vec{K}j}^2 Q_{\vec{K}j}^2}{2T} \right\} dP_{\vec{K}j} dQ_{\vec{K}j} = \prod_{\vec{K}} \prod_{j=1}^{2} \frac{2\pi T}{\omega_{\vec{K}j}} \qquad (5.275)$$

It is interesting to note, of course, that historically the EM field was first analyzed this way, before the development of the quantum theory. Let us see the result of this path. For the free and internal energies we have, respectively

$$F = -T \ln Z = -T \sum_{\vec{K},j} \ln \frac{2\pi T}{\omega_{\vec{K}j}} \qquad (5.276)$$

$$E = -\frac{\partial}{\partial \beta} \ln Z = \sum_{\vec{K}} \sum_{j=1}^{2} T = 2 \sum_{\vec{K}} T \qquad (5.277)$$

The Fourier expansion in Equation 5.240 has an infinite number of terms. For each wave vector $\vec{K}$, there are two independent polarizations, and each polarization is associated with a mean energy of T. This is the usual equipartition rule for classical oscillators. Once the sum over the infinite set of all wave vectors is considered, the mean energy in Equation 5.277 is seen to be infinite. Note that the divergence is coming from the unlimited larger wave vectors, that is, the sum includes radiation with arbitrarily short wavelengths. This difficulty with the classical theory for radiation was known as the *ultraviolet catastrophe*. Nevertheless, with the uniform distribution of thermal energy over the accessible degrees of freedom, every degree of freedom in the radiation is assigned the mean energy T, including those with arbitrarily short wavelengths.

Although the total energy is infinite, we can evaluate the mean energy assigned to a finite number of oscillators, corresponding to those with frequencies in some interval from ω to $\omega + d\omega$. This is closely related to the frequency spectrum of the thermal radiation. From the relations (5.243) together with the dispersion relation (5.249), we have

$$\omega^2 = K^2 c^2 = (2\pi)^2 \left(\frac{n_x^2}{A^2} + \frac{n_y^2}{B^2} + \frac{n_z^2}{C^2} \right) c^2 \tag{5.278}$$

This is the equation of a sphere in the space of coordinates $x' = n_x/A$, $y' = n_y/B$, $z' = n_z/C$, with the radius $R' = (\omega/2\pi c)$ and the volume $V' = (4\pi/3)R'^3$. However, the volume of such a sphere scaled by factors of A, B, C for each axis is approximately equal to the number of combinations of the three integral numbers n_x, n_y, n_z that fall within the sphere:

$$\frac{n_x^2}{A^2} + \frac{n_y^2}{B^2} + \frac{n_z^2}{C^2} \le R'^2 \tag{5.279}$$

(The reader could also try to justify this by thinking about how the volume of a sphere is modified when the axes are stretched. For instance, in 2D, how does the area of an ellipse compare to the area of an undistorted circle?) These choices of the integers give modes with frequencies less than the selected ω. Thus, the number of wave vectors for which the frequency is less than or equal to ω is

$$N_K \approx \frac{4\pi}{3} R'^3 ABC = \frac{V\omega^3}{6\pi^2 c^3} \tag{5.280}$$

Here, we used the fact that the system containing the radiation has volume $V = ABC$. For every wave vector there are two modes at the same frequency but with different polarizations. Also note that modes with $+K_x$ and $-K_x$ are independent; they are waves traveling in opposite directions. Thus, the total number of oscillators with the frequencies less than or equal to ω is

$$N(\omega) = 2N_K \approx \frac{V\omega^3}{3\pi^2 c^3} \tag{5.281}$$

Then by differentiation, the number of oscillators with frequencies lying in the interval from ω to $\omega + d\omega$ is

$$dN(\omega) = \frac{V\omega^2}{\pi^2 c^3} d\omega \tag{5.282}$$

As each mode carries an average thermal energy equal to T, the mean radiation energy in a frequency interval $d\omega$ is

$$dE(\omega) = TdN(\omega) = T\frac{V\omega^2}{\pi^2 c^3}d\omega \qquad (5.283)$$

This leads to the spectral energy density, defined from $u(\omega)d\omega = dE(\omega)V$,

$$u(\omega)d\omega = T\frac{\omega^2 d\omega}{\pi^2 c^3} \qquad (5.284)$$

Relation (5.283) or (5.284) is the *Rayleigh–Jeans law* as we have seen earlier (Equation 5.214). These formulas are true in experiments only for exceedingly high and low temperatures. But as we already know, they do not apply in the high-frequency limit. For ever-increasing frequency, these formulas yield a diverging energy density, and for the total radiation energy density, we have an infinite value

$$u = \int_0^\infty u(\omega)d\omega = \frac{T}{\pi^2 c^3}\int_0^\infty \omega^2 d\omega \qquad (5.285)$$

As mentioned above, the integral is divergent due to the high frequency oscillations of short wavelengths, hence the difficulty is called the *ultraviolet catastrophe*. It must be corrected by applying quantum theory.

In the quantum theory for this problem, the oscillator energies are used to construct the Hamiltonian and Equation 5.268 is replaced by a sum of quantum harmonic oscillators

$$W = H = \sum_{\vec{K}j} \hbar\omega_{\vec{K}}\left(n_{\vec{K}j} + \frac{1}{2}\right) = \sum_{\vec{K}j} H_{\vec{K}j} \qquad (5.286)$$

The number of quanta $n_{\vec{K}j}$ in each oscillator can now take on only the discrete values, 0, 1, 2, 3, and so on. Ignoring the divergent but irrelevant zero point energy, this leads to the grand canonical quantum partition function instead of Equation 5.275

$$\Xi = \prod_{\vec{K}}\prod_{j=1}^2 \sum_{n_{\vec{K}j}=0}^\infty \exp\left\{\beta(\mu - \hbar\omega_{\vec{K}})n_{\vec{K}j}\right\} = \prod_{\vec{K}}\prod_{j=1}^2 \frac{1}{1 - \exp\left\{\beta(\mu - \hbar\omega_{\vec{K}})\right\}} \qquad (5.287)$$

This is specified in terms of the chemical potential, μ. The canonical partition function was found earlier, where Planck radiation has already been discussed concerning other details. There is a grand thermodynamic potential associated with each independent mode of the EM field, according to

$$\Omega_{\vec{K}j} = -T\ln\Xi_{\vec{K}j} = -T\ln\left(\frac{1}{1 - \exp\left\{\beta(\mu - \hbar\omega_{\vec{K}})\right\}}\right) = T\ln\left(1 - \exp\left\{\beta(\mu - \hbar\omega_{\vec{K}})\right\}\right)$$

$$(5.288)$$

This can be used to find all the thermodynamic properties. Let us just review the spectrum of the radiation. In particular, from the definition of $\Omega_{\vec{K}j}$, we know the average number of quanta present in each mode via the expression

$$\bar{n}_{\vec{K}j} = -\frac{\partial \Omega_{\vec{K}j}}{\partial \mu} = \frac{\exp\{\beta(\mu - \hbar\omega_{\vec{K}})\}}{1 - \exp\{\beta(\mu - \hbar\omega_{\vec{K}})\}} \rightarrow \frac{1}{\exp\{\beta\hbar\omega_{\vec{K}}\} - 1} \tag{5.289}$$

The last step results because the chemical potential for photons as quasi-particles is zero. This is the Bose–Einstein distribution encountered earlier. Then, the population of the quantum high-frequency modes is limited by this factor, compared to that in the classical theory.

To get the radiation energy per unit frequency interval, we can use Equation 5.282 for the number of oscillators per unit frequency interval, and then multiply that by the average occupation numbers times the photon energy $\hbar\omega$. This produces

$$dE(\omega) = \bar{n}\hbar\omega\, dN(\omega) = \frac{\hbar\omega}{\exp(\beta\hbar\omega) - 1} \frac{V\omega^2}{\pi^2 c^3}\, d\omega \tag{5.290}$$

This is the Planck radiation law that was derived earlier in this chapter (see Equation 5.201). Instead of increasing indefinitely for large frequency, it has a peak in the distribution, corresponding to the Wien displacement law. The wavelength at which the peak occurs shifts to shorter and shorter values as the temperature is increased. Hence, with increasing temperature, a heated object first tends to glow in the infrared, then in the red, then through the other colors of the visible, even passing to become bluish, and eventually glowing more toward the ultraviolet. This effect can be used to estimate the surface temperatures of stars. For any temperature, the spectral distribution goes to zero both for very high and for very low frequencies, unlike the prediction of classical statistical mechanics. It has been verified to incredibly high precision, for example, when used to analyze the temperature of the cosmic microwave background radiation (about 2.74 K). This was a tremendous success for Planck's postulate of discrete energy levels, and serves very much as a basis for the credibility of the quantum theory.

6 Electron Gas in a Magnetic Field

6.1 EVALUATION OF DIAMAGNETISM OF A FREE ELECTRON GAS; DENSITY MATRIX FOR A FREE ELECTRON GAS

A crystal lattice is a geometrical entity with certain symmetry properties. Atoms occupy lattice sites but vibrate around those sites, leading to many of the mechanical and thermal properties of solids. Here, we consider this atomic model for matter to include magnetic effects, concentrating on the contributions due to the free electrons.

An electron has an intrinsic magnetic moment that is associated with its spin angular momentum. It also has an orbital magnetic moment associated with its orbital angular momentum. The atomic nuclei also possess a magnetic moment and an angular momentum.

Magnetic phenomena in solids may be described in terms of macroscopic magnetic quantities such as the magnetic field $\vec{H}$, magnetic induction $\vec{B}$, the magnetic susceptibility χ_M, the permeability μ_M, magnetization $\vec{M}$, and the magnetic moment $\vec{m}$. The magnetic susceptibility χ_M gives the linear response of the magnetization when a magnetic field is applied to a material. It may be either negative or positive. If χ_M is positive then the material displays *paramagnetism* and if χ_M is negative the material displays *diamagnetism*. For paramagnetism, the magnetization is parallel to the applied magnetic field; for diamagnetism, the magnetization is opposite to the direction of the applied magnetic field. There are other magnetic effects where a material can possess a magnetic dipole moment in the absence of an applied field, including *ferromagnetism* (aligned permanent dipoles), *antiferromagnetism* (anti-aligned permanent dipoles), and *ferrimagnetism* (antialigned permanent dipoles of different magnitudes).

Each electron in every atom has an induced diamagnetic response when affected by an externally applied magnetic field. The induced magnetic moment is usually very weak. In this way, there exists diamagnetism for all materials. Usually it is overshadowed by other magnetic effects present due to permanent magnetic dipoles (such as paramagnetism and ferromagnetism). When each atom within a crystal has all of its electrons in paired states of opposite spins, so as to exclude permanent dipoles, diamagnetic response can be observed.

In order that we may investigate diamagnetism we have to study a model atom for which all electrons are in paired states. Let us consider N free electrons moving in a crystalline lattice placed in a magnetic field $\vec{H}$ at absolute temperature T. It should be noted that the magnetic field $\vec{H}$ is the local field acting on an atom. Since it also includes the magnetic field produced by all other atoms, it is not quite the same as the external field. When the concentration of the atoms is small the distinction between the external and local fields becomes immaterial. If that is not the case, the local field that is truly responsible for magnetic forces is the magnetic induction, $\vec{B}$.

Each electron has a mass m. Recall the quantum treatment of an electron acted on by an electromagnetic field. The Hamiltonian $\hat{H}$ of a charge e in a magnetic field $\vec{H}$ is

$$\hat{H} = \frac{1}{2m}\left(\hat{\vec{p}} - \frac{e}{c}\hat{\vec{A}}\right)^2 + e\Phi \tag{6.1}$$

where $\hat{\vec{p}} = -i\hbar\vec{\nabla}$ is the momentum operator, Φ is the scalar potential, and $\vec{A}$ is the vector potential that determines the magnetic induction

$$\vec{B} = \vec{\nabla} \times \vec{A} \tag{6.2}$$

Because the vector potential depends on position, we have the commutation relation

$$\hat{\vec{p}} \cdot \vec{A} - \vec{A} \cdot \hat{\vec{p}} = -i\hbar\vec{\nabla} \cdot \vec{A} \tag{6.3}$$

Then, Equation 6.1 can be expressed as

$$\hat{H} = \frac{\hat{\vec{p}}^2}{2m} + \frac{ie\hbar}{mc}\vec{A} \cdot \vec{\nabla} + \frac{ie\hbar}{2mc}\vec{\nabla} \cdot \vec{A} + \frac{e^2}{2mc^2}\vec{A}^2 + e\Phi \tag{6.4}$$

Let $X = X(\vec{r})$ be an arbitrary function of the coordinates; then the fields are invariant under a gauge transformation of the potentials

$$\vec{A} = \vec{A}' - \vec{\nabla}X, \quad \Phi = \Phi' + \frac{1}{c}\frac{\partial X}{\partial t} \tag{6.5}$$

But $\vec{\nabla} \times \vec{\nabla}G = 0$, where G is any potential field, then

$$\vec{B} = \vec{\nabla} \times \vec{A} = \vec{\nabla} \times \vec{A}' \tag{6.6}$$

$$\vec{E} = -\vec{\nabla}\Phi - \frac{1}{c}\frac{\partial \vec{A}}{\partial t} = -\vec{\nabla}\Phi' - \frac{1}{c}\frac{\partial \vec{A}'}{\partial t} \tag{6.7}$$

If we consider Equations 6.1 through 6.7, then the Hamiltonian does change under this gauge transformation. We get

$$\hat{H} = \hat{H}' - \frac{ie\hbar}{mc}\vec{\nabla}X \cdot \vec{\nabla} - \frac{ie\hbar}{2mc}\Delta X - \frac{e^2}{2mc^2}\vec{A}' \cdot \vec{\nabla}X + \frac{e^2}{2mc^2}\left(\vec{\nabla}X\right)^2 + \frac{e}{c}\frac{\partial X}{\partial t} \tag{6.8}$$

If for the first gauge we have

$$ i\hbar \frac{\partial \Psi}{\partial t} = \hat{H}\Psi \tag{6.9} $$

then for the second we have

$$ i\hbar \frac{\partial \Psi'}{\partial t} = \hat{H}\Psi' \tag{6.10} $$

Equation 6.9 is not equal to Equation 6.10 but the field remains unchanged. It follows that the wave functions Ψ and Ψ' differ by a phase factor as their moduli are equal. That is, for the Schrödinger equation to remain invariant, every wave function must be multiplied by a phase factor

$$ \Psi = \Psi' \exp\left\{-i\frac{e}{\hbar c} X\right\} \tag{6.11} $$

then the wave functions in Equations 6.9 and 6.10 describe one and the same state.

The electron gas as charged particles has diamagnetism, that is, a magnetization which is in the opposite sense to the field. The diamagnetism is connected to the *Larmor precession*. Classically, a uniform magnetic field $\vec{H}$ causes charges to rotate about $\vec{H}$ with a constant angular velocity. The Larmor frequency is given as

$$ \omega_L = \frac{eH}{2mc} \tag{6.12} $$

In classical physics, the Larmor precession for electrons leads us to the diamagnetic susceptibility

$$ \chi_d = -\frac{e^2}{6mc^2} \sum_i \overline{r_i^2} \tag{6.13} $$

The electrons precess around the applied field, and the current associated with their motion generates a magnetic moment whose direction opposes the applied field. By this expression, classical electrodynamics enables us to explain the diamagnetism of electrons coupled to the nucleus.

To obtain the magnetic moment, we first need to evaluate the partition function

$$ Z = \int \exp\left\{-\frac{E(q,p)}{T}\right\} d\Gamma, \quad d\Gamma = dqdp \tag{6.14} $$

(where p is the generalized momentum and q the generalized coordinates) and if we substitute into this the energy expression (6.1) without operators, we see that in classical physics the diamagnetism of a gas of free electrons is equal to zero. Consider a reversible change at constant temperature. Then the work done on a system is equal to the change in its free energy F and the magnetic moment of a system is evaluated through the relation

$$M = -\frac{\partial F}{\partial H} \qquad (6.15)$$

Then, if no extra work is done on the system when the field changes, we get immediately $M = 0$. This argument is discussed further below.

In 1930, Landau understood that the diamagnetism of an electronic gas is a quantum effect and he began to calculate not the statistical integral but the statistical sum, that is

$$Z = \sum_i \exp\left\{-\frac{E_i}{T}\right\} \qquad (6.16)$$

For this, it necessary to know the eigen energies E_i by solving the stationary Schrödinger equation

$$\hat{H}\Psi_I = E_I \Psi_I \qquad (6.17)$$

The energy of an electron in a magnetic field is quantized. Let us examine first the operator

$$\exp\left\{-\frac{\hat{H}}{T}\right\} = \sum_{n=0}^{\infty} \frac{(-1)^n}{n! T^n} \hat{H}^n \qquad (6.18)$$

The eigenstates of powers of $\hat{H}$ satisfy

$$\hat{H}^n \Psi_i = E_i^n \Psi_i \qquad (6.19)$$

Now, consider the sum over diagonal matrix elements of the operator (6.18), that is

$$\sum_{i=1}^{n} \int \Psi_i^* \exp\left\{-\frac{\hat{H}}{T}\right\} \Psi_i d\Gamma = \sum_i \exp\left\{-\frac{E_i}{T}\right\} = Z \qquad (6.20)$$

It follows that partition function is a trace

$$Z = \mathrm{Tr}\left(\exp\left\{-\frac{\hat{H}}{T}\right\}\right) \equiv \mathrm{Tr}(\exp\{-\beta\hat{H}\}), \quad \beta = \frac{1}{k_B T} \qquad (6.21)$$

We can also introduce the Bloch density matrix or the statistical operator

$$\rho(\vec{r}',\vec{r},\beta) = \sum_i \Psi_i^*(\vec{r}')\exp\{-\beta\hat{H}\}\Psi_i(\vec{r}) \tag{6.22}$$

that suffices to calculate the free energy in Fermi–Dirac statistics. Then, in coordinate representation, the partition function is the trace of ρ in real space:

$$Z = \int \rho(\vec{r},\vec{r},\beta) \, d\vec{r} \tag{6.23}$$

The density matrix (6.22), as is readily shown, satisfies the Bloch equation

$$\frac{\partial\rho}{\partial\beta} = -\hat{H}\rho \tag{6.24}$$

The boundary condition is that at $\beta = 0$

$$\rho(\vec{r}',\vec{r},0) = \delta(\vec{r}' - \vec{r}) \tag{6.25}$$

This expresses the completeness condition on the eigenfunctions. What is the advantage of working with ρ rather than the Landau wave functions $\Psi(\vec{r})$? The advantage is that we can more easily treat an infinite crystal. This avoids the complications of the boundaries of the system. In order to find Z by explicit calculation of E_i, we have to take a large but finite volume. Here, we have to show that the susceptibility is independent of the shape and the nature of the volume, as the volume is increased without limit.

For a uniform field, we assume that $\vec{B} = \vec{H} = (0,0,H)$, along the z-axis. One vector potential that leads to this field is

$$\vec{A} = \frac{1}{2}\vec{H} \times \vec{r} \tag{6.26}$$

The Cartesian components are simple:

$$A_x = -\frac{1}{2}Hy, \quad A_y = \frac{1}{2}Hx, \quad A_z = 0 \tag{6.27}$$

Following Sondheimer and Wilson (1951) [41], we arrive at the form of the solution by the invariance properties of the Schrödinger equation with respect to the change of coordinates. It should be noted that the electrons are free to move through the entire space. An infinite crystal is assumed.

No physical property is dependent on the choice of origin of the coordinate system. Consider if $\vec{r} \to \vec{r} + \vec{a}$, where $\vec{a}$ is a translation vector, then

$$\vec{A}(\vec{r}) \to \vec{A}' = \vec{A}(\vec{r} + \vec{a}) = \vec{A}(\vec{r}) + \frac{1}{2}\vec{H} \times \vec{a} = \vec{A}(\vec{r}) + \frac{1}{2}\vec{\nabla}[\vec{r} \cdot (\vec{H} \times \vec{a})] \quad (6.28)$$

If we compare Equation 6.28 with Equation 6.5, then we have the function that generated the gauge transformation

$$X = \frac{1}{2}[\vec{r} \cdot (\vec{H} \times \vec{a})] \quad (6.29)$$

and considering Equation 6.11, we have

$$\Psi' = \Psi \exp\left\{\frac{ie}{2\hbar c}[\vec{r} \cdot (\vec{H} \times \vec{a})]\right\} \quad (6.30)$$

Thus, from the above, the effect of the translation is

$$\hat{T}_{\vec{a}}\Psi(\vec{r}) = \Psi(\vec{r} + \vec{a})\exp\left\{\frac{ie}{2\hbar c}[\vec{r} \cdot (\vec{H} \times \vec{a})]\right\} \quad (6.31)$$

where $\hat{T}_{\vec{a}}$ is the translation operator by the vector $\vec{a}$. Thus, for the invariance of the Schrödinger equation, every shifted wave function is multiplied by the phase factor

$$\exp\left\{\frac{ie}{2\hbar c}[\vec{r} \cdot (\vec{H} \times \vec{a})]\right\} \quad (6.32)$$

Thus, there follows

$$\rho' = \rho'(\vec{r}',\vec{r},\beta) = \rho(\vec{r}' + \vec{a},\vec{r} + \vec{a},\beta) = \rho(\vec{r}',\vec{r},\beta)\exp\left\{\frac{ie}{2\hbar c}(\vec{r} - \vec{r}') \cdot (\vec{H} \times \vec{a})\right\}$$

$$= \rho(\vec{r}',\vec{r},\beta)\exp\left\{\frac{ie}{2\hbar c}\vec{H} \cdot [\vec{a} \times (\vec{r} - \vec{r}')]\right\} \quad (6.33)$$

Here, the translation on both indices of the density matrix was used in the first step, and then a phase factor appears for both the shift in $\vec{r}$ and the shift in $\vec{r}'$.

Suppose $\rho'(\vec{r}',\vec{r},\beta)$ is of the form

$$\rho'(\vec{r}',\vec{r},\beta) = F(\vec{r}' - \vec{r},\beta)\exp\left\{\frac{ie}{2\hbar c}\vec{H} \cdot (\vec{r}',\vec{r})\right\} \quad (6.34)$$

Our next problem is to find the function $F(\vec{r}' - \vec{r})$ by choosing $\vec{H}$ such that the vector potential is as shown in Equation 6.27.

We suppress the temperature argument. If we consider the case $\vec{H} = 0$, then the Hamiltonian is proportional to the Laplacian operator only

$$\hat{H} = -\frac{\hbar^2}{2m}\Delta, \quad \frac{\partial\rho}{\partial\beta} = \frac{\hbar^2}{2m}\Delta\rho \tag{6.35}$$

Consider the theory of thermal conductivity, where the diffusion equation describes the flow of heat, using diffusion constant, D

$$\frac{\partial U}{\partial t} = D\Delta U \tag{6.36}$$

Our equation for ρ and the diffusion equation for U have the same mathematical structure. The solution for heat diffusion, starting from a unit delta-function pulse of energy initiated at point $\vec{r}'$, is

$$U = \frac{\exp\left\{-\dfrac{(\vec{r} - \vec{r}')^2}{4Dt}\right\}}{(4\pi Dt)^{3/2}} \tag{6.37}$$

Then the density matrix follows this solution, also with a delta-function initial condition, with replacements $D \to (\hbar^2/2m)$ and $t \to \beta$

$$\rho = \left(\frac{m}{2\pi\hbar^2\beta}\right)^{3/2} \exp\left\{-\frac{m}{2\hbar^2\beta}(\vec{r} - \vec{r}')^2\right\} \tag{6.38}$$

This could have alternatively been obtained from the Dirac density matrix for free electrons. If we consider projection on the z-axis, then Equation 6.38 is a solution for $\vec{H}_z = 0$.

If $\vec{H} \neq 0$, then the Hamiltonian has more terms

$$\hat{H} = \frac{\hbar^2}{2m}\Delta + \frac{ie\hbar}{mc}\vec{A}\cdot\vec{\nabla} + \frac{ie\hbar}{2mc}\vec{\nabla}\cdot\vec{A} + \frac{e^2}{2mc^2}\vec{A}^2 \tag{6.39}$$

Using the vector potential (6.1.9) to describe a uniform field along z

$$\vec{A}\cdot\vec{\nabla} = \frac{H}{2}\left(x\frac{\partial}{\partial y} - y\frac{\partial}{\partial x}\right), \quad \vec{\nabla}\cdot\vec{A} = 0, \quad \vec{A}^2 = \frac{H}{4}(x^2 + y^2) \tag{6.40}$$

and as $H = H_z$, then it follows that

$$\vec{H} \cdot (\vec{r}' \times \vec{r}) = H(x'y - y'x) \tag{6.41}$$

From here, it follows that in the magnetic field, the density matrix satisfies

$$\frac{\partial \rho}{\partial \beta} = \left[\frac{\hbar^2}{2m}\Delta + \frac{ie\hbar H}{2mc}\left(y\frac{\partial}{\partial x} - x\frac{\partial}{\partial y} \right) - \frac{e^2 H^2}{8mc^2}(x^2 + y^2) \right]\rho \tag{6.42}$$

We solve Equation 6.42 using the boundary condition (6.38) and trying a solution in the form (6.34)

$$\rho(\vec{r}', \vec{r}, \lambda) = f(\beta)\exp\left\{ -\frac{m}{2\hbar^2\beta}(z - z')^2 - g(\beta)[(x - x')^2 + (y - y')^2] \right.$$
$$\left. + \frac{ieH}{2\hbar c}(x'y - y'x) \right\} \tag{6.43}$$

Let us express this as

$$\rho(\vec{r}', \vec{r}, \beta) \equiv f(\beta)G(\vec{r}', \vec{r}, \beta) \tag{6.44}$$

If we substitute Equation 6.44 into Equation 6.42, then we have

$$\frac{\partial \rho}{\partial \beta} = G(\vec{r}', \vec{r}, \beta)\left[\frac{\partial f}{\partial \beta} + \frac{mf}{2\hbar^2\beta^2}(z - z')^2 - f\frac{\partial g}{\partial \beta}((x - x')^2 + (y - y')^2) \right]$$
$$= \left[\frac{\hbar^2}{2m}\Delta + \frac{ie\hbar H}{2mc}\left(y\frac{\partial}{\partial x} - x\frac{\partial}{\partial y} \right) - \frac{e^2 H^2}{8mc^2}(x^2 + y^2) \right]\rho \tag{6.45}$$

Let us evaluate the RHS of Equation 6.45. One term involves a Laplacian of a phase like

$$\Delta \exp\{\Phi\} = \nabla\nabla\exp\{\Phi\} = \nabla(\nabla\exp\{\Phi\}) = [\Delta\Phi + (\nabla\Phi)^2]\exp\{\Phi\} \tag{6.46}$$

then it follows that

$$\frac{\hbar^2}{2m}\Delta\rho = \frac{\hbar^2}{2m}fG(\vec{r}', \vec{r}, \beta)\left[-\frac{m}{\hbar^2\beta} - 4g + \frac{m^2}{\hbar^2\beta^2}(z - z')^2 \right.$$
$$\left. + \left(2g(x - x') + \frac{ieH}{2\hbar c}y' \right) + \left(2g(y - y') + \frac{ieH}{2\hbar c}x' \right)^2 \right]$$

$$= G(\vec{r}',\vec{r},\beta)\left[-\frac{f}{2\beta} - \frac{2\hbar^2 fg}{m} + \frac{fm}{2\hbar^2\beta^2}(z-z')^2 + \frac{2\hbar^2 fg^2}{m}h\right] \quad (6.47)$$

$$h = \frac{2\hbar^2 fg^2}{m}\left[(x-x')^2 + (y-y')^2 - \frac{e^2H^2}{8mc^2}f(x'^2+y'^2) + \frac{ie\hbar H}{mc}fg(xy'-yx')\right]$$

$$\frac{ie\hbar H}{2mc}\left(y\frac{\partial}{\partial x} - x\frac{\partial}{\partial y}\right)\rho = \frac{ie\hbar}{2mc}fG(\vec{r}',\vec{r},\beta)\left[y\left(-2g(x-x') - \frac{ieH}{2\hbar c}y'\right)\right.$$

$$\left. -x\left(-2g(y-y') + \frac{ieH}{2\hbar c}x'\right)\right]$$

$$= G(\vec{r}',\vec{r},\beta)\left[\frac{ie\hbar H}{2mc}fg(x'y-y'x) + \frac{e^2H^2}{4mc^2}f(y'y+xx')\right] \quad (6.48)$$

$$-\frac{e^2H^2}{8mc^2}(x^2+y^2)\rho = -\frac{e^2H^2}{8mc^2}(x^2+y^2)fG(\vec{r}',\vec{r},\lambda) \quad (6.49)$$

The sum of Equations 6.47, 6.48, and 6.49 yields Equation 6.45. This leads to

$$\frac{df}{d\beta} + \frac{f}{2\beta} + \frac{2\hbar^2 fg}{m} = \left[(x-x')^2 + (y-y')^2\right]\left[f\frac{dg}{d\beta} + \frac{2\hbar^2 fg^2}{m} - \frac{e^2H^2f}{8mc^2}\right] \quad (6.50)$$

The equation applies for any values of the position coordinates. The functions f and g depend only on inverse temperature β. Then the only way to satisfy the equation is if both sides are zero. This leads to two independent equations

$$\frac{df}{d\beta} + \frac{f}{2\beta} + \frac{2\hbar^2 fg}{m} = 0 \quad (6.51)$$

and

$$f\frac{dg}{d\beta} + \frac{2\hbar^2 fg^2}{m} - \frac{e^2H^2f}{8mc^2} = 0 \quad (6.52)$$

We cannot have $f = 0$ or the density matrix would be equal to zero. Hence, we cancel out f in Equation 6.52 and we have

$$d\beta = \frac{dg}{\dfrac{e^2H^2}{8mc^2} - \dfrac{2\hbar^2 g^2}{m}} = \frac{d\left(\dfrac{4\hbar cg}{eH}\right)}{\dfrac{e^2H^2}{8mc^2}\left(1 - \dfrac{16\hbar^2c^2g^2}{e^2H^2}\right)\dfrac{4\hbar c}{eH}} \quad (6.53)$$

The factor $(e\hbar/2mc) = \mu_B$ is the Bohr magneton. Then

$$\frac{4\hbar c}{eH} \frac{e^2H^2}{8mc^2} = \frac{e\hbar}{2mc}H = \mu_B H \tag{6.54}$$

This simplifies the differential equation for g into

$$\mu_B H \, d\beta = \frac{d\left(\dfrac{4\hbar cg}{eH}\right)}{1 - \left(\dfrac{4\hbar cg}{eH}\right)^2} \tag{6.55}$$

which has the form

$$\alpha \, d\beta = \frac{dx}{1 - x^2} \tag{6.56}$$

The solution is

$$\alpha\beta = \frac{1}{2}\ln\left(\frac{x+1}{x-1}\right), \quad x > 1 \tag{6.57}$$

or

$$x = \frac{\exp\{2\alpha\beta\} + 1}{\exp\{2\alpha\beta\} - 1} = \coth\alpha\beta \tag{6.58}$$

The constant of integration is taken to be zero. From Equation 6.55, $x = 4\hbar cg/eH$, which implies that

$$g = \frac{eH}{4\hbar c}\coth(\mu_B H\beta) \tag{6.59}$$

and

$$\lim_{H\to 0} g = \frac{m}{2\hbar^2\beta} \tag{6.60}$$

which is consistent with the limiting solution (6.38).
    For $x < 1$, we have the solution

$$g = \frac{eH}{4\hbar c}\tanh(\mu_B H\beta) \tag{6.61}$$

and this gives

$$\lim_{H \to 0} g = 0 \qquad (6.62)$$

which does not satisfy the limiting transformation.

Let us next solve Equation 6.51, that is, the equation for $f(\beta)$

$$\frac{df}{d\beta} + \frac{f}{2\beta} + \frac{2\hbar^2 fg}{m} = 0 \qquad (6.63)$$

This may be written in the form

$$\frac{df}{f} + \frac{d\beta}{2\beta} + \coth(\mu_B H\beta)d(\mu_B H\beta) = 0 \qquad (6.64)$$

Then it follows that

$$f = \frac{C}{\sqrt{\beta} \sinh(\mu_B H\beta)} \qquad (6.65)$$

Let us find the constant C. The limiting solution (6.38) shows that

$$\lim_{H \to 0} f = \left(\frac{m}{2\pi\hbar^2\beta}\right)^{3/2} \qquad (6.66)$$

whereas here we need to use

$$\sinh(\mu_B H\beta) \approx \mu_B H\beta \quad \text{for } \mu_B H\beta \ll 1 \qquad (6.67)$$

Thus

$$\frac{C}{\sqrt{\beta}\mu_B H\beta} = \left(\frac{m}{2\pi\hbar^2}\right)^{3/2} \frac{1}{\beta^{3/2}} \qquad (6.68)$$

and the constant is

$$C = \mu_B H \left(\frac{m}{2\pi\hbar^2}\right)^{3/2} \qquad (6.69)$$

(Note that this is the normalization for the density matrix corresponding to one electron within the system volume.) Hence, the density matrix modulus function is

$$f = \left(\frac{m}{2\pi\hbar^2\beta}\right)^{3/2} \frac{\mu_B H\beta}{\sinh(\mu_B H\beta)} \tag{6.70}$$

Consequently, the density matrix for an electron in a magnetic field along z is

$$\rho(\vec{r}',\vec{r},\beta) = \left(\frac{m}{2\pi\hbar^2\beta}\right)^{3/2} \frac{\mu_B H\beta}{\sinh(\mu_B H\beta)} \exp[\gamma(\vec{r}',\vec{r},\beta)] \tag{6.71}$$

where

$$\gamma(\vec{r}',\vec{r},\beta) = -\frac{m}{2\hbar^2\beta}\Big[(z - z')^2 + \mu_B H\beta\coth(\mu_B H\beta)((x - x')^2 + (y - y')^2)$$
$$- 2i\mu_B H\beta(x'y - y'x)\Big] \tag{6.72}$$

One can check that it reduces to Equation 6.38 when the field is allowed to vanish. The function represents both the dynamic and thermal effects on electrons.

The density matrix is applied to the problem of diamagnetism in the next heading.

6.2 EVALUATION OF FREE ENERGY

Considering the statistical physics (Boltzmann statistics), for the electron gas, the free energy F per unit volume is given by the relation

$$F = -n_0 T \ln Z \tag{6.73}$$

where $n_0 = (N/V)$ is the number density of particles and Z is the partition function

$$Z = \int \rho(\vec{r},\vec{r},\beta)d\vec{r} \tag{6.74}$$

which is the trace of the density matrix, that is, evaluated for $\vec{r}' = \vec{r}$. Thus, if we consider Equation 6.71, we have the partition function:

$$Z = \left(\frac{m}{2\pi\hbar^2\beta}\right)^{3/2} \frac{\mu_B H\beta}{\sinh(\mu_B H\beta)} \tag{6.75}$$

The magnetic moment M per unit volume is given by

$$M = -\frac{\partial F}{\partial H} = -n_0 \mu_B L(\mu_B H \beta) \tag{6.76}$$

where $L(a)$ is the Langevin function:

$$L(a) = \coth a - \frac{1}{a}, \qquad a = \mu_B H \beta \tag{6.77}$$

Because $M \neq 0$, there exists a magnetic moment. This is diamagnetism as $M \propto -L(a)$ and $\coth a$ is always greater than $1/a$. This effect is a pure quantum effect, at small field, proportional to the square of the Bohr magneton, $\mu_B = e\hbar/2mc$, itself dependent on Planck's constant. In the classical case where $\mu_B = 0$ ($\hbar = 0$), the result would be $M = 0$ and diamagnetism does not exist.

We can analyze the limiting cases for weak and strong magnetic fields.

1. If the magnetic field $\vec{H}$ is weak and the temperature T is high, that is, $a \ll 1$, then $L(a) \approx a/3$ and from Equation 6.76 we have

$$M = -\frac{n_0 \mu_B^2 H \beta}{3} = -\frac{CH}{T} \tag{6.78}$$

The constant of proportionality is

$$C = \frac{n_0 \mu_B^2}{3} \tag{6.79}$$

2. If the magnetic field $\vec{H}$ is strong and the temperature T is low, that is, $a \gg 1$, then $L(a) \approx 1$ and the magnetization is $M \approx -n_0 \mu_B$.

A graph of $L(a)$ versus a is shown in Figure 6.1. The slope of the curve at the origin is equal to 1/3. The curve asymptotically attains a maximum value at $L(a \to \infty) = 1$.

The expression for the diamagnetic susceptibility per unit volume of the system results from

$$\chi_d = \frac{dM}{dH} = -\frac{3C}{T}\left[\frac{1}{a^2} - \operatorname{cosech}^2 a\right] \tag{6.80}$$

Suppose the temperature T is high, that is, $a \ll 1$, then we may expand (6.80):

$$X_d \cong -\frac{C}{T}\left[1 - \frac{1}{12}a^2 + \cdots\right] \tag{6.81}$$

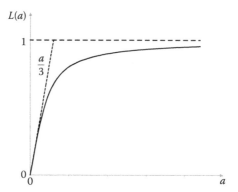

FIGURE 6.1   Variation of the Langevin function versus its argument.

Consider the case where H → 0 or also T → ∞, then from Equation 6.80, we have

$$\chi_d = -\frac{C}{T} \tag{6.82}$$

It is the correct high-temperature limit; however, it is not applicable to a degenerate electron gas (Boltzmann statistics was applied here).

6.3   APPLICATION TO A DEGENERATE GAS

Now, we consider the case when the particles are described by Fermi–Dirac statistics. The free energy per unit volume is given by the relation

$$\Omega = F - n_0\mu = -2T \sum_i \ln(1 + \exp\{\beta(\mu - E_i)\}) = 2\sum_i f(E_i) \tag{6.83}$$

where μ is the Fermi energy and the factor of 2 accounts for spin degeneracy. The contribution from each degree of freedom is

$$f(E_i) = -T \ln(1 + \exp\{\beta(\mu - E_i)\}) \tag{6.84}$$

Now define the standard function for the Laplace transformation of $f(E_i)$:

$$\tilde{F}(\beta) = \int_0^\infty f(E)\exp\{-\beta E\}dE \tag{6.85}$$

and the inverse Laplace transformation will be

$$f(E) = \frac{1}{2\pi i} \int_{C-i\infty}^{C+i\infty} \tilde{F}(\beta)\exp\{\beta E\}d\beta \tag{6.86}$$

The constant C is selected in such a way that all the singular points should be to the left of the integration contour in the complex plane. From Equations 6.85 and 6.86, we may express the statistical sum through f:

$$\sum_i f(E_i) = \sum_i \frac{1}{2\pi i} \int_{C-i\infty}^{C+i\infty} \tilde{F}(\beta) \exp\{\beta E_i\} d\beta = \frac{1}{2\pi i} \int_{C-i\infty}^{C+i\infty} \tilde{F}(\beta) Z(-\beta) d\beta \quad (6.87)$$

Keep in mind that the canonical partition function is

$$Z(\beta) = \sum_i \exp\{-\beta E_i\} \quad (6.88)$$

Similarly from Equation 6.85, for a different function represented as a Laplace transform pair

$$\frac{Z(\beta)}{\beta^2} = \int_0^\infty z(E) \exp\{-\beta E\} dE \quad (6.89)$$

and from Equation 6.86

$$z(E) = \frac{1}{2\pi i} \int_{C-i\infty}^{C+i\infty} \exp\{\beta E\} \frac{Z(\beta)}{\beta^2} d\beta \quad (6.90)$$

From Equation 6.87

$$\sum_i f(E_i) = \frac{1}{2\pi i} \int_{C-i\infty}^{C+i\infty} \tilde{F}(\beta)\beta^2 \int_0^\infty z(E) \exp\{\beta E\} dE\, d\beta$$

$$= \int_0^\infty z(E) dE \frac{1}{2\pi i} \int_{C-i\infty}^{C+i\infty} \tilde{F}(\beta)\beta^2 \exp\{\beta E\} d\beta \quad (6.91)$$

and from Equation 6.86, this last integral contains

$$\frac{1}{2\pi i} \int_{C+i\infty}^{C-i\infty} \tilde{F}(\beta)\beta^2 \exp\{\beta E\} d\beta = \frac{\partial^2 f}{\partial E^2} \quad (6.92)$$

from which it follows that

$$\sum_i f(E_i) = \int_0^\infty z(E) \frac{\partial^2 f}{\partial E^2} dE \quad (6.93)$$

We can make the definition

$$\frac{\partial f}{\partial E} = \frac{1}{1 + \exp\{\beta(E - \mu)\}} = f_0(E) \tag{6.94}$$

Relation (6.94) is the Fermi occupation number function. From the above, considering Equation 6.83, it follows that the free energy per volume is given from

$$F - n_0\mu = 2\int_0^\infty z(E)\frac{\partial f_0}{\partial E}\,dE \tag{6.95}$$

This is the standard Fermi–Dirac equation.

6.4 EVALUATION OF CONTOUR INTEGRALS

From Equation 6.90

$$z(E) = \frac{1}{2\pi i}\int_{C-i\infty}^{C+i\infty} \exp\{\beta E\}\frac{Z(\beta)}{\beta^2}\,d\beta \tag{6.96}$$

It should be noted that for a free electron gas, we already found the canonical partition function

$$Z(\beta) = \left(\frac{m}{2\pi\hbar^2\beta}\right)^{3/2}\frac{\mu_B H\beta}{\sinh(\mu_B H\beta)} \tag{6.97}$$

Then the function related via Laplace transform is

$$z(E) = \left(\frac{m}{2\pi\hbar^2}\right)^{3/2}\frac{\mu_B H}{2\pi i}\int_{C-i\infty}^{C+i\infty}\frac{\exp\{\beta E\}d\beta}{\beta^{5/2}\sinh(\mu_B H\beta)} \tag{6.98}$$

The integral requires us to take β as a complex valued parameter. For the poles of the integrand, we have $\beta = 0$ and $\sinh(\mu_B H\beta) = -i\sin(i\mu_B H\beta) = 0$, which implies that $\mu_B H\beta = \pm in\pi$ (where n is an integer); the latter gives poles of the first order, that is

$$\beta_n = \pm\frac{in\pi}{\mu_B H} \tag{6.99}$$

However, the singularity at $\beta = 0$ is a branch point due to the 5/2 power. A contour (AB) that places all singular points to its left is shown in Figure 6.2.

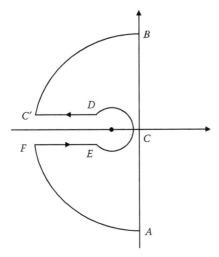

FIGURE 6.2   A contour (AB) that places all singular points to its left.

In order to find the integral (6.98), it is necessary to form a closed contour but the branch point and a branch cut to $-\infty$ should be excluded. It is necessary that all the poles should be on the left of C. Then this closed contour encloses all the poles of Equation 6.99. From the theory of residues, the integral around the whole contour is the sum over all its residues

$$\int_{AB} + \int_{BC'DEFA} = 2\pi i \sum_{n=1}^{\infty} \left[\text{Res}\left(\frac{in\pi}{\mu_B H} \right) + \text{Res}\left(-\frac{in\pi}{\mu_B H} \right) \right] \tag{6.100}$$

Let us check whether the arcs (BC′ and FA) at infinity contribute to the integral. β is a complex valued parameter, written with a modulus and a phase

$$\beta = |\beta| \exp\{i\phi\}, \quad \frac{\pi}{2} \le \phi \le \frac{3\pi}{2} \tag{6.101}$$

On the left of C, $\cos\phi < 0$ and $|\beta| > 0$. For $|\beta| \to \infty$, the hyperbolic sine function diverges faster than the exponential in the numerator. Then the integral at infinity along arcs BC′ and FA are both equal to zero. What remains is

$$J = \frac{\mu_B H}{2\pi i} \int_{C-i\infty}^{C+i\infty} \frac{\exp\{\beta E\} d\beta}{\beta^{5/2} \sinh(\mu_B H \beta)}$$

$$= \frac{\mu_B H}{2\pi i} \int_{FEDC'} \frac{\exp\{\beta E\} d\beta}{\beta^{5/2} \sinh(\mu_B H \beta)} + \mu_B H \sum_{n} \text{Res}\left(\frac{\exp\{\beta E\}}{\beta^{5/2} \sinh(\mu_B H \beta)} \right) \tag{6.102}$$

In the neighborhood of a pole β_n, we have

$$\sinh(\mu_B H\beta) = \sinh(\mu_B H [\beta - \beta_n] + \mu_B H\beta_n) = \mu_B H [\beta - \beta_n](-1)^n \quad (6.103)$$

Considering the fact that

$$\cosh(in\pi) = \cos(n\pi) = (-1)^n \quad (6.104)$$

we have

$$\mu_B H \mathrm{Res}\left(\frac{in\pi}{\mu_B H}\right) = (-1)^{n+1}\left(\frac{\mu_B H}{n\pi}\right)^{5/2} \exp\left\{\frac{in\pi E}{\mu_B H} - \frac{i\pi}{4}\right\} \quad (6.105)$$

which is $\mathrm{Res}(+\beta_n)$, which is not equal to $\mathrm{Res}(-\beta_n)$. Consequently

$$\mu_B H \sum_n \mathrm{Res}\left(\frac{\exp\{\beta E\}}{\beta^{5/2}\sinh(\mu_B H\beta)}\right) = \sum_n (-1)^{n+1}\left(\frac{\mu_B H}{n\pi}\right)^{5/2} \exp\left\{\frac{in\pi E}{\mu_B H} - \frac{i\pi}{4}\right\} \quad (6.106)$$

Let us consider now the contour in Figure 6.3.

$$\frac{1}{y^{5/2}\sinh y} = \frac{1}{y^{5/2}\left(y + \dfrac{y^3}{6} + 0(y^5)\right)} = \frac{1}{y^{7/2}}\left(1 + \frac{y^2}{6} + 0(y^4)\right)^{-1}$$

$$\underset{y \to 0}{=} \frac{1}{y^{7/2}} - \frac{1}{6y^{3/2}} + 0\left(y^{1/2}\right) \quad (6.107)$$

$$I_1 = \frac{\mu_B H}{2\pi i}\int_G \exp\left\{\frac{Ey}{\mu_B H}\right\}\left[\frac{1}{y^{5/2}\sinh y} - \left(\frac{1}{y^{7/2}} - \frac{1}{6y^{3/2}}\right)\right](\mu_B H)^{3/2}\,dy \quad (6.108)$$

$$I_2 = \frac{\mu_B H}{2\pi i}\int_{CDEF} \exp\left\{\frac{Ey}{\mu_B H}\right\}\left[\left(\frac{1}{y^{7/2}} - \frac{1}{6y^{3/2}}\right)\right](\mu_B H)^{3/2}\,dy, \quad I = I_1 + I_2 \quad (6.109)$$

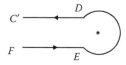

FIGURE 6.3   Contour diagram for the evaluation of Equation 6.102.

and as $y \to 0$, $I_1 = 0$ and thus $I_2 = I$. In the complex plane

$$\frac{1}{\Gamma(z)} = \frac{1}{2\pi i}\int t^{-z}\exp\{t\}dt \tag{6.110}$$

and thus

$$I_2 = \frac{8}{15\sqrt{\pi}}E^{5/2} - \frac{(\mu_B H)^2}{3\sqrt{\pi}}E^{1/2} \tag{6.111}$$

As $y \to 0$, $I_1 = 0$ which follows that I_1 is taken on the paths FE and C′D when in FE, $y = x\exp\{-i\pi\}$ and in C′D, $y = x\exp\{i\pi\}$, that is, they differ only in phase. On the lower edge of FE

$$x = |y|, \quad \sinh y = \sinh(x\exp\{-i\pi\}) = -\sinh x \tag{6.112}$$

$$\frac{1}{y^{7/2}} = \frac{1}{ix^{7/2}}, \quad \frac{1}{y^{5/2}\sinh y} = -\frac{(\exp\{i\pi\})^{5/2}}{x^{5/2}\sinh x} = -\frac{i}{x^{5/2}\sinh x} \tag{6.113}$$

Thus

$$I_1 = \frac{(\mu_B H)^{5/2}}{2\pi i}\frac{1}{i}\int_0^{+\infty}\left(\frac{1}{x^{7/2}} - \frac{1}{6x^{3/2}} - \frac{1}{x^{5/2}\sinh x}\right)\exp\left\{-\frac{Ex}{\mu_B H}\right\}dx \tag{6.114}$$

The integral does not contain the unit imaginary number i which implies that

$$I_1^{lower} = I_1^{upper} \tag{6.115}$$

Thus, the function $z(E)$ is found as

$$z(E) = \left(\frac{m}{2\pi\hbar^2}\right)^{3/2}\left\{\begin{array}{l}\dfrac{8}{15\sqrt{\pi}}E^{5/2} - \dfrac{(\mu_B H)^2}{3\sqrt{\pi}}E^{1/2} - 2(\mu_B H)^{5/2}\displaystyle\sum_{n=1}^{\infty}\dfrac{(-1)^n}{(n\pi)^{5/2}} \\[2ex] \times\cos\left[\dfrac{n\pi E}{\mu_B H} - \dfrac{\pi}{4}\right] + \dfrac{(\mu_B H)^{5/2}}{\pi}\displaystyle\int_0^{\infty}\left(\dfrac{1}{x^{7/2}} - \dfrac{1}{6x^{3/2}} - \dfrac{1}{x^{5/2}\sinh x}\right) \\[2ex] \times\exp\left\{-\dfrac{Ex}{\mu_B H}\right\}dx\end{array}\right. \tag{6.116}$$

6.5 DIAMAGNETISM OF A FREE ELECTRON GAS; OSCILLATORY EFFECT

It is known already that

$$\Omega = F - n_0\mu = 2\int_0^\infty z(E)\frac{\partial f_0}{\partial E}dE \tag{6.117}$$

and for its evaluation it is supposed that for strong degeneracy, we have $T \ll \mu$ and $\partial f_0/\partial E$ behaves as a δ-function. Then, it may be seen that (6.116) yields the desired susceptibility. Substituting Equation 6.116 into 6.117, $\partial f_0/\partial E$ behaves as a δ-function only in the first two terms. The fourth term may be exactly evaluated and the first three approximately.

Let us evaluate the fourth term:

$$J = \int_0^\infty \cos\left(\frac{n\pi E}{\mu_B H} - \frac{\pi}{4}\right)\frac{\partial f_0}{\partial E}dE = \mathrm{Re}\int_0^\infty \exp\left\{i\left(\frac{n\pi E}{\mu_B H} - \frac{\pi}{4}\right)\right\}\frac{\partial f_0}{\partial E}dE$$

$$= \mathrm{Re}\exp\left\{i\left(\frac{n\pi E}{\mu_B H} - \frac{\pi}{4}\right)\right\}\int_0^\infty \exp\left\{i\left(\frac{n\pi}{\mu_B H}(E-\mu)\right)\right\}\frac{\partial f_0}{\partial E}dE \tag{6.118}$$

It should be noted that

$$f_0(E) = \frac{1}{\exp\{\beta(E-\mu)\}+1} \tag{6.119}$$

The following change of variable is helpful:

$$\beta(E-\mu) = x \tag{6.120}$$

Then, considering

$$\frac{\partial f_0}{\partial x} = -\frac{1}{4\cosh^2(x/2)}, \quad \beta = \frac{n\pi}{\mu_B H}T \tag{6.121}$$

we have

$$J_1 = -\frac{1}{4}\int_{-\frac{\mu}{T}\to-\infty}^\infty \frac{\exp\{i\beta x\}}{\cosh^2(x/2)}dx \tag{6.122}$$

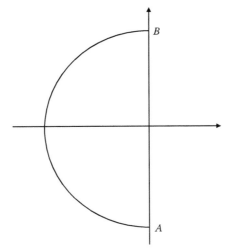

FIGURE 6.4 Contour diagram for the evaluation of Equation 6.122.

which is convenient to integrate in the complex plane. Use the contour in the plane as in Figure 6.4.

Let us consider

$$\cosh i\phi = \cos\phi \tag{6.123}$$

The poles are located at

$$x_m = i(2m + 1)\pi \tag{6.124}$$

The contour should be enclosed so that the point x_m is an interior point

$$\cosh\frac{x}{2} = \cosh\left(\frac{x - x_m}{2} + \frac{x_m}{2}\right) = \frac{i}{2}(-1)^m(x - x_m) \tag{6.125}$$

The expansion in the neighborhood of the point x_m, using (6.125), gives

$$\frac{\exp\{i\beta x\}}{\cosh^2(x/2)} = \frac{\exp\{i\beta x_m\} + i\beta\exp\{i\beta x_m\}(x - x_m)}{-1/4(x - x_m)^2} \tag{6.126}$$

Considering Equation 6.124, we find that

$$J_1 = -\frac{\pi\beta}{2\sinh(\pi\beta)} \tag{6.127}$$

and

$$J = -\frac{n\pi}{\mu_B H} \frac{T}{\sinh\left(\dfrac{\pi^2 nT}{\mu_B H}\right)} \cos\left(\frac{n\pi\mu_0}{\mu_B H} - \frac{\pi}{4}\right) \tag{6.128}$$

Thus

$$F = F_0(H,T) + \int_0^\infty \Phi(E)\,dE \tag{6.129}$$

where

$$\Phi(E) = 2\left(\frac{m}{2\pi\hbar^2}\right)^{3/2}\frac{\partial f_0}{\partial E}\left\{\frac{8}{15\sqrt{\pi}}E^{5/2} - \frac{(\mu_B H)^2}{3\sqrt{\pi}}E^{1/2} + \frac{(\mu_B H)^{5/2}}{\pi}\right.$$

$$\left.\int_0^\infty\left(\frac{1}{y^{7/2}} - \frac{1}{6y^{3/2}} - \frac{1}{y^{5/2}\sinh y}\right)\exp\left\{-\frac{Ey}{\mu_B H}\right\}dy\right\} \tag{6.130}$$

$$F_0(H,T) = n_0\mu + 2\left(\frac{m}{2\pi\hbar^2}\right)^{3/2}(\mu_B H)^{3/2}\frac{T}{\sqrt{\pi}}\sum_{n=1}^{\infty}\frac{(-1)^n}{n^{3/2}}\frac{\cos\left[\dfrac{n\pi}{\mu_B H} - \dfrac{\pi}{4}\right]}{\sinh\left(\dfrac{\pi^2 nT}{\mu_B H}\right)} \tag{6.131}$$

We explore the case of a weak field and low temperatures. In this situation

$$\frac{(\mu_B H)^{5/2}}{\pi}\int_0^\infty\left(\frac{1}{y^{7/2}} - \frac{1}{6y^{3/2}} - \frac{1}{y^{5/2}\sinh y}\right)\exp\left\{-\frac{Ey}{\mu_B H}\right\}dy < \frac{(\mu_B H)^2}{3\sqrt{\pi}}E^{1/2} \tag{6.132}$$

From thermodynamics, the magnetic moment is

$$M = -\frac{\partial F}{\partial H} = \frac{4}{3\sqrt{\pi}}\left(\frac{m}{2\pi\hbar^2}\right)^{3/2}\mu_B^2 H\int_0^\infty E^{1/2}\frac{\partial f_0}{\partial E}\,dE = \frac{1}{3}\frac{\sqrt{2m^3}}{\pi^2\hbar^3}\mu_B^2 H\int_0^\infty E^{1/2}\frac{\partial f_0}{\partial E}\,dE \tag{6.133}$$

where we have kept only the linear term.
But

$$\chi_d = \frac{dM}{dH} \tag{6.134}$$

Then

$$\chi_d = \frac{1}{3}\frac{\sqrt{2m^3}}{\pi^2\hbar^3}\mu_B^2\int_0^\infty E^{1/2}\frac{\partial f_0}{\partial E}dE \qquad (6.135)$$

and for $\chi_d < 0$, then on basis of quantum statistics

$$\chi_d = -\frac{1}{3}\chi_{para} \qquad (6.136)$$

where χ_d and χ_{para} are the diamagnetic and paramagnetic susceptibility. From Equation 6.135, we have

$$\chi_d = -\frac{1}{3}\frac{\sqrt{2m^3}}{\pi^2\hbar^3}\mu_B^2 n_0\mu^{1/2}\left[1-\frac{\pi^2}{12}\left(\frac{T}{\mu}\right)^2\right]$$

$$= -\frac{n_0\mu_B^2}{2\mu}\left[1-\frac{\pi^2}{12}\left(\frac{T}{\mu}\right)^2\right], \quad \mu = \frac{\hbar^2}{2m}(3\pi n_0)^{2/3} \qquad (6.137)$$

The Fermi energy has been related to the number density of electrons, n_0. As χ_d is one third of the paramagnetic susceptibility, it is easy for the paramagnetic term to dominate over the diamagnetic term.

As the electron is moving in a medium (crystal) subjected to a magnetic field and not in a vacuum, the mass of the electron m should be changed to the effective mass m_{eff}. Then in μ_B there appears a change as $\mu_B = (e\hbar/2mc)$ becomes $\mu_B^* = \mu_B(m/m_{eff})$, and the Fermi level changes to $\mu^* = \mu(m/m_{eff})$. With these modifications, we have

$$\chi_d = \frac{n_0\mu_B^{*2}}{2\mu^*}\left[1-\frac{\pi^2}{12}\left(\frac{T}{\mu^*}\right)^2\right] \qquad (6.138)$$

There is a corresponding change in the paramagnetic term, so

$$\chi_{para} = \frac{3n_0\mu_B^2}{2\mu^*}\left[1-\frac{\pi^2}{12}\left(\frac{T}{\mu^*}\right)^2\right] \qquad (6.139)$$

$$\chi_d - \chi_{para} = A\mu_B^2\left[1-\frac{1}{3}\left(\frac{m}{m_{eff}}\right)^2\right] \qquad (6.140)$$

The potential energy $-\vec{\mu}_B \cdot \vec{H}$ does not change while only the translational energy $E = (\hbar^2 K^2/2m_{eff})$ in the crystal changes.

Let us check the situation of a strong magnetic field. A strong magnetic field means

$$T \lesssim \mu_B H \ll \mu \qquad (6.141)$$

In such a case, the paramagnetic and diamagnetic portions of the magnetization may not be separated. In strong fields the magnetization of the metal shows oscillation dependence with respect to the magnetic field. This is called *De Haas–van Alphen effect*. This phenomenon is a powerful tool for the investigation of Fermi surfaces. It is a direct consequence of the underlying Fermi–Dirac statistics. The De Haas–van Alphen effect can be observed in pure specimens at low temperatures in strong magnetic fields.

The sum of all oscillatory terms in Equation 6.129 is a decreasing function of n_0. As a consequence, the most pronounced term will be $n_0 = 1$, which has $\cos((\pi\mu_0/\mu_B H) - (\pi/4))$. It is necessary that $\sinh(\pi^2 nT/\mu_B H)$ is small in Equation 6.129 so that the effect should be pronounced. In this way $(\pi^2 nT/\mu_B H) \approx 1$ is equivalent to Equation 6.141.

7 Magnetic and Dielectric Materials

7.1 THERMODYNAMICS OF MAGNETIC MATERIALS IN A MAGNETIC FIELD

We already know that the magnetic field does no work on charges on its own. This is connected with the fact that the Lorentz force is always perpendicular to the velocity of the particle it is acting on. We may see that magnetic forces acting on electric currents $\vec{j}$ do not perform any work directly on the charges. The elementary work of the magnetic force may be written as follows:

$$\delta W = \vec{j} \cdot \vec{F}_L \, dV \tag{7.1}$$

where

$$\vec{F}_L = \frac{1}{c} \vec{j} \times \vec{B} \tag{7.2}$$

is the Lorentz force per unit volume, $\vec{j}$ is the current density, and $\vec{B}$ is the magnetic induction. In a stationary magnetic field, the elementary work is found to be zero:

$$\delta W = \frac{1}{c} \vec{j} \cdot (\vec{j} \times \vec{B}) dV = 0 \tag{7.3}$$

If the magnetic field strength changes with time, however, then the total electromagnetic work is not zero, $\delta W \neq 0$. This is due to the fact that the change of the magnetic induction $\vec{B}(t)$ induces an electric field, by Faraday's law. This implies that there appear electric forces, described by Coulomb's law

$$\vec{F}_E = \rho_E \vec{E} \tag{7.4}$$

This equation gives the force per unit volume on the volume charge density ρ_E. When a material is exposed to a changing magnetic induction, then, a work will be involved, which technically, is associated only with the forces due to the electric field. For instance, there will be a work involved when a material is magnetized by

329

placing it in an increasing magnetic induction. Consider Maxwell equations (i.e., Faraday's law):

$$-\frac{1}{c}\frac{d\vec{B}}{dt} = \vec{\nabla} \times \vec{E} \tag{7.5}$$

This gives the change of the magnetic induction within the time interval dt. The magnetic work done on the medium per unit time is considered to be the power associated with the induced electric force, given as

$$\vec{F}_E \cdot \vec{v} = Ne\vec{E} \cdot \vec{v} = (\vec{j} \cdot \vec{E})dV \tag{7.6}$$

Here, N is the number of electrons within volume dV and $\vec{v}$ represents their average velocity. Considering the fact that the work done is the power multiplied by the time dt, then an increment of work due to the electric force is

$$\eth W = -dt \int (\vec{j} \cdot \vec{E})dV \tag{7.7}$$

The negative sign is due to the fact that the work is done by external forces. Ampere's law gives the relation between the free current density and the magnetic field it generates (the displacement current is ignored here, so that only the magnetic energy is found)

$$\frac{c}{4\pi}\vec{\nabla} \times \vec{H} = \vec{j} \tag{7.8}$$

Then, the magnetic work done is expressed as

$$\eth W = -dt \int \frac{c}{4\pi}(\vec{\nabla} \times \vec{H}) \cdot \vec{E}\, dV \tag{7.9}$$

Consider the identity

$$\vec{\nabla} \cdot [\vec{E} \times \vec{H}] = \vec{E} \cdot (\vec{\nabla} \times \vec{H}) - \vec{H} \cdot (\vec{\nabla} \times \vec{E}) \tag{7.10}$$

We select some closed surface, then the term involving the Poynting vector $\vec{S} = (c/4\pi)\vec{E} \times \vec{H}$ corresponds to radiation out of the system:

$$\int \vec{\nabla} \cdot \left[\frac{c}{4\pi}\vec{E} \times \vec{H} \right] dV = \oiint_A \vec{S} \cdot d\vec{A} = P_{rad} \tag{7.11}$$

We do not account for the radiant energy as a magnetic energy. It would correspond, rather, to degrees of freedom outside the magnetic system under investigation. The term that remains is identified as the magnetic part, corresponding to the rate at which energy is stored in the magnetic fields

$$-\frac{c}{4\pi}\int \vec{E}\cdot(\vec{\nabla}\times\vec{H})dV = -\frac{c}{4\pi}\int \vec{H}\cdot(\vec{\nabla}\times\vec{E})dV - P_{rad} = \frac{\partial W_M}{dt} - P_{rad} \qquad (7.12)$$

From here we arrive at the effective work that stores energy in a magnetic form

$$\partial W_M = -dt\frac{c}{4\pi}\int \vec{H}\cdot(\vec{\nabla}\times\vec{E})dV = \frac{dt}{4\pi}\int \vec{H}\cdot\frac{\partial \vec{B}}{\partial t}dV \qquad (7.13)$$

The change of the free energy per unit volume now takes the form

$$dF = -SdT + \xi d\rho + \frac{1}{4\pi}(\vec{H}\cdot d\vec{B}) \qquad (7.14)$$

and the internal energy E change is

$$dE = TdS + \xi d\rho + \frac{1}{4\pi}(\vec{H}\cdot d\vec{B}) \qquad (7.15)$$

Here, we have the dependence of the free energy on the entropy per unit volume, the mass density ρ of the medium, and the chemical potential per unit mass, ξ. This formula for the free energy (7.14) may be exactly transformed to that of the electric field if we do the changes $\vec{H}\leftrightarrow\vec{E}$ and $\vec{B}\leftrightarrow\vec{D}$; see Section 7.2.

Note that Equation 7.14 contains the general form for the magnetic term, even for nonlinear magnetic media. If the medium has a linear relationship between $\vec{B}$ and $\vec{H}$ (true only at small fields), such as $\vec{B} = \mu\vec{H}$, where μ is the magnetic permeability, then the stored magnetic energy per unit volume takes the familiar form

$$W_M = \frac{1}{8\pi}\vec{B}\cdot\vec{H} \qquad (7.16)$$

From here, we can also transform to applied field as the independent variable

$$d\tilde{F} = -SdT + \xi d\rho - \frac{1}{4\pi}(\vec{B}\cdot d\vec{H}), \quad d\tilde{E} = TdS + \xi d\rho - \frac{1}{4\pi}(\vec{B}\cdot d\vec{H}) \quad (7.17)$$

that is, where

$$\tilde{F} = \tilde{F}(T,\rho,\vec{H}) \quad \text{and} \quad \tilde{E} = \tilde{E}(T,\rho,\vec{H}) \qquad (7.18)$$

The definition of the function $d\tilde{F}$ is as follows:

$$d\tilde{F} \equiv dF - \frac{1}{4\pi}d(\vec{H} \cdot \vec{B}) \tag{7.19}$$

This is the free energy without the work done by the field. Then, from here, we have the stated result

$$d\tilde{F} = -SdT + \xi d\rho - \frac{1}{4\pi}(\vec{B} \cdot d\vec{H}) \tag{7.20}$$

and furthermore

$$\frac{\partial \tilde{F}}{\partial \vec{H}} = -\frac{1}{4\pi}\vec{B} \tag{7.21}$$

Then it follows for linear magnetic media that

$$\tilde{E} = E_0(S,\rho) - \frac{\mu}{8\pi}H^2, \quad \tilde{F} = F_0(T,\rho) - \frac{\mu}{8\pi}H^2 \tag{7.22}$$

The terms with a zero subscript correspond to the energies when no magnetic field is applied. Considering $B = \mu H$, then, from Equations 7.14 and 7.6, we have

$$E = E_0(S,\rho) + \frac{1}{8\pi\mu}B^2, \quad F = F_0(T,\rho) + \frac{1}{8\pi\mu}B^2 \tag{7.23}$$

From here one can find the entropy, which depends on the temperature variation of the permeability

$$S = -\left(\frac{\partial F}{\partial T}\right)_{\rho,B} = S_0(T,\rho) + \frac{1}{8\pi\mu^2}\left(\frac{\partial \mu}{\partial T}\right)_{\rho,B}B^2 \tag{7.24}$$

When the density is fixed, we have a relation

$$dS = \left(\frac{\partial S}{\partial T}\right)_B dT + \left(\frac{\partial S}{\partial B}\right)_T dB = 0 \tag{7.25}$$

This leads to

$$dT = -\frac{\left(\frac{\partial S}{\partial B}\right)_T dB}{\left(\frac{\partial S}{\partial T}\right)_B} = \frac{1}{4\pi\mu^2}\frac{\left(\frac{\partial \mu}{\partial T}\right)}{\frac{C_B}{T} + \frac{1}{8\pi}\frac{d}{dt}\left\{\frac{1}{\mu^2}\left(\frac{\partial \mu}{\partial T}\right)\right\}B^2}BdB \tag{7.26}$$

Here,

$$\frac{C_B}{T} = \left(\frac{\partial S}{\partial T}\right)_B \tag{7.27}$$

7.2 THERMODYNAMICS OF DIELECTRIC MATERIALS IN AN ELECTRIC FIELD

Suppose that a dielectric material with dielectric permittivity ε is found in an electric field. The field strongly affects the properties of the dielectric, principally, by inducing an electric polarization. Near the surface of the dielectric, electrostatic theory shows that the polarization leads to induced surface charges. At the depth $l_D = (\sqrt{\varepsilon_0\varepsilon/eN})$ within the dielectric, there appear electric dipoles responsible for the electric polarization. These dipoles screen the external field. If the dielectric has a concentration of electrons equal to $N \approx 10^{23}\,\mathrm{cm}^{-3}$, then this screening depth is about $l_D \approx 1.0$ nm.

Let us analyze a dielectric surface f with the potential Φ. We are interested in studying the thermodynamic properties of the dielectric. This may be done by examining the work done on it due to an infinitesimal variation of the field at its interior. We consider the field in which the dielectric is located as if it is created by some charged external conductors. We may then calculate the variation of the field as a consequence of the variation of the conductors' charges. Let us for convenience consider a single conductor with a charge q and electrostatic potential Φ. Consider the infinitesimal increase of the conductor's charge by dq. The work needed to produce this increase is given by

$$\eth W = \Phi\,dq = \Phi\int d\sigma = -\frac{1}{4\pi}\int_f \Phi\,d\mathrm{D}_n\,df \tag{7.28}$$

Here, $\eth W$ is the mechanical work done by the field produced by the charge dq that is brought from infinity (where the potential is zero) onto the surface of the conductor df. This charge increment produces the surface charge density $d\sigma = -(1/4\pi)d\mathrm{D}_n$. This charge passed through a potential difference of Φ; that potential is constant on the entire surface of the conductor. In our relationship, the quantity $d\vec{\mathrm{D}}_n$ is the component of the electric displacement vector normal to the surface of the conductor and pointing into the conductor (it is the component pointing out of the dielectric). That is the reason for the minus sign, in contrast to what is expected from Gauss' law, $\vec{\nabla}\cdot\vec{\mathrm{D}} = 4\pi\rho_E$.

If we use the divergence theorem on the above expression (7.28), then we have

$$\eth W = -\frac{1}{4\pi}\int_V \vec{\nabla}\cdot(\Phi d\vec{\mathrm{D}})\,dV = -\frac{1}{4\pi}\int [\vec{\nabla}\Phi\cdot d\vec{\mathrm{D}} + \Phi\vec{\nabla}\cdot d\vec{\mathrm{D}}]\,dV \tag{7.29}$$

Here, V is the entire volume exterior to the conductor, that is, it is the volume inside the dielectric. Since there are no bulk (free) charges there, the last term is zero. By using the definition of electric field for electrostatic situations, $\vec{E} = -\vec{\nabla}\Phi$, we obtain the electric field energy

$$\delta W_E = \frac{1}{4\pi} \int (\vec{E} \cdot d\vec{D}) dV \qquad (7.30)$$

This is the same as what we would have obtained from Equation 7.14 with the changes suggested there, $\vec{H} \leftrightarrow \vec{E}$ and $\vec{B} \leftrightarrow \vec{D}$. Furthermore, if the dielectric is linear, with a relation, $\vec{D} = \varepsilon\vec{E}$, then we get the familiar result for energy per unit volume stored in the electric field

$$W_E = \frac{1}{8\pi} \vec{E} \cdot \vec{D} \qquad (7.31)$$

The free energy of the system per unit volume may now be evaluated as follows:

$$dF = -SdT + \xi d\rho + \frac{1}{4\pi}(\vec{E} \cdot d\vec{D}) \qquad (7.32)$$

Here, ρ is the mass per unit volume, S is the entropy per unit volume, and ξ is the chemical potential per unit mass of the substance. T is the absolute temperature. It follows from Equation 7.32 that

$$\frac{\partial F}{\partial \vec{D}} = \frac{\vec{E}}{4\pi} \qquad (7.33)$$

and if we consider a linear medium, $\vec{D} = \varepsilon\vec{E}$, then one can write

$$F = F_0(T,\rho) + \frac{1}{8\pi\varepsilon}D^2 \qquad (7.34)$$

where F_0 is the free energy when no field is present. Similar to magnetic problems, we define the function $d\tilde{F}$ as follows, transforming to a function $\tilde{F}(T,\rho,\vec{E})$:

$$d\tilde{F} \equiv dF - \frac{1}{4\pi}d(\vec{E} \cdot \vec{D}) \qquad (7.35)$$

Then, from here, we have

$$d\tilde{F} = -SdT + \xi d\rho - \frac{1}{4\pi}\vec{D} \cdot d\vec{E} \qquad (7.36)$$

and

$$\frac{\partial \tilde{F}}{\partial \vec{E}} = -\frac{1}{4\pi}\vec{D} \tag{7.37}$$

There also follows

$$\tilde{F} = F_0(T,\rho) + \frac{\varepsilon}{8\pi}E^2 \tag{7.38}$$

If we consider Equations 7.32 and 7.36, then the entropy is obtained equivalently in two alternative ways (for linear materials):

1. $S = -\left(\dfrac{\partial F}{\partial T}\right)_{\rho,D} = S_0 + \dfrac{1}{8\pi\varepsilon^2}\dfrac{d\varepsilon}{dT}D^2$ (7.39)

2. $S = -\left(\dfrac{\partial \tilde{F}}{\partial T}\right)_{\rho,E} = S_0 + \dfrac{1}{8\pi}\dfrac{d\varepsilon}{dT}E^2$ (7.40)

We may conclude that the entropy differs from zero only inside the dielectric.

Let the variation in the free energy density dF that results from an infinitesimal change in the field which occurs at constant temperature and does not destroy the thermodynamic equilibrium of the medium be expressed as

$$dF = \frac{1}{4\pi}\int(\vec{E} \cdot d\vec{D})dV \tag{7.41}$$

Now, let us consider $\vec{G}$ as being the field of an external source and $\vec{E}'$ as the field due to polarization of the medium. Then, the total field is

$$\vec{E} = \vec{E}_{ext} + \vec{E}' \tag{7.42}$$

The external field is generated by some charges outside the dielectric, hence it is the same as the electric displacement: $\vec{E}_{ext} = \vec{D}$. This is because the displacement depends only on free charges, not on any bound charges within the dielectric,

$$\vec{D} = \vec{E} + 4\pi\vec{P}, \quad \vec{\nabla} \cdot \vec{D} = 4\pi\rho_{free} \tag{7.43}$$

This shows that the field produced by electric polarization is

$$\vec{E}' = \vec{E} - \vec{D} = -4\pi\vec{P} \tag{7.44}$$

Then the free energy density with terms due to the source fields removed is

$$dF' = \frac{1}{4\pi} \int \left[\vec{E} \cdot d\vec{D} - \vec{D} \cdot d\vec{D} \right] dV = \frac{1}{4\pi} \int \left[\left(\vec{E} - \vec{D} \right) \cdot d\vec{D} \right] dV \qquad (7.45)$$

Equation 7.44 can be applied, leading to the free energy change associated only with the dielectric material,

$$dF' = -\int_V (\vec{P} \cdot d\vec{D}) dV \qquad (7.46)$$

which is not interpreted as the variation of the free energy density in the same way as in formula (7.41). It should be noted that the energy density at any point in the body can depend only on the field actually present there, and not on the field which would be present if the body were removed. As $\vec{P}$ is zero in vacuum and different from zero only in a dielectric, the integral is evaluated only using the volume V of the dielectric:

$$dF' = -\int_V (\vec{P} \cdot d\vec{D}) dV = -\vec{p}_E \cdot d\vec{D} \qquad (7.47)$$

Thus, the thermodynamic identity for the free energy density may now be evaluated as

$$dF = dF_0(\rho, T) - \vec{p}_E \cdot d\vec{D} \qquad (7.48)$$

Considering the fact that all the field equations are linear when $\vec{D} = \varepsilon\vec{E}$, the electric dipole moment $\vec{p}_E$ must be a linear function of the uniform field $\vec{D}$ through the relation $\vec{p}_E = \alpha\vec{D}$, where α is the polarization coefficient. Then

$$F = F_0(\rho, T) - \frac{\alpha D^2}{2} \qquad (7.49)$$

is the thermodynamic identity for the free energy density.

7.3 MAGNETIC EFFECTS IN MATERIALS

Earlier we found a basic expression for the work done to produce a magnetic field in a system. Assuming the system has a fixed number N of particles, the change in internal energy is

$$dE = TdS + \frac{1}{4\pi}\vec{H} \cdot d\vec{B} \tag{7.50}$$

Note, however, that with the relation between magnetic field and magnetic induction in a magnetic material

$$\vec{B} = \vec{H} + 4\pi\vec{M} \tag{7.51}$$

then

$$dE = TdS + \frac{1}{4\pi}\vec{H} \cdot d\vec{H} + \vec{H} \cdot d\vec{M} \tag{7.52}$$

The first magnetic term in Equation 7.52, $(1/4\pi)\vec{H} \cdot d\vec{H}$, involves the creation of the fields, regardless of the presence of a magnetic material. Therefore, it makes sense to isolate only the other part that relates to the magnetic material, as the part of most interest for the thermodynamics of the matter system. Hence, most authors define the internal energy change for the magnetic material system itself at fixed magnetic field as only from

$$dE = TdS + \vec{H} \cdot d\vec{M} \tag{7.53}$$

One can ask whether Equation 7.53 makes sense, physically. The last term is the energy cost to create a net magnetic moment in a material. In a typical paramagnetic material, for instance, $\vec{M}$ will be generated in the same direction as $\vec{H}$; hence, the magnetic work is positive. This implies an energy input is necessary to magnetize the material. Compare the work required to establish a current in an inductor, $W = (1/2)Li^2$, which is also positive. In an inductor with some magnetic core, the inductance L is enhanced compared to that for an air-filled inductor, to store even more energy. The term $\vec{H} \cdot d\vec{M}$ in Equation 7.53 corresponds to the extra energy that is stored in the inductor due to the presence the magnetically polarizable material. The term that we have dropped, $(1/4\pi)\vec{H} \cdot d\vec{H}$, would correspond to only the energy associated with an air-filled core, which is of lesser interest in the current discussion.

By the usual procedure, one can do a double Legendre transformation from $S(E,\vec{M})$ to a Helmholtz free energy $F(T,\vec{H}) = E - TS - \vec{M} \cdot \vec{H}$, whose differential is seen to be

$$dF = -SdT - \vec{M} \cdot d\vec{H} \tag{7.54}$$

Thus, one can obtain the entropy and magnetic moment from

$$S = -\left(\frac{\partial F}{\partial T}\right)_{\vec{H}}, \quad \vec{M} = -\left(\frac{\partial F}{\partial \vec{H}}\right)_{T} \tag{7.55}$$

Compare to the free energy $F(T,V)$ for a gas

$$dF = -SdT - pdV \tag{7.56}$$

with the relations

$$S = -\left(\frac{\partial F}{\partial T}\right)_V, \quad p = -\left(\frac{\partial F}{\partial V}\right)_T \tag{7.57}$$

In a mathematical sense, the magnetic moment is analogous to pressure, and the magnetic field is analogous to the volume. This correspondence can be useful for thinking about various thermodynamic processes applied to the magnet.

Note that to apply this formalism, the Hamiltonian used to find the canonical partition function Z, with $F = -T\ln Z$, should include any internal magnetic interactions, and also the interaction $-\vec{M} \cdot \vec{H}$ with the external field. One could consider this term as part of the internal energy of the system.

To illustrate this, consider the application of these equations to a simple problem, that of N two-state paramagnetic ions with dipole moments $\vec{m}_i$ of fixed magnitudes m, whose directions are random unless a magnetic field is applied. We suppose the ions are fixed at lattice sites, and they have only an up state and a down state, where the component of $\vec{m}_i$ along $\vec{H}$ can be $m_i = \pm m$. Any phonon vibrations in the system are initially ignored; only the magnetic degrees of freedom are being considered. We take the Hamiltonian for the system to be only the potential energy of the dipoles in the field

$$H = -\sum_{i=1}^{N} \vec{m}_i \cdot \vec{H} = -\sum_{i=1}^{N} m_i H \tag{7.58}$$

Then, because the dipoles are considered noninteracting, the canonical partition function is obtained from that for an individual dipole

$$Z = Z_1^N, \quad Z_1 = \sum_{m_i = \pm m} \exp(-\beta m_i H) = 2\cosh \beta m H \tag{7.59}$$

The free energy is then

$$F(T,H) = -NT \ln Z_1 = -NT \ln(2\cosh \beta m H) \tag{7.60}$$

The expectation value of the total magnetic moment component along the field can be found easily from Equation 7.55:

$$M = -\left(\frac{\partial F}{\partial H}\right)_T = NT\frac{\beta m \sinh \beta m H}{\cosh \beta m H} = Nm \tanh \beta m H \tag{7.61}$$

This is the same as would be obtained from the statistical average

$$M = N \frac{m \exp\{\beta m H\} - m \exp\{-\beta m H\}}{\exp\{\beta m H\} + \exp\{-\beta m H\}} = Nm \tanh \beta m H \qquad (7.62)$$

The expectation value of the internal energy is also simple

$$E = -MH = -Nm \, H \tanh(\beta m H) \qquad (7.63)$$

Alternatively, the formula

$$E = -\left(\frac{\partial}{\partial \beta}\right)_H (\beta F) = \left(\frac{\partial}{\partial \beta}\right)_H \ln Z \qquad (7.64)$$

leads to an identical result.

The entropy is obtained also by a simple derivative as in Equation 7.55

$$S = -\left(\frac{\partial F}{\partial T}\right)_H = N[\ln(2 \cosh \beta m H) - \beta m H \tanh \beta m H] \qquad (7.65)$$

An alternative route to the entropy that gives the same result is $S = (E - F)/T$. The behavior of S with temperature in Equation 7.65 for different magnetic fields in shown in Figure 7.1. As can be expected, the higher the magnetic field, the lower the entropy becomes. One can check the behavior of S in the low and high temperature limits. As $T \to 0$, corresponding to $\beta \to \infty$, one has the correct limit

$$S \approx N[\ln\{\exp(\beta m H)\} - \beta m H] = 0 \qquad (7.66)$$

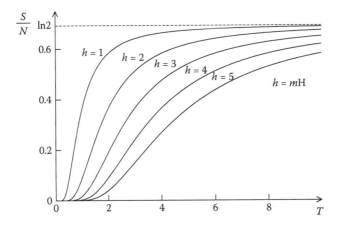

FIGURE 7.1 Behavior of the entropy of a paramagnet with temperature, for different magnetic field strengths. Shown is the entropy per atom in units of Boltzmann's constant versus thermal energy in arbitrary units. The magnetic fields scaled with the magnetic dipoles, $h = mH$, are in the same units as the thermal energy.

At the other extreme, for $T \rightarrow \infty$ or $\beta \rightarrow 0$

$$S \approx N \ln 2 \tag{7.67}$$

which simply reflects the complete accessibility of the two states per dipole for high temperature. In the limit of zero magnetic field, $H \rightarrow 0$, it is interesting that the entropy also goes to the same result, $S \approx N \ln 2$. This reflects the zero-field mac-rostate of the dipoles, that have become completely randomized, the same as they become at very high temperature.

7.4 ADIABATIC COOLING BY DEMAGNETIZATION

Entropic effects in a paramagnetic can be used in a dramatic fashion to produce refrigeration by an adiabatic demagnetization process. This is a phenomenon that resembles in a close way the adiabatic cooling of a gas by expansion. It is also known as the magnetocaloric effect and can reach temperatures well below 1 K. Typically, the process can be carried out in some paramagnetic salts, such as cerium magne-sium nitrate. Alloys with gadolinium, a rare-earth magnet paramagnet, have also been used for designing magnetic refrigerators. A gas of paramagnetic atoms is also a good medium. The general idea is fairly simple, which can be described by the following steps:

1. The magnetic medium is initially at some temperature T_1 and has no net magnetic moment, $M = 0$, without an applied field, $H = 0$. A convenient ini-tial temperature might be a few kelvin, close to the boiling point of liquid helium.
2. A magnetic field $H_1 > 0$ is applied to the medium, producing a nonzero magnetization within it. This causes the sample to heat up, due to the work done on it in producing the magnetization.
3. Left in contact with the original thermal reservoir at temperature T_1, heat flows out of the sample until it is allowed to cool close to the temperature T_1. The magnetic field H_1 is still turned on. After the sample has cooled down, it has a magnetic moment M_1 and an entropy S_1. The alignment of the microscopic magnetic moments means that the magnetic entropy has been reduced compared to the initial state where M was zero.
4. Now, with the medium thermally insulated from the reservoir, the magnetic field is turned off. This is the adiabatic demagnetization step, which takes place without the flow of heat. Thus, it takes place at constant entropy. By turning off the field, the microscopic magnetic dipoles within the medium become directionally disorganized, and the only way for the entropy to remain constant is if the temperature falls. One can expect that in fact it could even be that the entropy in the magnetic degrees of freedom increases at the expense of a reduction of the entropy in the lattice vibrations. In any case, the adiabatic demagnetization causes the temperature to drop, much in the same way as an adiabatic expansion of a gas causes it to cool. The final temperature could be typically a 10th of the initial temperature.

The procedure can be repeated to enhance the effect. If only the magnetic degrees of freedom are considered, the entropy considered as $S(T,H)$ could be similar to that shown in Figure 7.1 for the simple paramagnetic model discussed above. The curves of entropy at higher magnetic field fall below and to the right of those at lower field. When a previously established field is reduced toward zero, the state point in an adiabatic process moves directly to the left, toward lower temperature. With only magnetic degrees of freedom present, the temperature would go very close to zero as $H \to 0$.

Consider how to find the final temperature, starting from some field H_1 at temperature T_1. The demagnetization process at fixed entropy must go to a final temperature T_2 given from the equation:

$$S(T_2,0) = S(T_1,H_1) \qquad (7.68)$$

The magnetic entropy (7.65) of our simple paramagnetic model by itself does not give a solution to this equation. In reality, there is also at the very least, a lattice entropy, and perhaps some electronic or other contributions. So the removal of the magnetic field cannot bring the temperature so close to zero. Some qualitative approximations can be made, however. One approximation is to assume that the magnetization is linearly proportional to the magnetic field, with some susceptibility $\chi(T) = C/T$, which is true at least at weak field:

$$M = \chi(T)H \qquad (7.69)$$

Also we have the relation to the Helmholtz free energy

$$M = -\left(\frac{\partial F}{\partial H}\right)_T \qquad (7.70)$$

Combining these leads to a relation

$$F(T,H) = F(T,0) - \frac{1}{2}\chi(T)H^2 \qquad (7.71)$$

Now, this leads to an expression for the field dependence of the entropy

$$S(T,H) = -\left(\frac{\partial F}{\partial T}\right)_H = S(T,0) + \frac{1}{2}\frac{\partial\chi}{\partial T}H^2 \qquad (7.72)$$

Here, the quantity $S(T,0)$ includes both the magnetic and nonmagnetic forms of entropy at zero magnetic field. As the paramagnetic susceptibility follows a Curie–Weiss law, $\chi(T)T = C$, the derivative of the susceptibility is negative

$$\frac{\partial\chi}{\partial T} = -\frac{\chi}{T} = -\frac{C}{T^2} < 0 \qquad (7.73)$$

One can check if this makes sense, by comparing with the low-field expansion of the entropy (7.65) in the simple model paramagnet above. There results for $\beta m H \ll 1$

$$S_{\text{para}} \approx N\{\ln[2 + (\beta m H)^2] - (\beta m H)^2\} \approx N\left\{\ln 2 - \frac{1}{2}(\beta m H)^2\right\} \qquad (7.74)$$

This matches the expression (7.72) for $S(T, H)$ derived from the free energy relation, with the Curie–Weiss constant being $C = N m^2$. The same value of C is obtained from the expression for the magnetization. One also sees the limiting nonzero magnetic entropy when the field is turned off. Now, when applied to an adiabatic process, matching the initial and final entropies from Equation 7.72 gives

$$S(T_2, 0) = S(T_1, 0) - \frac{1}{2}\frac{C}{T_1^2}H_1^2 \qquad (7.75)$$

If the entropy function for the whole system is known, this can be solved for the final temperature T_2. Even without that information, one sees that the final entropy of the nonmagnetic degrees of freedom is less than its initial value, which corresponds to a drop in temperature. The entropy is a monotonically increasing function of the temperature, so we conclude that $T_2 < T_1$. By turning off the applied magnetic field, the magnetic entropy increases, at the expense of a simultaneous equivalent decrease in the entropy of the other degrees of freedom.

8 Lattice Dynamics

8.1 PERIODIC FUNCTIONS OF A RECIPROCAL LATTICE

In many applications of statistical mechanics, the system is composed of atoms or molecules in a regular crystalline arrangement, on a lattice in three-dimensional space, rather than the random positions found in a gas or liquid. When the lattice arrangement repeats periodically, and can be described by the repetition of the points in a fundamental unit cell, it is known as a Bravais lattice [5,14]. Here we discuss some of the implications of having atoms in these kinds of regular arrangements.

8.2 RECIPROCAL LATTICE

We introduce the notion of the *reciprocal lattice vectors* of a Bravais lattice that we use further in the chapter. The lattice has primitive vectors $\vec{a}_i, i = 1,2,3$, that are dependent on the crystal symmetry. These vectors cannot all be in a single plane for a three-dimensional lattice. Then we can select a (possibly nonorthogonal) Cartesian coordinate system $\varsigma_1, \varsigma_2, \varsigma_3$ whose axes are aligned with the primitive vectors. Vectors $\vec{a}_i$ and $\hat{\varsigma}_i$, $i = 1,2,3$, are in the same direction (Figure 8.1). Then a point of the lattice at coordinates $(\varsigma_1, \varsigma_2, \varsigma_3)$ can be denoted by a triple of integers, $\boldsymbol{n} = (n_1, n_2, n_3)$ and is displaced from the origin by the vector

$$\vec{a}_n = \varsigma_1 \hat{e}_1 + \varsigma_2 \hat{e}_2 + \varsigma_3 \hat{e}_3 = n_1 \vec{a}_1 + n_2 \vec{a}_2 + n_3 \vec{a}_3 \tag{8.1}$$

The lattice is generated by the choice of only integer coordinates, (n_1, n_2, n_3). Any point of the lattice can be reached from another one by an integral number of primitive vector displacements.

To define the reciprocal lattice we consider the microscopic physical space values (say, of potential $V(r)$) in the crystal to be periodic. Periodicity implies that the function is unchanged when shifted by any lattice vector, $\vec{a}_n$,

$$V(\vec{r} + \vec{a}_n) = V(\vec{r}) \tag{8.2}$$

We can ask, in what physical situation can this arise, if ever? One way would be to imagine an infinite crystal with no boundaries, but that is unphysical. When we wish to deal with a crystal of finite size, the question of boundary conditions usually arises. In a microscopic crystal, the boundaries are free surfaces, and atoms near the surfaces have a different environment than atoms in the interior, so how can there be a true periodicity? Suppose, however, that the crystal is very large. Then we expect

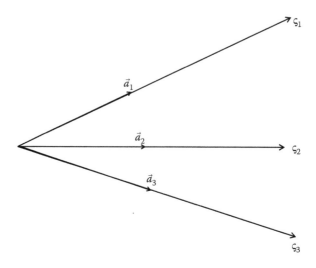

FIGURE 8.1 Primitive vectors $\vec{a}_i$, and along the same directions, the nonorthogonal Cartesian coordinate system, with axes $\xi_i, i = 1, 2, 3$.

the precise form of boundary conditions should not affect the physical description of properties within the bulk of the crystal, far from the surface. We can artificially impose what are called *cyclic* or *periodic boundary conditions*. Here the value of any wave function is constrained to be the same at equivalent points on opposite sides of the crystal. This periodicity is represented mathematically by Equation 8.2, with the appropriate choice of the $\vec{a}_n$ being displacements across the whole crystal. Now for any lattice displacement, a translation operator exists such that

$$V(\vec{r} + \vec{a}_n) = \hat{T}_{\vec{a}_n} V(\vec{r}) \tag{8.3}$$

Let N be the number of primitive displacements to span across the crystal, counting along the direction of each primitive translation $\vec{a}_j, j = 1, 2, 3$. Suppose a translation $N\vec{a}_j$ brings us to a point not only physically but also mathematically identical to the original point. Thus we may have N unit translations across the sample lead to the original wave function,

$$\hat{T}_{\vec{a}_j}^N \Psi(\vec{r}) = t_{\vec{a}_j}^N \Psi(\vec{r}) = \Psi(\vec{r}) \tag{8.4}$$

Then it follows that the translation operator eigenvalues satisfy

$$t_{\vec{a}_j}^N = 1 \tag{8.5}$$

so that every eigenvalue, $t_{\vec{a}_j}$ of $\hat{T}_{\vec{a}_j}$, is a root of unity of the form $\exp\{i\alpha\}$, where $N\alpha = 2\pi \times$ integer. As the translation operators along the different axes commute among themselves and with the Hamiltonian $\hat{H}$, then a complete set of simultaneous eigenfunctions of $\hat{H}$ and an operator $\hat{T}_{\vec{a}_n}$ for all lattice vectors $\vec{a}_n$ as in Equation 8.1 may be found such that

$$\hat{T}_{\vec{a}_n} \Psi = (\hat{T}_{a_1})^{n_1} (\hat{T}_{a_2})^{n_2} (\hat{T}_{a_3})^{n_3} \Psi = \exp\{i(n_1\alpha_1 + n_2\alpha_2 + n_3\alpha_3)\}\Psi \qquad (8.6)$$

We can note that each factor in the exponent involves a coordinate like $n_1 = \varsigma_1/a_1$, and so on.

Now suppose we expand Equation 8.2 in a Fourier series, where the basis states are plane waves. This is an expression in the skew coordinates,

$$V(\vec{r}) = \sum_{k_1 k_2 k_3 = -\infty}^{\infty} v_{k_1 k_2 k_3} \exp\left\{2\pi i\left(\frac{\varsigma_1}{a_1}k_1 + \frac{\varsigma_2}{a_2}k_2 + \frac{\varsigma_3}{a_3}k_3\right)\right\} \qquad (8.7)$$

where k_i, $i = 1, 2, 3$ are some integers representing wave vectors. Let us do the transformation to orthogonal Cartesian coordinates x_i via,

$$\varsigma_i = \alpha_{ik} x_k, \quad i,k = 1, 2, 3 \qquad (8.8)$$

where (α_{ik}) is the transformation matrix. Then

$$V(\vec{r}) = \sum_{\beta_1 \beta_2 \beta_3} V_{\beta_1 \beta_2 \beta_3} \exp 2\pi i(\beta_1 x_1 + \beta_2 x_2 + \beta_3 x_3) = \sum_{\vec{b}} V_{\vec{b}} \exp(i\vec{b} \cdot \vec{r}) \qquad (8.9)$$

Here

$$\vec{r} = (x_1, x_2, x_3), \quad \vec{b} = 2\pi(\beta_1, \beta_2, \beta_3) = (b_1, b_2, b_3) \qquad (8.10)$$

Thus,

$$V(\vec{r}) = \sum_{\vec{b}} V_{\vec{b}} \exp(i\vec{b} \cdot \vec{r}) \qquad (8.11)$$

is the desired expansion using the reciprocal lattice, that is, a set of fundamental wave vectors $\vec{b}$. In order to find the allowed wave vectors $\vec{b}$ we use the periodicity property (8.2), for any lattice displacement $\vec{a}_n$:

$$V(\vec{r} + \vec{a}_n) = \sum_{\vec{b}} V_{\vec{b}} \exp(i\vec{b} \cdot \vec{r})\exp(i\vec{b} \cdot \vec{a}_n) \qquad (8.12)$$

In order that this should be identical to (8.11), the reciprocal lattice vectors $\vec{b}$ must satisfy

$$\exp(i\vec{b} \cdot \vec{a}_n) = 1 \tag{8.13}$$

or

$$(\vec{b} \cdot \vec{a}_n) = 2\pi\kappa \tag{8.14}$$

where κ is some integer and from Equation 8.1, in terms of the primitive vectors,

$$(\vec{b} \cdot \vec{a}_n) = n_1(\vec{b}_1 \cdot \vec{a}_1) + n_2(\vec{b}_2 \cdot \vec{a}_2) + n_3(\vec{b}_3 \cdot \vec{a}_3) \tag{8.15}$$

The vectors $\vec{b}_1$, $\vec{b}_2$, $\vec{b}_3$ are considered the primitive vectors of the reciprocal lattice. All reciprocal lattice vectors can be composed from them as linear combinations, for example,

$$\vec{b} = m_1\vec{b}_1 + m_2\vec{b}_2 + m_3\vec{b}_3 \tag{8.16}$$

$\vec{b}_1$, $\vec{b}_2$, $\vec{b}_3$ can be found by supposing that each primitive vector of the real space lattice is perpendicular to two of the primitive vectors of the reciprocal lattice. Then, suppose the individual factors in Equation 8.15 are

$$n_1(\vec{b}_1 \cdot \vec{a}_1) = g_1, \quad n_2(\vec{b}_2 \cdot \vec{a}_2) = g_2, \quad n_3(\vec{b}_3 \cdot \vec{a}_3) = g_3 \tag{8.17}$$

$\vec{a}_1$, $\vec{a}_2$, $\vec{a}_3$ are non-coplanar vectors by assumption. These are necessary and significant conditions to define the reciprocal lattice vectors, $\vec{b}$, which have a multitude of solutions.

From Equations 8.15 and 8.17, it is obvious that $\vec{b}$ has the dimensions of inverse length. Thus it is a reciprocal vector to the direct lattice vectors, $\vec{a}_n$. A lattice constructed from reciprocal lattice vectors is called a *reciprocal lattice*.

Consider how to express the primitive reciprocal vectors. In order to force $\vec{b}_1$ to be perpendicular to two of the real space primitive vectors, we can write it as a vector cross product, as in $\vec{b}_1 = c_1\vec{a}_2 \times \vec{a}_3$. Now also the scalar product of a primitive reciprocal vector with its paired real space primitive vector can be taken to be $\vec{b}_1 \cdot \vec{a}_1 = 2\pi$, that is, using $\kappa = 1$. This means that the constant is $c_1 = 2\pi/[\vec{a}_1 \cdot (\vec{a}_2 \times \vec{a}_3)]$. The same approach applies for finding the other two primitive reciprocal vectors, so one gets these as the basis of the reciprocal lattice:

$$\vec{b}_1 = \frac{2\pi\,\vec{a}_2 \times \vec{a}_3}{\vec{a}_1 \cdot (\vec{a}_2 \times \vec{a}_3)}, \quad \vec{b}_2 = \frac{2\pi\,\vec{a}_3 \times \vec{a}_1}{\vec{a}_1 \cdot (\vec{a}_2 \times \vec{a}_3)}, \quad \vec{b}_3 = \frac{2\pi\,\vec{a}_1 \times \vec{a}_2}{\vec{a}_1 \cdot (\vec{a}_2 \times \vec{a}_3)}, \tag{8.18}$$

For any given direct lattice, these can be applied to find the resulting structure of the reciprocal lattice. The vectors $\vec{b}$ that can be constructed with Equation 8.16 represent the wave vectors of plane waves that have the desired periodicity of the original real space lattice. The reciprocal space, therefore, is important for the Fourier expansion of any physical quantities defined at the lattice sites.

8.3 VIBRATIONAL MODES OF A MONATOMIC LATTICE

In investigating real crystals there can be two types of imperfections:

1. Dynamic imperfections due to the displacements of the atoms as a result of thermal motions.
2. Static imperfections that include vacancies, interstitial atoms, dislocations, stacking faults, and so on. We consider impurities as static imperfections. The impurity as we know it has a twofold effect: (1) it substitutes a host atom at the atom's site; (2) it displaces neighboring host atoms from their equilibrium positions.

In this chapter only dynamic imperfections in crystals are analyzed. Atoms in a solid are always in constant thermally induced motion. Models in which to discuss this motion are required. We may conveniently illustrate many of the critical concepts in one dimension. In this heading the dynamics of a linear chain of atoms is investigated in detail. For the sake of convenience, we assume only one atom per unit cell of the given crystalline solid. Here we consider only classical motion of atoms in the chain with no reference to quantization:

1. We assume the atomic arrangement with minimum energy to be a periodic array of atoms. In order to avoid surface effects we consider the familiar periodic boundary conditions. The crystal has N lattice vectors $\vec{a}_n$, and the last one is coupled to the first one of the chain. The potential energy of the system is expanded in powers of atomic displacements from the equilibrium positions. The *harmonic approximation* is used in the expansion of the potential energy where orders higher than quadratic (in displacements) are ignored. The approximation is good for most solids at room temperature. This can be further refined using the perturbation method. The harmonic approximation leads to a classical equation of motion that is conveniently solved by introducing special linear combinations of atomic displacement coordinates called *normal coordinates*. We express the potential and kinetic energy of the system in terms of the normal coordinates. This expression contains no cross terms coupling different normal coordinates. For each we have only quadratic terms. This reduces the equation of motion to a set of independent harmonic oscillator equations that are easily solvable. The solutions of the system are called *normal modes*.
2. Despite the large size and apparent complexity of the system, the ideal crystal's perfect periodicity enables us to find the normal mode frequencies. Generally, atomic motion within the crystal is described as a linear superposition of normal modes. These normal modes correspond to traveling plane wave solutions.

8.3.1 LINEAR MONATOMIC CHAIN

We study first the vibration of a one-dimensional chain of atoms. Suppose we have a finite chain of N identical atoms of mass M and interatomic distances a (lattice spacing or lattice constant) as in Figure 8.2. To find normal modes, we impose boundary conditions at the end of the finite chain. Neighboring atoms are coupled by springs with a force constant α.

In Figure 8.2 the un-heavy atoms show the equilibrium positions and the heavy ones the displaced positions of atoms (nonequilibrium positions), where U_k is the displacement of the kth atom in the chain from its equilibrium position. In the above system, each atom has a degree of freedom, and the entire system has N degrees of freedom. The model is well described by a linear elementary Bravais lattice, where the positions of the atoms are determined by a translation vector $\vec{T}_{\bar{a}_n} = n\vec{a}$, where integral values of n give the equilibrium positions of the atoms in the chain.

The atoms are linked by coupling forces such that when there is an excitation, it is propagated in the chain in the form of a wave that displaces all atoms from their equilibrium positions. We assume that the coupling between near neighbors is very much larger than that between more distant neighbors.

If we consider a plane wave propagating along the ox-axis of the chain, then the displacement at position x and time t is U: $U = U(x, t)$. For the linear array of atoms the equation $U = U(x, t)$ only has meaning at the lattice sites $x = na$. The equation is made more precise: $U_n = U(na, t)$ where U_n is considered as the displacement of the nth atomic site at time t. The positions of the atoms are determined by a translation vector

$$\vec{T}_{\bar{a}_n} = n\vec{a}: \quad \vec{a} = a\hat{e}_x \qquad (8.19)$$

The thermodynamic properties of crystals are linked with the vibrational motion of atoms (or ions) about their equilibrium positions, which are sites of the crystal lattice. The relative displacement of atoms is small compared to the distance between them. Further, in quantum crystals, where the amplitudes of the zero-point oscillations are high, the mean square deviation of atoms from their sites is about 30% of the inter-atomic distances. However in the majority of solids they are not over 10% of the value at the melting temperature.

If we assume that the displacements U_n of atoms from their equilibrium positions are small, then the potential energy $W(U_n)$, that is a univalent function of U_n, is small, such that we may expand in atomic displacements U_n.

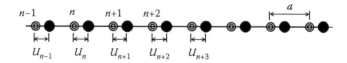

FIGURE 8.2 A one-dimensional chain of N identical atoms of mass M and inter-atomic distances a, where U_n is considered as the displacement of the nth atomic site from its equilibrium position.

$$W(U_n) = W(0) + \sum_n \left[\frac{\partial W}{\partial U_n}\right]_0 U_n + \frac{1}{2}\sum_{nn'}\left[\frac{\partial^2 W}{\partial U_n \partial U_{n'}}\right]_0 U_n U_{n'} + \cdots \qquad (8.20)$$

The subscript 0 implies that the derivatives are evaluated at the equilibrium configuration. The ground state potential energy $W(0) = W_0$ is the equilibrium value and may always be set to zero and thus made a reference point. We ignore W_0 as it will not affect our answers. As the potential energy has a minimum at the equilibrium position, then

$$\left[\frac{\partial W}{\partial U_n}\right]_0 = 0 \qquad (8.21)$$

This is because in the equilibrium configuration there is no net force on the nuclei. At equilibrium W is minimum.

Thus the remaining term in Equation 8.20 is quadratic in displacements and describes the harmonic vibrations of the lattice atoms:

$$W(U_n) = \frac{1}{2}\sum_{nn'}\left[\frac{\partial^2 W}{\partial U_n \partial U_{n'}}\right]_0 U_n U_{n'} \qquad (8.22)$$

When all higher-order terms are dropped, such an approximation is called *harmonic*. Here U_n and $U_{n'}$ refer to the displacements of two neighboring atoms. Surface effects can be avoided by appropriate boundary conditions. A crystal with N lattice vectors $\vec{a}_n$ can be thought of as part of an infinite crystal that is made up of smaller crystals vibrating identically. This can be approximated with periodic boundary conditions.

In order that the chain should have the property of translational invariance, it is necessary that

$$\left[\frac{\partial^2 W}{\partial U_n \partial U_{n'}}\right]_0 = \left[\frac{\partial^2 W}{\partial U_{n'} \partial U_n}\right]_0 = P(|n - n'|) \qquad (8.23)$$

where $P(|n - n'|)$ is the stress tensor. The argument $|n - n'|$ gives the distance between two sites, when multiplied by the lattice constant a.

In addition to potential energy, the kinetic energy of the lattice is

$$E_{kin} = \sum_n \frac{1}{2}M|\dot{U}_n|^2 \qquad (8.24)$$

Let us find the equations of motion for the system. For that we write the Lagrangian of the system;

$$L(U_n, \dot{U}_n) = \sum_n \frac{1}{2} M |\dot{U}_n|^2 - \frac{1}{2} \sum_{nn'} P(|n - n'|) U_n U_{n'}. \tag{8.25}$$

and the Lagrangian equations of motion are

$$\frac{\partial}{\partial t} \frac{\partial L}{\partial \dot{U}_m} - \frac{\partial L}{\partial U_m} = 0 \tag{8.26}$$

When the derivatives are evaluated, the index m can match with n or n'. This leads to two equivalent terms in the potential part:

$$\frac{\partial L}{\partial U_m} = \frac{1}{2} \sum_{nn'} \left[\frac{\partial^2 W}{\partial U_n \partial U_{n'}} \right]_0 U_n \delta_{mn'} + \frac{1}{2} \sum_{nn'} \left[\frac{\partial^2 W}{\partial U_n \partial U_{n'}} \right]_0 U_{n'} \delta_{mn}$$

$$= \sum_n \left[\frac{\partial^2 W}{\partial U_n \partial U_m} \right]_0 U_n = \sum_n P(|n - m|) U_n \tag{8.27}$$

Changing the dummy indexes, this gives the basic equation of motion,

$$M \ddot{U}_n = -\sum_{n'} P(|n - n'|) U_{n'} \tag{8.28}$$

The RHS is the total force on the atom n. An individual factor $P(|n - n'|)$ is the force constant (or spring constant) associated with the displacement of the n' atom. The solution of this equation now proceeds in the usual way. Normal coordinates are sought, whose equations of motion are decoupled.

The translational symmetry of the lattice with one atom per unit cell guarantees that the tensor $P(|n - n'|)$ depends only on the difference $|n - n'|$. For crystals with more than one atom per unit cell, we require separate equations for each of the atoms in the primitive cell.

Equation 8.28 is a large number of coupled differential equations. It has no direct solution. If we suppose that the wavelength $\lambda \gg a$ then we have a continuous distribution of the given wave as along a string. The $P(|n - n'|)$ are interaction constants representing coupling between pairs of atoms. Coupling between near neighbors should be much larger than that between more distant neighbors. Hence if Equation 8.28 represents point particles coupled by springs, then we expect the spring constants to be much larger for nearest neighbors. Hence we are concerned mostly with short-range interactions.

Let us find the normal modes of the vibrations, that is, types of movements such that all the atoms oscillate with time with the same frequency according to the factor $\exp\{-i\omega t\}$. We find the solution of Equation 8.28 in the form of a progressive wave,

$$U_n(t) = U_n^0 \exp\{-i\omega t\} \tag{8.29}$$

Then considering Equation 8.28 gives

$$MU_n^0\omega^2 = \sum_{n'} P(|n - n'|)U_{n'}^0 \tag{8.30}$$

The 0 superscript refers to an amplitude or constant. The stress tensor $P(|n - n'|)$ has the property of translational symmetry. If we do the translation $n \to n + 1$ and $n' \to n' + 1$, then the difference $|n - n'|$ is invariant relative to this translation. Consequently, U_n^0 and U_{n+1}^0 should be enough to describe a given defined system. Suppose the quantities U_n^0 and U_{n+1}^0 are connected through the relation (assuming equal amplitudes everywhere):

$$U_{n+1}^0 = \Gamma U_n^0 \tag{8.31}$$

and $|\Gamma| = 1$. This works out if Γ is a complex valued function, represented:

$$\Gamma = \exp\{iqa\} \tag{8.32}$$

Here q has the dimension of the inverse length. Including an amplitude A, we should have

$$U_n^0 = A\exp\{i\vec{q} \cdot \vec{a}_n\} \tag{8.33}$$

and $\vec{a}_n$ is the nth lattice vector and $\vec{q}$ is a wave vector. Considering Equations 8.29 and 8.33, the wave solution is expressed as

$$U_n(t) = A\exp\{i(\vec{q} \cdot \vec{a}_n - \omega t)\} \tag{8.34}$$

Here A is the displacement of the atom for $n = 0$ at the instant $t = 0$ and $|\vec{q}| = 2\pi/\lambda$ is the number of waves per unit length. λ is the wavelength of the wave. If we substitute Equation 8.34 into Equation 8.30, then the dynamic equation becomes

$$M\omega^2 = \sum_{n'} P(|n - n'|)\exp\{iqa(n' - n)\} \tag{8.35}$$

If we let $l = n' - n$ represent separation of lattice sites, then we have

$$M\omega^2 = \sum_{l=-\infty}^{\infty} P(|l|)\exp\{iqal\} = P(0) + 2\sum_{l=1}^{\infty} P(l)\cos(qal) \tag{8.36}$$

In the simplification, the sine functions cancel out as they are odd functions with respect to l and the cosine functions remain as they are even functions with respect to l.

Equation 8.36 shows that the wave frequencies $\omega(q)$ for different q are different. This phenomenon is called dispersion and Equation 8.36 is the dispersion relation. It should be noted that

$$\omega(-q) = \omega(q) \tag{8.37}$$

which implies that the system has inversion symmetry. Waves traveling toward positive x or towards negative x have the same dispersion properties.

Considering Equation 8.36, the frequency ω of vibration of an atom does not depend on n. All the atoms in the chain vibrate with the same frequency, as assumed to get this solution.

The wave frequencies cannot be completely determined without specifying the allowed wave vectors q. The solution for U_n depends on the factor $\exp(i\vec{q} \cdot \vec{a}_n) = \exp(iqna)$. But one can see that different values of q can give identical atomic motions, because the wave function is only defined at the lattice sites. If the phase factor qa is shifted by any multiple of 2π, an equivalent wave solution results, because this shift leaves U_n unchanged. Therefore, all possible wave solutions can be described by using a range of q whose width is $2\pi/a$. The simplest choice would be to take the range, $0 \leq q < 2\pi/a$. While this is perfectly acceptable, it does not allow for consideration of the symmetry in the dispersion relation, $\omega(q)$. Instead, it makes sense to take a range symmetrical around $q = 0$, that is, $-\pi/a \leq q \leq \pi/a$. This range of q is usually called the first Brillouin zone, as part of the central zone scheme. Technically, the two outer end points give identical wave solutions, because they differ by $2\pi/a$. We only need to use q within this range to specify all possible wave solutions.

The first Brillouin zone, $-\pi/a \leq q \leq \pi/a$, is defined in physics as the region of q closest to the origin, that gives all of the distinct allowed wave solutions. Consider a wave vector near the center of the Brillouin zone at an infinitesimal value of q. We may expand $\omega(q)$ to obtain a general expression,

$$\omega(q) = \omega(0) + \left.\frac{d\omega}{dq}\right|_0 q + \cdots \tag{8.38}$$

If we consider Equation 8.36 and take the limit $q \to 0$ then it follows that the frequency must be zero,

$$\omega(0) \approx \sum_{n'} P(|n - n'|) = 0 \tag{8.39}$$

This is because the net force on any atom is zero if each atom is displaced from its equilibrium position by the same amount (e.g., all $U_n = 1$). Note that the remaining leading term in the dispersion exhibits the group velocity of the waves,

$$v_g = \frac{d\omega}{dq} \tag{8.40}$$

If that is the dominant term, the waves have an acoustic dispersion relation, and v_g evaluated at $q = 0$ is the speed of sound in the medium.

The phase velocity v_{ph} of a wave is the ratio of frequency and wave vector:

$$v_{ph} = \frac{\omega}{q} \tag{8.41}$$

The significance of the phase velocity should be questioned because many distinct q values may be used to describe the same lattice wave. If we consider a periodic medium, say a crystal, in contrast with a homogeneous medium, then the phase velocity is clearly not a uniquely defined concept, and a maximum wave vector exists,

$$q_{max} = \frac{2\pi}{\lambda_{min}} = \frac{\pi}{a} \tag{8.42}$$

Here λ_{min} is the minimum wavelength in the chain. At this wavelength, the entire chain is made up of two sub-lattices, where all the even atoms vibrate in out-of-phase with respect to the odd atoms. Wavelengths less than $\lambda_{min} = 2a$ cannot propagate down a linear chain lattice. We know that the wave motions are periodic in space and time. If $\lambda_{min} = 2a$, adjacent atoms are always 180° out of phase. For $\lambda < 2a$ the atomic motions would reproduce the same motions possible at a longer wavelength, corresponding to some other q within the Brillouin zone.

The actual allowed values of q depend on the boundary conditions. Suppose that we have a chain of N atoms as in Figure 8.3.

Using periodic boundary conditions, we close this chain so that the distance between the first and the Nth atom is a (they are considered as nearest neighbors). The closed chain has N intervals. There is no preferred position in the chain, which still possesses translational symmetry. Then the atomic displacements satisfy the condition:

$$U_{n\pm N} = U_n \tag{8.43}$$

(Born-von Karman condition). Condition (8.43) implies that n and $n \pm N$ refer to one and the same atom. This implies that

$$\exp\{\pm iqaN\} = 1 \tag{8.44}$$

FIGURE 8.3 A one-dimensional chain of N identical atoms with inter-atomic distances a.

and q is allowed to satisfy

$$qaN = 2\pi m, \quad m = 0, \pm 1, \pm 2, \ldots \tag{8.45}$$

$$q = \frac{2\pi m}{aN} \tag{8.46}$$

This implies that q is quantized and has only these discrete values. Recall that there is no change in the solution by adding multiples of $2\pi/a$ to the wave vector $\vec{q}$. Considering the inequality

$$-\frac{\pi}{a} \le q \le \frac{\pi}{a} \tag{8.47}$$

and Equation 8.46 then we have the range of the index m,

$$-\frac{N}{2} \le m \le \frac{N}{2} \tag{8.48}$$

Thus the allowable wave vector q takes countable discrete values. For a chain of N sites, there are N allowed q values ($m = -N/2$ and $m = +N/2$ are equivalent).

The same set of atomic displacements may be represented by wave vector q and by $q' = q + m(2\pi/a)$. It should be noted that $b = 2\pi/a$ is the primitive vector of the reciprocal lattice for a one-dimensional crystal. This is discussed further in Chapter 9: the reciprocal lattice vectors are those whose plane waves give a unit exponential on the real space lattice sites, $\exp(i\vec{b} \cdot \vec{a}_n) = 1$. This redundancy implies that the dispersion relation for these wave solutions is fully specified over the limited range of q: $-\pi/a \le q \le \pi/a$. The redundancy in q, often expressed as periodicity in reciprocal space of certain physical quantities, is a consequence of the periodicity in real space of the physical system.

Consider now a particular case, where only nearest neighbor interactions are present. This would be a model where springs connect nearest-neighbor atoms. In Equation 8.33 we consider only three values:

$$n' = n - 1, n, n + 1 \tag{8.49}$$

that correspond to P(1), P(0), and P(|-1|), respectively.

If we consider the fact that

$$\sum_{n'} P(|n - n'|) = 0 \tag{8.50}$$

then these couplings must satisfy

$$\frac{1}{2}P(0) = -P(1) = \alpha \tag{8.51}$$

Here α is the spring constant of an individual spring between two atoms.
   Considering Equations 8.28 and 8.51 we have

$$M\ddot{U}_n = -\alpha[(U_n - U_{n-1}) + (U_n - U_{n+1})] \tag{8.52}$$

On the RHS of Equation 8.52 in the square brackets we see the displacements relative to the neighboring atoms. Equation 8.52 involves the expected Hooke's Law
spring force F in classical mechanics:

$$F = -\alpha\Delta x \tag{8.53}$$

From the general law of dispersion in Equation 8.36 we have

$$M\omega^2 = P(0) + 2\sum_{\ell=1}^{\infty} P(\ell)\cos qa\ell \tag{8.54}$$

and considering Equation 8.51 we have

$$M\omega^2 = P(0) + 2P(1)\cos qa = 4\alpha \sin^2 \frac{qa}{2} \tag{8.55}$$

and

$$\omega = \pm\omega_{max}\left|\sin \frac{qa}{2}\right| \tag{8.56}$$

where

$$\omega_{max} = \sqrt{\frac{4\alpha}{M}} \tag{8.57}$$

The plus and minus signs in Equation 8.56 denote waves traveling either to the right
or to the left, for a chosen q. This reflects the inversion symmetry of the chain. To
completely count all possible solutions, however, we keep only the positive solution
for $\omega(q)$. Then we associate positive q with waves moving to the right and negative
q with waves moving to the left, but still with positive frequency. Therefore, for a
chosen (positive) frequency, there are always two modes.
   The frequency and wave vector may not be chosen independently. Specifying one,
the other is determined from the dispersion relation (8.56) for the chain. Whatever

the choice of the wave vector $\vec{q}$ or wavelength λ, the angular frequency ω may never be greater than that in Equation 8.57. This is in contrast to an ideal continuous string that will propagate a wave of any frequency.

We see in Figure 8.4 that the dispersion curve flattens gradually as the edge of the Brillouin zone is approached ($q \rightarrow \pm \pi/a$). This flattening may be understood as due to a reflected wave (or wave in the opposite direction) that gradually increases its amplitude until at $q = \pm\pi/a$ the reflected amplitude is equal to the incident component.

For N atoms with periodic boundary conditions, we have N allowed values of q. These N atoms have N degrees of freedom. Thus all the modes are accounted for. In relation (8.56) we keep only the positive branch of the vibration spectrum. Formula (8.56) is the dispersion law and its graph is shown in Figure 8.4. The maximum frequency a wave may have in the lattice is expression (8.57). This frequency corresponds to the zone edges, $q_{max} = \pm\pi/a$. Values of q outside this range are unnecessary.

In the case of long waves, that is, for small $q : |q| = 2\pi/\lambda$, we have approximately

$$\sin \frac{qa}{2} \approx \frac{qa}{2} \tag{8.58}$$

In this case Equation 8.56 takes the form

$$\omega \approx \omega_{max} \frac{a}{2} q = cq \tag{8.59}$$

which is a linear dependence of ω on q, with some wave speed c. The speed is the phase velocity v_{ph}, which takes the form

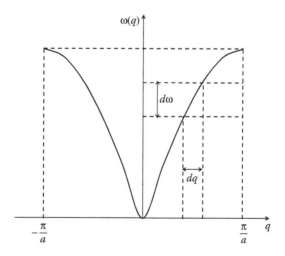

FIGURE 8.4 The dispersion relation for a one-dimensional chain of atoms.

$$v_{ph} = \frac{\omega}{q} = \omega_{max} \frac{a}{2} = a\sqrt{\frac{\alpha}{M}} \tag{8.60}$$

and the group velocity is the same, in this case,

$$v_g = \frac{d\omega}{dq} = \omega_{max} \frac{a}{2} \cos\frac{qa}{2} \approx \omega_{max} \frac{a}{2} \tag{8.61}$$

This implies that for long wavelengths,

$$\rho(\omega) = \frac{2N}{\pi} \frac{1}{\sqrt{\omega_{max}^2 - \omega^2}}, \quad 0 < \omega < \omega_{max} \tag{8.62}$$

This should be expected, as waves of long wavelength are not sensitive to the discreteness of the lattice. In other words, a large number of atoms will participate in all the displacements as on a homogeneous string. Thus long waves correspond to the propagation of sound waves. To the contrary, it is puzzling that the group velocity goes to zero at $q = \pm\pi/a$. These modes of vibration, however, characterize stationary waves. If we consider that two equivalent wave vectors are

$$q' = q + \frac{2\pi m}{a} \tag{8.63}$$

Then we may expand $\omega (q')$ in a Taylor series:

$$\omega(q') = \omega(q) + (q' - q)\frac{d\omega}{dq} + \frac{1}{2}(q' - q)^2 \frac{d^2\omega}{dq^2} + \cdots \tag{8.64}$$

Using only the first two terms of the expansion, with powers higher than $(q' - q)$ infinitesimally small, then

$$\omega(q') = \omega(q) + (q' - q)\frac{d\omega}{dq} \tag{8.65}$$

This shows formally that the group velocity of the wave packet is $v_G = d\omega/dq$. As we have seen this velocity is zero at the boundary of the Brillouin zone, $q = \pi/a$. There is energy associated with the wave packet and located within it. The group velocity is the velocity with which energy is carried and that velocity goes to zero at $q = \pm\pi/a$. If we consider v_{ac} to be the velocity of an acoustic wave then

$$\omega = \omega_{max} \left| \sin\frac{qa}{2} \right|_{q\to 0} \cong \omega_{max} \frac{qa}{2} = v_{ac}q \tag{8.66}$$

and

$$\omega_{max} = \frac{2}{a} v_{ac} \tag{8.67}$$

Thus the phase velocity is

$$v_{ph} = \frac{\omega}{q} = \omega_{max} \left| \frac{\sin(qa/2)}{q} \right| = v_{ac} \frac{\sin(qa/2)}{(qa/2)} \tag{8.68}$$

and the group velocity is

$$v_g = \left| \frac{\partial \omega}{\partial q} \right| = v_{ac} \left| \cos \frac{qa}{2} \right| \tag{8.69}$$

It is seen for small q that

$$v_{ph} = v_g = v_{ac} \tag{8.70}$$

8.3.2 DENSITY OF STATES

Consider q restricted to the first Brillouin zone for convenience. The notion of a normal mode with q' chosen outside the first Brillouin zone is physically indistinguishable from one associated with q within the first zone with $\vec{q}' - \vec{q} = \vec{b} =$ (reciprocal lattice vector).

We evaluate how many modes have frequencies in the range from ω to $\omega + d\omega$. Define the number of modes to be $\rho(\omega)d\omega$ and call the distribution function $\rho(\omega)$ the *density of modes*. The density of modes refers to the number of modes dZ per unit frequency interval $d\omega$:

$$\rho(\omega) = \frac{dZ}{d\omega} \tag{8.71}$$

In the quantum picture, the different modes correspond to the different possible states for phonons. The modal frequency ω corresponds to an allowed phonon state of energy $E = \hbar\omega$. Then the density of modes $\rho(\omega)$ may be referred to as a density of states for phonons. We already know that the allowed wave vectors on a chain of N atoms are $q = (2\pi m/aN)$. Then in the limit of large N, and supposing then that the allowed wave vectors go over to a continuum of modes, we can write, $dq = (2\pi/aN)dm$. Note that m is assumed to take values from $-N/2$ to $N/2$. For every chosen frequency, there are two modes ($\pm q$ values for waves moving to the right/left). Then from Equation 8.56 we have

$$\rho(\omega) = 2 \frac{dm}{d\omega} = \frac{2N}{\pi} \frac{1}{\sqrt{\omega_{max}^2 - \omega^2}}, \quad 0 < \omega < \omega_{max} \tag{8.72}$$

It should be noted that $\rho(\omega)$ is infinite at the cutoff frequency ω_{max}. The divergence at $\omega = \omega_{max}$ is not physically unrealistic. For this integral, this singularity presents no problem and the integral is finite. The formula recovers the total number of modes equal to N:

$$\int \rho(\omega)d\omega = \frac{2N}{\pi} \int_0^{\omega_{max}} \frac{d\omega}{\sqrt{\omega_{max}^2 - \omega^2}} = N \qquad (8.73)$$

The density of states is important later when considering the statistics of oscillator or phonon modes.

8.4 VIBRATIONAL MODES OF A DIATOMIC LINEAR CHAIN

We generalize these discussions by considering the propagation of lattice vibrations in crystals with more than one atom per unit cell. We assume that the mass of nearest neighbors differs but the mass of next nearest neighbors is the same. An example in three dimensions is the sodium chloride structure.

In this model masses on alternate planes are different, as in Figure 8.5. The propagation of the wave remains one-dimensional. Our analyses are carried out in analogy with those in the previous heading. For the case of two types of atoms per primitive cell, there arises a new phenomenon. In the dispersion law there arises two branches, called the *acoustic* and *optical* branches. As $q \to 0$, the acoustic frequency goes to zero, whereas, the optical branch frequency tends to a nonzero value. The optical modes appear at infrared frequencies and can be excited by infrared radiation.

Consider for convenience a linear Bravais lattice for which the elementary linear cell of size a has two atoms. Such a system has $2N$ degrees of freedom for N unit cells, each labeled by an index n. Figure 8.5 shows a linear chain with two types of atoms arranged alternately. A distance $a_0 = a/2$ separates two neighboring atoms. Suppose we have light and heavy atoms of masses m' and m'' respectively, with $m' < m''$. The force of interaction between atoms is considered to be of short range. This implies that the atoms interact only with their neighbors, and as a result, the equations of motion are

$$m'\ddot{U}_n' = -\alpha(2U_n' - U_n'' - U_{n-1}'')$$
$$m''\ddot{U}_n'' = -\alpha(2U_n'' - U_n' - U_{n+1}') \qquad (8.74)$$

where α is a force constant related with the elasticity (spring constant).

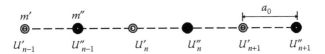

FIGURE 8.5 A linear diatomic chain with model masses on alternate planes being different.

We can assume that the oscillations of the atoms with different masses may have different amplitudes. Then we find the solutions of Equation 8.74 in the form of progressive waves in the crystal:

$$U_n' = A' \exp\{i(qna - \omega t)\}, \quad U_n'' = A'' \exp\{i(qna - \omega t)\} \qquad (8.75)$$

In U_n' and U_n'' we assume the same ω and q for the motion of the two types of atoms. We assume a simple wave motion for which both types of atom participate. Although the displacement of both types of atoms is different, they both contribute to one and the same wave disturbance. Note also that Equation 8.75 contains the unit cell size a, which is twice the nearest-neighbor separation: $a = 2a_0$. If we substitute Equation 8.75 into Equation 8.74 then we have

$$(m'\omega^2 - 2\alpha)A' + \alpha(\exp\{-iqa\} + 1)A'' = 0,$$
$$(m''\omega^2 - 2\alpha)A'' + \alpha(1 + \exp\{-iqa\})A' = 0 \qquad (8.76)$$

This system of homogeneous equations has a solution only if the determinant is equal to zero:

$$\begin{vmatrix} m'\omega^2 - 2\alpha & \alpha(1 + \exp\{-iqa\}) \\ \alpha(1 + \exp\{iqa\}) & m''\omega^2 - 2\alpha \end{vmatrix} = 0 \qquad (8.77)$$

This determinant gives immediately the dispersion law from the roots of the equation,

$$\omega^4 - \omega^2 2\alpha\left(\frac{1}{m'} + \frac{1}{m''}\right) + \frac{4\alpha^2}{m'm''}\sin^2\frac{qa}{2} = 0 \qquad (8.78)$$

Making some definitions,

$$\omega_o^2 = 2\alpha\left(\frac{1}{m'} + \frac{1}{m''}\right), \quad \gamma^2 = \frac{4m'm''}{(m' + m'')^2}, \quad \gamma^2 \le 1 \qquad (8.79)$$

the roots of Equation 8.78 are

$$\omega^2 = \frac{\omega_o^2}{2}\left[1 \pm \sqrt{1 - \gamma^2 \sin^2\frac{qa}{2}}\right] \qquad (8.80)$$

Negative values of ω^2 would not give progressive waves, however, Equation 8.80 will give only positive values because $\gamma \le 1$. We use only the positive solutions for ω, as in Section 8.1.

Recall for the monatomic chain that each value of q corresponds to one value of ω (univalent dependence of ω on q). For the diatomic chain, with dispersion (8.80), each value of q corresponds to two values of ω (bivalent dependence of ω on q as a result of the $\pm$ signs). As a consequence, there are two modes of vibrations of the type (8.75). The unit cell in this case has size $a = 2a_0$ and the first zone for q runs from $-\pi/2a_0$ to $\pi/2a_0$. This is the same as restricting q to the range, $-\pi/a \leq q \leq \pi/a$, using the unit cell size. We have N values of q in this interval and thus for the two complete branches there are $2N$ normal modes. These correspond to the $2N$ degrees of freedom of the $2N$ atoms in the lattice.

Now we determine what differentiates the two branches in Equation 8.80. The lower frequency,

$$\omega_{ac}^2 = \omega_1^2 = \frac{\omega_o^2}{2}\left[1 - \sqrt{1 - \gamma^2 \sin^2 \frac{qa}{2}}\right] \tag{8.81}$$

is called the *acoustic branch* and the higher frequency,

$$\omega_{op}^2 = \omega_2^2 = \frac{\omega_o^2}{2}\left[1 + \sqrt{1 - \gamma^2 \sin^2 \frac{qa}{2}}\right] \tag{8.82}$$

is called the *optical branch*. The splitting into acoustic and optical branches occurs for any crystal with two atoms per primitive cell. In three dimensions, the atoms can vibrate either parallel or perpendicular to $\vec{q}$. Thus the modes have one longitudinal and two transverse polarizations. Splitting into acoustic and optical branches occurs for both polarizations. For a general 3D crystal with two atoms per primitive cell, then there are six modes in all.

Let us investigate the two branches within the Brillouin zone $-\pi/a \leq q \leq \pi/a$. If we consider the fact that ω_{ac} and ω_{op} are even functions of q then we limit ourselves to the domain: $0 \leq q \leq \pi/a$. For $q = 0$ we have

$$\omega_{ac}(0) = 0, \quad \omega_{op}(0) = \omega_o \tag{8.83}$$

and for $q = \pi/a$ we have

$$\omega_{ac}\left(\frac{\pi}{a}\right) = \sqrt{\frac{2\alpha}{m''}}, \quad \omega_{op}\left(\frac{\pi}{a}\right) = \sqrt{\frac{2\alpha}{m'}}, \quad m'' > m' \tag{8.84}$$

The mass difference is seen to lead to a difference or gap between these frequencies, at the edge of the Brillouin zone.

For long wavelength, that is, $q \to 0$, we may expand Equation 8.82 into a MacLaurin series, limiting ourselves to only the first term of the series:

$$\omega_{ac} \cong \frac{\omega_o}{4}\gamma a q \tag{8.85}$$

$$\omega_{op} \cong \omega_o\left[1 - \frac{\gamma^2 q^2 a^2}{8}\right] \tag{8.86}$$

The acoustic branch is linear, as expected. The dispersion of the optical branch is weak in the neighborhood of $q = 0$. Relations (8.85) and (8.86) are shown in Figure 8.6. The optical branch frequency decreases parabolically near $q = 0$.

Let us examine the physical sense of the difference between the acoustic and the optical modes of vibration of atoms in the chain. In this connection we compare the amplitudes and phases of vibration of neighboring atoms for each of the branches.

From Equation 8.75 define the ratio:

$$\frac{U'_n}{U''_n} = \frac{A'}{A''} \tag{8.87}$$

Get this ratio from the system of Equation 8.76:

$$(m'\omega^2 - 2\alpha)A' + \alpha(\exp\{-iqa\} + 1)A'' = 0 \tag{8.88}$$

This gives

$$\frac{U'_n}{U''_n} = \frac{A'}{A''} = \frac{\alpha(\exp\{-iqa\} + 1)}{2\alpha - m'\omega^2} \tag{8.89}$$

The amplitudes are complex valued. However, to get the physical displacements, one takes the real part. At the points $q = 0$ and $q = \pm\pi/a$ (limiting vibrations) the amplitudes are real.

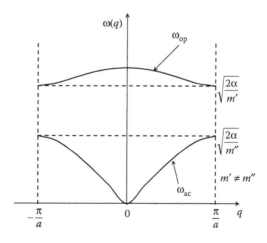

FIGURE 8.6 Dispersion relation, showing the optical and acoustic branches for a diatomic chain.

Look first at the case $qa \ll 1$, that is, long wavelengths:

$$q = \frac{2\pi}{\lambda} \ll \frac{1}{a}, \quad \lambda \gg 2\pi a \tag{8.90}$$

For the acoustic vibrations, with $\omega_{ac}(0) \approx 0$, see Equation 8.85, we have

$$\left.\frac{U'_n}{U''_n}\right|_{ac} \approx 1 \tag{8.91}$$

For long wavelength acoustic vibrations, the displacements of light and heavy atoms coincide. The elementary cell vibrates as a whole and the vibrations of neighboring atoms are in phase. Both types of atoms move together in a long wavelength pressure wave. We have the vibration of a density, and as the vibration of a density is a sound vibration, thus the name acoustic vibration. For optical modes, consider Equations 8.79, 8.80, and 8.82, and then for $q = 0$ we have

$$\left.\frac{U'_n}{U''_n}\right|_{op} = \left.\frac{\alpha(\exp\{-iqa\} + 1)}{2\alpha - m'\omega_{op}^2}\right|_{q=0} = \exp\{\pm i\pi\}\frac{m''}{m'} = -\frac{m''}{m'} \tag{8.92}$$

This implies that one vibration lags the other by a phase of $\pm\pi$. The heavy and light atoms vibrate in opposite directions (out-of-phase). In addition, their amplitudes are inversely proportional to their masses, which is reasonable under the requirements of Newton's Third Law.

The center of mass of each cell that has two types of atoms is at rest. We have

$$m'U'_n + m''U''_n = 0 \tag{8.93}$$

Dividing Equation 8.93 by $m' + m''$ shows that the center of mass is conserved. In an optical mode, the heavy and light atoms have displacement in opposite directions for long wavelengths. If there is a charge difference of the atoms, this causes a polarization wave with wave vector $\bar{q}$. Such a system may either absorb or emit light, hence the name optical vibration. For most solids this frequency is similar to that of infrared light and thus the name optical vibration. They can be excited with infrared radiation.

Consider next the case, $q = \pi/a$, at the Brillouin zone edge:

$$\left.\frac{U'_n}{U''_n}\right|_{ac} = 0, \quad m' \neq m'' \tag{8.94}$$

Then, for short acoustic waves, the light (m') atoms are at rest and the heavy ones are displaced. As a consequence, the density changes considerably as the heavy atoms move.

For the optical branch at $q = \pi/a$, the ratio is indeterminate,

$$\left.\frac{U'_n}{U''_n}\right|_{op} = \frac{0}{0} \tag{8.95}$$

It is known from Equation 8.84 that

$$\omega_{ac}\left(\frac{\pi}{a}\right) = \sqrt{\frac{2\alpha}{m'}}, \quad \omega_{op}\left(\frac{\pi}{a}\right) = \sqrt{\frac{2\alpha}{m''}} \tag{8.96}$$

Let us remove the indeterminacy in Equation 8.95. Suppose $q = \frac{\pi}{a} - \varepsilon$, then

$$\left.\frac{U'_n}{U''_n}\right|_{op} = \frac{\alpha(1 - \exp\{ia\varepsilon\})}{2\alpha - m'\omega^2_{op}} \tag{8.97}$$

$$\omega^2_{op} = \frac{\omega^2_o}{2}\left\{1 + \sqrt{1 - \gamma^2\cos^2\frac{\varepsilon a}{2}}\right\} \tag{8.98}$$

and as a consequence

$$\left.\frac{U'_n}{U''_n}\right|_{op} \approx \frac{\varepsilon}{\varepsilon^2} \to \infty \quad \text{as } \varepsilon \to 0 \tag{8.99}$$

This implies that $U''_n = 0$. The heavy atoms are at rest and with $U'_n \neq 0$, the light atoms are vibrating. In this sense, the physical distinction between acoustic and optical modes at $q = \pi/a$ is minimal. In both cases, one of the sublattices (odd or even lattice sites) is vibrating, while the other is at rest.

8.5 VIBRATIONAL MODES IN A THREE-DIMENSIONAL CRYSTAL

We consider the lattice dynamics in a three-dimensional crystal based on a harmonic interaction potential with one free parameter, the force constant of the nearest-neighbor interaction. The atoms are assumed to undergo small oscillations around their equilibrium positions. Here we find solutions to the equations of motion of the atoms in the form of plane waves with phonon frequency ω that is a function of the phonon wave vector q. This is an extension of the model for a chain of atoms.

We consider a crystal whose unit cells have s atoms with masses m_k, $k = 1, 2, 3, ..., s$. The crystal with volume $V = N\Omega_o$ has $N = G^3$ elementary cells with Ω_o being the volume of an elementary cell. Each atom in the crystal has three degrees of freedom. Then for the N atoms there are $3N$ degree of freedom. Including the s atoms per elementary cell, there are $3Ns$ degrees of freedom in the whole crystal.

We consider a crystal defined by a unit cell with primitive lattice vectors $\vec{a}_i$, $i = 1, 2, 3$. Suppose the crystal to be a parallelepiped with edges of length $G|\vec{a}_i|$, $i = 1, 2, 3$, that are parallel to the corresponding primitive lattice vectors. The equilibrium primitive cell positions are denoted by the vectors $\vec{a}_n$:

$$\vec{a}_n = n_1\vec{a}_1 + n_2\vec{a}_2 + n_3\vec{a}_3 \tag{8.100}$$

where n_1, n_2, n_3 are integers ranging from 1 to G. The displacement of the kth atom from the its equilibrium position is denoted by $\vec{U}_n^k$ so that the position vector $\vec{r}_n^k$ of the kth atom is

$$\vec{r}_n^k = \vec{a}_n + \vec{U}_n^k \tag{8.101}$$

The displacement $\vec{U}_n^k$ has α-Cartesian components, $U_{n\alpha}^k$, $\alpha = (x, y, z)$. We proceed classically, and then the total kinetic energy may be written as

$$E_{kin} = \frac{1}{2}\sum_{nk\alpha} m_k \dot{U}_{n\alpha}^k \dot{U}_{n\alpha}^k \tag{8.102}$$

Here m_k is the mass of the kth atom in a unit cell.

Since the vibration of the lattice involves small excursions from the equilibrium positions, we may expand the many-body potential energy W in a Taylor series of atomic displacements $U_{n\alpha}^k$.

$$W = W_0 + \sum_{nk\alpha} U_{n\alpha}^k \left[\frac{\partial W}{\partial U_{n\alpha}^k}\right]_0 + \frac{1}{2}\sum_{nk\alpha n'k'\alpha'} U_{n\alpha}^k U_{n'\alpha'}^{k'}\left[\frac{\partial^2 W}{\partial U_{n\alpha}^k \partial U_{n'\alpha'}^{k'}}\right]_0 + \cdots \tag{8.103}$$

Here the subscript zero "0" implies that the derivatives are evaluated at the equilibrium configuration. The derivatives here are merely constants. The ground state potential energy (the potential energy in the equilibrium configurations of the atoms) $W_0 = W(0)$ may always be set equal to zero and thereby made a reference point. This will not affect our answers. We assume that the electrons in the crystal always have time to adjust themselves to the configuration with lowest energy even during crystal vibration. Since in the equilibrium configuration the potential energy is a minimum it follows that

$$\left\{\frac{\partial W}{\partial U_{n\alpha}^k}\right\}_0 = 0 \tag{8.104}$$

This is because the force acting on any atom must vanish in the equilibrium configuration.

If we limit ourselves to the harmonic approximation, in which terms of order higher than quadratic are ignored, then W becomes

$$W = \frac{1}{2} \sum_{nk\alpha n'k'\alpha'} W_{\alpha\alpha'}\left(\begin{matrix} kk' \\ nn' \end{matrix}\right) U^k_{n\alpha} U^{k'}_{n'\alpha'} \qquad (8.105)$$

where

$$\left[\frac{\partial^2 W}{\partial U^k_{n\alpha} \partial U^{k'}_{n'\alpha'}}\right]_0 \equiv W_{\alpha\alpha'}\left(\begin{matrix} kk' \\ nn' \end{matrix}\right) \qquad (8.106)$$

This constant depends on the atomic arrangement of the crystal and the inter-atomic potential. The harmonic potential W of the three-dimensional crystal is written in terms of the displacement vector $\vec{U}^k_n$ of each atom from its equilibrium position n as in Equation 8.105. The sum extends over all pairs of atoms located at cells n and n'.

At low temperatures the harmonic approximation is very reasonable. The separation between atoms is of the order of 0.1–1 nm and at room temperature vibrations have amplitudes of the order 0.01 nm. Experiments may show this idealization to be false. The general approach is to make idealizations and then find corrections to this approach to give better results. Even so, the harmonic approximation is a good approximation for most solids at room temperature. Within the harmonic approximation, classical mechanics gives an exact solution to the equations of motion. This can be done by first finding special linear combinations called *normal coordinates* of the atomic displacement coordinates. The potential and kinetic energies of the system contain no cross terms if expressed in normal coordinates. They contain only quadratic terms. We then find equations of motion that reduce to a set of independent harmonic oscillator equations that are easily solvable. The solutions are called *normal modes* of the system. The Lagrangian L of the system is

$$L\left\{U^k_{n\alpha}, \dot{U}^k_{n\alpha}\right\} = E_{\text{kin}} - W \qquad (8.107)$$

and the equations of motion are

$$\frac{d}{dt}\frac{\partial L}{\partial \dot{U}^k_{n\alpha}} - \frac{\partial L}{\partial U^k_{n\alpha}} = 0 \qquad (8.108)$$

or

$$m_k \ddot{U}^k_{n\alpha} = -\sum_{n'k'\alpha'} W_{\alpha\alpha'}\left(\begin{matrix} kk' \\ nn' \end{matrix}\right) U^{k'}_{n'\alpha'} \qquad (8.109)$$

This gives a system of $3Ns$ equations. These equations are true for each Cartesian component of the displacement $\vec{U}_n^k$. Equation 8.109 is a large number of coupled differential equations and thus may not be solved directly.

Consider the properties of the potential in Equation 8.106.

$$W_{\alpha'\alpha}\begin{pmatrix} k'k \\ n'n \end{pmatrix} = W_{\alpha\alpha'}\begin{pmatrix} kk' \\ nn' \end{pmatrix} \tag{8.110}$$

1. It is invariant relative to interchange of all indices. That corresponds to interchanging the two atoms. The differential equation (8.105) is the same if the atoms are interchanged.

$$W_{\alpha\alpha'}\begin{pmatrix} kk' \\ nn' \end{pmatrix} = W_{\alpha\alpha'}\begin{pmatrix} kk' \\ n-n' \end{pmatrix}, \quad n-n' = \begin{pmatrix} n_1 - n'_1 \\ n_2 - n'_2 \\ n_3 - n'_3 \end{pmatrix} \tag{8.111}$$

2. The expression in (8.105) is invariant under translation by any lattice vector $\vec{a}_n$ (see Equation 8.100). These force constants depend only on the separation of the cells.

3. We suppose that, at some initial moment, the crystal is at the temperature $T = 0$.

Then there is no vibration and consequently the displacement is a constant. If all the atoms in the crystal have the same displacement, say, $U_{\alpha'}^o$ from equilibrium, that is, $U_{\alpha'}^{k'} \equiv U_{\alpha'}^o$, then the entire system is simply displaced without distortion. Then the RHS of Equation 8.109 will have the same value as when all $U_{n'\alpha'}^{k'}$ vanish:

$$\sum_{\alpha'} U_{\alpha'}^o \sum_{n'k'} W_{\alpha\alpha'}\begin{pmatrix} kk' \\ nn' \end{pmatrix} = 0 \tag{8.112}$$

From this it follows that

$$\sum_{n'k'} W_{\alpha\alpha'}\begin{pmatrix} kk' \\ nn' \end{pmatrix} = 0 \tag{8.113}$$

which is the condition that (8.112) vanishes for an arbitrary choice of $U_{\alpha'}^o$. Equation 8.113 expresses the fact that the forces on any atom are zero, if each atom is displaced from the equilibrium position by the same amount say, $U_{\alpha'}^o$.

Expression (8.105) is invariant relative to the translation by any lattice vector defined in (8.100). Then it satisfies the Bloch theorem, giving a plane wave of amplitude $\vec{A}^k$ for the kth atoms:

$$U_{\alpha n}^k = A_\alpha^k \exp\{i(\vec{q} \cdot \vec{a}_n - \omega t)\} = \exp\{i\vec{q} \cdot \vec{a}_n\} U_{\alpha 0}^k(t) \tag{8.114}$$

Here ω is a phonon frequency associated with the wave vector $\vec{q}$. Substitute this in the dynamical equation, (8.109), to get

$$\omega^2 m_k A_\alpha^k = \sum_{n'k'\alpha'} W_{\alpha\alpha'}\left(\begin{matrix} kk' \\ nn' \end{matrix}\right) \exp\{i\vec{q} \cdot (\vec{a}_{n'} - \vec{a}_n)\} A_{\alpha'}^{k'} \qquad (8.115)$$

Call

$$B_{\alpha\alpha'}^{kk'}(\vec{q}) = \sum_{n'} W_{\alpha\alpha'}\left(\begin{matrix} kk' \\ n-n' \end{matrix}\right) \exp\{i\vec{q} \cdot (\vec{a}_{n'} - \vec{a}_n)\} \qquad (8.116)$$

the *dynamical matrix*. Thus the normal modes can be determined from the dynamical matrix $B_{\alpha\alpha'}^{kk'}$.

If we move from the sum with respect to n' to the sum with respect to $l = n - n'$, then we see that the RHS of Equation 8.115 is independent of n. We expect that as the LHS is independent of n, it follows that

$$\omega^2 m_k A_\alpha^k = \sum_{k'\alpha'} B_{\alpha\alpha'}^{kk'}(\vec{q}) A_{\alpha'}^{k'} \qquad (8.117)$$

Thus to describe the oscillations, Equation 8.117 may be transformed:

$$\sum_{k'\alpha'} \left\{ B_{\alpha\alpha'}^{kk'} - m_{k'}\,\omega^2 \delta_{kk'} \delta_{\alpha\alpha'} \right\} A_{\alpha'}^{k'} = 0 \qquad (8.118)$$

This is a system of homogeneous equations for the amplitudes $A_{\alpha'}^k$. This system has nontrivial solutions only if the determinant is equal to zero. This condition determines the phonon frequencies $\omega(\vec{q})$. Thus

$$\left| B_{\alpha\alpha'}^{kk'} - m_{k'}\omega^2 \delta_{kk'} \delta_{\alpha\alpha'} \right| = 0 \qquad (8.119)$$

It is a determinant of the type $3s \times 3s$ with $3s$ independent eigenvalues. One can show there are $3s - 3$ optical modes.

The number of equations in Equation 8.118 is obtained from the number of possible polarization directions (three of them), that is, two transverse and one longitudinal polarization, times the number of atoms in the unit cell.

Considering the fact that there may be special symmetries within the unit cell, and the $3s \times 3s$ system is of finite size, the solution of Equation 8.118 is quite possible, especially with modern computation. In principle, the system should be solvable for all possible values of q (we have N of them). This would be performed within the first Brilluoin zone.

Thus, the equation of the determinant (8.119) gives the phonon spectrum of the crystal. This gives us information about the role of the electron–electron, electron–ion, and electron–phonon interactions. It explains phonon characteristics responsible for atomic properties.

8.5.1 Properties of the Dynamical Matrix

If we consider Equation 8.116 we see that the dynamical matrix is Hermitian. In fact, it takes real values

$$B_{\alpha'\alpha}^{k'k}(\vec{q}) = B_{\alpha\alpha'}^{kk'\,*}(\vec{q}) \tag{8.120}$$

The characteristic equation has a real spectrum. This implies that ω^2 is real. Physically, it is obvious that ω^2 may not be less than zero, otherwise from Equation 8.114 there would appear the term $\exp\{\pm \omega_o t\}$, which implies instability of the lattice. If we solve Equation 8.119 then we have $3s$ different frequency eigenvalues

$$\omega_1^2(\vec{q}),\ldots,\omega_1^2(\vec{q}),\ldots,\omega_{3s}^2(\vec{q}), \quad j = 1,\, 2,\ldots,3s \tag{8.121}$$

$\vec{q}$ is a three-dimensional vector and thus the surface $\omega^2(\vec{q})$ is a four dimensional hyper-surface. If we do the change $\vec{q} \to -\vec{q}$ then it is easy to see that

$$B_{\alpha\alpha'}^{kk'}(-\vec{q}) = \left\{B_{\alpha'\alpha}^{k'k}(\vec{q})\right\}^* = B_{\alpha'\alpha}^{k'k} \tag{8.122}$$

and if we substitute in Equation 8.119 we see that the determinant changes rows for columns and vice versa and thus

$$\omega_j^2(-\vec{q}) = \omega_j^2(\vec{q}) \tag{8.123}$$

and as $\omega_j^2 \geq 0$ then using only the positive solutions,

$$\omega_j(-\vec{q}) = \omega_j(\vec{q}) \tag{8.124}$$

If in Equation 8.115 we let $\vec{q} \to -\vec{q}$ then

$$A_{j\alpha}^k(-\vec{q}) = \left(A_{j\alpha}^k\right)^*(-\vec{q}) \tag{8.125}$$

Let the wave vector transform,

$$\vec{q} \to \vec{q}' = \vec{q} + \vec{b} \tag{8.126}$$

where $\vec{b}$ is any vector of the reciprocal lattice. If we substitute Equation 8.126 into Equation 8.114 then

$$\begin{aligned}
\left(U_{\alpha n}^k\right)' &= A_\alpha^k \exp\{i(\vec{q}\cdot\vec{a}_n - \omega t)\}\exp\{i\vec{b}\cdot\vec{a}_n\} \\
&= U_{\alpha n}^k \exp\{i\vec{b}\cdot\vec{a}_n\} = U_{\alpha n}^k \exp\{i2\pi l\}
\end{aligned} \tag{8.127}$$

Here $\vec{b} \cdot \vec{a}_n = 2\pi l$, where l is an integer and $\exp\{i2\pi l\} = 1$. Shifting the wave vector by any reciprocal lattice vector gives a mathematically equivalent solution.

It should also be noted from the definition of $B_{\alpha\alpha'}^{kk'}(\vec{q})$ that

$$B_{\alpha\alpha'}^{kk'}(\vec{q} + \vec{b}) = B_{\alpha\alpha'}^{kk'}(\vec{q}) \tag{8.128}$$

This is the confirmation that $\omega_j^2(\vec{q})$ is periodic in the reciprocal lattice. The minimum value of $\vec{b}$ in order to have the minimum domain of $\vec{q}$ is the reciprocal value of $\vec{a}_n = n_1\vec{a}_1 + n_2\vec{a}_2 + n_3\vec{a}_3$. A shift of $\vec{q}$ by any primitive $\vec{b}$-vector gives an irrelevant 2π shift in the argument of the wave function. This is a result of the fundamental relation for the reciprocal lattice vectors, $\vec{b}_i \cdot \vec{a}_j = 2\pi\delta_{ij}$. This result can be made symmetrical around the origin, and we see that it makes sense to limit $\vec{q}$ according to constraints for each primitive direction,

$$-\pi \le (\vec{q} \cdot \vec{a}_i) \le \pi, \quad i = 1, 2, 3 \tag{8.129}$$

which defines a domain of physically distinct values of $\vec{q}$. For a cubic crystal we have

$$-\frac{\pi}{a_\alpha} \le q_\alpha \le \frac{\pi}{a_\alpha}, \quad \alpha = (x, y, z) \tag{8.130}$$

where q_α is one of the Cartesian components. In a more general case, these constraints define a three-dimensional region of the reciprocal space, known as the first Brilluoin zone.

If Ω_o is the elementary volume of the cell then for a cubic crystal we have

$$\frac{(2\pi)^3}{a^3} = \frac{(2\pi)^3}{\Omega_o} \tag{8.131}$$

We show the validity of Equation 8.131 for any crystal structure. Suppose

$$\phi_i = (\vec{q} \cdot \vec{a}_i) = q_x a_{ix} + q_y a_{iy} + q_z a_{iz} \tag{8.132}$$

then

$$\iiint dx\, dy\, dz = \int_{-\pi}^{} \iiint d\phi_1\, d\phi_2\, d_1\phi_3\, \frac{1}{|J|} \tag{8.133}$$

where J is a Jacobian:

$$J = \left| \frac{\partial \phi_i}{\partial q_\alpha} \right| = |a_{i\alpha}| = \vec{a}_1 \cdot (\vec{a}_2 \times \vec{a}_3) = \Omega_o \tag{8.134}$$

Thus

$$\frac{1}{\Omega_o} \iiint_{-\pi} d\phi_1 d\phi_2 d\phi_3 = \frac{(2\pi)^3}{\Omega_o} \qquad (8.135)$$

For the dispersion law in place of the graph of $\omega_j^2(\vec{q})$ we have $\omega_j(\vec{q}) = const$ there are points in the Brillouin zone where the dynamical matrix is real. That is where $\exp\{i\vec{q} \cdot (\vec{a}_n - \vec{a}_{n'})\}$ should go to 1, that is, where

1. $\vec{q} = 0$,
2. $(\vec{q} \cdot \vec{a}_i) = \pm\pi$

We examine the case $\vec{q} = 0$, (long wavelengths). If we consider Equation 8.115 then

$$\omega_j^2(0)A_{j\alpha}^k(0) = \sum_{n'k'\alpha'} W_{\alpha\alpha'}\begin{pmatrix} kk' \\ nn' \end{pmatrix} A_{j\alpha'}^{k'}(0) \qquad (8.136)$$

For $\omega_j^2(\vec{q}) = 0$ then the RHS of Equation 8.136 is equal to zero. This is a particular case of the one-dimensional lattice for $\vec{q} = 0$, for which $\omega_{ac} = 0$. If we consider Equation 8.113 then the RHS of Equation 8.136 is equal to zero and for that

$$A_{\alpha'}^{k'}(0) = A_{\alpha'}^{(0)}(0) \qquad (8.137)$$

There exists three directions for which $\omega_j^2(0) = 0$, independent of the symmetry of the crystal. This is because $\vec{q} = 0$ corresponds to identical displacements (see Equation 8.137) of all atoms of the crystal. Hence,

$$\omega_j^2(q \to 0) \to 0 \qquad (8.138)$$

We see that the three modes are degenerate at $q = 0$. Notwithstanding this, if there is more than one type of atom in the case where atoms occupy physically inequivalent sites, the different types of atoms may vibrate out of phase with one another. Thus Equation 8.138 need not hold for all modes. However it must hold for three modes. These three modes are called *acoustic modes*. This is due to the fact that at low q there are frequencies of the order of those of sound. The remaining modes where Equation 8.138 does not hold have frequencies in the infrared region of the spectrum. They are the *optical modes*.

From Equation 8.137 it follows that the amplitude (for acoustic modes) is independent of the type of atom and from Equation 8.115 we have

$$U_{n\alpha}^k = A_\alpha^o \exp\{-i\omega t\} \qquad (8.139)$$

The entire elementary cell vibrates as a whole but with different phases. If we assume that $\lambda \to \infty$ ($q \to 0$) then such a wave does not feel the discreteness of the lattice in an elastic medium.

If the crystal is isotropic then as far as the polarization is concerned, we have one longitudinal polarization and two transverse polarizations. The atomic motions can either be aligned with the wave vector, or transverse to it. In an anisotropic crystal it is not possible to seek out longitudinal and transverse waves. A crystal as a whole is anisotropic. We examine a cubic crystal for a longitudinal and transverse wave, simply for convenience.

The index $j = 1, 2, 3, \ldots, 3s$ labels the normal modes. For the index $j = 1, 2, 3$ we have $\omega_j(0) = 0$ (acoustic modes) and for the other $3s - 3$ cases we have $\omega_j(0) \neq 0$ ($3s - 3$ optical modes). Then from Equation 8.115 we have

$$\omega^2(0)\sum_k m_k A_\alpha^k(0) = \sum_{k'\alpha'} A_{\alpha'}^{k'}(0)\sum_{n'k} W_{\alpha\alpha'}\binom{kk'}{nn'}$$

(8.140)

As $W_{\alpha\alpha'}$ is symmetric with respect to the indices n and n', then from Equation 8.140

$$\sum_{n'k} W_{\alpha\alpha'}\binom{kk'}{nn'} = \sum_{nk} W_{\alpha\alpha'}\binom{kk'}{nn'} = 0$$

(8.141)

and as $\omega_j(0) \neq 0$ then

$$\sum_k m_k A_\alpha^k(0) = 0$$

(8.142)

If we multiply Equation 8.142 by a factor $\exp\{i(\vec{q} \cdot \vec{a}_n - \omega t)\}$ then

$$\sum_k m_k U_{n\alpha}^k(0)_{\vec{q}=0} = 0, \quad \alpha = (x, y, z)$$

(8.143)

by which it follows that

$$\sum_k m_k U_n^k(0)\Big|_{\vec{q}=0} = 0$$

(8.144)

This gives us the character of the vibration for the $3s - 3$ cases. In particular, for the elementary cell the center of mass remains at rest. Such a vibration is called optical.

For any mode, from $\omega(-\vec{q}) = \omega(\vec{q})$, we expand $\omega(\vec{q})$ in a series around $\vec{q} = 0$:

$$\omega(\vec{q}) = \omega(0) + \sum_{\alpha\alpha'} C_{\alpha\alpha'} q_\alpha q_{\alpha'}$$

(8.145)

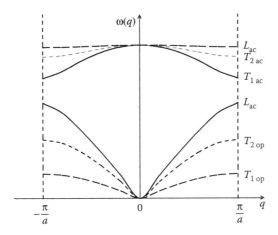

FIGURE 8.7   A dispersion relation, showing the transverse and longitudinal components of the optical and acoustic branches.

We may bring the sum $\sum_{\alpha\alpha'} C_{\alpha\alpha'} q_\alpha q_{\alpha'}$ to the diagonal form $\sum_\alpha C_\alpha q_\alpha^2$, and that leads us to $C\vec{q}^2$.

All three acoustic frequencies are equal to zero for $\vec{q} = 0$. Hence we have degeneracy. The dispersion relation for $\vec{q} \to 0$ is of the form,

$$\omega_j(\vec{q}) = C|\vec{q}| \tag{8.146}$$

as expected for acoustic vibrations.

In a crystal of the sodium chloride type we have $s = 2$ and $j = 1 - 6$. There are three branches which are obviously acoustic, see Figure 8.7. The lower branches that linearly tend to zero for small $\vec{q}$ are acoustic branches, and the remaining three branches are optical. For both the acoustic and optical types, there are branches with longitudinal and transverse vibrations. The longitudinal waves have stronger restoring forces and hence higher frequencies and propagation speeds ($\omega_L > \omega_{T_2} > \omega_{T_1}$). Here and in Figure 8.7, T_1 and T_2 correspond to the transverse modes and L to the longitudinal modes. It is left as an exercise for the reader to set up the normal mode problem for the simple cubic sodium chloride structure and verify the general results in Figure 8.7.

8.5.2   Cyclic Boundary for Three-Dimensional Cases

8.5.2.1   Born–Von Karman Cyclic Condition

Consider a crystal (or cell) as illustrated in Figure 8.8. We have a cyclic boundary condition if an integral shift in the lattice sites (G is a whole number),

$$\vec{a}_{\vec{n}} \to \vec{a}_{\vec{n}} + G\vec{a}_i, \quad i = 1,2,3. \tag{8.147}$$

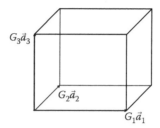

FIGURE 8.8   A cubic crystal or cell.

makes the displacements in Equation 8.114,

$$U_{n\alpha}^k = A_\alpha^k \exp\{i(\vec{q} \cdot \vec{a}_n - \omega t)\} \tag{8.148}$$

invariant or unchanged. This requires

$$\exp\{i\vec{q} \cdot G\vec{a}_i\} = 1 \tag{8.149}$$

where

$$G(\vec{q} \cdot \vec{a}_i) = 2\pi m_i \tag{8.150}$$

and m_i is a whole number. This gives

$$(\vec{q} \cdot \vec{a}_i) = \frac{2\pi m_i}{G} \tag{8.151}$$

Referring to Equation 8.17 and the basic identity for reciprocal lattice vectors, $\vec{b}_1 \cdot \vec{a}_1 = 2\pi$, it can be seen that this holds only if $\vec{q}$ is proportional to a reciprocal lattice vector:

$$\vec{q} = \frac{1}{G}\vec{b}_m \tag{8.152}$$

where $\vec{b}_m$ is a reciprocal lattice vector and $\vec{q}$ is now quasi discrete.

As shown earlier, the wave vector $\vec{q}$ is bounded by the limits of the Brillouin zone (Wigner–Seitz cell) that may be found from

$$-\pi \le (\vec{q} \cdot \vec{a}_i) \le \pi \tag{8.153}$$

Using Equation 8.152, we have

$$-\frac{G}{2} \le m_i \le \frac{G}{2} \tag{8.154}$$

where m_i takes G values and $\vec{q}$ takes G^3 values. Note that $G^3 = N$ is the number of elementary cells in the crystal. If the crystal has unequal edges (in units of the primitive vectors) then $G_1 \neq G_2 \neq G_3$ and $\vec{q}$ has $G_1 G_2 G_3 = N$ values, where each G_i is a whole number. We know that the volume of the Brillouin zone is equal to $((2\pi)^3/\Omega_o)$. We may calculate the number of different vibrational modes per unit volume of the given crystal, which is equal to the density of the modes. This is

$$\frac{N}{((2\pi)^3/\Omega_o)} = \frac{N\Omega_o}{(2\pi)^3} = \frac{V}{(2\pi)^3} \tag{8.155}$$

where V is the volume of the given crystal.

If we have to evaluate any sum with respect to $\vec{q}$, that is, $\Sigma_{\vec{q}}$, then as $\vec{q}$ is quasi-continuous. Supposing that G is large, we can write a differential relationship,

$$\frac{dN}{d\vec{q}} = \frac{V}{(2\pi)^3} \quad \text{or} \quad dN = \frac{V}{(2\pi)^3} d\vec{q} \tag{8.156}$$

Then a sum over discrete states goes over to a continuous integral (approximately) by the transformation,

$$\sum_{\vec{q}} \to \int dN \to \frac{V}{(2\pi)^3} \int d\vec{q} \tag{8.157}$$

8.6 NORMAL VIBRATION OF A THREE-DIMENSIONAL CRYSTAL

The kinetic energy of the lattice is

$$E_{\text{kin}} = \sum_{nk\alpha} \frac{1}{2} m_k \left|\dot{U}_{n\alpha}^k\right|^2 \tag{8.158}$$

where m_k is the mass of the kth atom in a cell. The potential energy is

$$W = \frac{1}{2} \sum_{nk\alpha n'k'\alpha'} W_{\alpha\alpha'}\binom{kk'}{nn'} U_{n\alpha}^k U_{n'\alpha'}^{k'} \tag{8.159}$$

It is convenient for us to transform to normal modes that are decoupled. This implies that instead of the system of the vibrating atoms we deal with a set of waves (modes). The amplitudes of these are called normal (mode) coordinates. These amplitudes $a_j(\vec{q},t)$ are dependent on the wave vector $\vec{q}$ and the particular mode, indexed by wave vector $\vec{q}$ and polarization state j. We have an expression for the total atomic displacement of the kth atom in cell n, as a sum over contributions from all the normal modes:

$$U^k_{n\alpha} = \frac{1}{\sqrt{N}} \sum_{\bar{q}j} a_j(\bar{q},t) A^k_{j\alpha}(\bar{q})\exp\{i\bar{q} \cdot \bar{a}_n\} \tag{8.160}$$

where $A^k_{j\alpha}(\bar{q})\exp\{i\bar{q} \cdot \bar{a}_n\}$ are plane wave normal modes in the crystal and a_j is the amplitude of the jth polarization state of normal mode at wave vector $\bar{q}$. The factor of $1/\sqrt{N}$ is for convenient normalization. If $a_j(\bar{q})$ is complex valued and

$$a_j(-\bar{q}) = a_j^*(\bar{q}) \tag{8.161}$$

then we can show that the expression in Equation 8.160 is real:

$$U^{k\;*}_{n\alpha} = \frac{1}{\sqrt{N}} \sum_{\bar{q}j} a_j(-\bar{q}) A^k_{j\alpha}(-\bar{q})\exp\{-i\bar{q} \cdot \bar{a}_n\} = U^k_{n\alpha} \tag{8.162}$$

Thus, Equation 8.161 must hold for the lattice displacements to be real. The normal modes are assumed to have the following normalization:

$$\sum_{nk\alpha} m_k A^k_{j\alpha}(\bar{q})\exp\{i\bar{q} \cdot \bar{a}_n\}\left\{A^k_{j'\alpha}(\bar{q})\exp\{i\bar{q}' \cdot \bar{a}_n\}\right\}^* = NM\delta_{\bar{q}\bar{q}'}\delta_{jj'} \tag{8.163}$$

where M represent the mass of all the atoms in a primitive cell, that is,

$$\sum_{k=1}^{s} m_k = M \tag{8.164}$$

Let us define

$$S = \sum_n \exp\{i(\bar{q} - \bar{q}') \cdot \bar{a}_n\} \tag{8.165}$$

where the $\bar{a}_n$ are the lattice positions,

$$\bar{a}_n = n_1\bar{a}_1 + n_2\bar{a}_2 + n_3\bar{a}_3, \quad \bar{n} = (n_1,n_2,n_3) \tag{8.166}$$

If $\bar{q} = \bar{q}'$ then $S = N$ is the number of the lattice vectors $\bar{a}_n$. If $\bar{q} \neq \bar{q}'$ then $\bar{q}$ and

$$\bar{q}' = \bar{q} + \bar{b} \tag{8.167}$$

where $\vec{b}$ is a reciprocal lattice vector, are physically equivalent. If we consider an arbitrary direct lattice vector $\vec{a}_m$ then we can do the transformation

$$\vec{a}_n \rightarrow \vec{a}_n + \vec{a}_m \tag{8.168}$$

in Equation 8.165. The crystal is the closure of the entire space. Then S is invariant relative to such a translation and thus

$$S = \sum_n \exp\{i(\vec{q} - \vec{q}') \cdot \vec{a}_n\} = \sum_n \exp\{i(\vec{q} - \vec{q}') \cdot (\vec{a}_n + \vec{a}_m)\}$$

$$= \exp\{i(\vec{q} - \vec{q}') \cdot \vec{a}_m\} \, S \tag{8.169}$$

and

$$[1 - \exp\{i(\vec{q} - \vec{q}') \cdot \vec{a}_m\}]S = 0 \tag{8.170}$$

But $\vec{a}_m$ is an arbitrary vector of the direct lattice. Thus, when $\vec{q} \neq \vec{q}'$ (or if $\vec{q}$ and $\vec{q}'$ differ by a reciprocal lattice vector as in Equation 8.167), it follows that $S = 0$. Thus we get in general

$$S = \sum_n \exp\{i(\vec{q} - \vec{q}') \cdot \vec{a}_n\} = N\delta_{\vec{q}\vec{q}'} \tag{8.171}$$

The quantity S is called the *lattice sum*. Here if $\vec{q} - \vec{q}'$ is zero or any reciprocal lattice vector, the two waves are physically equivalent, and Equation 8.171 is still true.

Let us find an orthogonality condition for the normal modes, by looking at the equations for the amplitude of two different waves, j and j', that is,

$$m_k \omega_j^2 A_{j\alpha}^k = \sum_{k'\alpha'} B_{\alpha\alpha'}^{kk'} A_{j\alpha'}^{k'}, \quad m_k \omega_{j'}^2 \left(A_{j'\alpha}^k \right)^* = \sum_{k'\alpha'} \left(B_{\alpha\alpha'}^{kk'} \right)^* \left(A_{j'\alpha'}^{k'} \right)^* \tag{8.172}$$

Let us multiply the first and second equation (8.172) by $A_{j'\alpha}^{*k}$ and $A_{j\alpha}^k$, respectively. Then take the sum with respect to the indices k and α, after which we subtract the second from the first, leading to

$$\left(\omega_j^2 - \omega_{j'}^2\right) \sum_{k\alpha} m_k A_{j\alpha}^k A_{j'\alpha}^{*k} = \sum_{kk'\alpha\alpha'} \left[B_{\alpha\alpha'}^{kk'} A_{j\alpha'}^{k'} A_{j'\alpha}^{*k} - B_{\alpha\alpha'}^{*kk'} A_{j\alpha}^k A_{j'\alpha'}^{*k'} \right] = 0 \tag{8.173}$$

Zero is obtained if we do the change of indices $k \overset{\rightarrow}{\leftarrow} k'$ and $\alpha \overset{\rightarrow}{\leftarrow} \alpha'$ on the RHS and use the fact that the dynamical matrix $B_{\alpha\alpha'}^{kk'}$ is Hermitian. If $j \neq j'$ and there is no degeneracy then the different branches do not intersect and thus $\omega_j^2 \neq \omega_{j'}^2$. Then we obtain the mode orthogonality condition,

$$\sum_{k\alpha} m_k A^k_{j\alpha} A^{*k}_{j'\alpha} = 0, \quad \omega_j \neq \omega_{j'} \tag{8.174}$$

If $j = j'$ then $\omega_j^2 = \omega_{j'}^2$ and now this last expression need not be equal to zero. What should the sum $\sum_{k\alpha} m_k A^k_{j\alpha} A^{*k}_{j'\alpha}$ be equal to for $j = j'$ (or any two nondegenerate modes)?

We evaluate for a particular case. Consider a lattice with only one type of atom, $k = 1$:

$$\sum_\alpha m A^k_{j\alpha} A^{*k}_{j'\alpha} = m \sum_\alpha A_{j\alpha} A^*_{j'\alpha} \tag{8.175}$$

If the amplitudes are taken in such a way that they are unit polarization vectors, that is, $|U^k_{n\alpha}| = 1$, where an individual normal mode is

$$u_j = U^k_{n\alpha} = A^k_{j\alpha} \exp\{i(\vec{q} \cdot \vec{a}_n - \omega t)\} \tag{8.176}$$

then it follows from the condition, $U^{*k}_{n\alpha} U^k_{n\alpha} = 1$ that

$$\sum_{k\alpha} m_k A^k_{j\alpha} A^{*k}_{j'\alpha} = M\delta_{jj'} \tag{8.177}$$

We take this as the standard normalization of the normal modes.

Now consider the energy of vibrations. From the definition,

$$E_{kin} = \sum_{nk\alpha} \frac{1}{2} m_k \left| \dot{U}^k_{n\alpha}(t) \right|^2 \tag{8.178}$$

if we consider Equation 8.160 then we have

$$E_{kin} = \frac{1}{2N} \sum_{nk\alpha} m_k \sum_{\vec{q},j} \dot{a}_j(\vec{q},t) A^k_{j\alpha} \exp\{i\vec{q}\cdot\vec{a}_n\} \sum_{\vec{q}',j'} \dot{a}^*_{j'}(\vec{q}',t) \left[A^k_{j'\alpha}(\vec{q}) \exp\{i\vec{q}\cdot\vec{a}_n\} \right]^*$$

$$= \frac{1}{2N} \sum_{\vec{q}j\vec{q}'j'} \dot{a}_j(\vec{q},t) \dot{a}^*_{j'}(\vec{q}',t) NM\delta_{\vec{q}\vec{q}'}, \delta_{jj'} = \frac{M}{2} \sum_{\vec{q},j} |\dot{a}_j(\vec{q},t)|^2 \tag{8.179}$$

Thus, as we could have concluded due to the orthogonality of different modes, there results a sum over them, without cross terms:

$$E_{kin} = \frac{M}{2} \sum_{\vec{q},j} |\dot{a}_j(\vec{q},t)|^2 \tag{8.180}$$

Let us also evaluate the expression (8.105) for the potential energy:

$$W = \frac{1}{2} \sum_{nk\alpha n'k'\alpha'} W_{\alpha\alpha'}\binom{kk'}{nn'} U_{n\alpha}^{k} U_{n'\alpha'}^{k'} = \frac{1}{2N} \sum_{nk\alpha n'k'\alpha'} W_{\alpha\alpha'}\binom{kk'}{nn'}$$

$$\sum_{\vec{q},j} a_j^*(\vec{q},t)\left[A_{j\alpha}^k(\vec{q})\exp\{i\vec{q}\cdot\vec{a}_n\}\right]^* \sum_{\vec{q}',j'} a_{j'}(\vec{q}',t)A_{j'\alpha'}^{k'}(\vec{q}')\exp\{i\vec{q}'\cdot\vec{a}_{n'}\}$$

$$= \frac{1}{2N} \sum_{nk\alpha n'k'\alpha'} \sum_{\vec{q}j\vec{q}'j'} W_{\alpha\alpha'}\exp\{i\vec{q}\cdot(\vec{a}_{n'}-\vec{a}_n)\}$$

$$\exp\{i(\vec{q}'-\vec{q})\cdot\vec{a}_{n'}\}a_{j'}(\vec{q}',t)a_j^*(\vec{q},t)A_{j'\alpha'}^{k'}(\vec{q})A_{j\alpha}^{*k}(\vec{q})$$

$$= \frac{1}{2N} \sum_{nk\alpha k'\alpha'jj'\vec{q}\vec{q}'} B_{\alpha\alpha'}^{kk'}(\vec{q}')A_{j'\alpha'}^{k'}(\vec{q})A_{j\alpha}^{*k}(\vec{q})\exp\{i(\vec{q}'-\vec{q})\cdot\vec{a}_n\}a_{j'}(\vec{q}',t)a_j^*(\vec{q},t)$$

$$\tag{8.181}$$

From the eigenvalue problem, (8.172), we have

$$m_k\omega_j^2 A_{j'\alpha}^k = \sum_{k'\alpha'} B_{\alpha\alpha'}^{kk'} A_{j'\alpha'}^{k'} \tag{8.182}$$

then

$$W = \frac{1}{2N} \sum_{nk\alpha jj'\vec{q}\vec{q}'} m_k\omega_{j'}^2 A_{j'\alpha}^k A_{j\alpha}^{*k}(\vec{q})\exp\{i(\vec{q}'-\vec{q})\cdot\vec{a}_n\}a_{j'}(\vec{q}',t)a_j^*(\vec{q},t) \tag{8.183}$$

and from Equation 8.177

$$\sum_{k\alpha} m_k A_{j'\alpha}^k A_{j\alpha}^{*k} = M\delta_{jj'} \tag{8.184}$$

Then

$$W = \frac{1}{2N} \sum_{njj'\vec{q}\vec{q}'} M\omega_{j'}^2\delta_{jj'}\exp\{i(\vec{q}'-\vec{q})\cdot\vec{a}_n\}a_{j'}(\vec{q}',t)a_j^*(\vec{q},t) \tag{8.185}$$

and using the lattice sum (8.171) then we have

$$W = \frac{1}{2N} \sum_{jj'\vec{q}\vec{q}'} M\omega_{j'}^2\delta_{jj'}a_{j'}(\vec{q}',t)a_j^*(\vec{q},t)N\delta_{\vec{q}\vec{q}'} \tag{8.186}$$

$$W = \frac{M}{2} \sum_{\vec{q},j} \omega_j^2(\vec{q})|a_j(\vec{q},t)|^2 \tag{8.187}$$

The total energy of vibration of the crystal is rather simple,

$$E = \frac{M}{2} \sum_{\vec{q},j} \left\{ |\dot{a}_j(\vec{q},t)|^2 + \omega_j^2(\vec{q})|a_j(\vec{q},t)|^2 \right\} \tag{8.188}$$

This expression is the energy of a system of independent harmonic oscillators with frequencies ω_j. Here we exhibit explicitly the sums over wave vector and polarization.

To represent the energy of the system in the form of independent harmonic oscillators we have to express the complex valued functions $a_j(\vec{q},t)$ through real normal coordinates such that the symmetry condition (8.161) should be automatically satisfied.

Let us start from the Lagrangian

$$L = E_{kin} - W \tag{8.189}$$

then the equation of motion for the normal coordinates is found to be

$$\ddot{a}_j(\vec{q},t) + \omega_j^2(\vec{q})a_j(\vec{q},t) = 0, \quad j = 1, 2, 3 \tag{8.190}$$

$\vec{q}$ has N values and the system (8.190) has $3N$ equations ($k = 1, 2, 3 \ldots s$ is not present any more). The normal coordinates $a_j(\vec{q},t)$ are complex valued. We want to move from complex valued normal coordinates to real-valued normal coordinates.

The wave in a crystal should be traveling waves and not standing waves, and thus the real normal coordinates should be canonical. Thus we try the canonical transformation:

$$a_j(\vec{q}) = Q_j(\vec{q}) + \frac{i}{\omega_j(\vec{q})} \dot{Q}_j(\vec{q}) \tag{8.191}$$

As a result of the symmetry in Equation 8.161 for $a_j(\vec{q},t)$, then we have

$$Q_j(-\vec{q}) = Q_j^*(\vec{q}) \tag{8.192}$$

The difficulty is that this gives $3N/2$ equations and not $3N$ as expected. Thus we select $a_j(\vec{q})$ symmetrically, such as

$$a_j(\vec{q}) = \frac{1}{2} \left\{ Q_j(\vec{q}) + Q_j(-\vec{q}) + \frac{i}{\omega_j(\vec{q})} [\dot{Q}_j(\vec{q}) - \dot{Q}_j(-\vec{q})] \right\} \tag{8.193}$$

Now introduce

$$\alpha_j(\vec{q}) = \frac{1}{2} \left[Q_j(\vec{q}) + \frac{i}{\omega_j(\vec{q})} \dot{Q}_j(\vec{q}) \right] \tag{8.194}$$

which has the symmetry,

$$\alpha_j(-\vec{q}) = \frac{1}{2}\left[Q_j(-\vec{q}) + \frac{i}{\omega_j(\vec{q})}\dot{Q}_j(-\vec{q})\right] \tag{8.195}$$

Thus from Equation 8.193 to Equation 8.195 this transformation gives a real coordinate,

$$a_j(\vec{q}) = \alpha_j(\vec{q}) + \alpha_j^*(\vec{q}) \tag{8.196}$$

We show that the $Q_j(\vec{q})$ are normal coordinates. For this we classify the dependence of α_j on the time t. Let us find the derivative of α_j with respect to time:

$$\dot{\alpha}_j = \frac{1}{2}\left[\dot{Q}_j(\vec{q}) + \frac{i}{\omega_j(\vec{q})}\ddot{Q}_j(-\vec{q})\right] \tag{8.197}$$

It is necessary to get

$$\ddot{Q}_j(\vec{q}) + \omega_j^2(\vec{q})Q_j(\vec{q}) = 0 \tag{8.198}$$

which imposes the restriction,

$$\dot{\alpha}_j = \frac{1}{2}\left[\dot{Q}_j + \frac{i}{\omega_j(\vec{q})}\left(-\omega_j^2(\vec{q})\right)Q_j\right] = i\omega_j(\vec{q})\alpha_j(\vec{q}) \tag{8.199}$$

It follows from here that

$$\alpha_j \approx \exp\{i\omega_j t\}, \quad \alpha_j^* \approx \exp\{-i\omega_j t\} \tag{8.200}$$

For a given $\vec{q}$ and j from (8.160) considering (8.196) we have

$$U_{n\alpha}^k = \frac{1}{\sqrt{N}}\sum_{\vec{q}j}\alpha_j(\vec{q})A_{j\alpha}^k(\vec{q})\exp\{i\vec{q}\cdot\vec{a}_n\} + \frac{1}{\sqrt{N}}\sum_{\vec{q}j}\alpha_j^*(\vec{q})A_{j\alpha}^k(-\vec{q})\exp\{-i\vec{q}\cdot\vec{a}_n\} \tag{8.201}$$

and from

$$A_{j\alpha}^k(-\vec{q}) = A_{j\alpha}^{*k}(\vec{q}) \tag{8.202}$$

the two terms are of similar argument, and we have

$$U_{n\alpha}^k \approx A\cos[(\vec{q}\cdot\vec{a}_n) - \omega_j(\vec{q})t + \theta] \tag{8.203}$$

which shows that in a crystal we have a traveling wave.

If now we substitute Equation 8.196 into Equation 8.188 then that the total energy is

$$E = \frac{M}{2} \sum_{\vec{q},j} \left\{ \left| \dot{\alpha}_j(\vec{q}) + \dot{\alpha}_j^*(\vec{q}) \right|^2 + \omega_j^2(\vec{q}) \left| \alpha_j(\vec{q}) + \alpha_j^*(\vec{q}) \right|^2 \right\} \tag{8.204}$$

From condition (8.199) we have a cancellation of terms, and

$$E = M \sum_{\vec{q},j} \omega_j^2(\vec{q}) \left\{ \left| \alpha_j(\vec{q}) \right|^2 + \left| \alpha_j(-\vec{q}) \right|^2 \right\} \tag{8.205}$$

As the Brillouin zone has a center of inversion then

$$E = 2M \sum_{\vec{q},j} \left| \alpha_j(\vec{q}) \right|^2 \omega_j^2(\vec{q}) \tag{8.206}$$

and from here considering Equation 8.194 for $\alpha_j(\vec{q})$ this can be expressed as

$$E = \frac{M}{2} \sum_{\vec{q},j} \left[\dot{Q}_j^2(\vec{q}) + \omega_j^2 Q_j^2(\vec{q}) \right] \tag{8.207}$$

This is the expected form with Q_j as normal coordinates. Let us find the Hamiltonian function H of the vibration of the crystal. The generalized momentum is given by

$$P_j(\vec{q}) = \frac{\partial E}{\partial \dot{Q}_j(\vec{q})} = m' \dot{Q}_j \tag{8.208}$$

Thus,

$$H(P,Q) = \sum_{\vec{q}j} \left\{ \frac{P_j^2(\vec{q})}{2m'} + \frac{m' \omega_j^2(\vec{q}) Q_j^2(\vec{q})}{2} \right\} \tag{8.209}$$

From here the dynamics is described by usual Hamiltonian equations,

$$\dot{Q}_j = \frac{\partial H}{\partial P_j} = \frac{P_j(\vec{q})}{m'}, \quad \dot{P}_j = -\frac{\partial H}{\partial Q_j} = -m\omega_j^2(\vec{q}) Q_j(\vec{q}), \quad m\ddot{Q}_j + m\omega_j^2(\vec{q}) Q_j(\vec{q}) = 0 \tag{8.210}$$

One can then verify that the correct dynamics results,

$$\ddot{Q}_j(\vec{q}) + \omega_j^2(\vec{q}) Q_j(\vec{q}) = 0 \tag{8.211}$$

For high temperatures T in a crystal, the kinetic energy is much higher than the potential energy. If the temperatures are lower then the potential energy is much higher than the kinetic energy. When we examine the interactions of the electrons with the crystal lattice vibrations then the above notions are very convenient to take note of.

In the classical mechanics representation the Hamiltonian of lattice normal modes has the form of decoupled harmonic oscillators:

$$H(P,Q) = \sum_{\vec{q}j} \left\{ \frac{P_j^2(\vec{q})}{2M} + \frac{M\omega_j^2(\vec{q})Q_j^2(\vec{q})}{2} \right\} \tag{8.212}$$

and as a quantum operator it becomes,

$$\hat{H}(P,Q) = \sum_{\vec{q}j} \left\{ \frac{\hat{P}_j^2(\vec{q})}{2M} + \frac{M\omega_j^2(\vec{q})\hat{Q}_j^2(\vec{q})}{2} \right\} \tag{8.213}$$

In the coordinate representation the momentum operator is the usual derivative,

$$\hat{P}_j(\vec{q}) = -i\hbar \frac{\partial}{\partial Q_j(\vec{q})} \tag{8.214}$$

Here we consider the quantization of lattice modes. Let us solve the Schrödinger equation:

$$\hat{H}\Psi = E\Psi \tag{8.215}$$

We find the eigenvalues E and eigenvectors Ψ of the operator (8.213). The eigenvectors Ψ describe the small vibrations of the crystal lattice and E is the eigenvalue of the energy. As the Hamiltonian Equation 8.213 represents the sum of independent identical summands then the solution of Equation 8.215 may be found in the form of a product of wave functions:

$$\Psi = \prod_{\vec{q}j} \Psi_{N_{\vec{q}j}}[Q_j(\vec{q})] \tag{8.216}$$

where $\Psi_{N_{\vec{q}j}}$ is the wave function of a single oscillator (normal mode), for which Equation 8.215 becomes

$$\hat{H}\Psi_{N_{\vec{q}j}} = E_{N_{\vec{q}j}}\Psi_{N_{\vec{q}j}} \tag{8.217}$$

and

$$E = \sum_{\vec{q}j} E_{N_{\vec{q}j}} \tag{8.218}$$

where

$$E_{N_{\vec{q}j}} = \hbar\omega_j(\vec{q})\left(N_{\vec{q}j} + \frac{1}{2}\right) \tag{8.219}$$

Equation 8.217 for the square-integrable function $\Psi_{N_{\vec{q}j}}$ and its eigenvalues is a standard problem in quantum mechanics for a harmonic oscillator. The solution of this problem is well known. The eigen energy of the whole system is just the sum of the energies for each quantized normal mode, which are referred to as phonons. Each phonon state has a quantized energy of vibration, given by Equation 8.219. The quantum numbers $N_{\vec{q}j}$ can be 0 or positive integers. These eigenvalues indicate the number of vibrational quanta at each normal mode of oscillation for the lattice. Hence they can be considered as the number of phonons present of each mode. Because of quantization, we see that the lattice can vibrate only in certain discrete ways. It cannot vibrate in a way that would correspond to a fractional number of phonons.

Let us introduce the dimensionless canonical coordinate

$$\xi_{\vec{q}j} = \xi_j(\vec{q}) = \left(\frac{M\omega_j(\vec{q})}{\hbar}\right)^{1/2} Q_j(\vec{q}) \tag{8.220}$$

then the coordinate space wave functions are

$$\Psi_{N_{\vec{q}j}}(\xi_j) = \left(\frac{M\omega_j(\vec{q})}{\hbar}\right)^{1/4} \exp\left\{-\frac{\xi_j}{2}\right\} H_{N_{\vec{q}j}}(\xi_j) \tag{8.221}$$

The function $H_{N_{\vec{q}j}}(\xi_j)$ is a Hermite polynomial:

$$H_{N_{\vec{q}j}}(\xi_j) = \frac{(-1)^{N_{\vec{q}j}}}{\sqrt{N_{\vec{q}j}!2^{N_{\vec{q}j}}\sqrt{\pi}}} \exp\{\xi_j^2\} \frac{d^{N_{\vec{q}j}}}{d\xi_j^{N_{\vec{q}j}}}\left[\exp\{-\xi_j^2\}\right] \tag{8.222}$$

We know that

$$\hat{P}_j(\vec{q}) = -i\hbar\frac{\partial}{\partial Q_j(\vec{q})} = -i\hbar\left(\frac{M\omega_j(\vec{q})}{\hbar}\right)^{1/2}\frac{\partial}{\partial\xi_{\vec{q}j}} = -\left(M\omega_j(\vec{q})\hbar\right)^{1/2} i\frac{\partial}{\partial\xi_{\vec{q}j}} \tag{8.223}$$

Thus

$$\hat{P}_j(\vec{q}) = \left(M\omega_j(\vec{q})\hbar\right)^{1/2}\hat{P}_{\xi_{\vec{q}j}} \tag{8.224}$$

Here we have the dimensionless operator

$$\hat{P}_{\xi_{\bar{q}j}} = -i\frac{\partial}{\partial \xi_{\bar{q}j}}$$

(8.225)

Dirac was the first person who introduced that. In the case of the electron–phonon interaction it is convenient to use the method of the second field quantization to describe the particles. Then the field is represented as some quasi-classical particles. This method is used to investigate the field with variable number of particles, as we have when a system has a set of phonon excitations.

One can introduce second quantization for harmonic oscillators, by defining an annihilation operator and a creation operator:

$$\hat{a}_{\bar{q}j} = \frac{1}{\sqrt{2}}\left(\xi_{\bar{q}j} - i\hat{P}_{\xi_{\bar{q}j}}\right), \quad \hat{a}_{\bar{q}j}^{\dagger} = \frac{1}{\sqrt{2}}\left(\xi_{\bar{q}j} + i\hat{P}_{\xi_{\bar{q}j}}\right)$$

(8.226)

Then one can show

$$\alpha_j^{\dagger}(\vec{q}) = \frac{1}{2}\left(\frac{\hbar}{M\omega_j(\vec{q})}\right)^{1/2} a_{\bar{q}j}^{\dagger}$$

(8.227)

In the same manner we can find that

$$\alpha_j(\vec{q}) = \frac{1}{2}\left(\frac{\hbar}{M\omega_j(\vec{q})}\right)^{1/2} a_{\bar{q}j}$$

(8.228)

We can find the commutation of these operators. Hence we let then act on a state of the oscillator. From the properties of the Hermite polynomials it follows that

$$\xi\Psi_N = \sqrt{\frac{N}{2}}\Psi_{N-1} + \sqrt{\frac{N+1}{2}}\Psi_{N+1}, \quad \frac{\partial\Psi_N}{\partial\xi} = \sqrt{\frac{N}{2}}\Psi_{N-1} - \sqrt{\frac{N+1}{2}}\Psi_{N+1}$$

(8.229)

then from here and Equation 8.226 we have

$$\hat{a}_{\bar{q}j}\Psi_{N_{\bar{q}j}} = \sqrt{N_{\bar{q}j}}\Psi_{N_{\bar{q}j}-1}, \quad \hat{a}_{\bar{q}j}^{\dagger}\Psi_{N_{\bar{q}j}} = \sqrt{N_{\bar{q}j}+1}\Psi_{N_{\bar{q}j}+1}$$

(8.230)

When $\hat{a}_{\bar{q}j}\left(\hat{a}_{\bar{q}j}^{\dagger}\right)$ acts on the state $|N_{\bar{q}j}\rangle = \Psi_{N_{\bar{q}j}}$ (this is Dirac notation) of the given oscillator it reduces (increases) the phonon number $N_{\bar{q}j}$ by 1. The energy of an oscillator was given above,

$$E_{N_{\bar{q}j}} = \hbar\omega_j(\vec{q})\left(N_{\bar{q}j} + \frac{1}{2}\right)$$

(8.231)

The state of the oscillator for which $N_{\bar{q}j} = 0$ is a state with no phonons (of the indicate wave vector and polarization). The expression $\hbar\omega_j(\vec{q})N_{\bar{q}j}$ gives the excitation energy of the system. This energy depends on $N_{\bar{q}j}$ and increases with an increase in phonon number $N_{\bar{q}j}$. The excitation energy has the form of the total energy of an ideal gas. The phonons are the particles of this gas.

In Dirac notation we have

$$\hat{a}_{\bar{q}j}\left|N_{\bar{q}j}\right\rangle = \sqrt{N_{\bar{q}j}}\left|N_{\bar{q}j}-1\right\rangle, \quad \hat{a}_{\bar{q}j}^\dagger\left|N_{\bar{q}j}\right\rangle = \sqrt{N_{\bar{q}j}+1}\left|N_{\bar{q}j}+1\right\rangle \quad (8.232)$$

Here $\hat{a}_{\bar{q}j}\left(\hat{a}_{\bar{q}j}^\dagger\right)$ is called the *annihilation* (*creation*) operator. It follows that

$$\hat{a}_{\bar{q}j}\hat{a}_{\bar{q}j}^\dagger\left|N_{\bar{q}j}\right\rangle = \left(N_{\bar{q}j}+1\right)\left|N_{\bar{q}j}\right\rangle, \quad \hat{a}_{\bar{q}j}^\dagger\hat{a}_{\bar{q}j}\left|N_{\bar{q}j}\right\rangle = N_{\bar{q}j}\left|N_{\bar{q}j}\right\rangle \quad (8.233)$$

Thus from Equation 8.233 we have

$$\left[\hat{a}_{\bar{q}j},\hat{a}_{\bar{q}j}^\dagger\right] = 1 \quad (8.234)$$

Also

$$\left[\hat{a}_{\bar{q}j},\hat{a}_{\bar{q}'j'}\right] = 0, \quad \left[\hat{a}_{\bar{q}j},\hat{a}_{\bar{q}'j'}^\dagger\right] = \delta_{\bar{q}\bar{q}'}\delta_{jj'}, \quad \left[\hat{a}_{\bar{q}j}^\dagger,\hat{a}_{\bar{q}'j'}^\dagger\right] = 0 \quad (8.235)$$

The commutation in Equation 8.234 is true for Bose particles. In the quantum state there may exist an arbitrary number of bosons (in this case, phonons). The wave function is symmetric with respect to the interchange of two identical phonons.

The operator $\hat{a}_{\bar{q}j}\left(\hat{a}_{\bar{q}j}^\dagger\right)$ may be represented in the matrix form, that is, by indicating their matrix elements between states:

$$a_{N'N} = \sqrt{N}\delta_{N',N-1}, \quad a_{N'N}^* = \sqrt{N+1}\delta_{N',N+1} \quad (8.236)$$

From here follows the related matrix elements

$$\alpha_{N'N} = \left(\frac{\hbar}{2M\omega}\right)^{1/2}\sqrt{N}\delta_{N',N-1}, \quad \alpha_{N'N}^* = \left(\frac{\hbar}{2M\omega}\right)^{1/2}\sqrt{N+1}\delta_{N',N+1} \quad (8.237)$$

Let us find the Hamiltonian of the vibrating crystal in the representation of second field quantization. Considering Equations 8.213 and 8.224, the dimensionless Hamiltonian of the system has the form:

$$\hat{H} = \sum_{\bar{q}j}\frac{\hbar\omega_{\bar{q}j}}{2}\left(\hat{P}_{\xi_{\bar{q}j}}^2 + \hat{\xi}_{\bar{q}j}^2\right) \quad (8.238)$$

where

$$\hat{\xi}_{\bar{q}j} = \frac{1}{\sqrt{2}}\left(\hat{a}_{\bar{q}j} + \hat{a}_{\bar{q}j}^{\dagger}\right), \quad \hat{P}_{\xi_{\bar{q}j}} = \frac{-i}{\sqrt{2}}\left(\hat{a}_{\bar{q}j} - \hat{a}_{\bar{q}j}^{\dagger}\right) \tag{8.239}$$

Thus,

$$\hat{H} = \sum_{\bar{q}j} \hbar\omega_{\bar{q}j}\left(\hat{a}_{\bar{q}j}^{\dagger}\hat{a}_{\bar{q}j} + \frac{1}{2}\right) \tag{8.240}$$

The eigenvectors of Equation 8.240 are related to those of the operator $\hat{a}_{\bar{q}j}^{\dagger}\hat{a}_{\bar{q}j}$. Since the operator $\hat{a}_{\bar{q}j}^{\dagger}\hat{a}_{\bar{q}j}$ has the sense of $N_{\bar{q}j}$ and does not change the state $|N_{\bar{q}j}\rangle = \Psi_{N_{\bar{q}j}}$, it is referred to as the *phonon number operator in the mode* $\bar{q}_j$. The eigenvalues of $\hat{a}_{\bar{q}j}^{\dagger}\hat{a}_{\bar{q}j}$ are $N_{\bar{q}j}$, the number of phonons in the state of wave vector $\vec{q}$, polarization j, which ranges from 0 to infinity. Even when there is no phonon, there is still energy. This can be expected from the Heisenberg uncertainty principle. The value

$$E_0 = \frac{\hbar\omega_{\bar{q}j}}{2} \tag{8.241}$$

describes the *ground state (zero-point)* or the *vacuum state energy* with no phonon, that is $|0_{\bar{q}j}\rangle$. This energy implies that even at the temperature of 0 K the atoms in a crystal have an oscillatory motion. This energy is a considerable quantity. The localization of the atoms closely around their exact equilibrium positions leads to uncertainty in their velocities, according to Heisenberg uncertainty principle. There is no state of frozen atomic positions.

The excited state with just one phonon is produced from the ground state by using the creation operator:

$$\hat{a}_{\bar{q}j}^{\dagger}|0_{\bar{q}j}\rangle = \sqrt{1}|1_{\bar{q}j}\rangle = |1_{\bar{q}j}\rangle \tag{8.242}$$

This can be repeated $N_{\bar{q}j}$ times to get a state with $N_{\bar{q}j}$ phonons,

$$\left[\hat{a}_{\bar{q}j}^{\dagger}\right]^{N_{\bar{q}j}}|0_{\bar{q}j}\rangle = \sqrt{N_{\bar{q}j}!}|N_{\bar{q}j}\rangle \quad \text{or} \quad |N_{\bar{q}j}\rangle = \frac{\left[\hat{a}_{\bar{q}j}^{\dagger}\right]^{N_{\bar{q}j}}}{\sqrt{N_{\bar{q}j}!}}|0_{\bar{q}j}\rangle \tag{8.243}$$

One also realizes the effect of the annihilation operator on the ground state,

$$\hat{a}_{\bar{q}j}|0_{\bar{q}j}\rangle = 0 \text{ (no state)} \tag{8.244}$$

Then it follows that the total energy is

$$E = \sum_{\bar{q}j} \hbar\omega_{\bar{q}j}\left(N_{\bar{q}j} + \frac{1}{2}\right) \tag{8.245}$$

This is the energy of normal vibrations of the oscillators. The associated state has a direct product of Dirac kets corresponding to each allowed mode, $\vec{q}_j$.

$$\left| N_1 N_2 N_3 \ldots \right\rangle = \left| N_1 \right\rangle \left| N_2 \right\rangle \left| N_3 \right\rangle \ldots \tag{8.246}$$

Here each subscript refers to a distinct mode, for example, $1 \equiv \vec{q}_1 j_1$, $2 \equiv \vec{q}_2 j_2$, $3 \equiv \vec{q}_3 j_3$, and so on, with an occupation number for every allowed mode.

The above describes an approximate solution (harmonic approximation of the expansion of the potential energy) for the lattice vibrations. Phonons are quanta of elementary excitations. Phonons may also be called quanta of sound vibrations just as photons are quanta of light or electromagnetic radiation. However, for the optical case, especially in a "perfect" vacuum, the photon field is an exact representation of the optical field.

In the process of interaction with each other (due to the ignored nonlinear terms in the potential energy) phonons may appear and disappear. As a result there arises the question of the mean value of the phonon number per normal vibration, in thermal equilibrium. Planck solved this problem. The mean phonon number for one mode is equal to

$$\overline{N}_{\vec{q}j} = \frac{1}{\exp\{\hbar\omega_{\vec{q}j}/T\} - 1} \tag{8.247}$$

This is also called the Bose–Einstein distribution. In the formula

$$\overline{E}_{\vec{q}j} = \left(\overline{N}_{\vec{q}j} + \frac{1}{2} \right) \hbar\omega_{\vec{q}j} \tag{8.248}$$

the quantity $\overline{N}_{\vec{q}j}$ is the mean phonon number. It should be noted that at any finite temperature a crystal has a distribution of phonons. They contribute significantly to the warming of the crystal only at very low temperatures or high intensities. If optical branches are excited, the result is optical phonons. If the temperature is raised then more phonons are created, otherwise they are destroyed. There is no conservation law for phonon number, because they can be created and destroyed, similar to photons. Because they are Bose quasi-particles, any phonon state can hold an arbitrary number of quanta. There is no problem of the conservation of energy in the process of creation and annihilation as energy flow into or out of a crystal is by conduction of heat, which involves phonon propagation.

We have seen that the spectrum of phonons (or quantized lattice vibrations) in a crystal depends directly on the type of lattice, the number of atoms in the primitive cell, the atomic masses, and so on. This spectrum is determined from an eigenvalue problem. Once $\omega_j(\vec{q})$ has been found, however, the energy of lattice vibrations is quantized according to Equation 8.245. The distinction between different crystalline lattices fades away. All that matters in the end is the occupation numbers $N_{\vec{q}j}$ for numbers of phonons in each fundamental mode. Further, in thermal equilibrium, the mean occupation numbers (8.247) are most important for the statistical properties of the vibrations.

9 Condensed Bodies

Condensed matter refers to materials that are neither gaseous nor liquid, where the atoms arrange themselves in some kind of structure. The atoms come together due to their attractive interactions, and can form ordered arrays (crystals) and apparently disordered or less-ordered patterns (soft-matter, fractal aggregates, etc.). The periodicity of ordered arrays not only makes their analysis simpler, but leads to the possibility of well-defined and long-lived lattice vibrations or phonons that would not be present in soft-matter materials.

Crystalline solids in thermal equilibrium support a spectrum of phonons as discussed in the previous chapter. Based on knowledge of the phonon spectrum, we can consider how those modes are populated at a given temperature, and how that population affects the basic properties of a solid, such as its specific heat and thermal expansion.

9.1 APPLICATION OF STATISTICAL THERMODYNAMICS TO PHONONS

We still consider in this heading as in the previous heading, elastic vibrations of a crystal. Here we assume that the elastic response of the crystal is a linear function of the forces. This is equivalent to the assumption that the elastic energy is a quadratic function of the relative displacement of any two points in the crystal (harmonic approximation). In statistics we begin from a Hamiltonian, from which we find the spectrum of phonon excitations (quantized lattice vibrations). The partition function can then be used to determine, for example, the phonon contribution to specific heat and other thermal properties.

Let us write the Hamiltonian of a system of N atoms in a crystal as follows:

$$\hat{H} = K(p) + W(q) + \tilde{W}(q) \tag{9.1}$$

The first term in Equation 9.1 is the kinetic energy of thermal motion, the second is the potential energy of interaction between atoms, and the third is a local potential energy (such as an externally applied field). If we neglect the interaction term $W(q)$ in Equation 9.1 then we have the Hamiltonian of an ideal gas. However, the interactions can be so strong, such that we may not neglect that term in Equation 9.1, and the particles come together and form a condensed body. In that case the interaction potential energy forms minimum energy states that can correspond to ordered crystalline arrays of the atoms, that is, a crystal lattice.

The energy of interaction may dominate over the energy of thermal motion, especially at low temperature. This results in the atoms being localized at the lattice sites of a crystal lattice (at zero temperature). For finite temperature, there can be small displacements from the equilibrium locations. The displacements U of the atoms from their equilibrium positions are much less than a lattice constant a. Thus, we can analyze the system in terms of these small (lattice) vibrations about the equilibrium positions.

A solid can be considered as a system of particles held together by springs. The displacements of the particles are dependent on time t, that is, $U = U(t)$. Let U be some implicit harmonic function of time t. The equations of motion for the particles are as follows:

$$m_k \ddot{U}_{n\alpha}^k = -\sum_{n'k'\alpha'} W_{\alpha\alpha'} \left(\begin{smallmatrix} kk' \\ nn' \end{smallmatrix} \right) U_{n'\alpha'}^{k'} \tag{9.2}$$

We consider elementary cells with S atoms that have masses m_k, $k = 1,2,\ldots,s$. The crystal has N elementary cells. Each atom in the crystal has three degrees of freedom. Then for the N atoms there are $3N$ degrees of freedom; for the whole crystal then there are $3Ns$ degrees of freedom. In Equation 9.2 the displacement U_n^k of the kth atom from the equilibrium position has the α-Cartesian component $U_{n\alpha}^k$, $\alpha = (x,y,z)$. Equation 9.2 is a system of $3Ns$ equations. The solution of Equation 9.2 can be written:

$$U_{n\alpha}^k = A_\alpha^k \exp\{i(\vec{q} \cdot \vec{a}_n - \omega t)\} \tag{9.3}$$

Here ω is the phonon frequency associated with the wave vector $\vec{q}$, A_α^k is the amplitude of the wave, and $\vec{a}_n$ are the vectors that defines the equilibrium positions:

$$\vec{a}_{\bar{n}} = n_1\vec{a}_1 + n_2\vec{a}_2 + n_3\vec{a}_3 \tag{9.4}$$

Here $\vec{a}_i$ are basis vectors and n_i are integers with $i = 1,2,3$.

If we substitute Equation 9.3 into Equation 9.2 then we arrive at a secular equation whose solutions gives the phonon frequencies $\omega_{\bar{q}j}$, where j identifies the polarization. These phonon frequencies correspond to the phonon energies:

$$E_{\bar{q}j} = \hbar\omega_{\bar{q}j}\left(v_{\bar{q}j} + \frac{1}{2}\right), \quad v_{\bar{q}j} = 0,1,2,\ldots \tag{9.5}$$

when the modes are excited to the quantum number $v_{\bar{q}j}$. That is when the modes are occupied by $v_{\bar{q}j}$ phonons. If we denote $\bar{q}j$ by the symbol α, then we may write the total phonon energy as

$$E = \sum_\alpha \hbar\omega_\alpha\left(v_\alpha + \frac{1}{2}\right), \quad v_\alpha = 0,1,2,\ldots \tag{9.6}$$

9.2 FREE ENERGY OF CONDENSED BODIES IN THE HARMONIC APPROXIMATION

We have already solved the problem of the spectrum. Now we describe the phonons as a condensed phase with the number v_α of particles with energy $\hbar\omega_\alpha$. The Helmholtz free energy is,

$$F = -T \ln Z \tag{9.7}$$

where Z is the partition function. The partition function for an individual oscillator (or normal mode) is a sum over states,

$$Z_\alpha = \sum_{v_\alpha=0}^{\infty} \exp\left\{ -\frac{\hbar\omega_\alpha(v_\alpha + 1/2)}{T} \right\} = \frac{\exp\{-\hbar\omega_\alpha/2T\}}{1 - \exp\{-\hbar\omega_\alpha/T\}} \tag{9.8}$$

This is the partition function for one oscillator with frequency ω_α. From Equation 9.8 we have

$$F_\alpha = -T \ln Z_\alpha = \frac{\hbar\omega_\alpha}{2} + T \ln\left(1 - \exp\left\{ -\frac{\hbar\omega_\alpha}{T} \right\} \right) \tag{9.9}$$

Then the free energy of the entire lattice is a sum over all modes,

$$F = \sum_\alpha F_\alpha = 3N\varepsilon_0 + T \sum_\alpha \ln\left(1 - \exp\left\{ -\frac{\hbar\omega_\alpha}{T} \right\} \right) \tag{9.10}$$

The first term is the zero point energy,

$$\sum_\alpha \frac{\hbar\omega_\alpha}{2} = 3Ns\varepsilon_0 \tag{9.11}$$

where ε_0 is a mean energy per atom, and Ns is the number of particles. The spectrum ω_α depends on the details of the lattice. If ω_α is considered continuous then we may change the summation to integration.

Let us evaluate the vibrational internal energy E_α of the oscillators in thermal equilibrium. For a single oscillator the partition function is Z_α. The average vibrational internal energy of the single oscillator is E_α:

$$E_\alpha = -T^2 \frac{\partial}{\partial T}\left(\frac{F_\alpha}{T} \right) = T^2 \frac{\partial}{\partial T}(\ln Z_\alpha) = \frac{\hbar\omega_\alpha}{2} + \frac{\hbar\omega_\alpha}{\exp\left\{ \frac{\hbar\omega_\alpha}{T} \right\} - 1} \tag{9.12}$$

The quantity

$$n(\omega_\alpha) = \frac{1}{\exp\left\{\dfrac{\hbar\omega_\alpha}{T}\right\} - 1} \tag{9.13}$$

is the *Planck function* encountered earlier. The average energy for the entire system is

$$E = \sum_\alpha E_\alpha = \sum_\alpha \left[\frac{\hbar\omega_\alpha}{2} + \frac{\hbar\omega_\alpha}{\exp\left\{\dfrac{\hbar\omega_\alpha}{T}\right\} - 1} \right] \tag{9.14}$$

When $\hbar\omega_\alpha \ll T$, the temperature is so high that the thermal energy T is large compared to the separation $\hbar\omega_\alpha$ between energy levels. Then one expects that a classical approximation is acceptable. Expanding the mean energy E_α in a Taylor's series, we have

$$E_\alpha = \hbar\omega_\alpha \left[\frac{1}{2} + \frac{1}{\left(1 + \dfrac{\hbar\omega_\alpha}{T} + \cdots\right) - 1} \right] \approx T \tag{9.15}$$

This is in agreement with the classical result (equi-partition of energy). Thus for high temperatures all modes are excited to approximately the same energy.

At low temperatures, $\hbar\omega_\alpha \gg T$, and then

$$\exp\left\{\frac{\hbar\omega_\alpha}{T}\right\} \gg 1 \tag{9.16}$$

The mean energy becomes

$$E_\alpha \approx \hbar\omega_\alpha \left(\frac{1}{2} + \exp\left\{-\frac{\hbar\omega_\alpha}{T}\right\} \right) \tag{9.17}$$

This differs from $E_\alpha = T$. As $T \to 0$ it approaches properly the zero-point energy $\hbar\omega_\alpha/2$. The contribution to the free energy F_α of the αth oscillator is negligible if $\hbar\omega_\alpha \gg T$, except for the $(\hbar\omega_\alpha/2)$ term. At low temperatures, the high-frequency modes are *frozen out* and do not contribute to the heat capacity. This effect was discussed already for the vibration of diatomic molecules, see Figures 4.8 and 4.9.

The specific heat capacity may be evaluated:

$$C = \left(\frac{\partial E}{\partial T}\right)_V \qquad (9.18)$$

This gives

$$C = \sum_\alpha \left(\frac{\hbar\omega_\alpha}{T}\right)^2 \frac{\exp\left\{\dfrac{\hbar\omega_\alpha}{T}\right\}}{\left(\exp\left\{\dfrac{\hbar\omega_\alpha}{T}\right\} - 1\right)^2} = \sum_\alpha \left(\frac{\hbar\omega_\alpha}{2T}\right)^2 \frac{1}{\sinh^2\left(\dfrac{\hbar\omega_\alpha}{2T}\right)} \qquad (9.19)$$

As $T \to 0$, the phonon heat capacity approaches zero exponentially fast. In experimental measurements of low-temperature heat capacity for crystals, a power law in T has been observed, however, those experiments must include other contributions, such as electronic degrees of freedom. Further, one must more carefully look at the frequency distribution before drawing any conclusions from expression (9.19).

The phonon entropy is evaluated as follows:

$$S = -\left(\frac{\partial F}{\partial T}\right)_V = \sum_\alpha \left[\frac{\hbar\omega_\alpha}{2T}\coth\left(\frac{\hbar\omega_\alpha}{2T}\right) - \ln\left\{2\sinh\left(\frac{\hbar\omega_\alpha}{2T}\right)\right\}\right] \qquad (9.20)$$

In the harmonic approximation, the thermodynamic functions are additive functions of the normal mode frequencies ω_α. This has the consequence that all these functions are expressible as averages over the frequency distribution function $f(\omega)$ defined such that $f(\omega)d\omega$ is the number of allowed frequencies (normal modes) in the interval from ω to $\omega + d\omega$. For example, the free energy (9.10) can be expressed as a frequency integral:

$$F = 3N\varepsilon_0 + T \int_0^{\omega_{max}} f(\omega)d\omega \ln\left(1 - \exp\left\{-\frac{\hbar\omega}{T}\right\}\right) \qquad (9.21)$$

Various bodies differ in their $f(\omega)$. $f(\omega)$ can be considered the density of states in frequency space.

The evaluation of F, E, C, and S gives the thermodynamic properties in terms of the frequency distribution. It is clear that in order to evaluate the thermodynamic properties we must have the precise information on the frequency spectrum $f(\omega)$ (see Figure 9.1 for a typical example). It is possible for the spectrum to have an upper limit, ω_{max}, beyond which there are no more modes.

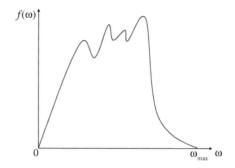

FIGURE 9.1   An example of the frequency distribution function $f(\omega)$.

9.3   CONDENSED BODIES AT LOW TEMPERATURES

Low temperatures are defined as those for which

$$T \ll \hbar\omega_{max} \tag{9.22}$$

where T is the energy of thermal motion. This means

$$\exp\left\{-\frac{\hbar\omega_{max}}{T}\right\} \ll 1 \tag{9.23}$$

and

$$\ln\left(1 - \exp\left\{-\frac{\hbar\omega_{max}}{T}\right\}\right) \approx \exp\left\{-\frac{\hbar\omega_{max}}{T}\right\} \ll 1 \tag{9.24}$$

In this case the upper limit of Equation 9.21 can be considered infinite. For small T in the sum with respect to α only terms with small frequencies play a role:

$$\hbar\omega_{\alpha} < T \tag{9.25}$$

But the vibrations of small frequencies are ordinary sound waves, and their wavelength is related to the frequency by

$$\lambda \approx \frac{2\pi U}{\omega} \tag{9.26}$$

where U is the speed of sound. For sound waves, the wavelengths λ are much larger than the lattice constant a;

$$\lambda \gg a \tag{9.27}$$

This implies that sound has

$$\omega \ll \frac{2\pi U}{a} \tag{9.28}$$

Then in order that only acoustic vibrations contribute significantly to the thermodynamics, the temperature should satisfy the condition:

$$T \ll \frac{2\pi\hbar U}{a} \tag{9.29}$$

In an appropriate limit, we can consider a crystalline solid as an elastic medium. As we have already seen, each normal mode of vibration of the elastic continuum is characterized by its wavelength λ. For the condition $\lambda \gg a$ the neighboring atoms in the solid are displaced by nearly the same amount. Here there is no significance to the fact that atoms are at a finite separation a from each other. The normal modes of vibration of a continuous elastic medium are very near those of the actual solid. The modes of vibrations of the elastic continuum for which $\lambda \approx a$ correspond, however, to markedly different displacements of neighboring atoms. In that case, the discrete spacing of atoms is very important. Thus the actual normal modes of vibrations of the atoms are quite different from those of the elastic continuum.

If we describe a real solid as if it were an elastic continuum, that is a good approximation for modes of long wavelengths, for which $\lambda \gg a$. That is to say, for low frequencies ω, the density of modes $f_c(\omega)$ for a continuous elastic medium should be very close to that of $f(\omega)$ for a real solid. For the case of short wavelengths or high frequencies there is a considerable difference between $f(\omega)$ and $f_c(\omega)$. A real solid has no modes for such high frequencies, that is, $f(\omega) = 0$ for $\omega > \omega_{max}$, whereas, there is no limiting value for short wavelength (or high frequency) for a continuous medium.

Consider a model for a solid as an isotropic elastic continuous medium with volume V. In an isotropic solid state, there are the possibilities of longitudinal sound waves (with velocity U_l) and transverse waves (with velocity U_t) with two independent polarization directions, see Figure 9.2. The longitudinal modes have compressions (contractions) and rarefactions (expansions) along the direction of the wave vector. On the other hand, the transverse modes are somewhat comparable to the vibrations of a string under tension, where the displacements are perpendicular to the wave vector. The frequencies of these waves are related through acoustic dispersion relations for an elastic continuum,

$$\omega = U_{l,t}q \tag{9.30}$$

where $\vec{q}$ is a wave vector.

The number of possible wave modes $d\Gamma$ of one polarization, with frequency in the interval from ω to $\omega + d\omega$ is

$$d\Gamma = \frac{dx\,dy\,dz\,dp_x\,dp_y\,dp_z}{(2\pi\hbar)^3} = \left.\frac{V4\pi p^2\,dp}{(2\pi\hbar)^3}\right|_{p=\hbar q} = \frac{Vq^2\,dq}{2\pi^2} = f_l(\omega)d\omega \tag{9.31}$$

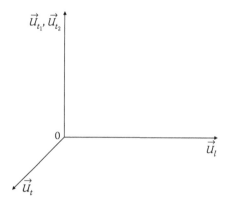

FIGURE 9.2 Illustrates the possibilities of longitudinal sound waves (with velocity U_l) and transverse waves (with velocity U_t) with two independent polarization directions.

The total number of modes in a desired frequency interval, for all polarizations is

$$f_c(\omega)\,d\omega = (f_l(\omega) + f_{t_1}(\omega) + f_{t_2}(\omega))\,d\omega = \frac{V\omega^2 d\omega}{2\pi^2}\left(\frac{1}{U_l^3} + \frac{1}{U_{t_1}^3} + \frac{1}{U_{t_2}^3}\right)$$

$$= \frac{V\omega^2 d\omega}{2\pi^2}\left(\frac{1}{U_l^3} + \frac{2}{U_t^3}\right) = \frac{V\omega^2 d\omega}{2\pi^2}\frac{3}{\overline{U}^3} \tag{9.32}$$

where $\overline{U}$ is some mean value of the velocity of sound.

Considering condition (9.22), then as $T \to 0$ we have $\omega_{max} \to \infty$ and the upper limit of integration in Equation 9.21 may be set to infinity instead of cutting it off at ω_{max}. It is immaterial whether we set the upper limit at ω_{max} or at infinity, as there is no area under the curve in Figure 9.1 from ω_{max} to infinity. Almost none of the modes in the neighborhood of ω_{max} are excited.

It follows that the free energy F for the continuum model of an elastic solid is

$$F = 3N\varepsilon_0 + \frac{3VT}{2\pi^2\overline{U}^3}\int_0^\infty \omega^2 d\omega \ln\left(1 - \exp\left\{-\frac{\hbar\omega}{T}\right\}\right) \tag{9.33}$$

If we integrate by parts and use the fact that

$$\ln(1 - q) \approx q, \quad q \ll 1 \tag{9.34}$$

then

$$F = 3N\varepsilon_0 - \frac{\pi^2 VT^4}{30\hbar^3\overline{U}^3} \tag{9.35}$$

then the entropy is

$$S = -\left(\frac{\partial F}{\partial T}\right)_V = \frac{2\pi^2 V T^3}{15\hbar^3 \bar{U}^3} \tag{9.36}$$

and the total energy is

$$E = F + TS = 3N\varepsilon_0 + \frac{\pi^2 V T^4}{10\hbar^3 \bar{U}^3} \tag{9.37}$$

From the classical theorem of equipartition we would expect to get

$$E = 3NT \tag{9.38}$$

But our formula (9.37) is quantum mechanical and applies to low temperature, where modes are less strongly excited. We may also evaluate the heat capacity from Equation 9.37:

$$C_V = \left(\frac{\partial E}{\partial T}\right)_V = \frac{2\pi^2 V T^3}{5\hbar^3 \bar{U}^3} \tag{9.39}$$

This is called the law of Debye cubes. This takes into account only the acoustic phonon spectrum, for a continuum elastic solid.

9.4 CONDENSED BODIES AT HIGH TEMPERATURES

Now consider temperatures exceeding the phonon level spacings,

$$T \gg \hbar\omega \tag{9.40}$$

For either the continuum elastic solid, or, a solid state material with some Bravais lattice, we can write down the free energy in Equation 9.33 as follows:

$$F = 3N\varepsilon_0 + T\sum_\alpha \ln\left[1 - \exp\left\{-\frac{\hbar\omega_\alpha}{T}\right\}\right]\Bigg|_{\frac{\hbar\omega_\alpha}{T} \ll 1} \approx 3N\varepsilon_0 + T\ln\left(\prod_{\alpha=1}^{3N}\frac{\hbar\omega_\alpha}{T}\right) \tag{9.41}$$

Define the geometrical mean frequency $\bar{\omega}$ from

$$\prod_{\alpha=1}^{3N}\frac{\hbar\omega_\alpha}{T} = \left(\frac{\hbar\bar{\omega}}{T}\right)^{3N} \tag{9.42}$$

Then the high-temperature free energy is approximated by

$$F = 3N\varepsilon_0 + 3NT \ln \frac{\hbar\bar{\omega}}{T} \tag{9.43}$$

The entropy is

$$S = -\left(\frac{\partial F}{\partial T}\right)_V = 3N - 3N \ln \frac{\hbar\bar{\omega}}{T} \tag{9.44}$$

It follows that the mean internal energy obeys equipartition,

$$E = 3N\varepsilon_0 + 3NT \tag{9.45}$$

The specific heat is a constant, following the *Dulong and Petit law*

$$C = 3N \tag{9.46}$$

Then the specific heat per degree of freedom is

$$\frac{C}{N} = 3 \tag{9.47}$$

which is just the result of the three possible polarizations. This is in total disagreement with experiments on lattice vibrations around normal room temperatures. Obviously this high-temperature analysis does not apply in the limit $T \to 0$ where experimentally and using the quantum theory one finds $C_V \to 0$. This shows the importance of quantum effects at low temperature.

9.5 DEBYE TEMPERATURE APPROXIMATION

We study now the general case of a crystal lattice at any temperature. We assume that the vibrations are quantized. The quantum of vibration or excitation of a crystal lattice is called a *phonon* as seen earlier. As we already know, a phonon travels at the speed of sound in a crystal lattice, as sound waves are elastic in nature. In this way we do not consider the internal energy as belonging to vibrations of individual atoms, but assume the energy to reside in elastic waves, which like electromagnetic waves in an enclosure, have a quantized energy.

In Debye's approximation we assume that a solid is a continuous elastic body, where there is no dispersion of the sound waves. Stated otherwise, the velocity of sound is assumed to be a constant, independent of the wave vector and polarization: $\omega = \bar{U}q$. Since the frequency can increase without limit, for a continuum the frequency is cut off at a maximum frequency below which we have $3N$ normal modes.

Here we have N allowed values of q in the Brillouin zone where N is the number of atoms. The 3 corresponds to the three independent polarizations.

Debye assumed that the mean energy of an oscillator (phonon) was obtained by multiplying ω by the mean occupation of that mode, (9.13),

$$n(\omega) = \frac{1}{\exp\left\{\dfrac{\hbar\omega}{T}\right\} - 1} \tag{9.48}$$

The mean energy of one phonon mode may be represented as

$$\overline{E} = \frac{\hbar\omega}{\exp\left\{\dfrac{\hbar\omega}{T}\right\} - 1} \tag{9.49}$$

Debye used the expression in Equation 9.21 (sum over modes replaced by sum over frequency) for the evaluation of all formulas. He decided to extrapolate the graph in Figure 9.3 as shown in Figure 9.4.

In the Debye approximation the true frequency distribution function $f(\omega)$ is approximated to $f_D(\omega)$, a distribution for a continuous elastic medium that is cut off at an upper limiting frequency, ω_D. This is not only at exceedingly low frequencies (where $f(\omega)$ and $f_D(\omega)$ are nearly the same) but also for all $3N$ lowest frequency modes of the elastic continuum. Thus in the Debye approximation $f(\omega)$ is replaced by $f_D(\omega)$, defined partly by

$$f_D(\omega) = \begin{cases} f_c(\omega), & \omega \leq \omega_D \\ 0, & \omega > \omega_D \end{cases} \tag{9.50}$$

where ω_D is the *Debye frequency (the cut-off frequency)*. The function $f_c(\omega)$ is the frequency distribution found earlier (9.32) for a continuous solid. The Debye

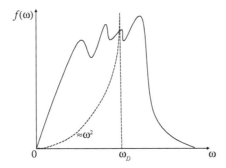

FIGURE 9.3 Variation of the true frequency distribution function $f(\omega)$ together with its approximation $f_D(\omega)$ versus the frequency ω.

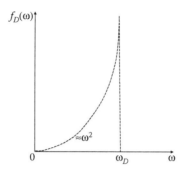

FIGURE 9.4 The true frequency distribution function $f(\omega)$ approximated to $f_D(\omega)$, a distribution for a continuous elastic medium that is cut off at an upper limiting frequency, ω_D.

frequency ω_D is chosen such that the following integral should yield the correct total number of $3N$ normal modes:

$$\int_0^{\omega_{max}} f(\omega)\,d\omega \tag{9.51}$$

That is,

$$3N = \int_0^\infty f_D(\omega)\,d\omega = \int_0^{\omega_D} f_c(\omega)\,d\omega = \int_0^{\omega_D} \frac{3V\omega^2}{2\pi^2 \bar{U}^3}\,d\omega = \frac{3V}{2\pi^2 \bar{U}^3}\frac{\omega_D^3}{3} \tag{9.52}$$

It follows that the Debye frequency is determined by

$$\omega_D^3 = 6\pi^2 \bar{U}^3 \frac{N}{V} \tag{9.53}$$

N/V is the density of the matter or the number of atoms per unit volume. The Debye frequency is dependent only on the mean velocity of sound and on the number of atoms per unit volume. The Debye frequency corresponds to a Debye wavelength, $\lambda_D = 2\pi\bar{U}/\omega_D$, that is of the order of the interatomic spacing $a = (N/V)^{-1/3}$. At typical values $\bar{U} \approx 5 \times 10^3$ m/s, $a \approx 10^{-10}$ m, then $\omega_D \approx 10^{14}$ rad/s, which lies in the infrared region of the electromagnetic spectrum.

The Debye frequency marks the boundary between high $T \gg \hbar\omega_D$ and low temperatures $T \ll \hbar\omega_D$. Thus we may have

$$F = 3N\varepsilon_0 + \frac{9TN}{\omega_D^3}\int_0^{\omega_D} \omega^2\,d\omega\,\ln\left(1 - \exp\left\{-\frac{\hbar\omega}{T}\right\}\right) \tag{9.54}$$

If we integrate by parts, using that the definitions,

$$\frac{\hbar\omega}{T} = z, \quad \hbar\omega_D = \theta \tag{9.55}$$

and also considering the *Debye function* from the following definition:

$$D(x) = \frac{3}{x^3}\int_0^x \frac{z^3 dz}{\exp\{z\} - 1} \tag{9.56}$$

then we have

$$F = 3N\varepsilon_0 + 3NT\ln\left(1 - \exp\left\{-\frac{\theta}{T}\right\}\right) - NTD\left(\frac{\theta}{T}\right) \tag{9.57}$$

Here θ is known as *the Debye temperature.*
   The entropy is evaluated as follows:

$$S = -\left(\frac{\partial F}{\partial T}\right)_V = -3N\ln\left(1 - \exp\left\{-\frac{\theta}{T}\right\}\right) + 4ND\left(\frac{\theta}{T}\right) \tag{9.58}$$

Then the internal energy in the Debye approximation is

$$E = F + TS = 3N\varepsilon_0 + 3NTD\left(\frac{\theta}{T}\right) \tag{9.59}$$

The constant-volume specific heat is found as

$$C_V = \left(\frac{\partial E}{\partial T}\right)_V = 3ND\left(\frac{\theta}{T}\right) - 3N\frac{\theta}{T}D'\left(\frac{\theta}{T}\right) \tag{9.60}$$

Let us describe the following limiting cases:

1. Low temperatures: $T \ll \theta$. The Debye function is approximated as

$$D(x) = \frac{T^3}{\theta^3}\frac{\pi^4}{5} \tag{9.61}$$

Thus,

$$E = 3N\varepsilon_0 + \frac{3}{5}N\pi^4\frac{T^4}{\theta^3} \tag{9.62}$$

This is characteristic of a gas of massless bosons. Also we find

$$C_V = \left(\frac{\partial E}{\partial T}\right)_V = \frac{12}{5} N\pi^4 \frac{T^3}{\theta^3} \tag{9.63}$$

This is the *Debye cubic law* that is fundamentally related to the three-dimensional nature of real crystals. This result is in total agreement with experiments, except for exceedingly low temperatures where

$$C_V \cong aT + bT^3 \tag{9.64}$$

The linear term in C_V is due to electronic contributions to the specific heat (in metals).

For the condition $\hbar\omega/T \gg 1$ and for relative frequencies $\omega \ll \omega_D$, what should be the physical meaning of the dependence of specific heat on the temperature cubed? It means that only oscillators of low frequency ω are thermally excited to an appreciable extent and contribute to the heat capacity C_V. Hence, the knowledge of $f(\omega)$ for low frequencies is sufficient for the evaluation of C_V (at low temperatures). It is in this low-temperature region that the Debye approximation in which a solid is replaced by an elastic continuum is most appropriate. For the low-temperature region we may now conveniently replace the upper limit θ/T in our integrals by ∞. It should be noted that at exceedingly low temperatures, normal modes of frequencies as high as ω_D are not excited. This situation is analogous to that of electromagnetic waves where all frequencies are permissible, but the short wavelength modes do not obtain any appreciable occupation.

2. High temperatures: $T \gg \theta$. As $z \ll 1$ then $\exp\{z\} - 1 \approx z$ and it follows that $D(x) \approx 1$. Then the internal energy is

$$E = 3N\varepsilon_0 + 3NT \tag{9.65}$$

and

$$C_V = 3N \tag{9.66}$$

This is the *Dulong and Petit law*, as we saw earlier. It verifies the applicability of the Debye approximation. However, real materials do not follow this relation even at very high temperatures. The reason is that when the temperature is well into a range where classical mechanics holds, the typical atomic displacements become large, and the harmonic approximation is not accurate. If the atomic displacements are appreciable compared to the lattice spacings, then anharmonic effects must be included. This is just one example where the nonlinear interactions between neighboring atoms must be taken into account. Expansion of a crystal with increasing temperature is another example.

9.6   VOLUME COEFFICIENT OF EXPANSION

The harmonic theory of linear phonons (restoring forces proportional to displacements) does not lead to any prediction for the thermal expansion of solids. For theory to predict the expansion of a solid, one needs to include anharmonic terms in the potential between atoms.

In general, we can see the thermodynamic conditions that determine thermal expansion. A material has a coefficient of linear expansion α and a coefficient of volume expansion $\beta = 3\alpha$. In order to evaluate the volume coefficient of expansion β:

$$\beta \equiv \frac{1}{V}\left(\frac{\partial V}{\partial T}\right)_p \tag{9.67}$$

it is necessary to know $V = V(p,T)$, which can be derived from the Gibbs thermodynamic potential $G = G(p,V)$ according to

$$V(p,V) = \left(\frac{\partial G}{\partial p}\right)_T \tag{9.68}$$

From definition

$$G(p,T) = F(V(p,T),T) + pV(p,T) \tag{9.69}$$

If we consider a solid state, then

$$V(p,T) = V(p,0) + \Delta V(T) \tag{9.70}$$

and

$$
\begin{aligned}
G(p,T) &= F(V(p,0) + \Delta V(T),T) + p(V(p,o) + \Delta V(T)) \\
&\approx F(V(p,0),T) + \left(\frac{\partial F}{\partial V}\right)_T \Delta V(T) + pV(p,o) + p\Delta V(T) \\
&= F(V(p,0),T) - p\Delta V(T) + pV(p,o) + p\Delta V(T) \\
&= F(V(p,0),T) + pV(p,o) = G(p,0) + \Delta F(T)
\end{aligned} \tag{9.71}
$$

1. For low temperatures we insert the phonon contribution to free energy found above (this model is not entirely general), giving,

$$G = G_0 - \frac{\pi^2 V T^4}{30\hbar^3 \bar{U}^3} \tag{9.72}$$

and from Equation 9.67 we have the following:

$$V(p,T) = V(p,0) - \frac{\pi^2 T^4}{30\hbar^3} \frac{\partial}{\partial p}\left(\frac{V}{\bar{U}^3}\right) \tag{9.73}$$

The second term in this relation is the temperature-dependent part of the volume. Applying Equation 9.67 gives

$$\beta = -\frac{1}{V} \frac{2\pi^2 T^3}{15\hbar^3} \frac{\partial}{\partial p}\left(\frac{V}{\bar{U}^3}\right) \tag{9.74}$$

and

$$S = -\left(\frac{\partial G}{\partial T}\right)_p = \frac{2V\pi^2 T^3}{15\hbar^3 \bar{U}^3} \tag{9.75}$$

$$C_V = T\left(\frac{\partial S}{\partial T}\right)_V = \frac{2V\pi^2 T^3}{5\hbar^3 \bar{U}^3} \tag{9.76}$$

Then we can calculate the ratio,

$$\frac{\beta}{C_V} = -\frac{1}{3V^2}\left(\frac{\partial V}{\partial p}\right) \tag{9.77}$$

which is independent of temperature. The mean sound velocity $\bar{U}$ was assumed to be independent of the pressure to get this result. The result can also be expressed with the help of the bulk modulus, $B = -V(\partial p/\partial V)$, and the specific heat per unit volume, $c_v = (C_V/V)$. Then we have

$$\frac{\beta}{c_V} = \frac{1}{3B} \tag{9.78}$$

2. For high temperatures

$$G = G_0 + 3NT \ln\frac{\hbar\bar{\omega}}{T}, \quad V(p,T) = V_0 + 3NT\frac{\partial}{\partial p}\ln\frac{\hbar\bar{\omega}}{T} \tag{9.79}$$

Now we obtain

$$\beta = \frac{1}{V}3N\frac{\partial}{\partial p}\ln\bar{\omega} \tag{9.80}$$

and as $C_V \approx 3N$ in the harmonic approximation, then

$$\frac{\beta}{C_V} = \frac{1}{V}\frac{\partial}{\partial p}\ln\bar{\omega} \tag{9.81}$$

This is also independent of temperature and is called the *Grüneisen law*.

Let us examine the Debye case for low temperatures. In this case the Debye temperature $\theta = \theta(p)$:

$$G(p,T) = \theta f\left(\frac{T}{\theta}\right), \quad V(p,T) = \frac{d\theta}{dp}f\left(\frac{T}{\theta}\right) + \theta f'\left(\frac{T}{\theta}\right)\left(-\frac{T}{\theta}\right)\frac{d\theta}{dp} \tag{9.82}$$

It follows that

$$\beta = \frac{d\theta}{dp}f'\left(\frac{T}{\theta}\right)\frac{1}{\theta} - \frac{1}{\theta}\frac{d\theta}{dp}f'\left(\frac{T}{\theta}\right) - \frac{T}{\theta}\frac{d\theta}{dp}f''\left(\frac{T}{\theta}\right)\frac{1}{\theta} = -\frac{T}{\theta}\frac{d\theta}{dp}f''\left(\frac{T}{\theta}\right)\frac{1}{\theta} \tag{9.83}$$

$$S = -\frac{\partial G}{\partial T} = \theta f'\left(\frac{T}{\theta}\right)\frac{1}{\theta}, \quad C = T\frac{\partial S}{\partial T} = -f''\left(\frac{T}{\theta}\right)\frac{1}{\theta} \tag{9.84}$$

Thus,

$$\frac{\beta}{C} = \frac{1}{\theta}\frac{\partial\theta}{\partial p} \tag{9.85}$$

This is a function of temperature. To make these results more concrete, however, further theory would need to be developed to explain how the Debye temperature varies with the applied pressure. That can only take place when anharmonic terms in the potential between atoms are included in the theory. The Debye temperature depends on the phonon frequencies. In the harmonic approximation, the frequencies do not vary with the system size, for an elastic continuum model. When nonlinear terms are included, the frequencies will change with the system size, leading to nonzero thermal expansion coefficients.

9.7 EXPERIMENTAL BASIS OF STATISTICAL MECHANICS

Crystalline nonconductors (that are also nonmagnetic) make good candidates for experimental verification of the statistical predictions expected for phonons and their effects in thermal equilibrium. Even metals can be considered, because the phonon contributions should dominate the electronic contributions at sufficiently high temperatures. If the phonons are really the primary degrees of freedom present and they behave as expected for boson excitations, then according to Debye's theory the specific heat should vary as T^3 at low temperature and attain a constant value $C_v \approx 3Nk_B$ (the Dulong and Petit law) at moderate temperatures compared to the Debye temperature. As mentioned earlier, at very high temperatures, anharmonic effects develop, leading to considerable complications that invalidate the linear theory. Let us just consider some materials at room temperature (about 295 K). The prediction from Dulong and Petit law for the molar specific heat is $C_v \approx 3N_A k_B = 3R = 24.9 \text{ JK}^{-1}\text{mol}^{-1}$. Some experimentally measured molar specific heats of common crystalline materials are given in Table 9.1. The values for many of the elemental materials are surprisingly close to the Dulong–Petit value. Significant deviations occur for compounds, which obviously have a complex phonon spectrum, involving effectively more degrees of freedom than accounted for in the Debye theory. Except for the simplest of materials, the matching of theoretical phonon specific heat to experimental values is a complex problem.

TABLE 9.1
**Room Temperature Molar Specific Heats in $JK^{-1}mol^{-1}$
for Various Substances**

Al	Cu	Ag	Au	Pb	Zn	NaCl	KCl	MgF$_2$
24.3	24.5	24.9	25.6	26.4	25.2	36.8	50.7	61.6

---

10 Multiphase Systems

10.1 CLAUSIUS-CLAPEYRON FORMULA

Phases are physically homogeneous states of a system, characterized by different kinds of order (e.g., solid vs. liquid, ferromagnetic vs. paramagnetic, etc.). In many cases, two phases can exist simultaneously in equilibrium with each other, separated by a boundary. For example, a single-component system that consists of two phases may be a solid with its liquid or a liquid with its gas. It is supposed that the system is in equilibrium with a heat reservoir at a constant temperature T and pressure p, in order that the system itself should always have the temperature T and mean pressure p. The system may exist in either of its two possible phases or in a mixture of the two. We examine the equilibrium conditions of such a system. If we consider a system with two phases in equilibrium, apart from the equality in the temperature of the phases, $T_1 = T_2$, and the pressures, $p_1 = p_2$, the chemical potentials must be equal:

$$\mu_1(T, p) = \mu_2(T, p) \tag{10.1}$$

However, μ_1 and μ_2 are different functions. Relation (10.1) seems to be a paradox: for how can two functions be equal if their arguments do not coincide? For example, with $\mu_1 = \exp\{x^2\}$ and $\mu_2 = x + x^4$, the equality cannot hold for all values of x. The solution of the paradox is that the pressure and temperature are neither arbitrary nor independent, if the two different phases are to be in thermodynamic stability. In equilibrium, $p = f(T)$ or $T = g(p)$ (where f and g are inverses of each other). Two phases may exist in equilibrium with each other not for an arbitrary pressure p and temperature T but only along some *coexistence curve* in the pT-plane. The points lying on either side of the curve represent homogeneous states of the substance.

Thus, if the two phases are both present, specifying the value of p defines T or vice versa (see Figures 10.1 and 10.2). The intersection of μ_1 and μ_2 is at one point, that is, for a defined temperature T if the pressure p is given (see Figure 10.2). From the Gibbs–Duhem relation (Equation 2.345), we have

$$\frac{\partial \mu_1}{\partial T} = -s_1, \quad \frac{\partial \mu_2}{\partial T} = -s_2 \tag{10.2}$$

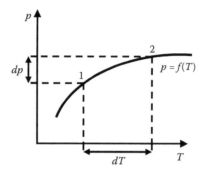

FIGURE 10.1   The pT-plane for a phase equilibrium.

where s_1 and s_2 are the entropies per particle of phases 1 and 2, respectively. This implies that the entropies of the two phases are different ($s_1 \neq s_2$). If the temperature is given, then at the intersection point, the Gibbs-Duhem relation also implies

$$\frac{\partial \mu_1}{\partial p} = v_1, \quad \frac{\partial \mu_2}{\partial p} = v_2 \qquad (10.3)$$

where v_1 and v_2 are the volumes per particle of phases 1 and 2, respectively. It follows that the volumes of the two phases are also different ($v_1 \neq v_2$). As this corresponds to some given mass in the system, the mass densities are different, which is often one of the most obvious distinguishing features of different phases.

Consider two points 1 and 2 infinitesimally near each other on the coexistence curve in Figure 10.1. Point 1 corresponds to temperature T and pressure p and point 2 to $T + dT$ and $p + dp$. For point 1 we have

$$\mu_1(T,p) = \mu_2(T,p) \qquad (10.4)$$

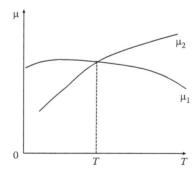

FIGURE 10.2   Relation between μ and T for the coexistence of two phases in phase equilibrium at constant pressure.

and for point 2 we have

$$\mu_1(T + dT, p + dp) = \mu_2(T + dT, p + dp) \quad (10.5)$$

Expanding on both sides leads to

$$\mu_1(T,p) + \frac{\partial\mu_1}{\partial p}dp + \frac{\partial\mu_1}{\partial T}dT = \mu_2(T,p) + \frac{\partial\mu_2}{\partial p}dp + \frac{\partial\mu_2}{\partial T}dT \quad (10.6)$$

Subtracting Equation 10.1 from Equation 10.6, then

$$\frac{\partial\mu_1}{\partial p}dp + \frac{\partial\mu_1}{\partial T}dT = \frac{\partial\mu_2}{\partial p}dp + \frac{\partial\mu_2}{\partial T}dT \quad (10.7)$$

Then using Equations 10.2 and 10.3, we have

$$v_1 dp - s_1 dT = v_2 dp - s_2 dT \quad (10.8)$$

and this gives

$$\frac{dp}{dT} = \frac{s_2 - s_1}{v_2 - v_1} \quad (10.9)$$

The subscripts refer to the two different phases. Then the slope of $p(T)$ is determined by the ratio of entropy change to volume change at the chosen temperature.

There is a change in entropy with a change in phase, or a *phase transformation*. The entropy change can be generally expressed as

$$dS = \frac{\delta Q}{T} \quad (10.10)$$

where δQ is the heat absorbed by the system. This is

$$s_2 - s_1 = \frac{q_{1\rightarrow 2}}{T} \quad (10.11)$$

where $q_{1\rightarrow 2}$ is the latent heat of transformation per particle at that temperature. During the transformation, the temperature remains constant. The latent heat is the heat absorbed by the substance when a given amount of phase 1 is transformed to phase 2. Generally, one phase is less ordered (i.e., higher entropy) than the other. If heat enters the system, some part of the system transforms into the less ordered phase, while the temperature does not change. If heat leaves the system, some part of it transforms into the more ordered phase, again with no change in temperature.

Latent heats might be called heat of fusion (for solid to liquid transformation) or heat of vaporization (liquid to gas), as some common examples. One could also have a transformation directly from solid to vapor (sublimation), which occurs in CO_2 at atmospheric pressure. This can be seen in the so-called dry ice, the white powdery solid CO_2 produced with a sudden burst from a fire extinguisher onto a thermally insulating material.

Combining the last equation with Equation 10.9, we arrive at another relation for the slope of $p(T)$

$$\frac{dp}{dT} = \frac{q_{1\to2}}{T(v_2 - v_1)} \tag{10.12}$$

Formulas (10.9) and (10.12) are called the Clausius–Clapeyron formulas. If v is the molar volume, then $q_{1\to2}$ is the latent heat per mole and if v is the volume per gram, then $q_{1\to2}$ is the latent heat per gram. The relationship gives interesting information about the coexistence curve of the two phases.

Let us examine and construct a curve of phase equilibrium. Consider, for example, phase equilibrium for phase 1, solid (S) and phase 2, liquid (L). There is some difference in the volumes of the two phases for a given mass

$$v_2 - v_1 = v_L - v_S \tag{10.13}$$

In the transformation from solid (S) to liquid (L), the entropy of the given substance almost always increases. (We may have an exception in solid helium ^3He for which in a certain temperature range the nuclear spins in the solid are oriented randomly, whereas, those in the liquid are aligned antiparallel to each other in order to conform to the quantum mechanical Fermi–Dirac statistics.) For the cases where the entropy increases, the corresponding latent heat $q_{S\to L}$ is positive: heat gets absorbed by the material in the transformation from solid (S) to liquid (L). In most cases, a solid expands on melting such that $v_L - v_S > 0$. Then the Clausius–Clapeyron formula 10.9 shows that the slope of the solid–liquid equilibrium line (melting curve) is positive. Some substances, such as water, contract upon melting, such that $v_L - v_S < 0$. In those cases, the slope of the melting curve is negative. For the case of the transformation from liquid (L) to vapor (G), the entropy always increases. As a consequence, the corresponding latent heat $q_{L\to G} > 0$ is positive and heat gets absorbed in the transformation. In these cases, $v_G - v_L > 0$ and the substance expands under vaporization, and thus the Clausius–Clapeyron formula shows that the slope of the liquid–vapor equilibrium line (vaporization curve) is positive. See Figure 10.3 for the pT-plane. This curve of phase equilibrium ends at a certain point called the *critical point* at a *critical temperature* T_c and the *critical pressure* p_c. For temperatures greater than T_c and pressures above p_c, there are no different phases; the system is always homogeneous. Liquid and vapor become indistinguishable; hence, the two-phase concept is unnecessary. At the critical point, the differences between the phases vanish. This notion of the critical point was introduced by D.I. Mendeleyev (1960). In Figure 10.3

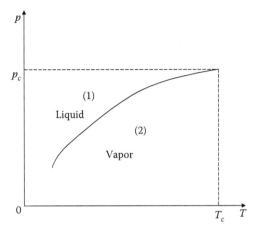

FIGURE 10.3 pT-plane for the coexistence of two phases in phase equilibrium. The curve of phase equilibrium ends at a certain point called the critical point at a critical temperature T_c and the critical pressure p_c.

for regions (1) and (2), we have homogeneous single-phase states of liquid and vapor, respectively. They only coexist in equilibrium at points on the phase boundary line.

In Figure 10.3, for the regions (1) and (2) there are homogeneous states of liquid and vapor, respectively. The volume of the vapor (gas) is often much greater than the volume of the liquid containing the same number of particles. This argument can be applied to Equation 10.12:

$$v_2 - v_1 \approx v_2 = v_G, \quad v_2 \gg v_1 \tag{10.14}$$

then

$$\frac{dp}{dT} = \frac{q_{1\rightarrow2}}{Tv_2} = \frac{q}{Tv} \tag{10.15}$$

It is assumed that the vapor may be adequately treated as an ideal gas, then $v = T/p$ is the volume per particle. Then from Equation 10.15 if q is the latent heat per particle, we have

$$\frac{dp}{dT} = \frac{pq}{T^2} \tag{10.16}$$

or

$$\frac{dp}{p} = d\ln p = q\frac{dT}{T^2} \tag{10.17}$$

Integration leads to a relation for the vapor pressure

$$\ln p = \ln p_0 - \frac{q}{T} \tag{10.18}$$

or

$$p = p_0 \exp\left\{-\frac{q}{T}\right\} \tag{10.19}$$

We assume that latent heat q is approximately temperature-independent. In Equation 10.19, p_0 is a constant that would correspond to the pressure at which $q = 0$. Relation (10.19) shows that the vapor pressure p changes with temperature according to an exponential law, with the temperature dependence determined by the magnitude of the latent heat of vaporization.

Ordinary substances are capable of existing in solid, liquid, and gaseous phases, with the phase equilibrium lines separating the respective phases as in Figure 10.4. There are three coexistence curves in Figure 10.4: A separates solid from gas, B separates solid from liquid, and C separates liquid from gas. The point denoted by (1) for which all the three lines intersect is called the triple point, which corresponds to a unique temperature T_{Tr} and pressure p_{Tr}. At this point, arbitrary amounts of all the three phases can coexist in equilibrium with each other. At the triple point, the equilibrium condition is defined by two equations:

$$\mu_G(T,p) = \mu_L(T,p), \quad \mu_G(T,p) = \mu_S(T,p) \tag{10.20}$$

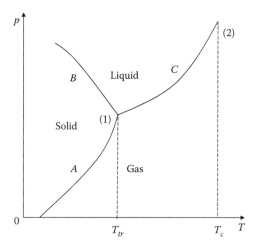

FIGURE 10.4 Phase diagram of ordinary substances capable of existing in solid, liquid, and gaseous phases, with the phase equilibrium lines separating the respective phases.

For each of the functions μ_S, μ_L, and μ_G, the solution at the triple point is unique. For the transformation from solid to liquid, we have

$$\frac{dp}{dT} = \frac{q_{S \to L}}{T(v_L - v_S)} \tag{10.21}$$

and if $v_L > v_S$ then $(dp/dT) > 0$ and if $v_L > v_S$ then $(dp/dT) < 0$ (contraction of a volume for the transformation of solid to liquid, an example is water). For the transformation from solid to gas, we have $v_S \ll v_G$ and $q_{S \to G} \cong q_{S \to L} + q_{L \to G}$. At the *critical point* (2), where the liquid–gas equilibrium curve C ends, we have $v_G - v_L \to 0$ and beyond (2) there is no further phase transformation as there exists only one homogeneous state (the dense gaseous state).

10.2 CRITICAL POINT

Suppose we have ether (a common organic compound) in an enclosed tube at room temperature. The less dense vapor occupies the space above the more dense liquid. The meniscus of the phase boundary can be seen due to the difference of the refractive indices of liquid versus gas. As a result of heating, the pressure increases and the meniscus become less notable. After passing through the critical temperature T_c, the meniscus completely disappears. This experiment was done by Landau in 1930. The experiment demonstrates changes of the properties of the substance connected with the critical point. For temperatures above the critical temperature, there is no boundary between the phases and the substance becomes homogeneous.

Figure 10.5 shows some isotherms of a liquid and a gas in the (p,V)-plane, that is, the dependence of p on V under isothermal expansion of a homogeneous substance $(1) \to (3)$ and $(4) \to (6)$. If we consider Chapter 2

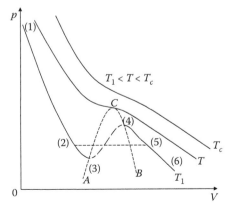

FIGURE 10.5 Dependence of p on V under isothermal expansion of a homogeneous substance.

$$-\left(\frac{\partial p}{\partial V}\right)_T \geq 0 \qquad (10.22)$$

then this is true for (1) → (2) and (5) → (6). The portions (2) → (3) and (4) → (5) of the isotherm correspond to the metastable state of the superheated liquid and the supercooled gas and Equation 10.22 is still true. (For the liquid phase between (2) and (3), the temperature is above the usual vaporization temperature. For the vapor phase between (4) and (5), the temperature is below the usual condensation temperature.) In Figure 10.5, points (2) and (5) have the same pressure p. Hence, there may be a discontinuity between the given points of the isotherms. At points (3) and (4), the thermodynamic inequality in Equation 10.22 is violated and

$$\left(\frac{\partial p}{\partial V}\right)_T = 0 \qquad (10.23)$$

Suppose we construct the locus of the points of termination of the isotherms of the liquid and gas, then we obtain a curve ACB in Figure 10.5 on which the thermodynamic inequalities are violated (for a homogeneous substance). This is the boundary of a region in which the substance can never exist in a homogeneous state. Equation 10.23 and its second derivative

$$\left(\frac{\partial^2 p}{\partial V^2}\right)_T = 0 \qquad (10.24)$$

characterize the critical points (saddle point).

Let us use the following notations for series expansion near the critical point:

$$V = V_c + v, \quad T = T_c + t \qquad (10.25)$$

Then the expansion of $\left(\dfrac{\partial p}{\partial V}\right)_T$ about V_c and T_c is

$$\left(\frac{\partial p}{\partial V}\right)_T = \left(\frac{\partial p}{\partial V}\right)_{T_c} + \left(\frac{\partial^2 p}{\partial V \partial t}\right)_{T_c} t + \left(\frac{\partial^2 p}{\partial V^2}\right)_{T_c} v + \frac{1}{2}\left(\frac{\partial^3 p}{\partial V \partial t^2}\right)_{T_c} t^2$$

$$+ \left(\frac{\partial^3 p}{\partial V^2 \partial t}\right)_{T_c} vt + \frac{1}{2}\left(\frac{\partial^3 p}{\partial V^3}\right)_{T_c} v^2 + \cdots \qquad (10.26)$$

If we apply Equations 10.22 and 10.23 and we limit ourselves to the first three terms of the expansion of Equation 10.26, then

$$-\left(\frac{\partial p}{\partial V}\right)_T = At + Bv^2, \quad A = -\left(\frac{\partial^2 p}{\partial V \partial t}\right)_{T_c}, \quad B = -\frac{1}{2}\left(\frac{\partial^3 p}{\partial V^3}\right)_{T_c} \qquad (10.27)$$

Let us look at the case for which $v = 0$; it follows that $V = V_c$; the system is at the critical density. Now if we take $t > 0$, then the requirement is

$$-\left(\frac{\partial p}{\partial V}\right)_{T>T_c} > 0 \tag{10.28}$$

This will insure that all points for which $T > T_c$ correspond to stable states. Here, there is no distinction of phases and this requires

$$A > 0 \tag{10.29}$$

Similarly, if $t = 0$ for $v \neq 0$, then it follows that we are in a nonhomogeneous domain, and thermodynamic stability requires

$$B > 0 \tag{10.30}$$

Let us integrate Equation 10.27 with respect to v:

$$p(v,t) = -Atv - \frac{1}{3}Bv^3 + f(t) \tag{10.31}$$

This equation defines the form of an isotherm of a homogeneous substance in the neighborhood of the critical point. If $t > 0$, then the isotherm p is a monotonically decreasing function of v. If $t = 0$ and $v = 0$, then we have $T = T_c$, $V = V_c$, and $p = p_c$ (critical point) and if $t < 0$, then there exists a maximum v and a minimum v between which there is a region with

$$\left(\frac{\partial p}{\partial V}\right)_t > 0 \tag{10.32}$$

That corresponds to a fictitious homogeneous substance. The above information is shown in Figure 10.6.

The line (2) → (5) is the line for which there coexists two phases of the system. In the interval (3) → (4), a homogeneous state is impossible (region of fictitious homogeneous substance). This implies that for $t < 0$, we have two curves, that is, (1) → (3) and (4) → (6). At the points (3) and (4), we have a minimum and a maximum, respectively. The portion (2) → (3) is the isotherm of a superheated liquid and (5) to (6) for a supercooled gas. It can be seen from Figure 10.6 that at the points (2) and (5) we have

$$p_{(2)} = p_{(5)}, \quad \mu_{(2)} = \mu_{(5)} \tag{10.33}$$

where p and μ are the pressure and chemical potential, respectively. The condition of phase equilibrium in Equation 10.33 may be written in the form

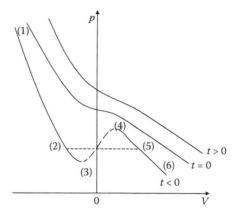

FIGURE 10.6   Dependence of p on V under isothermal expansion and showing the region for a fictitious homogeneous substance.

$$\int\limits_{(2)}^{(5)} d\mu = 0 \tag{10.34}$$

where the path of integration is from the liquid to the gaseous state. Thus

$$\int\limits_{(2)}^{(5)} d\mu = \int\limits_{(2)}^{(5)} \frac{\partial \mu}{\partial p} dp \tag{10.35}$$

Then with the definition

$$\frac{\partial \mu}{\partial p} = V \tag{10.36}$$

the integration becomes

$$\int\limits_{(2)}^{(5)} d\mu = \int\limits_{(2)}^{(5)} (V_c + v)dp = V_c(p_{(5)} - p_{(5)}) + \int\limits_{(2)}^{(5)} vdp = \int\limits_{(2)}^{(5)} v\left(\frac{\partial p}{\partial V}\right)_t dv \tag{10.37}$$

If we consider Equations 10.29, 10.30, and 10.31, then

$$\int\limits_{(2)}^{(5)} d\mu = \mu_{(5)} - \mu_{(2)} = v_c(p_{(5)} - p_{(2)}) + \int\limits_{v_{(5)}}^{v_{(5)}} v(-At - Bv^2)dv = 0 \tag{10.38}$$

It can be seen from Equation 10.38 that the integrand is an odd function and if the integral equals zero, then the limits of integration are symmetric, that is

$$v_{(5)} = -v_{(2)} \tag{10.39}$$

It follows from here that

$$V_{(2)} = V_c + v_{(2)}, \quad V_{(5)} = V_c + v_{(5)} \tag{10.40}$$

If we substitute $V_{(2)}$ in Equation 10.31, then

$$p_{(2)} = -A t v_{(2)} - \frac{1}{3} B v_{(2)}^3 + f(t) \tag{10.41}$$

and if we consider Equation 10.39, then

$$p_{(5)} = p_{(2)} = A t v_{(2)} + \frac{1}{3} B v_{(2)}^3 + f(t) \tag{10.42}$$

If we consider Equations 10.41 and 10.42, then it follows that

$$A t v_{(2)} + \frac{1}{3} B v_{(2)}^3 = 0 \tag{10.43}$$

Finally, this gives the result

$$v_{(2,5)} = \mp \sqrt{-\frac{3At}{B}} \tag{10.44}$$

and

$$\left| v_{(2)} \right| = \left| v_{(5)} \right| \approx (T_c - T)^{1/2} \tag{10.45}$$

Then the curve of the equilibrium phase in the (T,V)-plane has a simple maximum at the critical point. Now look at the density ρ:

$$\rho = \frac{\text{Mass}}{\text{Volume}} = \frac{M}{V} \tag{10.46}$$

Then

$$\rho_L = \frac{M}{V_L} = \frac{M}{V_c - \sqrt{-\frac{3At}{B}}} = \frac{M}{V_c}\left(\cfrac{1}{1 - \sqrt{-\frac{3At}{V_L^2 B}}}\right) \approx \frac{M}{V_c}\left(1 + \sqrt{-\frac{3At}{V_L^2 B}}\right)$$

$$= \rho_c\left(1 + \sqrt{-\frac{3At}{V_L^2 B}}\right) \tag{10.47}$$

where L stands for liquid. Thus, the liquid density is enhanced compared to the critical density

$$\rho_L = \rho_c\left(1 + \sqrt{\frac{-3At}{V_L^2 B}}\right) \tag{10.48}$$

In this aspect, the theory coincides with the experiment. Similarly, the vapor density is reduced compared to the critical density

$$\rho_G = \frac{M}{V_G} \approx \rho_c\left(1 - \sqrt{\frac{-3At}{V_G^2 B}}\right) \tag{10.49}$$

Equations 10.48 and 10.49 are shown in Figure 10.7.

In Figure 10.8, $\rho_{L0} = \rho_{L,T=0}$ and $\rho_{G0} = \rho_{G,T=0}$. The figure represents the densities ρ_{G0} and ρ_{L0} of a gas and liquid in mutual equilibrium at temperature T. Let us now consider the points (3) and (4) for which

$$-\left(\frac{\partial p}{\partial v}\right)_t = At + Bv^2 = 0 \tag{10.50}$$

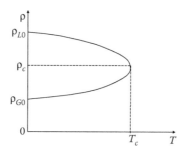

FIGURE 10.7   Densities ρ_{G0} and ρ_{L0} of a gas and liquid in mutual equilibrium at temperature T.

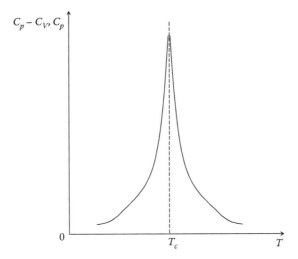

FIGURE 10.8 Behavior of the specific heat with temperature approaching T_c, where the strong increase in specific heat indicates the onset of a phase transition even before the transition temperature is reached

and

$$v'_{(3),(4)} = \pm\sqrt{-\frac{At}{B}} \qquad (10.51)$$

The volumes $v'_{(3)}$ and $v'_{(4)}$ are also symmetric and are proportional to $(T_c - T)^{1/2}$ and $\sqrt{3}$ times less than $v_{(2)}$ and $v_{(5)}$, respectively. The latent heat at the critical point equals zero. In the neighborhood of the critical point, we have

$$S_G - S_L = \frac{q_{L \to G}}{T} \qquad (10.52)$$

which is rearranged as

$$q = T\{S_G(t,V_c + v_{(s)}) - S_L(t,V_c - v_{(s)})\} \qquad (10.53)$$

If we expand in a series near the critical point, then

$$q = T\left\{\left(\frac{\partial S_G}{\partial v}\right)_{T_c} v_{(5)} + \left(\frac{\partial S_L}{\partial v}\right)_{T_c} v_{(5)}\right\} \approx T\left(\frac{\partial p}{\partial T}\right)_{V_c} \sqrt{\frac{A(T_c - T)}{B}}$$

$$\approx \sqrt{\frac{A}{B}(T_c - T)}\bigg|_{T \to T_c} \to 0 \qquad (10.54)$$

Then, if we consider Chapter 2 and the above relations, we have

$$C_p - C_v = -T \frac{\text{Const}}{-At - Bv^2} = -T \frac{\text{Const}}{-At - B\left(-\dfrac{3At}{B}\right)} = \frac{\text{Const}}{2A(T_c - T)} \quad (10.55)$$

from which it follows that $C_p - C_v \to \infty$ as $T \to T_c$ (see Figure 10.8).

The behaviors of $C_p - C_v$ and C_p in Figure 10.8 are similar and for all substances, $C_p > C_v$. Figure 10.8 is an indication that the onset of a phase transition can be antici-pated even before the transition temperature is reached due to a strong increase of the specific heat C_p or $C_p - C_v$, while for first-order phase transitions, C_p or $C_p - C_v$ diverges only when both phases coexist. The quantities C_p or $C_p - C_v$ may diverge in different ways for $T \to T_c$, depending on the side from which one approaches the critical temperature. Thus, we have seen that a fundamental aspect of phase transi-tions is the behavior of a system in the vicinity of the critical point. As seen earlier and from the preceding example, several thermodynamic quantities tend to diverge at this point, and the order parameter vanishes on the high-temperature side. This statement may be made more precise by finding the temperature dependence, say, for $C_p - C_v$ or other important thermodynamic quantities in the vicinity of the critical point. This is done by using power laws, the exponents of which are called critical indices or critical exponents. In our case, for $C_p - C_v \propto |T_c - T|^{-1}$ near the critical point, so the critical index is 1. A general theory for the complete analysis of systems near a critical point, known as the renormalization group theory, has been developed by Kadanoff, Wilson, Fisher, and others since the 1970s.

11 Macroscopic Quantum Effects

Superfluid Liquid Helium

11.1 NATURE OF THE LAMBDA TRANSITION

A study of a macroscopic system at exceedingly low temperatures gives an opportunity for investigating a system in its ground state and in the quantum states that lie close to it. The number of states accessible to the system or its entropy is quite small. The system exhibits much less randomness or a much greater degree of order than it could at higher temperatures. For low temperatures, some systems may exhibit a high degree of quantum order on a macroscopic scale. An example of such order is seen in a system of spins most of which at exceedingly low temperatures are aligned parallel to each other and giving rise to ferromagnetism.

However, a ferromagnet even at very low temperature still has spin wave excitations that create disorder. More interesting are examples of superconductors and superfluids [42–50], where a large fraction of the system condenses into a macroscopic quantum ground state, leading to impressive physical phenomena associated with an organized and frictionless particle motion. Here, we consider how such a macroscopic quantum order can appear in helium at low temperature.

For helium, there exist two isotopes, that is, He^3 and He^4[8, 47–50]. The isotope He^3 has spin 1/2 and thus obeys Fermi–Dirac statistics. The isotope He^4 has spin 0 and thus obeys Bose–Einstein statistics. The two helium isotopes remain liquid at absolute zero temperature under atmospheric pressure. Helium is the only element that conserves the liquid state at zero temperature and normal atmospheric pressure. Helium can be solidified, however, under a pressure above 25 atm as seen, for example, in experiments on super flow in solid helium by Moses Chan, University of Pennsylvania. Helium-4 is used extensively in cryogenics because of its low boiling point of 4.2 K at atmospheric pressure. The phase diagram for He^4 is sketched in Figure 11.1.

Helium remains liquid even at exceedingly low temperatures due to weak interatomic interactions enhanced by the small atomic mass of helium. A helium atom has two protons and two electrons. The two electrons in the filled 1 s electronic shell are of opposite spins and the protons are well screened by the electrons. The filled shell makes helium an inert gas. These are the reasons why the forces of interactions among the helium atoms are weak. The potential well between helium atoms is very wide compared to that of hydrogen (see Figure 11.2). As helium has large zero-point energy and weak molecular interactions, it is impossible to localize the

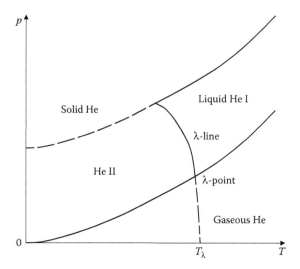

FIGURE 11.1   The phase diagram of helium-4 near the transition temperature T_λ.

atoms at well-defined lattice sites. Thus, helium does not solidify. It remains only in the liquid state. Compare argon and hydrogen. Due to argon's smaller zero-point energy, it can solidify at lower temperatures. Notwithstanding the fact that hydrogen is lighter than helium, it still solidifies at low temperatures due to its stronger molecular interactions.

The two liquids He^3 and He^4 have some significantly different physical properties. This difference arises from the fact that the constituent atoms of the two liquids obey different statistics. He^4 has particularly interesting properties.

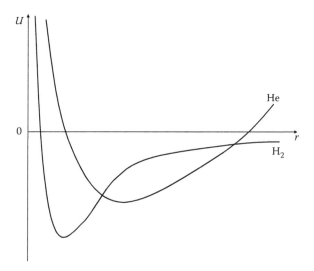

FIGURE 11.2   Comparison of the potential well of helium and hydrogen atoms.

He^4 exhibits two phases, that is, *normal* He I and *superfluid* He II. Figure 11.1 shows the phase diagram of helium near the transition temperature for this phase change, T_λ.

At a temperature of 2.17 K, He^4 undergoes a phase transition from He I to He II. This transition is called a λ-transition. This transition is seen in He^3 only at a much lower temperature (about 1 mK). As He I at its boiling point is cooled to 2.17 K, the boiling abruptly stops. This results in the low-temperature He II type that does not boil. This transition from He I to He II can be illustrated by heat capacity measurements (see Figure 11.3). In addition, the thermal conductivity rises substantially at 2.17 K—which at the same time explains the cessation of the boiling. Heat is conducted so fast through He II that the only vaporization is taking place at the outer surfaces, rather than in internal bubbles. Figure 11.3 shows the experimental heat capacity of liquid He^4 compared with that of an ideal boson gas (not to scale). The shape of the heat capacity curve resembles a λ; thus, the name of the point $T_\lambda = 2.17$ K is the λ-*point*. The λ-transition is considered to be due to Bose–Einstein condensation; a macroscopic fraction of the helium-4 atoms occupies the ground state in the condensed phase. This macroscopic fraction of atoms in the ground state is understood to be responsible for the amazing superfluid properties.

As He^3 atoms are fermions, no Bose–Einstein condensation or superfluidity is expected for them. However, that thinking is actually incorrect. At especially low temperatures around 1 mK, even He^3 can become superfluid due to a weak attractive interaction between atoms that pairs them into bosons. Superfluidity does appear in He^3 at a much lower temperature than that for He^4. A physically similar effect takes place in superconductors. Electrons clearly are fermions; however, the electron–phonon interaction produces an attractive interaction between electrons

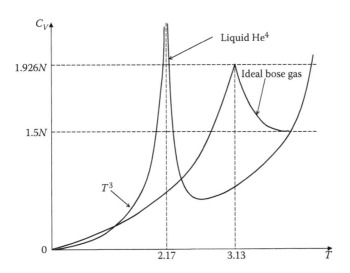

FIGURE 11.3 The phase transition from He I to He II, as indicated through heat capacity measurements.

that causes them to form into pairs of opposite spins, known as *Cooper pairs*. The Cooper pairs are bosons that can move around in the system like fundamental particles. Hence, under appropriate conditions, the Cooper pairs undergo a Bose–Einstein condensation, and being electrically charged with $q = -2e$, this leads to the phenomenon of superconductivity. Because many Cooper pairs can fall into the same low energy state, their motion together can lead to superconductivity: the flow of electrical current with essentially zero resistance. Thus, there are many similarities between superfluidity and superconductivity. The theory for superfluidity such as that in helium is simpler because the electromagnetic properties need not be considered.

11.2 PROPERTIES OF LIQUID HELIUM

Liquid helium above the λ-point, that is, He I, behaves as a fairly normal fluid. It has a normal gas–liquid critical point. On the other hand, helium below the λ-point, that is, (superfluid) He II, exhibits some strange properties such as zero entropy, zero viscosity (under certain conditions), and superfluidity. He II exhibits completely frictionless flow and can move through a thin capillary or *superleak* with zero resistance as long as its speed is below a critical velocity v_c. The critical velocity v_c increases as the capillary diameter decreases. Liquid He II placed in a container will easily creep up the walls and out of the container. It still has surface tension, but the zero viscosity allows it to flow easily in a thin layer even against gravity. He II used in cryogenics is famous for the problems caused by its *superleaks*, where it escapes the container through the smallest of apertures with ease.

In liquid He II, heat propagates very rapidly like a wave with a definite velocity. We call this *second sound*: this second sound is a temperature wave and is excited by heat rather than by a pressure pulse. The thermal conductivity of He II is extraordinarily high, as a result.

The peculiar properties of He II can be explained by the phenomenological two-fluid theory [46,47]. Imagine the following experiment: Place a cylinder in a liquid He II bath and rotate the cylinder. This leads to a momentum transfer from the rotating cylinder to the He II. This implies that the viscosity in this experiment is *nonzero*. This is explained using the two-fluid theory: Liquid He II is a mixture of two fluids—a normal fluid component and a superfluid component. The latter is a Bose condensate. The normal fluid has normal viscosity and density ρ_n and the superfluid, zero viscosity and density ρ_s. The two densities combine to make the total fluid density ρ. It is assumed that the normal fluid exhibits the properties of an ordinary liquid. Thus, it is the normal fluid that has viscosity which causes it to interact with the rotating cylinder. The superfluid has zero entropy, zero viscosity, and flows through capillaries without resistance. The fractions of normal and superfluid components, ρ_n/ρ and ρ_s/ρ, are dependent on temperature. Notwithstanding its utility to explain some physical properties, the two-fluid theory of liquid He II has a partial molecular basis.

Consider how molecular theory can explain the behavior of liquid He^4. Liquid He^4 behaves as a (nonideal) boson gas and obeys Bose–Einstein statistics. We have mentioned that liquid He^4 has weak intermolecular interactions. Assume that He^4 atoms move freely in a constant attractive central potential. Due to the interactions, the ideal Bose gas model should be inadequate to explain the properties of the liquid He II. Even so, this theory provides a partial understanding of the problem. From the prediction of Bose–Einstein condensation, an ideal gas at the density of liquid He^4 would undergo a sharp transition at $T_c = 3.13$ K, not far from the observed $T_\lambda = 2.17$ K. In the temperature range $0 \leq T \leq T_c$, the boson gas can be predicted theoretically to have some spectacular behavior.

For $T \leq T_c$, the boson gas model views liquid He^4 as a mixture of a gaseous phase with $N_{E>0}$ particles distributed over the excited states with the energy $E > 0$ and a condensed phase with $N_{E=0}$ particles occupying the ground state with energy $E = 0$. The accumulation of particles in the ground state is *Bose–Einstein condensation*. The Bose condensate corresponds to the *superfluid* component of liquid He II. The particles in the excited states correspond to the normal component of liquid He II. The number of atoms in the condensate is given from the Bose condensate theory, review Equation 5.193 and its derivation in Section 5.11

$$N_{E=0} = N\left[1 - \left(\frac{T}{T_c}\right)^{(3/2)}\right] \tag{11.1}$$

where N is the total number of helium atoms. From this expression, it is apparent that for $T = 0$, there exists only the superfluid component, whereas at $T = T_c$, there exists only the normal fluid component. At temperatures between 0 and T_c, the expression gives the fraction that is superfluid. For any temperature $T > T_c$, there is no macroscopic condensation and there exists only the normal fluid. Thus, the Bose condensation theory gives a physical description consistent with the ideas of the London/Tisza two-fluid model.

11.3 LANDAU THEORY OF LIQUID He II

From the thermodynamic functions, we may define the macroscopic properties of a body (in particular, liquid He II). Landau laid down the theory for liquid He II. He showed that the atoms of liquid He II were bosons [49,50]. In Figure 10.3, the ideal Bose gas is used to explain the thermodynamic behavior of liquid He II. From this model, the heat capacity behaves as $C_V \approx T^{3/2}$. Experimentally, the heat capacity of liquid He II is $C_V \approx T^{3/2}$ at low temperatures. This is characteristic of some type of excitations that may be regarded as phonons. It is a model that is analogous to a solid that consists of a background lattice plus phonon excitations. This is explained reasonably well by Landau's theory [49,50] that views the excitations as present in the normal component. From Landau's view, liquid He II is a perfect background fluid or superfluid (with zero entropy and viscosity) in which excitations (associated with the normal component) move.

Superfluidity and superconductivity appear to be physically similar effects. In superconductivity, the electrical resistance of some conductors becomes exceedingly small at low temperature, similar to the small viscosity of a superfluid at low temperature. There may be clues to the cause of these effects by examining other effects such as the discontinuity of the specific heat at the critical temperature.

Consider superfluids. At absolute temperature, $T = 0$ K, the motion of separate particles is absent (but there could, in principle, still be zero-point motion, yet it is suppressed). The entire liquid is in its ground state (superfluid phase) with internal energy, $E = E_0$. At nonzero temperature, the thermal motion in the liquid is due to elastic vibrations of the medium associated with longitudinal elastic waves. The frequencies are defined by $\omega(q) = U_l q$ and energy by $\varepsilon(p) = U_l p$, where, U_l, p, and q are respectively the longitudinal velocity of sound, the momentum, and the wave number. The relation $\omega(q) = U_l q$ shows a linear dispersion law relative to the wave number q. At exceedingly low temperatures, the phonon occupation numbers are sufficiently small such that there is no interaction among the phonons. Thus, we approximate them as an ideal boson gas. The energy levels of the quantum system are

$$E = E_0 + \sum_{\vec{p}} \varepsilon(\vec{p}) n(\vec{p})$$
(11.2)

Here, $\varepsilon(\vec{p})$ is the energy of a phonon with momentum $\vec{p}$ and $n(\vec{p})$ is its occupation number. According to Landau [15,49,50], there are two types of excitations. At low temperatures (and low wave numbers), the excitations are phonons or quantized sound waves with energy

$$\varepsilon(\vec{p}) = Up$$
(11.3)

where U is the velocity of sound. In the region around some momentum p_0, there is another type of excitation—a quantized classical vortex motion that Landau called *rotons*. Its energy is represented as

$$\varepsilon(p) = \Delta + \frac{(p - p_0)^2}{2m_{eff}}$$
(11.4)

Here, $\Delta = \varepsilon(p_0)$ is a constant (Landau's energy), m_{eff} is the effective mass of the excitation, and p_0 is the momentum of the excitation in the liquid. From experiments, these parameters are

$$\Delta \approx 8.7\,\text{K}, \quad m_{eff} \approx 0.16 m_{He}, \quad U = 2.4 \times 10^2\,\text{m·s}^{-1}, \quad \frac{p_0}{\hbar} = 1.9 \times 10^{10}\,\text{m}^{-1} \quad (11.5)$$

In liquid helium, the dispersion relation of the elementary excitations has the form shown in Figure 11.4. The majority of elementary excitations in liquid helium at thermal equilibrium have energies $\varepsilon(p)$ about the minima, that is, in the neighborhood of the points $p = 0$ and $p = p_0$. The expansion of $\varepsilon(p)$ about the point $p = p_0$ yields expression 11.4 for the roton energy. From Figure 11.4, phonons and rotons correspond to different parts of the same curve.

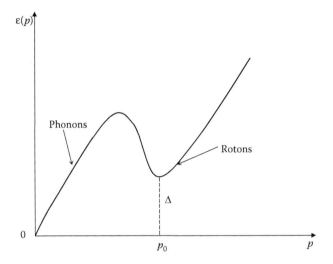

FIGURE 11.4   Dispersion relation of the elementary excitations in liquid helium.

We put the Landau two-fluid model on a more quantitative level by applying statistical mechanics to the phonon–roton gas. Consider that the phonon and roton numbers are not very large. Then we can take a two-fluid model that is a mixture of two ideal gases, that is, phonon and roton gases. The contribution to the thermodynamic properties of liquid He II is due to both phonons and rotons. The relative contributions of phonons and rotons are dependent on temperature. For exceedingly low temperatures (below 0.5 K), the phonon contribution is dominant over the roton contribution. Under Bose condensation, we know the free energy due to the phonons (see, e.g., Section 9.3) is

$$F = 3N\varepsilon_0 - \frac{\pi^2 VT^4}{90\hbar^3 U_l^3} \qquad (11.6)$$

and this gives the entropy

$$S = -\left(\frac{\partial F}{\partial T}\right)_V = \frac{2\pi^2 VT^3}{45\hbar^3 U_l^3} \qquad (11.7)$$

and the specific heat

$$C_V = T\left(\frac{\partial S}{\partial T}\right)_V = \frac{2\pi^2 VT^3}{15\hbar^3 U_l^3} \qquad (11.8)$$

This is the *law of Debye's cubes*. As $T \to 0$, $C_V \to 0$.

We may also evaluate the energy

$$E = F + TS = 3N\varepsilon_0 + \frac{\pi^2 V T^4}{30\hbar^3 U_l^3} \tag{11.9}$$

As we know that a phonon is a quantum mode, it then follows that S and C_V are pure quantum formulas. Let us evaluate the roton contribution to the free energy, F_r, in terms of an occupation number for rotons, N_r:

$$F_r = -T \ln Z_r, \quad Z_r = \frac{1}{N_r!} z_r^{N_r} \tag{11.10}$$

Using Stirling's formula

$$N_r! \approx \left(\frac{N_r}{e}\right)^{N_r} \tag{11.11}$$

we have

$$F_r = -N_r T \ln \frac{e z_r}{N_r} \tag{11.12}$$

It should be noted that N_r is not a fixed value and thus with respect to N_r at equilibrium the free energy F_r should have a minimum value. Thus

$$\frac{\partial F_r}{\partial N_r} = 0 = -T \ln \frac{e z_r}{N_r} + T \tag{11.13}$$

or

$$\overline{N_r} = z_r \tag{11.14}$$

Then, substituting this expression in Equation 11.12, we have

$$F_r = -N_r T \tag{11.15}$$

Since rotons can be considered as obeying Bose statistics, their total number may be evaluated as

$$N_r = \frac{V}{(2\pi\hbar)^3} \int \bar{n}(p) 4\pi p^2 \, dp \tag{11.16}$$

The quantity

$$\bar{n}(p) = \frac{1}{\exp\{\varepsilon(p)/T\} - 1} \tag{11.17}$$

is the phonon distribution and $\varepsilon(p)$ the energy in Equation 11.4. By ignoring the -1 in the boson occupation number (Boltzmann statistics), one obtains

$$N_r = \frac{4\pi V}{(2\pi\hbar)^3} \exp\left\{-\frac{\Delta}{T}\right\} p_0^2 \sqrt{2\pi m_{eff} T} \tag{11.18}$$

This is determined by momentum values near the roton peak, p_0, as in Figure 11.4. At low temperatures, the roton excitations are diminished in number by the Boltzmann factor $\exp\{-\Delta/T\}$. If we substitute Equation 11.18 into Equation 11.15, then we have

$$F_r = \frac{T^{(3/2)}}{2\pi^2\hbar^3} \exp\left\{-\frac{\Delta}{T}\right\} p_0^2 \sqrt{2\pi m_{eff} T} \tag{11.19}$$

The entropy S of the rotons is

$$S = -\left(\frac{\partial F_r}{\partial T}\right)_V \tag{11.20}$$

or

$$S = N_r \left(\frac{3}{2} + \frac{\Delta}{T}\right) \tag{11.21}$$

The internal energy of the rotons is as follows:

$$E = F + TS = N_r \left(\Delta + \frac{T}{2}\right) \tag{11.22}$$

Finally, the heat capacity that results is

$$C_V = \left(\frac{\partial E}{\partial T}\right)_V = N_r \left(\frac{3}{4} + \frac{\Delta}{T} + \frac{\Delta^2}{T^2}\right) \tag{11.23}$$

and as $T \to 0$ then $N_r \to 0$, $C_V \to 0$. These results have been found to agree with experiment. As seen above, the roton parts of the free energy, internal energy, entropy, and heat capacities have a temperature dependence that is essentially exponential. This is because the roton number diminishes exponentially with temperature (see Equation 11.18). At exceedingly low temperatures, the contribution of the roton part is less compared to the phonon part, and at exceeding high temperatures, a contrary behavior is observed.

11.4 SUPERFLUIDITY OF LIQUID HELIUM

Superfluidity is the phenomenon of frictionless flow of a liquid through extremely small holes (or capillaries) with great ease. P.L. Kapitza found this phenomenon in liquid helium in 1938. For a normal fluid such as hydrogen, the experiment for the measurement of friction (viscosity) is connected with the measurement of the decay time for a disk to stop rotating in the fluid. Rotating a disk in liquid hydrogen does this. This could not be done for liquid helium well below the lambda point.

Kapitza started to measure the velocity of flow of liquid helium through capillary tubes. He did not measure the decay time. No experiment allowed him to come to a general theoretical conclusion. P.L. Kapitza [48] showed that liquid helium was a superfluid and in general has no viscosity. Landau showed that the results were connected with the fact that the excitations above the ground state were not individual moving liquid helium atoms, or groups of atoms, but phonons. For $T = 0$, there are no such excitations and superfluidity results. In the superfluid state, with all the atoms in the same quantum state, there are no collisions between them, hence, no friction or viscosity. This changes if the fluid is set into motion.

Consider why motion of the helium leads to friction only above a critical speed. If liquid helium flows along a capillary with a constant velocity $\vec{v}$ relative to the laboratory reference frame, an excitation can be created in the fluid through its interaction with the walls of the container. The interaction can then dissipate its kinetic energy. The liquid flow will gradually slow down as a result of viscous effects in the liquid. In the reference frame of the fluid, the helium particles are initially at rest and the walls of the capillary move with velocity $-\vec{v}$. The motion of the capillary wall relative to the fluid generates an excitation whose wave vector then is opposite to $\vec{v}$. The liquid at rest will begin to move as a result of viscous forces in it. This motion is provoked by elementary excitations in the liquid helium. The energy of a single elementary excitation with momentum $\vec{p}$ appearing in the liquid is $\varepsilon(p)$. We consider now the coordinate system in which the capillary is at rest; then from the laws of classical mechanics, the total energy of the laboratory frame may be given as follows:

$$E = \varepsilon + \vec{p} \cdot \vec{v} + \frac{M\vec{v}^2}{2} \tag{11.24}$$

Here, the summand

$$\frac{M\vec{v}^2}{2} \tag{11.25}$$

represents the energy the initial fluid possesses if it moves with velocity $\vec{v}$ without creating excitations in the fluid that change its velocity. The appearance of an excitation changes the energy by the following amount:

$$\varepsilon + \vec{p} \cdot \vec{v} \tag{11.26}$$

For such an excitation to occur, this change must be negative, since the energy of the moving liquid must decrease:

$$\varepsilon + \vec{p} \cdot \vec{v} < 0 \tag{11.27}$$

The quantity $\varepsilon + \vec{p} \cdot \vec{v}$ achieves its minimum value when $\vec{p}$ and $\vec{v}$ are antiparallel (the wave vector of the quantum excitation is opposite to the velocity of the moving fluid) and thus

$$\varepsilon - pv < 0 \tag{11.28}$$

It follows from here that quantum excitations (phonons) are produced if

$$v > \frac{\varepsilon}{p} \tag{11.29}$$

That is, this is the velocity above which there is generation of quantum friction (viscosity). It should be noted that in principle, using the dispersion for the phonons

$$v > \frac{U_l p}{p} = U_l \neq 0 \tag{11.30}$$

Here, U_l is the longitudinal velocity of sound waves. Thus, it follows that

$$v > \left(\frac{\varepsilon}{p} \right)_{\min} \tag{11.31}$$

is the condition for excitations to appear in the fluid as it moves along the capillary. Geometrically, the ratio ε/p may be considered as a tangent line to a point of the curve $\varepsilon(p)$. The minimum value of ε/p is achieved when this tangent line passes through the origin in the $p\varepsilon$-plane. When this minimum is nonzero, excitations cannot appear in the liquid for velocities of flow below a critical value and the liquid exhibits the phenomenon of superfluidity. For superfluidity to occur, $\varepsilon(p)$ should not touch the abscissa at the origin. Confirming the validity of the above arguments, we consider liquid helium slightly above absolute zero temperature when the liquid contains excitations, though not in the ground state. Motion of the liquid relative to the walls of the capillary does not provoke new elementary excitations when the above condition (11.31) is unsatisfied.

Now we try to estimate the inertial mass density in the superfluid, isolating separately the contributions due to phonons and rotons. To do so requires a calculation of the momentum due to each, considered as bosonic gases. We have to consider the effect of excitations initially present in the liquid. Consider the bosonic gas (excitations) moving with a translational velocity $\vec{v}$ relative to the liquid to have the distribution function

$$\bar{n}(\varepsilon - \vec{p} \cdot \vec{v}) = \frac{1}{\exp\{\beta(\varepsilon - \vec{p} \cdot \vec{v})\} - 1}, \quad \beta = \frac{1}{T} \tag{11.32}$$

Consider the oz-axis to be the direction of motion of the background fluid. Then, each quasi-particle has a component of momentum $p \cos \theta$ along that axis. Thus, the energy of each quasi-particle is

$$\varepsilon - \vec{p} \cdot \vec{v} = \varepsilon - pv\cos\theta \tag{11.33}$$

The total momentum $\bar{P}$ of the gas of quasi-particles is the z-component of the momentum, that is, p_z:

$$\bar{P} = P_z = \sum_{\vec{p}} \bar{n}(\varepsilon - \vec{p} \cdot \vec{v})p_z = \frac{2\pi V}{(2\pi\hbar)^3} \int_0^\infty \int_0^\pi \bar{n}(\varepsilon - \vec{p} \cdot \vec{v})p^3 \, dp\cos\theta\sin\theta\, d\theta \tag{11.34}$$

Here, V is the volume. As v is small, then we expand $\bar{n}(\varepsilon - \vec{p} \cdot \vec{v})$ in a series:

$$\bar{n}(\varepsilon - \vec{p} \cdot \vec{v}) \cong \frac{1}{\exp\{\beta\varepsilon\} - 1} + \frac{\exp\{\beta\varepsilon\}}{(\exp\{\beta\varepsilon\} - 1)^2}\beta vp\cos\theta + O(v^2) \tag{11.35}$$

When $v \to 0$ then also $O(v^2) \to 0$. Thus

$$\bar{n}(p) \cong \frac{1}{\exp\{\beta\varepsilon\} - 1} + \frac{\exp\{\beta\varepsilon\}}{(\exp\{\beta\varepsilon\} - 1)^2}\beta vp\cos\theta \tag{11.36}$$

Substituting Equation 11.36 into Equation 11.34, we have

$$\bar{P} = \frac{2\pi V}{(2\pi\hbar)^3} \int_0^\infty \int_0^\pi \left[\frac{1}{\exp\{\beta\varepsilon\} - 1} + \frac{\beta vp\cos\theta\exp\{\beta\varepsilon\}}{(\exp\{\beta\varepsilon\} - 1)^2}\right]p^3 dp\cos\theta\sin\theta\, d\theta \tag{11.37}$$

If we do integration with respect to θ, then that of the first summand in the integrand is equal to zero and that of the second summand yields the total momentum in a gas of quasi-particles:

$$\bar{P} = \frac{4\pi V\beta v}{3(2\pi\hbar)^3} \int_0^\infty \frac{\exp\{\beta\varepsilon\}}{(\exp\{\beta\varepsilon\}-1)^2} p^4 \, dp \tag{11.38}$$

For the case of phonons, we substitute their acoustic dispersion relation, $\varepsilon(p) = Up$ in Equation 11.38

$$P_{ph} = \frac{4\pi V\beta v}{3(2\pi\hbar)^3 U} \int_0^\infty \frac{\exp\{\beta Up\} p^4 \, dp}{(\exp\{\beta Up\}-1)^2} = \frac{4}{3}\left(\frac{E_{ph}}{U^2}\right) \tag{11.39}$$

and we have

$$\bar{P} = P_{ph} = \frac{4}{3}\frac{E_{ph}}{U^2} \tag{11.40}$$

If the inertial mass of the phonon gas is $m_{ph} = P/v$, then the phonon mass density ρ_n^{ph} is

$$\rho_n^{ph} = \frac{P_{ph}}{vV} = \frac{2\pi^2}{45\hbar^3 U^5} T^4 \tag{11.41}$$

which is just a derived concept. Expression 11.9 without the zero-point energy was used to evaluate E_{ph} to obtain this result. The result is quite small at low temperatures, however, only according to a power law in the temperature.

For the roton contribution, considering that rotons can be described by a Boltzmann distribution, $\bar{n} \cong \exp\{-\beta\varepsilon(p)\}$, Equation 11.38 is then replaced by

$$P_2 \approx \frac{4\pi V\beta v}{3(2\pi\hbar)^3} \int_0^\infty \exp\{-\beta\varepsilon(p)\} p^4 \, dp \tag{11.42}$$

But for rotons, the energy dispersion is

$$\varepsilon(p) = \Delta + \frac{(p-p_0)^2}{2m_{eff}} \tag{11.43}$$

then

$$P_2 = \frac{4\pi V}{3(2\pi\hbar)^3} v p_0^4 \exp\left\{-\frac{\Delta}{T}\right\}\sqrt{\frac{2\pi m_{eff}}{T}} = \frac{vp_0^2}{3T} N_r \tag{11.44}$$

Thus, the roton mass density is

$$\rho_2 = \frac{P_2}{vV} = \frac{p_0^2}{3T}\frac{N_r}{V} \tag{11.45}$$

We see from Equations 11.45 and 11.41 that at exceedingly low temperatures, the phonon contribution to the density is large compared to the roton contribution. They are comparable at about 0.6 K and at higher temperatures, the roton contribution predominates. With increased temperature, an increased amount of the mass of the liquid becomes normal and at $\rho_n/\rho = 1$, superfluidity disappears completely. The point for which this occurs is the lambda point of the liquid. This is a phase transition of the second kind because there is no discontinuous change in the free energy or other macroscopic variables at the transition temperature.

12 Nonideal Classical Gases

12.1 PAIR INTERACTIONS APPROXIMATION

In previous chapters, we studied systems such as ideal gases where the statistical sum (partition function) Z may be evaluated exactly. For real gases, however, the evaluation of Z has some mathematical difficulties that we resolve via approximations. A real gas has weak interactions among the particles. The typical potential energy U between a pair of particles might have an attractive potential well at an intermediate distance, and a repulsive potential at very short distances. At large separations, the interparticle potential tends to go to zero, which takes one back to the situation where the ideal gas theory is applicable, at very low particle density.

Consider a monatomic gas of N identical particles of mass m in a containing vessel of volume V at a temperature T. The temperature is assumed to be sufficiently high so that the density $n \equiv N/V$ is low in a quantum sense. The typical separation of the particles is assumed to be much greater than their typical de Broglie wavelength, and the gas can be treated by classical mechanics.

The energy E of the gas may be written as a sum of kinetic and potential energy contributions (each p_K and q_K is a vector of three components):

$$E(p,q) = \sum_{K=1}^{N} \frac{p_K^2}{2m} + U(q_1,\ldots,q_N) \qquad (12.1)$$

U is the energy of interaction of the gas particles with each other. In principle, it would not have to be a sum of pair interactions. However, that is the simplest approximation and we apply it here.

Let us denote the potential energy of interaction between particles i and j by

$$U_{ij} \equiv U(\vec{r}_i - \vec{r}_j) = U(r_{ij}) \qquad (12.2)$$

We assume that the potential energy (12.2) is dependent only on the relative separations $\left(r_{ij} = \left| \vec{r}_i - \vec{r}_j \right| \right)$ of the i and j particles (a spherically symmetric potential). Consider also that the total U is simply the sum of all interactions between pairs of particles:

$$U = U_{12} + U_{13} + \cdots + U_{23} + U_{24} + \cdots + U_{N-1,N} \qquad (12.3)$$

or

$$U = \sum_{\substack{i,j=1 \\ i<j}}^{N} U_{ij} = \frac{1}{2} \sum_{\substack{i,j=1 \\ i\neq j}}^{N} U_{ij} \tag{12.4}$$

As mentioned above, the interactions are strongly repulsive when particles are very close together and weakly attractive for larger separations.

The free energy F of the system is evaluated with the usual procedure:

$$F = -T \ln Z_N \tag{12.5}$$

where

$$Z_N = \frac{1}{N!(2\pi\hbar)^{3N}} \int \exp\left\{-\frac{U(q_1,\ldots,q_N)}{T}\right\} \prod_{K=1}^{N} \exp\left\{-\frac{p_K^2}{2mT}\right\} dx_K \, dy_K \, dz_K \, dp_{x_K} \, dp_{y_K} \, dp_{z_K} \tag{12.6}$$

or

$$Z_N = \frac{(2\pi mT)^{3N/2}}{N!(2\pi\hbar)^{3N}} \int \exp\left\{-\frac{U(q_1,\ldots,q_N)}{T}\right\} dV_1 \cdots dV_N \tag{12.7}$$

The factor $N!$ takes into consideration the indistinguishability of the classical particles.

If $U = 0$ then the integration only over the momenta is important,

$$Z_N = \frac{(2\pi mT)^{3N/2}}{N!(2\pi\hbar)^{3N}} V^N = Z_{id} \tag{12.8}$$

where Z_{id} is the sum of states for an ideal gas. When $U \neq 0$ the partition function gets a modification. That modification is

$$\frac{1}{V^N} \int \exp\left\{-\frac{U(q_1,\ldots,q_N)}{T}\right\} dV_1 \cdots dV_N = Z_q \tag{12.9}$$

which is the configurational part of the statistical sum Z. Then the net partition function is the product

$$Z = Z_{id} Z_q \tag{12.10}$$

Hence, from Equation 12.5 we have two contributions to the free energy,

$$F = F_{id} + F_q \tag{12.11}$$

where the kinetic and configurational parts are

$$F_{id} = -T \ln Z_{id}, \quad F_q = -T \ln Z_q \tag{12.12}$$

The quantity F_q may be written in explicit form:

$$F_q = -T \ln \left[\frac{1}{V^N} \int \exp\left\{ -\frac{U(q_1,\ldots,q_N)}{T} \right\} dV_1 \cdots dV_N \right] \tag{12.13}$$

and F_{id} is the free energy of an ideal gas. Due to the nonindependence of the N integrations, the evaluation of Equation 12.13 is mathematically difficult. Thus, the evaluation of Z_q is the essential difficulty for the analysis of nonideal gases. Z_q may be easily evaluated for the following limiting cases:

1. As $U \to 0$ (ideal gas, which is trivial)
2. As $\beta \to 0$ (for high temperatures) then exp $\{-\beta U\} \to 1$ and $Z_q \to V^N$ where $\beta = (1/T)$.

Let us consider the integral needed in to get Z_q or F_q, in Equation 12.9:

$$\int \exp\left\{ -\frac{U(q_1,\ldots,q_N)}{T} \right\} dV_1 \cdots dV_N = \int \exp\left\{ -\beta \sum_{pair} U(r_{ij}) \right\} dV_1 \cdots dV_N \tag{12.14}$$

The integral in Equation 12.14 has all information about any phase transition or other nonideal behavior that might take place in the system. That is to say, the nature of a phase transition may be investigated by examining the integral in Equation 12.14. This is done on purely mathematical grounds without any physical knowledge. Let us make the denotation

$$\eta_{ij} = \exp\left\{ -\beta U_{ij} \right\} - 1 \tag{12.15}$$

It is evident that this quantity is different from zero when the ith and jth particles are at any finite distance from each other. In the ideal gas limit these parameters go to zero. If we let

$$\exp\left\{ -\beta U_{ij} \right\} = 1 + \eta_{ij} \tag{12.16}$$

then Equation 12.14 becomes

$$\int \exp\left\{-\beta\sum_{\text{pair}} U(r_{ij})\right\} dV_1 \cdots dV_N = \int dV_1 \int (1 + \eta_{12}) dV_2 \cdots$$

$$\int (1 + \eta_{1N}) \cdots (1 + \eta_{N-1,N}) dV_N \qquad (12.17)$$

Let us evaluate the integral over dV_N:

$$\int (1 + \eta_{1N}) \cdots (1 + \eta_{N-1,N}) dV_N = \int\left\{1 + \sum_{i=1}^{N-1} \eta_{iN} + \sum_{i \neq j} \eta_{iN}\eta_{jN} + \cdots\right\} dV_N \qquad (12.18)$$

If we neglect three-particle and higher-order collision terms (keeping only linear terms in the η) then we have

$$\int (1 + \eta_{1N}) \cdots (1 + \eta_{N-1,N}) dV_N = \int\left\{1 + \sum_{i=1}^{N-1} \eta_{iN}\right\} dV_i \qquad (12.19)$$

The quantity η_{ij} is dependent only on the separation of the particles. The integral

$$\int \eta_{iN} dV_N \qquad (12.20)$$

can be considered as an infinitesimal quantity. The quantity η_{iN} decreases quickly as r_{iN} increases. The integral (12.20) also is independent of the coordinate of the ith particle.

Then we can make the following definition:

$$\omega = -\int \eta_{iN} dV_N = \int (1 - \exp\{-\beta U_{iN}\}) dV_N \qquad (12.21)$$

Then

$$\int (1 + \eta_{1N}) \cdots (1 + \eta_{N-1,N}) dV_N = V - (N-1)\omega \qquad (12.22)$$

Similarly for the integrations over the other particles' coordinates,

$$\int (1 + \eta_{1,N-1}) \cdots (1 + \eta_{N-2,N-1}) dV_{N-1} = V - (N-2)\omega, \ldots, \int (1 + \eta_{12}) dV_2$$

$$= V - \omega \qquad (12.23)$$

Thus, from Equation 12.9 the configurational integral is

$$Z_q = \left(1 - \frac{\omega}{V}\right)\left(1 - \frac{2\omega}{V}\right)\cdots\left(1 - \frac{(N-1)\omega}{V}\right) \tag{12.24}$$

and

$$\ln Z_q = \sum_{k=0}^{N-1} \ln\left(1 - \frac{k\omega}{V}\right) \tag{12.25}$$

If the gas is of low density then the quantities

$$\frac{\omega}{V}, \frac{2\omega}{V}, \ldots, \frac{N\omega}{V} \tag{12.26}$$

are small. Then the logarithms on the RHS of Equation 12.25 can be expanded in a series, keeping only the first terms:

$$\ln Z_q = -\frac{\omega}{V}\sum_{k=0}^{N-1} k = -\frac{\omega}{V}\frac{N(N-1)}{2} \approx -\frac{\omega N^2}{2V} \tag{12.27}$$

where we used the approximation

$$\ln(1 + x) \approx x, \quad x \ll 1 \tag{12.28}$$

Thus, the configurational free energy is approximated as

$$F_q = -\frac{N^2 T}{2V}\int\left(\exp\{-\beta U(r)\} - 1\right)dV \tag{12.29}$$

and the total free energy is

$$F = F_{id} - \frac{N^2 T}{2V}\int\left(\exp\{-\beta U(r)\} - 1\right)dV \tag{12.30}$$

The pressure p of the nonideal gas is found to be

$$p = -\left(\frac{\partial F}{\partial V}\right)_T = \frac{NT}{V} - \frac{N^2 T}{2V^2}\int\left(\exp\{-\beta U(r)\} - 1\right)dV$$

$$= \frac{NT}{V}\left[1 - \frac{N}{2V}\int\left(\exp\{-\beta U(r)\} - 1\right)dV\right] \tag{12.31}$$

or

$$p = \frac{NT}{V}\left[1 + \frac{N}{V}B(T) + \frac{N^2}{V^2}C(T) + \cdots\right] \tag{12.32}$$

This is known as the *virial expansion*, where $B(T)$ is called the *second virial coefficient* and $C(T)$ the *third virial coefficient*. For an ideal gas

$$B = C = \cdots = 0 \tag{12.33}$$

Equation 12.32 is for the case when the gas is not dense. In Equation 12.32 the coefficient B is

$$B(T) = \frac{1}{2}\int\left(1 - \exp\{-\beta U(r)\}\right)dV \tag{12.34}$$

For further evaluation it is necessary to know the dependence of $B(T)$ on temperature, and this depends on the pair potential $U(r)$. The function $U(r)$ usually has a form like that in Figure 12.1. In domain I, due to the fact that $U \to \infty$, there results $\exp\{-\beta U\} \to 0$. This part of the integration gives

$$B_1(T) = \frac{1}{2}\int_0^{2r_0} dV = \frac{1}{2}\frac{4\pi}{3}(2r_0)^3 = 4v_0 \equiv b \tag{12.35}$$

Here, v_0 is some effective volume of a particle. In domain II we have another contribution,

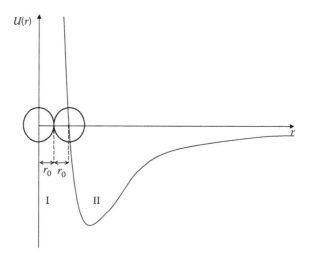

FIGURE 12.1 Dependence of the pair interaction energy $U(r)$ of gas particles on the particle's relative separation.

$$B_2(T) = \frac{1}{2} \int_{2r_0}^{\infty} \left(1 - \exp\{-\beta U(r)\}\right) dV \tag{12.36}$$

In order that there should be no condensation (the system remains gaseous), it is necessary that $\beta U < 1$. Thus, we can expand $\exp\{-\beta U\}$ in a series and we have

$$B_2(T) \approx \frac{1}{2T} \int U(r) dV \equiv -\frac{a}{T} \tag{12.37}$$

where $a > 0$ describes the interaction between the particles.
    Thus, the net virial coefficient is

$$B(T) = b - \frac{a}{T} \tag{12.38}$$

Hence, the equation of state under these approximations, ignoring the third virial coefficient, is

$$p = \frac{NT}{V}\left[1 + \frac{N}{V}\left(b - \frac{a}{T}\right)\right] \tag{12.39}$$

This is the equation of state in the approximation of pairwise interactions. The pressure is dependent on the number of collisions. For each collision, however, the distance of closest approach is $2r_0$, whereas point-like particles have a closest approach of zero. The factor b therefore represents an excluded volume not accessible by a particle. Thus, the particle velocities increase, the effective volume is reduced and the pressure consequently increases as a result of b. The other term, dependent on parameter a, has an opposite effect.

12.2   VAN DER WAALS EQUATION

If we consider Equation 12.39 with the total excluded volume $Nb \ll V$, then it can be rearranged as

$$\left(p + \frac{N^2}{V^2}a\right)(V - Nb) = NT \tag{12.40}$$

This is known as the Van der Waals equation. We may also get this equation via a different path, but still in the approximation of pair wise interactions. We consider again Equation 12.30:

$$F = F_{id} - \frac{N^2 T}{2V} \int \left(\exp\{-\beta U\} - 1\right) dV \tag{12.41}$$

which is approximated as

$$F = -NT \ln \frac{eV}{N} + Nf(T) + \frac{N^2 T}{V}\left(b - \frac{a}{T}\right) \tag{12.42}$$

Combine some terms,

$$-NT \ln \frac{eV}{N} + \frac{N^2 Tb}{V} = -NT\left[\ln \frac{eV}{N} - \frac{Nb}{V}\right] \cong -NT\left[\ln \frac{eV}{N} + \ln\left(1 - \frac{Nb}{V}\right)\right]$$

$$= NT \ln\left[\frac{eV}{N}\left(1 - \frac{Nb}{V}\right)\right] = -NT \ln \frac{e(V - Nb)}{N} \tag{12.43}$$

The resulting pressure is again found to be

$$p = \frac{NT}{V - Nb} - \frac{N^2}{V^2}a \tag{12.44}$$

This is the *Van der Waals equation*. In the form Equation 12.40, it resembles a modified ideal gas law. The volume factor is reduced by the excluded volume Nb; this tends to increase the pressure. The pressure is combined with a factor of squared number density times the interaction parameter a. One can see that the effect of the parameter a is to reduce the pressure compared to that for an ideal gas. These should be the primary modifications to the ideal gas law, due to a typical isotropic intermolecular potential. The more precise details will depend on the actual values of a and b, as determined by the potential $U(r)$.

12.3   COMPLETELY IONIZED GAS

The method for evaluation of thermodynamic averages for ideal gases seen in Chapter 4 is not applicable to a gas of charged particles. We study such a gas in this section. This is essentially the situation in a plasma, that is, a gas at very high temperatures, where electrons easily free themselves from atoms, leaving a gas of positive ions and negative electrons. These charges can have a significant interaction with each other. Let us try to evaluate the coefficient a from Equation 12.37,

$$a = -\frac{1}{2}\int_{2r_0}^{\infty} U(r)dV \tag{12.45}$$

For the case of a charged (singly ionized) gas

$$U = \frac{e^2}{\varepsilon_0 \varepsilon r} \tag{12.46}$$

which is the Coulomb potential energy of one pair of charges. Here, e is the electronic charge, ε_0 the permittivity of free space, ε the relative permittivity of the medium, and r the separation of two charged particles. For spherical symmetry, the volume element is

$$dV = 4\pi r^2 dr \qquad (12.47)$$

then from Equation 12.45 we have

$$a = -\left.\frac{e^2 \, 4\pi R^2}{2(2\varepsilon_0\varepsilon)}\right|_{R\to\infty} \to \infty \qquad (12.48)$$

This implies that it is impossible to take the limit of an infinite vessel under these approximations; we can only consider that $a = a(R)$. This is called the dimensional effect. In ordinary volumes of plasma there is no dimensional effect (the dependence of the parameters on the dimension of the vessel). There appears then the correlation of the collective effect [7]. Here, all the system interacts via Coulomb interactions. But there arises some screening of the Coulomb interaction. This is important, because the screening actually is sufficient to allow the integral in Equation 12.45 to be convergent. Let us examine the various approximations.

1. The Coulomb interaction is assumed to be weak. It is assumed that the mean potential energy is less than the kinetic energy. Thus

$$\left.\frac{e^2}{r}\right|_{r\approx n^{-\frac{1}{3}}} \approx e^2 n^{\frac{1}{3}} \ll T \qquad (12.49)$$

or

$$n \ll \left(\frac{T}{e^2}\right)^3 \qquad (12.50)$$

   where n is the concentration of the particles (ions).
2. We consider a neutral gas. Let e be the absolute value of the electronic charge, and let there be particles with charges $Z_a e$, where $Z_a = \pm 1, \pm 2, \ldots$ and n_{a0} is the mean number of ions of a selected charge state (the ath type of ion). This is a gas with singly ionized, doubly ionized, and so on, ions. Then from the condition of electroneutrality we have

$$\sum_a e Z_a n_{a0} = 0 \qquad (12.51)$$

We write the Coulomb interaction energy of a system of electrons. The potential energy of a particular charge $Z_a e$ is

$$E_p = \frac{1}{2} e Z_a \Phi_a \tag{12.52}$$

where Φ_a is the total electrostatic potential of the ath type of ion due to other charges. The factor of 1/2 eliminates double counting of pair-wise Coulomb interactions. The potential energy of all charges in a unit volume is

$$E_p = \frac{1}{2} \sum_a e Z_a \Phi_a n_{a0} \tag{12.53}$$

Within a volume V the electrostatic energy of the system (Coulomb energy) is as follows:

$$E_C = \frac{V}{2} \sum_a e Z_a \Phi_a n_{a0} \tag{12.54}$$

How do we calculate the potentials Φ_a? This can be done as follows: It should be noted that each ion creates around itself approximately a spherically symmetrical charged ion cloud. That is, selecting a particular ion in the gas (at the origin of coordinates) and considering the density of distribution of other ions relative to it, then this density will depend on a distance r from the center. Considering that n_a is the mean distribution of the ath type ion then the electrostatic potential energy of the ath type ion is given by $eZ_a \Phi(r)$, where $\Phi(\vec{r})$ is the total potential at the point $\vec{r}$ in space. The condition for finding $\Phi(\vec{r})$ is Poisson's equation, in SI units:

$$\Delta \Phi = -\frac{\rho}{\varepsilon_0 \varepsilon} \tag{12.55}$$

where ρ is the charge density.

We consider the fact that each ion creates in its neighborhood a nonequilibrium charge cloud. For example, a positive ion will draw negative charges into the surrounding space. But these negative charges themselves repel each other. So there is a limit to the amount of negative charge that can be attracted. Within a sphere centered on the positive ion, the total charge enclosed will tend towards zero as the size of the sphere increases. The ion becomes electrically invisible at large distance, because it is effectively shielded or screened from the outside world by its charge cloud. To understand this shielding, is necessary to imagine that ρ is a smooth distribution, so the Poisson equation can be solved.

Let n_a be the density of ions (of the ath type) in the charged ionic cloud around a chosen central ion. The potential energy of each ion of the ath type in the electric field around the given ion is $U_a(\vec{r}) = Z_a e \Phi(\vec{r})$ where the total electrostatic potential is $\Phi(\vec{r})$.

Recalling Boltzmann's formula (Equation 4.41) in an equilibrium situation,

$$n(\vec{r}) = n_0 \exp\left\{-\frac{U(\vec{r})}{T}\right\}$$
(12.56)

then the concentration follows:

$$n_a = n_{a0} \exp\left\{-\frac{Z_a e \Phi(r)}{T}\right\}$$
(12.57)

If we consider Equations 12.55 and 12.57 then we have

$$\Delta\Phi = -\frac{1}{\varepsilon_0 \varepsilon} \sum_a e Z_a n_a = -\frac{1}{\varepsilon_0 \varepsilon} \sum_a e Z_a n_{a0} \exp\left\{-\frac{Z_a e \Phi}{T}\right\}$$
(12.58)

As the interactions of the ions are weaker than the temperature, $Z_a e \Phi < T$ (this is a high-temperature theory) Equation 12.58 becomes

$$\Delta\Phi = -\frac{1}{\varepsilon_0 \varepsilon} \sum_a e Z_a n_a = -\frac{1}{\varepsilon_0 \varepsilon} \sum_a e Z_a n_{a0}\left(1 - \frac{Z_a e \Phi}{T}\right) = \frac{1}{\varepsilon_0 \varepsilon} \sum_a \frac{Z_a^2 e^2 n_{a0} \Phi}{T}$$
(12.59)

or

$$\Delta\Phi - \chi^2\Phi = 0$$
(12.60)

where

$$\chi^2 = \frac{e^2}{\varepsilon_0 \varepsilon T} \sum_a Z_a^2 n_{a0}$$
(12.61)

and χ has the dimensions of inverse length. The spherically symmetric solution of Equation 12.61 is

$$\Phi(r) = A\frac{\exp\{-\chi r\}}{r}$$
(12.62)

where A is a constant. For very small r (in the neighborhood of the central ion) we should have a pure unshielded Coulomb potential as a boundary condition,

$$\Phi \cong \frac{e Z_b}{r}$$
(12.63)

that is, the potential of the central ion with charge $Z_b e$. This requires $A = Z_b e$ in Equation 12.62, and as a consequence,

$$\Phi(r) = Z_b e \frac{\exp\{-\chi r\}}{r} \tag{12.64}$$

At distances large compared to $(1/\chi)$ the potential becomes very small—the central ion gets completely screened. The length $(1/\chi)$ defines the radius of the ionic cloud created around the central ion. It is called the screening radius or *Debye–Hückel radius*. The evaluation above implicitly assumes that this radius $(1/\chi) = r_D$ is much greater than the mean distance between ions. That is so the potential in Equation 12.64 is defined as an averaged potential, which ignores the fluctuations due to instantaneous motions of the charges. Or, the cloud of charges is assumed to be continuous. This condition coincides with condition Equation 12.50.

Now the screened potential might be used to evaluate the interaction parameter between screened singly charged ions,

$$a = -\frac{1}{2} \int\limits_{2r_0}^{\infty} U(r) e^{-\chi r}\, dV = -\frac{1}{2} \int\limits_{2r_0}^{\infty} \frac{e^2}{\varepsilon_0 \varepsilon r} e^{-\chi r} 4\pi r^2\, dr \tag{12.65}$$

The integral now is convergent, due to the exponential factor which causes the integrand to vanish at large r. The net result is found as follows:

$$a = -\frac{2\pi e^2}{\varepsilon_0 \varepsilon} \int\limits_{2r_0}^{\infty} r e^{-\chi r}\, dr = -\frac{2\pi e^2}{\varepsilon_0 \varepsilon} \left[\frac{r e^{-\chi r}}{-\chi} \Big|_{2r_0}^{\infty} + \frac{1}{\chi} \int\limits_{2r_0}^{\infty} e^{-\chi r}\, dr \right] = -\frac{2\pi e^2}{\varepsilon_0 \varepsilon} \left[\frac{2r_0}{\chi} + \frac{1}{\chi^2} \right] \tag{12.66}$$

However, this does not fully take into account the statistical distribution of charges of different types.

Consider another approach. In the region where χr is small, the potential due to a central ion is approximated by a power series:

$$\Phi = \frac{e Z_b}{r} - e Z_b \chi + \cdots \tag{12.67}$$

where the first term is the Coulomb field of the central ion and the second term is the potential created by the surrounding ionic cloud. Other higher order terms are ignored here. The second term is the potential that acts at the point where the given ion is located, and, that is, the value that should be substituted into Equation 12.50, such that

$$\Phi_a = -e Z_a \chi \tag{12.68}$$

Thus, for the Coulombic part of the energy Equation 12.54 we have

$$E_c = -\frac{V}{2}\chi e^2 \sum_a n_{a0}Z_a^2 = -\frac{V}{2}e^2 \sum_a n_{a0}Z_a^2 \left(\frac{e^2}{\varepsilon_0 \varepsilon T}\sum_a n_{a0}Z_a^2\right)^{1/2} \tag{12.69}$$

or

$$E_c = -\frac{V}{2}\frac{e^3}{(\varepsilon_0 \varepsilon T)^{1/2}}\left(\sum_a n_{a0}Z_a^2\right)^{3/2} \tag{12.70}$$

If we use the fact that the total number of type a ions in the gas is $N_a = n_{a0}V$, then

$$E_c = -\frac{e^3}{2(\varepsilon_0 \varepsilon T V)^{1/2}}\left(\sum_a N_a Z_a^2\right)^{3/2} \tag{12.71}$$

If we consider the relation between the energy and the free energy,

$$\frac{E}{T^2} = -\frac{\partial}{\partial T}\frac{F}{T} \tag{12.72}$$

then the free energy becomes

$$F = F_{id} - \frac{2e^3}{3(\varepsilon_0 \varepsilon T V)^{1/2}}\left(\sum_a N_a Z_a^2\right)^{3/2} \tag{12.73}$$

The constant of integration is equal to zero as $F = F_{id}$ for $T \to \infty$. The pressure is found from Equation 12.73 according to,

$$p = -\left(\frac{\partial F}{\partial V}\right)_T = \frac{T}{V}\sum_a N_a - \frac{e^3}{3(\varepsilon_0 \varepsilon T V^3)^{1/2}}\left(\sum_a N_a Z_a^2\right)^{3/2} \tag{12.74}$$

This consists of an ideal gas contribution and the Coulomb potential correction. The attraction between ions and electrons reduces the total pressure.

Next consider this charged gas in an externally applied electric field, whose potential is Φ. If the electronic gas is considered to be a Fermi gas, then $n_a = n_a(\mu_a)$, where μ_a is the chemical potential of the ath type particle and n_a is its concentration. The distribution of the particles may be defined either by Boltzmann or Fermi distributions (because it is a high-temperature limit). As a result of an external electric field the change in the distribution density is accompanied by the replacement

$$\mu_a \to \mu_{a0} - Z_a e\Phi \tag{12.75}$$

where μ_{a0} is the chemical potential at the point where $\Phi = 0$. If we expand n_a in powers of Φ then we have

$$n_a = n_{a0} - Z_a e\Phi \frac{\partial n_{a0}}{\partial \mu_a} \tag{12.76}$$

and substituting in Equation 12.57, considering that

$$n_a \approx \exp\left\{\frac{\mu_a}{T}\right\} \tag{12.77}$$

Then following the logic as in Equations 12.58 and 12.61, we arrive at

$$\chi^2 = \frac{e^2}{\varepsilon_0 \varepsilon} \sum_a Z_a^2 \frac{\partial n_{a0}}{\partial \mu_a} \tag{12.78}$$

The interaction with the external field is

$$U_c = -\frac{e^3}{2V^{1/2}} \left(\sum_a N_a Z_a^2 \right) \left(\sum_a Z_a^2 \frac{\partial N_a}{\partial \mu_a} \right)^{1/2} \tag{12.79}$$

This analysis gives an introduction to statistical effects in a gas of ions. The primary physical effect is the screening of the Coulomb interactions. Such effects would be important in the physical description of plasmas, such as in stars and in laboratory experiments. However, motion of charged particles also generates magnetic fields, which are of extreme importance in plasma physics. Thus, the reader is encouraged to consult more specialized books for a complete treatment of models for plasmas.

13 Functional Integration in Statistical Physics

13.1 FEYNMAN PATH INTEGRALS

Quantum mechanical problems may be examined using the wave, matrix, and variational methods. The functional integration method is principally based on the variational method. This method is convenient when we consider the transition from classical to quantum mechanics. The relation between classical and quantum mechanics is not so direct for the matrix method.

Here, we make an analysis of quantum systems, starting from considering their dynamics according to some Hamiltonian $\hat{H}$ in terms of propagation from some initial state to a final state. This involves the construction of a "transition amplitude" between those states, for the quantum problem. In the sense of path integrals, however, it is derived based on minimization of the action of a classical system. Thus, the initial discussion is about classical notions of action. We will take those notions and apply them subsequently to the quantum transition amplitude. Finally, it will be seen that the quantum transition amplitude is connected to the canonical partition function in statistical physics. The connection is intriguing. It will be shown that the partition function $Z(\beta)$ is equivalent to a trace of the transition amplitude operator $\hat{K}(t)$, but only when usual time t is mapped over into inverse temperature according to $t \rightarrow -i\hbar\beta$. We will see that this comes about because a transition amplitude involves time-evolution via the operator $\exp\{-i\hat{H}t/\hbar\}$, whereas the partition function depends on a kind of evolution via the operator $\exp\{-\beta\hat{H}\}$. Both the transition amplitude and the partition function can be expressed with path integrals, a mathematical device that includes the effects of quantum fluctuations.

It is obvious that the basic notion in classical mechanics is that of the path $q(t)$ of a particle (where the path $q(t)$ depends on time t). In quantum mechanics, the state of a particle is described by a state vector $|\Psi(t)\rangle$ that depends on time, and position $q(t)$ is the expectation value of some operator according to $q(t) = \langle\Psi(t)|\hat{q}|\Psi(t)\rangle$. The notion of a well-defined path, with coordinates and momenta defined to arbitrary precision, is absent. However, a somewhat fuzzy range of motion of a particle, similar to a path, can be allowed within the limits of the Heisenberg uncertainty principle. It is well known that for the transition from quantum mechanics to classical mechanics, it is necessary to take the limit as Planck's constant $\hbar \rightarrow 0$ (correspondence principle). This limit is equivalent to letting the "fuzzy path" of a quantum system become precisely defined.

The Feynman space–time approach to quantum mechanics carries in itself all classical notions and in particular, employs the notion of a path. The methodical advantage of the related variational method enables us to examine more profoundly the principles of classical mechanics, and in particular, the principle of least action. The Feynman path-integral approach, however, does not use a single precise path for a particle. Instead, Feynman considered the idea that a particle travels from some initial state to a final state by moving on all possible connecting paths, and these are able to interfere with each other, due to the wave nature of quantum motion. We will see that such a sum over paths relates to statistical equilibrium properties because those can be calculated via the trace property for an operator, which is a form of sum analogous to a path integral.

13.2 LEAST ACTION PRINCIPLE

To connect classical physics to quantum physics, we start with a review of the concept of *action* and its variation. The dynamics of a classical system with a finite number of degrees of freedom s may be characterized by a set of functions

$$q_1(t), q_2(t), \ldots, q_s(t) = \{q(t)\} \tag{13.1}$$

These are dependent on time t such that the values together with their derivatives at any time completely define its state. The quantities $q(t)$ are called the *generalized coordinates*. The space formed from the $q(t)$ values is called the *configuration space*. The number of degrees of freedom defines the dimension of the space. The configuration space may be infinite if the system contains an infinite number of degrees of freedom.

During a certain time interval the state of the system changes and the point $q(t)$ of the configuration space describes some curve which is the trajectory (or path) of the system. The differential equation defining the path of the system in the configuration space is called the *equation of motion* of the system.

Consider a particle that moves from an initial point to a final point within a time interval t. Usually, its dynamics could be described by Newton's law of motion:

$$\vec{F} = m\vec{a} \tag{13.2}$$

(where m is the mass of the given particle, $\vec{a}$ its acceleration, and $\vec{F}$ the force acting on it). The variational method, however, takes a different viewpoint. If the kinetic energy KE less the potential energy PE (at each moment of its path) is integrated over time for the entire path, then the result of that integration is an extremum for the actual motion. Paths that are near the actual path will have different values that all either increase or decrease away from the extremum path. Usually, the extremum is a minimum. Then Newtonian classical mechanics is replaced by this *principle of least action*. The result of the average kinetic energy less the potential energy is minimized for the actual path of a particle moving from one point to another.

The kinetic energy KE less the potential energy PE is a quantity called the *Lagrangian L* of the system. When L is integrated over time, the result is a functional S, called the *action* of the system, for that particular path:

$$\int_{t_a}^{t_b} (KE - PE)\, dt \equiv \int_{t_a}^{t_b} L(q, \dot{q}, t)\, dt = S[q] = S \tag{13.3}$$

Here, $q(t_a) \equiv q_a$ and $q(t_b) \equiv q_b$ are the starting and ending coordinates of the path. These points are fixed. Here, $\dot{q}(t)$ is the generalized velocity, where the dot denotes the time derivative d/dt.

The path $q(t)$ that is actually chosen by the system is called the *classical path*. Nearby paths which are not chosen by the system are denoted by $\bar{q}(t)$ (see Figure 13.1). The classical path has the property of extremizing the action S in comparison with all neighboring paths

$$\bar{q}(t) = q(t) + \delta q(t) \tag{13.4}$$

having the same end points $q(t_a) \equiv q_a$ and $q(t_b) \equiv q_b$. As the kinetic and potential energies are functions of time t, for each possible path we obtain a different action. The action of the classical path could be a minimum or a maximum when compared to neighboring paths. The mathematical problem here is to find the curve for which the action is an extremum. This is called the *calculus of variations*.

In mathematics, there are many such variational problems. For example, a circle is usually defined as the locus of all points at a constant distance from a fixed point. Alternatively, a variational definition may be given: a circle is that curve of a given length that encloses the largest area. Any other curve encloses less area for a given perimeter than does the circle. If we are supposed to find that curve that encloses the largest area for a given perimeter then we have a problem that can be solved by the calculus of variations.

In classical mechanics, the problem now is to find the curve connecting an initial and final state, which has the least action. This is not an ideal method but illustrates some fundamental deep property of mechanics. It is especially interesting that it will be carried over into quantum mechanical problems via path integrals.

A wide class of dynamic systems may be characterized by a Lagrangian L that is a function of $q(t)$ and $\dot{q}(t)$. The Lagrangian is at most a quadratic function of $q(t)$ and $\dot{q}(t)$.

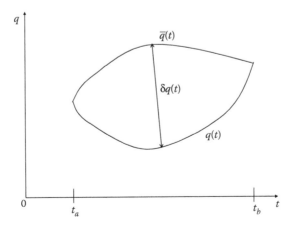

FIGURE 13.1   Path of a particle moving from a starting coordinate $q(t_a) \equiv q_a$ to an ending coordinate $q(t_b) \equiv q_b$ of a given path.

Let us express the property in Equation 13.4 formally. This may be done by introducing the *variation* of the action in Equation 13.3. The linear term in the Taylor expansion of $S[q(t)]$ in powers of $\delta q(t)$ is just the difference between the value for one path and a nearby one:

$$\delta S[q] = S[q + \delta q] - S[q] \tag{13.5}$$

The extremal principle for the classical path then says that the variation in the action is zero (the action is stationary) near the extremum:

$$\delta S[q] = 0 \tag{13.6}$$

This is for all variations of the path about the classical path such that the quantity $\delta q(t)$ vanishes at the end points:

$$\delta q(t_a) = \delta q(t_b) = 0 \tag{13.7}$$

As the action is the integral of the Lagrangian over the time, the extremal property of our problem may be phrased in terms of a differential equation. This may be done by calculating the variation of $S[q(t)]$:

$$\delta S[q] = S[q + \delta q] - S[q] = \int_{t_a}^{t_b} L(q + \delta q, \dot q + \delta \dot q, t)\,dt - \int_{t_a}^{t_b} L(q, \dot q, t)\,dt$$

$$= \int_{t_a}^{t_b} \left[\frac{\partial L}{\partial \dot q} \delta \dot q + \frac{\partial L}{\partial q} \delta q \right] dt = \int_{t_a}^{t_b} \left[\frac{d}{dt}\left(\frac{\partial L}{\partial \dot q} \delta q \right) - \frac{d}{dt}\frac{\partial L}{\partial \dot q} \delta q + \frac{\partial L}{\partial q} \delta q \right] dt$$

$$= \frac{\partial L}{\partial \dot q} \delta q \Big|_{t'} - \int_{t_a}^{t_b} \left[-\frac{d}{dt}\frac{\partial L}{\partial \dot q} + \frac{\partial L}{\partial q} \right] \delta q\, dt = 0 \tag{13.8}$$

The quantity $\delta q(t)$ achieves the value zero at the end points of the path $q(t)$ and between the end points it takes on any arbitrary value. As $\delta q(t)$ is arbitrary within the range of integration, the function in the brackets must vanish. Thus, the extremum yields the Euler–Lagrange equation for $q(t)$:

$$\frac{d}{dt}\frac{\partial L}{\partial \dot q} - \frac{\partial L}{\partial q} = 0 \tag{13.9}$$

To describe the dynamics of a system, it is necessary to set up its Lagrange function (Lagrangian). It may also be noted that the extremum of the action gives that curve along which condition (13.9) is satisfied. Equation 13.9 is the classical equation of motion.

We may formulate classical dynamics in an alternative way, which is reviewed here. Classical mechanics may be based on a Legendre-transformed function of the Lagrangian called the *Hamiltonian*:

$$H \equiv \frac{\partial L}{\partial \dot q} \dot q - L(q, \dot q, t) \tag{13.10}$$

This is the energy of the system at any time. The natural variables on which H depends are the generalized coordinate q and the generalized momentum p, according to the theory of Legendre transformations. The generalized momentum may be defined as follows:

$$p \equiv \frac{\partial}{\partial \dot{q}} L(q, \dot{q}, t) \tag{13.11}$$

In order that the Hamiltonian $H(p,q,t)$ be expressed in terms of its proper variables p and q, we solve Equation 13.11 for $\dot{q}$:

$$\dot{q} = v(p,q,t) \tag{13.12}$$

If we substitute this expression into Equation 13.10 then we have a Hamiltonian that is a function of p and q:

$$H(p,q,t) = pv(p,q,t) - L(q,v(p,q,t),t) \tag{13.13}$$

The action may then be expressed in terms of p and q through this Hamiltonian as follows:

$$S[p,q] = \int_{t_a}^{t_b} dt[\, p(t)\dot{q}(t) - H(p(t),q(t),t)] \tag{13.14}$$

This is called the *canonical form* of the action. The classical paths may be represented as $\bar{q}(t)$ and $\bar{p}(t)$. They extremize the action with respect to all neighboring paths that connect to desired fixed end points. However, the momentum is varied with no restriction. Then from

$$q(t) = \bar{q}(t) + \delta q(t), \quad p(t) = \bar{p}(t) + \delta p(t), \quad \delta q(t_a) = \delta q(t_b) = 0 \quad (13.15)$$

it follows that

$$\delta S[p,q] = \int_{t_a}^{t_b} dt \left[\delta p(t)\dot{q}(t) + p(t)\delta\dot{q}(t) - \frac{\partial H}{\partial p}\delta p(t) - \frac{\partial H}{\partial q}\delta q(t) \right] =$$

$$= \int_{t_a}^{t_b} dt \left\{ \left[\dot{q}(t) - \frac{\partial H}{\partial p} \right]\delta p(t) - \left[\dot{p}(t) + \frac{\partial H}{\partial q} \right]\delta q(t) \right\} + p(t)\delta q(t) \Big|_{t_a}^{t_b} \tag{13.16}$$

Considering that the variations should vanish for classical paths, we find that $\bar{q}(t)$ and $\bar{p}(t)$ must be the solutions of the *Hamilton equations* of motion:

$$\dot{q}(t) = \frac{\partial H}{\partial p}, \quad \dot{p}(t) = -\frac{\partial H}{\partial q} \tag{13.17}$$

These are the well-known canonical equations of classical mechanics, upon which we based the study of statistical mechanics.

Q13.1 Find the explicit form of the action for the motion of a free particle.

A13.1 The Lagrange function of a free particle of mass m contains only kinetic energy:

$$L = \frac{m\dot{q}^2}{2} \tag{13.18}$$

Let us write the equation of motion from

$$\frac{d}{dt}\frac{\partial L}{\partial \dot{q}} - \frac{\partial L}{\partial q} = 0 \tag{13.19}$$

Then we find

$$\frac{d}{dt}\frac{\partial L}{\partial \dot{q}} = \frac{d}{dt}(m\dot{q}) = m\ddot{q}, \quad \frac{\partial L}{\partial q} = 0 \tag{13.20}$$

This shows that the acceleration is zero

$$\ddot{q} = 0, \quad \dot{q} = v = \text{const}, \quad dq = vdt \tag{13.21}$$

and

$$(q_b - q_a) = v(t_b - t_a) \tag{13.22}$$

It follows that

$$v = \frac{q_b - q_a}{t_b - t_a} = \dot{q} \tag{13.23}$$

We may now evaluate the classical action of a free particle $S_{\text{classical}}$:

$$S_{\text{classical}} = \frac{m}{2}\int_{t_a}^{t_b} \dot{q}^2 dt = \frac{m\dot{q}^2}{2}\int_{t_a}^{t_b} dt = \frac{m}{2}\left(\frac{q_b - q_a}{t_b - t_a}\right)^2 (t_b - t_a) = \frac{m(q_b - q_a)^2}{2(t_b - t_a)} \tag{13.24}$$

Q13.2 Find the action of a harmonic oscillator.

A13.2 The Lagrangian of a harmonic oscillator contains both kinetic and potential energies:

$$L = \frac{m\dot{q}^2}{2} - \frac{m\omega^2 q^2}{2} \tag{13.25}$$

Let us write the equation of motion:

$$\frac{d}{dt}\frac{\partial L}{\partial \dot{q}} - \frac{\partial L}{\partial q} = 0 \tag{13.26}$$

Then, we obtain the required derivatives:

$$\frac{\partial L}{\partial \dot{q}} = m\dot{q}, \quad \frac{d}{dt}\frac{\partial L}{\partial \dot{q}} = m\ddot{q}, \quad \frac{\partial L}{\partial q} = -m\omega^2 q \tag{13.27}$$

The equation of motion is

$$m\ddot{q} + m\omega^2 q = 0 \tag{13.28}$$

We find the solution in the form:

$$q(t) = A\cos(\omega t + \alpha) \tag{13.29}$$

where α is the initial phase.

The action of the classical harmonic oscillator S_{CHO} can now be found:

$$S_{CHO} = \int_{t_a}^{t_b}\left[\frac{m\dot{q}^2}{2} - \frac{m\omega^2 q^2}{2}\right]dt = \int_{t_a}^{t_b}\left[\frac{m}{2}\left(\frac{d}{dt}(q\dot{q}) - \ddot{q}q\right) - \frac{m\omega^2 q^2}{2}\right]dt$$

$$= \frac{m}{2}q\dot{q}\Big|_{t_a}^{t_b} - \frac{m}{2}\int_{t_a}^{t_b} q[\ddot{q} + \omega^2 q]dt = \frac{m}{2}(q_b\dot{q}_b - q_a\dot{q}_a) \tag{13.30}$$

Let

$$A\cos\alpha = X, \quad A\sin\alpha = Y \tag{13.31}$$

Then

$$q(t) = X\cos\omega t - Y\sin\omega t \tag{13.32}$$

and

$$q_a = X\cos\omega t_a - Y\sin\omega t_a, \quad q_b = X\cos\omega t_b - Y\sin\omega t_b \tag{13.33}$$

and

$$X = \frac{q_a\sin\omega t_b - q_b\sin\omega t_a}{\sin\omega(t_b - t_a)}, \quad Y = \frac{q_a\cos\omega t_b - q_b\cos\omega t_a}{\sin\omega(t_b - t_a)} \tag{13.34}$$

Thus

$$S_{CHO} = \frac{m\omega}{2\sin\omega T}[(q_a^2 + q_b^2)\cos\omega T - 2q_a q_b], \quad T = t_b - t_a \qquad (13.35)$$

This is the action for the classical path of a harmonic oscillator.

13.3   REPRESENTATION OF TRANSITION AMPLITUDE THROUGH FUNCTIONAL INTEGRATION

Now we consider quantum dynamics of a system, which does not have a defined classical path. A quantum system can be thought to evolve as any wave system does, by propagation over a volume, with wave spreading and diffraction effects. This is analogous to propagation over multiple paths, à la Huygens' wavelet principle. We will be interested in how much each path contributes to the total amplitude to go from point a to point b. It is not just the path that corresponds to the extreme value of the action that contributes but rather all the paths that can contribute. We will see that quantum mechanics can be viewed in a new way, as shown by Feynman. All different paths contribute to the total amplitude for a transition from initial to final state with the same weights. However, the different paths contribute with different phases, allowing them to interfere. The phase for a given path is the action S for that path in units of the quantum of action, $\hbar$.

Consider the simplest quantum mechanical system with one degree of freedom that is described by the time-independent Hamiltonian operator $\hat{H}(\hat{p}, \hat{q})$:

$$\hat{H}(\hat{q}, \hat{p}) = \frac{\hat{p}^2}{2m} + U(\hat{q}) \qquad (13.36)$$

Here, $\hat{q}$ and $\hat{p}$ are, respectively, the operators of the coordinate and the momentum, m is the mass of the particle, and $U(\hat{q})$ is the operator of the potential energy. From the canonical quantization of the operators of the coordinate and momentum there follows the commutation relation:

$$[\hat{q}, \hat{p}] = i\hbar \qquad (13.37)$$

The eigenstates and eigenvalues of the operators $\hat{q}$ and $\hat{p}$ may be found from the equations:

$$\hat{q}|q\rangle = q|q\rangle, \quad \hat{p}|p\rangle = p|p\rangle \qquad (13.38)$$

where $|q\rangle$ and $|p\rangle$ are, respectively, the eigenvectors of the coordinate and momentum in the Dirac representation. The orthonormalization and closure relations of the eigenvectors $|q\rangle$ and $|p\rangle$ may be written in the following form:

$$\langle q||q'\rangle = \delta(q - q'), \quad \int_{-\infty}^{+\infty} dq|q\rangle\langle q| = \hat{I} \qquad (13.39)$$

$$\langle p \| p' \rangle = \delta(p - p'), \quad \int_{-\infty}^{+\infty} dp \, |p\rangle\langle p| = \hat{I} \tag{13.40}$$

where $\hat{I}$ is a unit or identity matrix.

It should be noted that

$$\langle p \| q \rangle = (2\pi\hbar)^{-1/2} \exp\left\{-\frac{i}{\hbar} pq\right\}, \quad \langle q \| p \rangle = \langle p \| q \rangle^* = (2\pi\hbar)^{-1/2} \exp\left\{\frac{i}{\hbar} pq\right\} \tag{13.41}$$

In order to define a transition amplitude, let us move from the Schrödinger operator $\hat{q}$ to the Heisenberg operator $\hat{q}(t)$, that is, change to the Heisenberg picture where the operators carry time dependence and the state vectors are time independent:

$$\hat{q}(t) = \exp\left\{\frac{i}{\hbar}\hat{H}t\right\}\hat{q}\exp\left\{-\frac{i}{\hbar}\hat{H}t\right\} \tag{13.42}$$

Let us denote by $|q_a,t_a\rangle$ the eigenvector of the Heisenberg operator $\hat{q}(t_a)$ at the time t_a:

$$\hat{q}(t_a)|q_a,t_a\rangle = q_a|q_a,t_a\rangle \tag{13.43}$$

From the rule of the transition from the Schrödinger to the Heisenberg picture, there follows

$$|q_a,t_a\rangle = \exp\left\{\frac{i}{\hbar}\hat{H}t_a\right\}|q_a\rangle \tag{13.44}$$

Let us suppose that at the time t_a, the particle is found in the localized state $|q_a, t_a\rangle$ that corresponds to the eigenvalue of the coordinate q_a. Then the *probability transition amplitude* (transition amplitude or propagator for the Schrödinger equation) to change to the localized state $\langle q_b,t_b|$ that corresponds to the eigenvalue of the coordinate q_b at time t_b is defined as

$$\langle q_b,t_b \| q_a,t_a \rangle \equiv K(q_b,t_b;q_a,t_a) = \langle q_b | \exp\left\{-\frac{i}{\hbar}\hat{H}(t_b - t_a)\right\}|q_a\rangle, \quad t_b > t_a \tag{13.45}$$

What is the physical sense of the transition amplitude of a particle in the Schrödinger representation? To answer this clearly, let us look again at the Schrödinger equation.

Let $|\Psi(t)\rangle$ be the Schrödinger state vector of the particle at time t. The evolution of the state of the particle is described by the time-dependent Schrödinger equation:

$$i\hbar\frac{\partial|\Psi(t)\rangle}{\partial t} = \hat{H}|\Psi(t)\rangle \tag{13.46}$$

The formal solution of Equation 13.46 starting from some initial state at time t_a has the form

$$\left|\Psi(t)\right\rangle = \exp\left\{-\frac{i}{\hbar}\hat{H}(t - t_a)\right\}\left|\Psi(t_a)\right\rangle = \hat{U}(t,t_a)\left|\Psi(t_a)\right\rangle, \quad t > t_a \quad (13.47)$$

Here, $\hat{U}(t,t_a)$ is an *evolution operator* with the initial condition

$$\hat{U}(t_a,t_a) = \hat{I} \quad (13.48)$$

where $\hat{I}$ is a unit or identity matrix. The state vector at a particular time $t = t_b$ can now be expressed by using a complete set of position eigenstates, inserting an identity operator $\hat{I} = \int dq_b \left|q_b\right\rangle\left\langle q_b\right|$:

$$\left|\Psi(t_b)\right\rangle = \int dq_b \left|q_b\right\rangle\left\langle q_b\right|\exp\left\{-\frac{i}{\hbar}\hat{H}(t_b - t_a)\right\}\left|\Psi(t_a)\right\rangle \quad (13.49)$$

If it was known that the system started at time t_a in a position eigenstate, $\left|\Psi(t_a)\right\rangle = \left|q_a\right\rangle$, then comparison with Equation 13.45 shows that the current state is expressed as a superposition

$$\left|\Psi(t_b)\right\rangle = \int dq_b \left|q_b\right\rangle K(q_b,t_b;q_a,t_a) \quad (13.50)$$

It is directly seen that the transition amplitude $K(q_b, t_b; q_a, t_a)$ is the amplitude for the system to be in position state $\left|q_b\right\rangle$ at time t_b, given that it started in position state $\left|q_a\right\rangle$ at time t_a. This is the physical interpretation of $K(q_b, t_b; q_a, t_a)$.

Next, consider a different viewpoint, by finding the evolution operator and breaking the time interval into small parts (path integral). If we substitute Equation 13.47 into the Schrödinger Equation 13.46, we obtain

$$i\hbar\frac{\partial\hat{U}(t,t_a)}{\partial t} = \hat{H}\hat{U}(t,t_a) \quad (13.51)$$

The first-order differential Equation 13.51 completely defines $\hat{U}(t,t_a)$, taking into consideration the initial condition (13.48). The formal solution of Equation 13.51, including the possibility of a time-dependent Hamiltonian, may be represented as a time-ordered product or T-product:

$$\hat{U}(t,t_a) = \hat{T}\exp\left\{-\frac{i}{\hbar}\int_{t_a}^{t}\hat{H}(t')dt'\right\} \quad (13.52)$$

The time-ordered product is defined as follows:

$$\hat{T}\prod_{i=1}^{N}\hat{F}(t_i) = \hat{F}(t_N)\hat{F}(t_{N-1})\ldots\hat{F}(t_2)\hat{F}(t_1), \quad t_N > t_{N-1} > \cdots t_2 > t_1 \quad (13.53)$$

Here, $\hat{F}(t)$ is a Heisenberg operator. The connection from Equation 13.52 to Equation 13.53 can be made by splitting the time interval of integration into N ordered smaller time intervals. In the Hamiltonian $\hat{H}(t)$, the time t plays the dual roles of an explicit argument and also the ordering parameter.

At time moment t_a the particle is localized at the point q_a, that is, the initial state is $|q,t_a\rangle \equiv |q_a\rangle$. It follows from Equation 13.50 that until the moment t_b, this state has evolved. The probability amplitude for this final state, given the initial state, is the overlap or scalar product

$$\langle q_b,t_b \,\|\, q_a,t_a \rangle = \langle q_b |\hat{U}(t_b,t_a)| q_a \rangle \quad (13.54)$$

This coincides exactly with relation 13.45. We can then make the following conclusions:

1. Relation (13.45) is the transition probability amplitude from the localized state at the point (q_a,t_a) to the localized state at the point (q_b,t_b). This is independent of the representation in which we are working. In this way, Equation 13.45 may also be represented as follows:

$$K(q_b,t_b;q_a,t_a) = \langle q_b |\hat{U}(t_b,t_a)| q_a \rangle \theta(t_b - t_a) \quad (13.55)$$

Since we are interested to use the expressions of the type (13.50) only for conditions $t_b > t_a$, then we can set $K = 0$ for $t_b < t_a$. This permits us to write the exact definition of K in exactly the form (13.55), where $\theta(t_b - t_a)$ is the *Heaviside step function*:

$$\theta(t_b - t_a) = \begin{cases} 1, & t_b > t_a \\ 0, & t_b < t_a \end{cases} \quad (13.56)$$

The introduction of $\theta(t_b - t_a)$ may have both a physical and a mathematical interest. The physical interest entails compelling the system at a starting point to be evolving toward the future. The mathematical interest as we see later (Section 13.9) is that $K(q_b,t_b; q_a,t_a)$, because of the factor $\theta(t_b - t_a)$, obeys a partial differential equation with the RHS (source term) being a delta function. That differential equation defines a *Green function* that in this case is $K(q_b,t_b;q_a,t_a)$. With the step function (13.56), this Green function will be the retarded solution to that differential equation. This is discussed further in Section 13.11.

2. The evaluation of the transition amplitude in Equation 13.45 is equivalent to the solution of the Schrödinger equation in Equation 13.46. This means that the transition amplitude completely gives the dynamics of the system. It may be seen from above that the transition amplitude is completely a quantum object.

13.3.1 TRANSITION AMPLITUDE IN HAMILTONIAN FORM

From here, let us now move into functional integration. We again derive Equation 13.45 using the classical Hamiltonian function $H(q,p)$ that corresponds to some quantum operator $\hat{H}(\hat{q}, \hat{p})$. We partition the time interval $(t_b - t_a)$ into N infinitesimal parts of equal widths, $\Delta t = (t_b - t_a)/N \equiv \varepsilon$. This gives a set of values t_i that are spaced a distance ε apart between the values t_a and t_b:

$$t_0 \equiv t_a < t_1 < \cdots < t_{N-1} < t_N \equiv t_b \qquad (13.57)$$

At each time t_i, we select some point q_i as the possible location of the quantum particle. We may construct a path by connecting all the points selected by straight lines.

We should recollect from mathematical analysis that an integral is defined through the Riemann sum, using points ξ_i at the centers of the time elements, as follows (see Figure 13.2):

$$\lim_{\substack{N \to \infty \\ \max|\Delta t_i| \to 0}} \sum_{i=0}^{N} f(\xi_i) \Delta t_i \equiv \int_{t_a}^{t_b} f(t) dt \qquad (13.58)$$

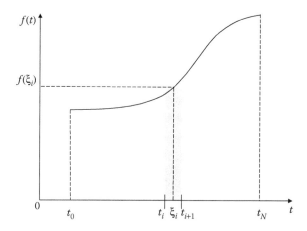

FIGURE 13.2 Graph of an arbitrary function $f(t)$, versus its argument t, to illustrate the evaluation of the Riemann integral.

Thus, in our case the argument of the exponential function in Equation 13.52 may be written as follows:

$$\lim_{\substack{N \to \infty \\ \max|\Delta t_i| \to 0}} \sum_{i=0}^{N} \hat{I}_N = \lim_{\substack{N \to \infty \\ \max|\Delta t_i| \to 0}} \sum_{i=0}^{N} \hat{H}(\xi_i) \Delta t_i \equiv \int_{t_a}^{t} \hat{H}(t') dt' \qquad (13.59)$$

Considering Equation 13.52, we may write a time-ordered product operator

$$\hat{U}_N(t, t_a) = \hat{T} \exp\left\{ -\frac{i}{\hbar} \hat{I}_N \right\} \qquad (13.60)$$

From the definition of the time-ordered product in Equation 13.53, we may write Equation 13.60 as follows:

$$\hat{U}_N(t, t_a) = \hat{F}(t, \xi_{N-1}) \dots \hat{F}(\xi_{i+1}, \xi_i) \dots \hat{F}(\xi_1, t_0) \qquad (13.61)$$

where

$$\hat{F}(t_{i+1}, t_i) = \exp\left\{ -\frac{i}{\hbar} \hat{H}(\xi_i) \Delta t_i \right\} \qquad (13.62)$$

From Equations 13.57 through 13.62, the evolution operator in Equation 13.52 may be obtained as a limiting value:

$$\hat{U}(t, t_a) = \lim_{N \to \infty} \hat{U}_N(t, t_a) \qquad (13.63)$$

Let us take the case of a Hamiltonian independent of time, then considering the partitioning with $\Delta t = (t_b - t_a)/N \equiv \varepsilon$ and Equation 13.50, we have

$$\exp\left\{ -\frac{i}{\hbar} \hat{H}(t_b - t_a) \right\} = \lim_{N \to \infty} \exp\left\{ -\frac{i}{\hbar} \hat{H} N \varepsilon \right\} = \lim_{N \to \infty} \left[\exp\left\{ -\frac{i}{\hbar} \hat{H} \varepsilon \right\} \right]^N \qquad (13.64)$$

Using the closure relation in Equation 13.7 and with the aid of Equation 13.64, it is possible to evaluate the matrix element (the transition amplitude) (13.50) by taking a multiple integral over all values of q_i for i between 1 and $N-1$:

$$\left\langle q_N \left| \exp\left\{ -\frac{i}{\hbar} \varepsilon N \hat{H} \right\} \right| q_0 \right\rangle = \left\langle q_N \left| \left(\exp\left\{ -\frac{i}{\hbar} \varepsilon \hat{H} \right\} \right)^N \right| q_0 \right\rangle$$

$$= \int_{-\infty}^{+\infty} dq_1 \int_{-\infty}^{+\infty} dq_2 \dots \int_{-\infty}^{+\infty} dq_{N-1} \left\langle q_N \left| \exp\left\{ -\frac{i}{\hbar} \varepsilon \hat{H} \right\} \right| q_{N-1} \right\rangle \left\langle q_{N-1} \left| \exp\left\{ -\frac{i}{\hbar} \varepsilon \hat{H} \right\} \right| q_{N-2} \right\rangle \dots$$

$$\times \left\langle q_2 \left| \exp\left\{ -\frac{i}{\hbar} \varepsilon \hat{H} \right\} \right| q_1 \right\rangle \left\langle q_1 \left| \exp\left\{ -\frac{i}{\hbar} \varepsilon \hat{H} \right\} \right| q_0 \right\rangle \qquad (13.65)$$

We do not integrate over $q_0 = q_a$ or $q_N = q_b$ because these are the fixed end points. We examine again (13.65), considering the limit $N \to \infty$, $\varepsilon \to 0$. We analyze in this limit one of the matrix elements in the integrand (13.65):

$$\langle q_{i+1}|\exp\left\{-i\frac{\varepsilon}{\hbar}\hat{H}\right\}|q_i\rangle \cong \langle q_{i+1}|\left(\hat{I} - i\frac{\varepsilon}{\hbar}\hat{H} + o(\varepsilon^2)\right)|q_i\rangle$$

$$= \int_{-\infty}^{+\infty} \frac{dp_i}{2\pi\hbar}\langle q_{i+1}\|p_i\rangle\langle p_i\|\left(\hat{I} - i\frac{\varepsilon}{\hbar}\hat{H} + o(\varepsilon^2)\right)|q_i\rangle \qquad (13.66)$$

In the last integral in Equation 13.66, we use the property of the closure relation in Equation 13.42. Further, considering Equation 13.7, we have as follows:

$$\langle q_{i+1}\|p_i\rangle = (2\pi\hbar)^{-1}\exp\left\{\frac{i}{\hbar}q_{i+1}p_i\right\} \qquad (13.67)$$

$$\langle p_i|\left(\hat{I} - i\frac{\varepsilon}{\hbar}\hat{H} + o(\varepsilon^2)\right)|q_i\rangle = \langle p_i\|q_i\rangle - i\frac{\varepsilon}{\hbar}\langle p_i|\hat{H}(\hat{p},\hat{q})|q_i\rangle + o(\varepsilon^2) \qquad (13.68)$$

Considering that $o(\varepsilon^2)$ is an infinitesimally small function, our evaluation is done up to the linear term in ε. From the formula

$$\langle p_i|\hat{H}(\hat{p},\hat{q})|q_i\rangle = \langle p_i|\left(\frac{\hat{p}^2}{2m} + V(\hat{q})\right)|q_i\rangle = \left(\frac{p_i^2}{2m} + V(q_i)\right)\langle p_i\|q_i\rangle \qquad (13.69)$$

there follows the expression

$$\langle p_i|\left(\hat{I} - i\frac{\varepsilon}{\hbar}\hat{H} + o(\varepsilon^2)\right)|q_i\rangle = \langle p_i\|q_i\rangle\left(1 - i\frac{\varepsilon}{\hbar}H(p_i,q_i) + o(\varepsilon^2)\right) \qquad (13.70)$$

If we consider

$$1 - i\frac{\varepsilon}{\hbar}H(p_i,q_i) + o(\varepsilon^2) \cong \exp\left\{-i\frac{\varepsilon}{\hbar}H(p_i,q_i)\right\} \qquad (13.71)$$

then to the approximation of $o(\varepsilon^2)$ it follows that

$$\langle q_{i+1}|\exp\left\{-i\frac{\varepsilon}{\hbar}H(p_i,q_i)\right\}|q_i\rangle \cong \int_{-\infty}^{+\infty} \frac{dp_i}{2\pi\hbar}\exp\left\{\frac{i\varepsilon}{\hbar}\left[p_i\left(\frac{q_{i+1} - q_i}{\varepsilon}\right) - H(p_i,q_i)\right]\right\}$$

$$(13.72)$$

If we substitute this expression into Equation 13.65 considering the fact that $q_a \equiv q_0$ and $q_b \equiv q_N$, then the transition amplitude to the approximation of $o(\varepsilon^2)$ is

$$K(q_b,t_b;q_a,t_a) \equiv \int\limits_{-\infty}^{+\infty}...\int\limits_{-\infty}^{+\infty}\prod_{i=1}^{N-1}\frac{dp_i dq_i}{2\pi\hbar}$$

$$\times \exp\left\{\frac{i\varepsilon}{\hbar}\sum_{i=0}^{N-1}\left[p_i\left(\frac{q_{i+1}-q_i}{\varepsilon}\right)-H(p_i,q_i)\right]\right\} \quad (13.73)$$

It should be noted that in the integrand, the argument of the exponent has the Riemann sum. We may do a limiting transformation of the sum into an integral if the functions $q(t)$ and $p(t)$ are continuous and the Hamilton function $H(p(t),q(t))$ is piecewise continuous.

Thus, if we consider that

$$\lim_{\substack{N\to\infty \\ \varepsilon\to 0}}\sum_{i=0}^{N-1}\left[p_i\left(\frac{q_{i+1}-q_i}{\varepsilon}\right)-H(p_i,q_i)\right]\varepsilon = \int\limits_{t_a}^{t_b}[p\dot{q}-H(p,q)]dt \quad (13.74)$$

where

$$\lim_{\varepsilon\to 0}\frac{q_{i+1}-q_i}{\varepsilon} = \dot{q}(t) \quad (13.75)$$

then from the Riemann limit, the transition amplitude (13.73) may be written in symbolic form as follows:

$$K(q_b,t_b;q_a,t_a) \equiv \int Dp(t)Dq(t)\exp\left\{\frac{i}{\hbar}\int\limits_{t_a}^{t_b}[p\dot{q}-H(p,q)]dt\right\} \quad (13.76)$$

where $q(t_a) \equiv q_a$, $q(t_b) \equiv q_b$, and the symbols $Dp(t)\, Dq(t)$ are functional measures that denote, respectively, a finite multiple integral:

$$Dp(t)Dq(t) \equiv \lim_{\substack{\varepsilon\to 0 \\ N\to\infty}}\prod_{i=1}^{N-1}\frac{dp_i dq_i}{2\pi\hbar} \quad (13.77)$$

The integral (13.76) is called the transition amplitude in the Hamiltonian form. It is known that the functional measure (13.77) and the integral (13.76) over it demands a concrete mathematical definition. For this one can use the method of functional analysis. We assume that the reader has knowledge of functional analysis and we do not examine the details of that question here.

13.3.2 TRANSITION AMPLITUDE IN FEYNMAN FORM

We further develop the transition amplitude using another approach. The path integral approach to quantum mechanics was developed by Richard Feynman in his PhD thesis [20,21]. This was in the mid-1940s, following a hint from an earlier paper by Dirac [22]. Apparently, Dirac's motivation was to formulate quantum mechanics starting from the Lagrangian rather than from the Hamiltonian formulation of classical mechanics. That Dirac was on the right track is evident above. It is the Lagrangian that appears within the integrand in the exponent in Equation 13.76.

Let us try again to make an attempt to understand the transition amplitude in a purely quantum mechanical sense. For the replacement $\hat{H}(\hat{p},\hat{q}) \rightarrow H(p,q)$ (13.69), the classical Hamiltonian function in Equation 13.76 has the form:

$$H(p,q) = \frac{p^2}{2m} + U(q) \tag{13.78}$$

Here, $U(q)$ is the potential energy associated with the particle. Our goal now is to integrate over the momenta in Equation 13.76, leaving only the integrations over coordinates. This is done with the aid of expression (13.73). We transform the finite multiple integral over the momenta. For the ith momentum the integral takes the form

$$\int_{-\infty}^{+\infty} \frac{dp_i}{2\pi\hbar} \exp\left\{\frac{i}{\hbar}\varepsilon\left[p_i\left(\frac{q_{i+1}-q_i}{\varepsilon}\right) - \frac{p_i^2}{2m} - U(q_i)\right]\right\}$$

$$= \exp\left\{-\frac{i}{\hbar}\varepsilon U(q_i)\right\}\int_{-\infty}^{+\infty} \frac{dp_i}{2\pi\hbar} \exp\left\{\frac{i\varepsilon}{\hbar}\left[p_i\left(\frac{q_{i+1}-q_i}{\varepsilon}\right) - \frac{p_i^2}{2m}\right]\right\} \tag{13.79}$$

The integral on the RHS of Equation 13.79 may be easily evaluated if we can transform it to the standard Gaussian integral:

$$\int_{-\infty}^{+\infty} dx \exp\{-ax^2 + bx\} = \left(\frac{\pi}{a}\right)^{1/2} \exp\left\{\frac{b^2}{4a}\right\}, \quad a > 0 \tag{13.80}$$

The integrand in Equation 13.79, however, is imaginary. Hence, it may be necessary to introduce a supplementary procedure for its evaluation. What should that supplementary procedure be? The procedure is the analytic continuation of the integral (13.79) by rotation of the integration variable through the angle $-\pi/2$ in the complex plane. This integral, with the aid of that rotation, is transformed into Gaussian form as in Equation 13.80. This can now be easily evaluated. The rotation of the integration variable through $\pi/2$ in the complex plane yields the same result. Using this analytic continuation, we arrive at the result from integrating out the momenta only:

$$\int_{-\infty}^{+\infty} \frac{dp_i}{2\pi\hbar} \exp\left\{\frac{i\varepsilon}{\hbar}\left[p_i\left(\frac{q_{i+1}-q_i}{\varepsilon}\right) - \frac{p_i^2}{2m} - U(q_i)\right]\right\}$$

$$= \left(\frac{m}{2\pi\hbar i\varepsilon}\right)^{1/2} \exp\left\{\frac{i\varepsilon}{\hbar}\left[\frac{m}{2}\left(\frac{q_{i+1}-q_i}{\varepsilon}\right)^2 - U(q_i)\right]\right\} \tag{13.81}$$

This result has complex amplitude $(m/2\pi\hbar i\varepsilon)^{1/2}$ modified by a phase factor. The dominant controlling factor is the phase argument of the exponential function in Equation 13.81. Considering Equation 13.45, the form of Equation 13.81 is not exactly like the transition amplitude:

$$K(q_{i+1},t_i + \varepsilon;q_i,t_i) = \langle q_{i+1} | \exp\left\{ -\frac{i\varepsilon}{\hbar}\hat{H} \right\} | q_i \rangle \tag{13.82}$$

If we consider the RHS of Equation 13.81, it follows that for $|q_{i+1} - q_i| \gg (\hbar\varepsilon/m)^{1/2}$, the phase of the transition amplitude oscillates extremely rapidly and the sum over neighboring trajectories may tend to cancel out by interference. Consequently, the type of the coordinates q_{i+1} and q_i contributing most to the transition amplitude in Equation 13.66 are those that satisfy the relation:

$$|q_{i+1} - q_i| \leq \left(\frac{\hbar\varepsilon}{m}\right)^{1/2} \tag{13.83}$$

This is the case for which the phase of the transition amplitude has an extremal (or stationary) value. In this case, the sum over neighboring trajectories will tend to interfere constructively because their phases are practically equal. This can be viewed differently when the phases of the transition amplitude are much larger than the quantum of the action $\hbar$, one is in a *quasi-classical situation*.

Consider the limit that the time step becomes infinitesimal, $\varepsilon = \Delta t \to dt$

$$\varepsilon \to 0, \quad N \to \infty \tag{13.84}$$

Then we have

$$\lim_{\varepsilon\to 0}\frac{q_{i+1} - q_i}{\varepsilon} = \dot{q}(t), \quad \lim_{\varepsilon\to 0}\frac{i\varepsilon}{\hbar}\left[\frac{m}{2}\left(\frac{q_{i+1} - q_i}{\varepsilon}\right)^2 - U(q_i)\right] = \frac{i}{\hbar}L(q,\dot{q},t)dt \tag{13.85}$$

where

$$L(q,\dot{q},t) = \frac{m}{2}\dot{q}^2 - U(q) \tag{13.86}$$

is the Lagrange function of the particle. It is apparent that the integral in Equation 13.76 involves the Lagrangian integrated over time, which is the action. Let us examine the classical action corresponding to one time step:

$$S[q_i,q_{i+1}] = \int_{t_i}^{t_i+\varepsilon} dt'L(q(t'),\dot{q}(t'),t') \tag{13.87}$$

$$q(t_i) \equiv q_i, \quad q(t_i + \varepsilon) \equiv q_{i+1} \tag{13.88}$$

The evaluation is done along some space–time path $q(t)$ passing from the point (q_i,t_i) to the point $(q_{i+1},t_i + \varepsilon)$. Considering Equation 13.83 as $\varepsilon \rightarrow 0$, the path $q(t)$ giving a considerable contribution to the transition amplitude should lie within the neighborhood of $(\hbar\varepsilon/m)^{1/2}$ of the coordinate q_i. For this reason, when evaluating the action (13.87) from the starting and end points, the condition (13.83) should be satisfied so that the path $q(t)$ should almost coincide with the line

$$q(t) \cong \left(1 - \frac{t - t_i}{\varepsilon}\right)q_i + \frac{t - t_i}{\varepsilon} q_{i+1} \tag{13.89}$$

If we substitute Equation 13.89 into Equation 13.87 while using Equation 13.86, then we have this step's contribution to the action:

$$S[q_i, q_{i+1}] \cong \frac{m}{2}\left(\frac{q_{i+1} - q_i}{\varepsilon}\right)^2 \varepsilon - \int_{t_i}^{t_i + \varepsilon} dt' U(q(t')) \cong \left[\frac{m}{2}\left(\frac{q_{i+1} - q_i}{\varepsilon}\right)^2 - U(q_i)\right]\varepsilon \tag{13.90}$$

In this way, the transition amplitude between the states $|q_i\,(t_i)\rangle$ and $|q_{i+1}\,(t_i + \varepsilon)\rangle$ as $\varepsilon \rightarrow 0$ may be written in the form:

$$K(q_{i+1}, t_i + \varepsilon; q_i, t_i) = \left(\frac{m}{2\pi i \hbar \varepsilon}\right)^{1/2} \exp\left\{\frac{i}{\hbar} S[q_i, q_{i+1}]\right\} \tag{13.91}$$

This is an important fundamental result, which gives the transition amplitude expressed only in terms of the particle's initial position and final position after an infinitesimal time step. The momentum degree of freedom has been integrated out. There is a normalization factor and a phase. We see from Equation 13.79 that the contribution to the phase from a given path is the action S for that path in units of the quantum of action, $\hbar$. In summary, the probability P_{ab} to go from a point (q_a,t_a) to the point (q_b,t_b) is the absolute square of the transition amplitude

$$P_{ab} = \left|K(q_b,t_b;q_a,t_a)\right|^2 \tag{13.92}$$

The argument is the amplitude $K(q_b,t_b; q_a,t_a)$ to go from point (q_a,t_a) to point (q_b,t_b). This amplitude is the sum of the contributions of the functions $K(q_{i+1},t_i + \varepsilon; q_i,t_i)$ from each path connecting the desired end points. The contribution of each path has a phase proportional to its action (13.79). In this way, quantum mechanics has been reduced to a simple procedure, in principle. Each amplitude for a chosen path can be thought to represent an event (the particular propagation along that path). The contributions of different paths are summed to generate the total amplitude, $K(q_b,t_b; q_a,t_a)$. The contributions of different paths interfere when converted into a probability by squaring the total amplitude. From the conceptual viewpoint, it is equivalent to saying that a quantum particle propagates along all possible (classical) paths—although the resulting interference effects tends to lead to a certain dominance of some paths over others. The quantum of action, $\hbar$, determines the practical range of paths that actually contribute. As expected, that range corresponds to the limits allowed within the Heisenberg uncertainty principle.

Thus, for a finite interval $(t_b - t_a)$ if we consider Equations 13.80 and 13.79, then the transition amplitude for a sequence of may time steps, with the momenta integrated out, is

$$K(q_b,t_b;q_a,t_a) \cong \lim_{\substack{\varepsilon \to 0 \\ N \to \infty}} \left(\frac{m}{2\pi i \hbar \varepsilon} \right)^{N/2} \int_{-\infty}^{+\infty} \cdots \int_{-\infty}^{+\infty} \prod_{i=1}^{N-1} dq_i \, \exp\left\{ \frac{i}{\hbar} \sum_{i=0}^{N-1} S[q_i, q_{i+1}] \right\} \qquad (13.93)$$

This is the Feynman path integral for the particle. It is expressed only in terms of the particle's coordinate, although it contains the effects of its momentum. The action in Equation 13.93, considering Equation 13.77, is evaluated in the direction of the piecewise line shown in Figure 13.3. We consider here a particle traveling from point (q_a,t_a) to (q_b,t_b) through a series of intermediate points with $q_1, q_2, ..., q_{N-1}$ that define a *path*. The transition amplitude $K(q_b,t_b;q_a,t_a)$ for the particle to start at (q_a,t_a) and end up at (q_b,t_b) is given by the sum over all possible paths. This implies that the particle seeks all possible values of the intermediate points (q_i,t_i). Taking the limits in expression (13.93) as the time increment ε approaches zero, the number of integrations over the

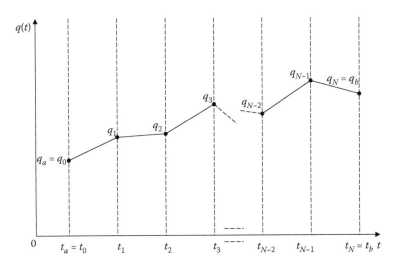

FIGURE 13.3   Illustration of the intermediate points t_i of the trajectory described by the function $q(t)$.

intermediate points become infinite and Equation 13.93 may be written symbolically as a path integral:

$$K(q_b, t_b; q_a, t_a) = \int Dq(t) \exp \left\{ \frac{i}{\hbar} \int_{t_a}^{t_b} L(q, \dot{q}, t) \, dt \right\}$$ (13.94)

where

$$Dq(t) = \frac{1}{A} \lim_{\substack{\varepsilon \to 0 \\ N \to \infty}} \prod_{i=1}^{N-1} \frac{dq_i}{A}$$ (13.95)

is the functional measure over the space of trajectories (or paths) and A is an appropriate normalizing factor as taken from Equation 13.93:

$$\frac{1}{A} = \left(\frac{m}{2\pi i \hbar \varepsilon} \right)^{1/2}$$ (13.96)

The transition amplitude (13.94) is called the *Feynman transition amplitude*. The RHS is expressed as a Feynman path integral. Considering Equation 13.83, the dominant contribution in Equation 13.4 is from a sufficiently small class of piecewise functions. These functions may be varied on a piecewise interval that is close to the classical path, on which the action achieves its extreme (or stationary) value. The paths that move close to the classical path will all have nearly the same phase; hence, they interfere constructively in the transition amplitude and make the major contribution. Paths farther from the classical path (in the sense of different values of the action), interfere destructively and their contributions tend to cancel out.

In the functional limit (13.94), a piecewise path is transformed to a smooth path. Thus, in evaluating the transition amplitude the dominant contribution in the action is from a sufficiently small class of functions that are all smooth—those near the classical path. From this reasoning, the most approximate method of the evaluation of Equation 13.94 is called the *stationary phase method*, which is discussed further below.

Thus, we have obtained two types of functional integrals for the evaluation of the transition amplitude:

1. The evaluation through the classical Hamiltonian $H(p,q)$ (the Hamilton transition amplitude in Equation 13.76). This involves integrations over coordinate and momenta.
2. The evaluation through the classical Lagrangian $L(q, \dot{q}, t)$ (Feynman transition amplitude (13.94)). This involves integrations only over coordinates.

It should be noted that Equation 13.94 is obtained from Equation 13.76 considering the fact that

$$\hat{H}(\hat{p}, \hat{q}) \to H(p, q)$$ (13.97)

From this argument the first evaluation of the transition amplitude is the Hamiltonian form. However, it is easily evaluated through the action using the notion of paths. This method is that of the stationary phase—the Feynman transition amplitude (13.94). Thus, in order to evaluate the transition amplitude, it is first written in the Hamiltonian form and then transformed into the Feynman form.

The transition amplitudes Equations 13.76 and 13.94 may be generalized for a problem with s degrees of freedom that may be described by a set of operators $\hat{q}_1, \hat{q}_2, \ldots, \hat{q}_s, \hat{p}_1, \hat{p}_2, \ldots, \hat{p}_s$:

$$K(q_1'', q_2'', \ldots, q_s'', t_b; q_1', q_2', \ldots, q_s', t_a) = \int \prod_{\alpha=1}^{s} Dp_\alpha Dq_\alpha$$

$$\exp\left\{\frac{i}{\hbar} \int_{t_a}^{t_b} \left[\sum_\alpha p_\alpha \dot{q}_\alpha - H(p,q)\right] dt\right\} \tag{13.98}$$

For the case of the Hamiltonian

$$H(p,q) = \sum_{\alpha=1}^{s} \frac{p_\alpha^2}{2m} + U(q_1, q_2, \ldots, q_s) \tag{13.99}$$

the integral (13.98) (evaluated over the momenta) is transformed to the Feynman transition amplitude:

$$K(q_1'', q_2'', \ldots, q_s'', t_b; q_1', q_2', \ldots, q_s', t_a) = \int \prod_\alpha Dq_\alpha \, \exp\left\{\frac{i}{\hbar} \int_{t_a}^{t_b} L(q_\alpha, \dot{q}_\alpha, t) dt\right\} \tag{13.100}$$

The functional measures in Equations 13.98 and 13.100 are defined, respectively, as in Equations 13.77 and 13.83.

Let us examine the transformation from the Hamiltonian to the Feynman transition amplitude when the Hamiltonian in Equation 13.99 is generalized. As an example, we examine the system with s degrees of freedom described by the following Lagrangian, quadratic in $\dot{q}_\alpha$:

$$L(q_\alpha, \dot{q}_\alpha) = \frac{1}{2} \sum_{\alpha,\beta=1}^{s} \dot{q}_\alpha C_{\alpha\beta}(q)\dot{q}_\beta + \sum_{\alpha=1}^{s} b_\alpha(q)\dot{q}_\alpha - U(q) \tag{13.101}$$

From here, we may construct the Hamiltonian of the system with the help of the relation that relates the canonically conjugate coordinates q_α to their canonical momenta:

$$p_\alpha = \frac{\partial L}{\partial \dot{q}_\alpha} = \sum_\beta C_{\alpha\beta}\dot{q}_\alpha + b_\alpha \tag{13.102}$$

Using this relation, we introduce the following matrices:

$$Q \equiv \begin{bmatrix} q_1 \\ q_2 \\ \vdots \\ q_s \end{bmatrix}, \quad P \equiv \begin{bmatrix} p_1 \\ p_2 \\ \vdots \\ p_s \end{bmatrix}, \quad B \equiv \begin{bmatrix} b_1 \\ b_2 \\ \vdots \\ b_s \end{bmatrix} \qquad (13.103)$$

which gives

$$P = C\dot{Q} + B \qquad (13.104)$$

and

$$\dot{Q} = C^{-1}(P - B) \qquad (13.105)$$

We consider the fact that $CC^{-1} = \hat{I}$. The quantity $C_{\alpha\beta}(q) = C_{\beta\alpha}(q)$ is assumed to be a real nondegenerate matrix. Degenerate matrices will not be considered here.

We may now conveniently through very simple transformations arrive at the following Hamiltonian:

$$H(P,Q) = P^T \dot{Q} - L = \frac{1}{2}\tilde{P}^T C^{-1}\tilde{P} + U(Q) \qquad (13.106)$$

where

$$\tilde{P} \equiv P - B(Q) \qquad (13.107)$$

We write the transition amplitude of the quantum system in the Hamilton form (see Equation 13.83):

$$K(Q_b,t_b;Q_a,t_a) = \int DPDQ \, \exp\left\{\frac{i}{\hbar}\int_{t_a}^{t_b}[P^T\dot{Q} - H(P,Q)]dt\right\} \qquad (13.108)$$

Here

$$DPDQ \equiv \prod_{\alpha=1}^{s} Dp_\alpha Dq_\alpha \qquad (13.109)$$

The evaluation of the functional integral may be done based on the functional analog of the change of variables in the integrand:

$$\tilde{P}(t) = P(t) - B(Q), \quad DP(t) = D\tilde{P}(t) \qquad (13.110)$$

This yields the following transition amplitude

$$K(Q_b,t_b;Q_a,t_a) = \int DQ \, \exp\left\{ \frac{i}{\hbar} \int_{t_a}^{t_b} [B^T \dot{Q} - V(Q)]dt \right\}$$

$$\times \int D\tilde{P} \, \exp\left\{ \frac{i}{\hbar} \int_{t_a}^{t_b} \left[-\frac{1}{2}\tilde{P}^T C^{-1}\tilde{P} + \tilde{P}^T \dot{Q} \right] dt \right\} \qquad (13.111)$$

We evaluate the integral over the momenta with the aid of the following:

$$\lim_{\substack{\varepsilon \to 0 \\ N \to \infty}} \prod_{k=1}^{N-1} \frac{d\tilde{P}_k}{(2\pi\hbar)^s} \, \exp\left\{ \frac{i\varepsilon}{\hbar} \sum_{k=1}^{N-1} \left[\tilde{P}_k^T \left(\frac{Q_{k+1} - Q_k}{\varepsilon} \right) - \frac{1}{2}\tilde{P}_k^T C^{-1}\tilde{P}_k \right] \right\} \qquad (13.112)$$

The evaluation of each of the above multiple integrals yields the following:

$$\int \frac{dP_k}{(2\pi\hbar)^s} \, \exp\left\{ \frac{i\varepsilon}{\hbar} \left[P_k^T \left(\frac{Q_{k+1} - Q_k}{\varepsilon} \right) - \frac{1}{2}P_k^T C^{-1}P_k \right] \right\}$$

$$= \left(\frac{1}{2\pi i\hbar\varepsilon} \right)^{s/2} [\det C^{-1}]^{-1/2} \, \exp\left\{ \frac{i\varepsilon}{\hbar} \left[\frac{1}{2}\left(\frac{Q_{k+1} - Q_k}{\varepsilon} \right)^T C \left(\frac{Q_{k+1} - Q_k}{\varepsilon} \right) \right] \right\} \qquad (13.113)$$

Thus, we arrive at the Feynman transition amplitude:

$$K(Q_b,t_b;Q_a,t_a) = \int DQ(t) \prod_{q_\alpha(t)} [\det C^{-1}(q_\alpha(t))]^{-1/2} \, \exp\left\{ \frac{i}{\hbar} \int_{t_a}^{t_b} L(Q,\dot{Q})dt \right\} \qquad (13.114)$$

where

$$DQ(t) = \frac{1}{A} \lim_{\substack{\varepsilon \to 0 \\ N \to \infty}} \prod_{i=1}^{N-1} \frac{dQ_i}{A} \qquad (13.115)$$

and

$$\frac{1}{A} = \left(\frac{1}{2\pi i\hbar\varepsilon} \right)^{s/2} \qquad (13.116)$$

To summarize, the Feynman path integral is used to represent the probability amplitude for a transition from an initial state to a final state, using only the coordinates, not the momenta. The particle can be considered to follow all connecting paths; however, interference takes place between all paths. Paths that are close to the classical path have nearly the same action and thus the same phase; hence, they interfere constructively and make the dominating contributions to the transition amplitude. One can expect that paths whose action is within about $\hbar$ of the classical path will contribute strongly. For paths whose action is more than $\hbar$ different from the classical path, interference becomes more and more destructive and there is lesser net contribution. It is a beautiful way to think about quantum dynamics, and at the same time, illustrates how the Heisenberg uncertainty limit comes about through wave (or path) interference.

13.3.3 EXAMPLE: A FREE PARTICLE

The Lagrangian for a nonrelativistic free particle is

$$L(\dot{q}) = \frac{m\dot{q}^2}{2} \tag{13.117}$$

The transition amplitude for a free particle may be written through Feynman path integration with the help of relation (13.77):

$$K(q_b, T; q_a, 0) = \int Dq \ \exp\left\{ \frac{i}{\hbar} \int_o^T \frac{m\dot{q}^2}{2} \, dt \right\}$$

$$= \frac{1}{A} \int \prod_{k=1}^{N-1} \frac{dq_k}{A} \ \exp\left\{ \frac{im\varepsilon}{2\hbar} \sum_{k=0}^{N-1} \left[\left(\frac{q_{k+1} - q_k}{\varepsilon} \right)^2 \right] \right\} \tag{13.118}$$

This represents a set of Gaussian integrals. Considering the fact that the integral of a Gaussian is again a Gaussian, the integrations can be carried out on one variable after the other. When the integrations are completed then the limit $N \to \infty$ can be applied. We carry out the calculations as follows. In the first step, we use the method of analytic continuation to evaluate the integrals:

$$\int_{-\infty}^{\infty} \left(\frac{m}{2\pi i \hbar \varepsilon} \right)^{1/2} dq_1 \ \exp\left\{ \frac{im}{2\hbar\varepsilon} [(q_2 - q_1)^2 + (q_1 - q_0)^2] \right\}$$

$$= \frac{1}{2^{1/2}} \exp\left\{ \frac{im}{2\hbar(2\varepsilon)} (q_2 - q_0)^2 \right\} \tag{13.119}$$

In the second step we have as follows:

$$\int_{-\infty}^{+\infty} \left(\frac{m}{2\pi i \hbar 2\varepsilon} \right)^{1/2} dq_2 \ \exp\left\{ \frac{im}{2\hbar\varepsilon} (q_3 - q_2)^2 \right\} \exp\left\{ \frac{im}{2\hbar(2\varepsilon)} (q_2 - q_0)^2 \right\}$$

$$= \frac{1}{3^{1/2}} \exp\left\{ \frac{im}{2\hbar(3\varepsilon)} (q_3 - q_0)^2 \right\} \tag{13.120}$$

If this procedure is continued, we see that a recursion process is established, that after $(N-1)$ steps yields

$$
\int_{-\infty}^{\infty} \left(\frac{m}{2\pi i \hbar (N-1)\varepsilon} \right)^{1/2} dq_{N-1} \exp\left\{ \frac{im}{2\hbar\varepsilon}(q_N - q_{N-1})^2 \right\}
$$

$$
\times \exp\left\{ \frac{im}{2\hbar(N-1)\varepsilon}(q_{N-1} - q_0)^2 \right\}
$$

$$
= \frac{1}{N^{1/2}} \exp\left\{ \frac{im}{2\hbar(N\varepsilon)}(q_N - q_0)^2 \right\} \tag{13.121}
$$

This result is multiplied by the normalization factor $(m/2\pi i\hbar\varepsilon)^{1/2}$. Then, considering that

$$
N\varepsilon = T, \quad q_N = q_b, \quad q_0 = q_a \tag{13.122}
$$

we have as follows:

$$
K(q_b, T; q_a, 0) = \left(\frac{m}{2\pi i \hbar T} \right)^{1/2} \exp\left\{ \frac{im}{2\hbar T}(q_b - q_a)^2 \right\}
$$

$$
= \left(\frac{m}{2\pi i \hbar T} \right)^{1/2} \exp\left\{ \frac{i}{\hbar} S_{\text{classical}}[0, T] \right\} \tag{13.123}
$$

This coincides with the quantum mechanical transition amplitude obtained by other methods, as expected. It is interesting that it depends on $S_{\text{classical}}[0,T]$, the classical action along the classical path, with the condition that $q(0) = q_a$ and $q(T) = q_b$. When absolute squared, K gives the probability per unit length that the particle will be found around position q_b, given that it was known to be at q_a at the start of the time interval T. This is simply a diffusive motion or spreading of a localized wave packet that began, in fact, as a Dirac delta function centered at q_a. One can note that the wave packet width (in probability) increases linearly with the time.

13.4 TRANSITION AMPLITUDES USING STATIONARY PHASE METHOD

13.4.1 MOTION IN POTENTIAL FIELD

It should be noted that the most effective method of evaluation of the transition amplitude is the stationary phase method. It is an approximate method. Let us do one example. We consider a particle moving through a potential $U(q)$. The Lagrangian for this particle is the difference of its kinetic and potential energies:

$$L(q,\dot{q}) = \frac{m}{2}\dot{q}^2 - U(q) \tag{13.124}$$

The transition amplitude may be written straight away through the Feynman path integral:

$$K(q_b,T;q_a,0) = \int Dq(t)\exp\left\{\frac{i}{\hbar}S[0,T]\right\} = \int Dq(t)\exp\left\{\frac{i}{\hbar}\int_0^T L(q,\dot{q})dt\right\} \tag{13.125}$$

The main contribution to this path integral must come from paths that are in the vicinity of the classical path, that is, those with an action within about $\hbar$ of the action of the classical path. From the least action principle $\delta S = 0$, we arrive at the Euler–Lagrange equation for the classical path:

$$\frac{d}{dt}\frac{\partial L}{\partial \dot{q}} - \frac{\partial L}{\partial q} = 0 \tag{13.126}$$

This results in the equation for the classical path:

$$m\ddot{\bar{q}} + \left.\frac{dU(q)}{dq}\right|_{q=\bar{q}(t)} = 0, \quad \bar{q}(0) = q_a, \quad \bar{q}(T) = q_b \tag{13.127}$$

In order to do functional integration in Equation 13.125, we consider that only those paths quite near to the classical path $\bar{q}(t)$ are important. Thus, for the new path we select the variation of the path, $\delta\, q(t) \equiv y(t)$ (see Figure 13.4):

$$q(t) = \bar{q}(t) + y(t), \quad y(0) = y(T) = 0, \quad Dq(t) = Dy(t) \tag{13.128}$$

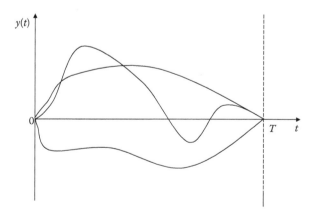

FIGURE 13.4 Variation of the distance $y(t)$ between the classical path and other paths with time t.

Let us expand the potential $U(q)$ in a Taylor series:

$$U(q) = U(\overline{q} + y) \cong U(\overline{q}) + U'(\overline{q})y + \frac{1}{2!}U''(\overline{q})y^2 + \frac{1}{3!}U'''(\overline{q})y^3 + \cdots \quad (13.129)$$

All differentials are evaluated along the classical path $\overline{q}(t)$, and the prime on U shows differentiation with respect to its coordinate argument. After some transformations, we arrive at the following:

$$L(q,\dot{q}) = \frac{m}{2}(\dot{\overline{q}} + y)^2 - U(\overline{q} + y) \cong \left(\frac{m}{2}\dot{\overline{q}}^2 - U(\overline{q})\right) + (m\dot{\overline{q}}\dot{y} - U'(\overline{q})y)$$

$$+ \left(\frac{m}{2}\dot{y}^2 - \frac{1}{2!}U''(\overline{q})y^2 - \frac{1}{3!}U'''(\overline{q})y^3 - \cdots\right) \quad (13.130)$$

The contribution of the second term in the action is seen to vanish:

$$\int_0^T (m\dot{\overline{q}}\dot{y} - U'(\overline{q})y)dt = m\int_0^T \frac{d}{dt}(\dot{\overline{q}}y)dt - \int_0^T (m\ddot{\overline{q}} + U'(\overline{q}))ydt = m\dot{\overline{q}}y\Big|_0^T = 0 \quad (13.131)$$

The contribution of the first term to the action is equal to the action of a classical path:

$$S_{classical}[0,T] = \int_0^T \left[\frac{m}{2}\dot{\overline{q}}^2 - U(\overline{q})\right]dt \quad (13.132)$$

Substituting these results into the expression of the transition amplitude Equation 13.125 and considering Equation 13.40, we obtain the following transition amplitude:

$$K(q_b,T;q_a,0) = \exp\left\{\frac{i}{\hbar}S_{classical}[0,T]\right\}$$

$$\times \int_0^0 Dy(t)\exp\left\{\frac{i}{\hbar}\int_0^T \left(\frac{m}{2}\dot{y}^2 - \frac{1}{2!}U''(\overline{q})y^2 - \frac{1}{3!}U'''(\overline{q})y^3 - \cdots\right)dt\right\} \quad (13.133)$$

This is the approximate formula using the stationary phase method. What is the physical sense of (13.133)? The first factor, having the classical action, is the dominant contribution to the transition amplitude from the classical path. The second factor, which is the functional integral, describes the contribution from the quantum corrections of the order $o(\hbar)$. In (13.133), if we consider

$$\int_0^T \left(\frac{1}{3!}U'''(\overline{q})y^3 + \cdots\right)dt \cong o(\hbar^2) \quad (13.134)$$

then the integrand in

$$\int_0^T \left(\frac{m}{2} \dot{y}^2 - \frac{1}{2!} U''(\bar{q}) y^2 \right) dt \qquad (13.135)$$

is quadratic in y. Thus, the integral in Equation 13.133 is a Gaussian and may be evaluated conveniently. In this way, the stationary phase method evaluates properly the first quantum correction $\hbar$ in the transition amplitude. We see next that the stationary phase method may conveniently be applied to an oscillator.

13.4.2 HARMONIC OSCILLATOR

Let us apply the results of the previous section to a simple example. We take $U(q)$ as the potential of a harmonic oscillator:

$$U(q) = \frac{m}{2} \omega^2 q^2 \qquad (13.136)$$

and

$$U'' = m\omega^2, \quad U^{(n)}(q) = 0, \quad n \geq 3 \qquad (13.137)$$

It follows that the transition amplitude of a harmonic oscillator is

$$K(q_b, T; q_a, 0) = F(T) \exp\left\{ \frac{i}{\hbar} S_{CHO}[0, T] \right\} \qquad (13.138)$$

where

$$F(T) = \int_0^0 Dy(t) \, \exp\left\{ \frac{i}{\hbar} \int_0^T \frac{m}{2} (\dot{y}^2 - \omega^2 y^2) \, dt \right\} \qquad (13.139)$$

This integral is evaluated according to the paths in Figure 13.4.

All the paths begin from point q_a at time moment t_a and end at point q_b at time moment t_b. The oscillator motion is periodic with some period T. Then the function $y(t)$ may be expanded as a Fourier sine series with the fundamental harmonic frequency equal to $2\pi/T$:

$$y(t) = \sqrt{\frac{2}{T}} \sum_n a_n \sin \frac{\pi n t}{T} \qquad (13.140)$$

Such a formula enables us to have $y = 0$ at ends points of the path. Then this calculation considers the motion over some integer number of periods. For convenience, we
let $t_a = 0$. In order to have a defined path it is necessary to select a_n:

$$a_n = \int y(t) \sin \frac{\pi n t}{T} dt \tag{13.141}$$

Each path has its own coefficients a_n and thus we have sets of coefficients $\{a_n\}$. Let
us evaluate the term proportional to the kinetic energy:

$$\int_0^T \dot{y}^2(t) dt = \int_0^T \frac{2}{T} \sum_{nn'} a_n a_{n'} \left(\frac{\pi n}{T} \right) \left(\frac{\pi n'}{T} \right) \cos \frac{\pi n t}{T} \cos \frac{\pi n' t}{T} dt \tag{13.142}$$

If $n \neq n'$, then the integral is equal to zero, whereas for $n = n'$, we have

$$\int_0^T \dot{y}^2(t) dt = \sum_n \frac{2}{T} (a_n)^2 \left(\frac{\pi n}{T} \right)^2 \int_0^T \cos^2 \frac{\pi n t}{T} dt \tag{13.143}$$

As

$$\int_0^T \cos^2 \frac{\pi n t}{T} dt = \int_0^T \sin^2 \frac{\pi n t}{T} dt \tag{13.144}$$

then

$$\int_0^T \sin^2 \frac{\pi n t}{T} dt = \frac{1}{2} \int_0^T \left(\sin^2 \frac{\pi n t}{T} + \cos^2 \frac{\pi n t}{T} \right) dt = \frac{1}{2} \int_0^T dt = \frac{T}{2} \tag{13.145}$$

Thus

$$\int_0^T \dot{y}^2(t) dt = \sum_n (a_n)^2 \left(\frac{\pi n}{T} \right)^2 \tag{13.146}$$

Similarly, the term proportional to the potential energy is obtained:

$$\int_0^T y^2(t) dt = \int_0^T \frac{2}{T} \sum_{nn'} a_n a_{n'} \sin \frac{\pi n t}{T} \sin \frac{\pi n' t}{T} dt$$

$$= \int_0^T \frac{2}{T} \sum_n (a_n)^2 \sin^2 \frac{\pi n t}{T} dt = \sum_n (a_n)^2 \tag{13.147}$$

and

$$\int_0^T \frac{m}{2}(\dot{y}^2 - \omega^2 y^2)\,dt = \frac{m}{2}\sum_n \left[\left(\frac{\pi n}{T}\right)^2 - \omega^2\right]a_n^2 \tag{13.148}$$

Then, we have to evaluate

$$F(T) = \int_0^0 Dy(t)\,\exp\left\{\frac{i}{\hbar}\int_0^T \frac{m}{2}(\dot{y}^2 - \omega^2 y^2)\,dt\right\} \tag{13.149}$$

This means we need a sum over all paths, which is a sum over the coefficients $\{a_n\}$. As $y \to a_n$, we use the transformation Jacobian J, obtaining

$$
\begin{aligned}
F(T) &= \int_0^0 Dy(t)\,\exp\left\{\frac{i}{\hbar}\int_{t_a}^{t_b} \frac{m}{2}(\dot{y}^2 - \omega^2 y^2)\,dt\right\} \\
&= J\prod_n \int_{-\infty}^{\infty} da_n\,\exp\left\{-\frac{m}{2\hbar i}\left[\left(\frac{\pi n}{T}\right)^2 - \omega^2\right]a_n^2\right\} \\
&= J\prod_n \left(\frac{2\pi i\hbar}{m}\right)^{1/2}\left[\left(\frac{\pi n}{T}\right)^2 - \omega^2\right]^{-1/2}
\end{aligned} \tag{13.150}
$$

For motion over an integer number of periods, we have $q_b = q_a$ and the Jacobian is

$$J = \left(\frac{m}{2\pi i\hbar}\right)^{1/2}\frac{1}{\prod_n((2\pi i\hbar/m)^{1/2}(\pi n/T)^{-1})} \tag{13.151}$$

Then

$$
\begin{aligned}
F(T) &= \int_0^0 Dy(t)\,\exp\left\{\frac{i}{\hbar}\int_0^T \frac{m}{2}(\dot{y}^2 - \omega^2 y^2)\,dt\right\} \\
&= J\prod_n \left(\frac{2\pi i\hbar}{m}\right)^{1/2}\left(\frac{\pi n}{T}\right)^{-1}\left[1 - \frac{\omega^2 T^2}{n^2\pi^2}\right]^{-1/2}
\end{aligned} \tag{13.152}
$$

and as

$$\prod_n\left[1 - \frac{\omega^2 T^2}{n^2\pi^2}\right]^{-1/2} = \left(\frac{\omega T}{\sin \omega T}\right)^{1/2} \tag{13.153}$$

then

$$F(T) = \left(\frac{m\omega}{2\pi i\hbar \sin \omega T} \right)^{1/2} \qquad (13.154)$$

Thus, finally we get for the oscillator

$$K(q_b, T; q_a, 0) = F(T) \exp\left\{ \frac{i}{\hbar} S_{CHO}[0, T] \right\} \qquad (13.155)$$

This corresponds to a purely quantum mechanical result. The absolute square of K or effectively, of $F(T)$, gives the probability per unit length of finding the mass back where it started at the beginning of an oscillation with period T. We see that this gives

$$|K(T)|^2 = |F(T)|^2 = \frac{m\omega}{2\pi\hbar \sin \omega T} \qquad (13.156)$$

At long times, that is, over many periods, the particle returns to where it started, with a constant probability because that is the case of $\omega T = 2\pi,\ 4\pi,\ 6\pi,\ \ldots$, and so on.

It should be recalled that for a free particle we have

$$\int Dq(t)\ \exp\left\{ \frac{i}{\hbar} \int_0^T \left(\frac{m\dot{q}^2}{2} \right) dt \right\} = \left(\frac{m}{2\pi i\hbar T} \right)^{1/2} \exp\left\{ \frac{im}{2\hbar T}(q_b - q_a)^2 \right\} \qquad (13.157)$$

We can think of a free particle as an oscillator with a vanishing frequency. If we take the limit as $\omega \to 0$ in Equation 13.155 and especially in the expression for $F(T)$, then we obtain

$$K(q_b, T; q_a, 0) \to \left(\frac{m}{2\pi i\hbar T} \right)^{1/2} \exp\left\{ \frac{im(q_b - q_a)^2}{2\hbar T} \right\} \qquad (13.158)$$

which recovers the transition amplitude for a free particle, as expected. The harmonic oscillator particle acts as a free particle when its binding force is very weak.

13.5 REPRESENTATION OF MATRIX ELEMENT OF PHYSICAL OPERATOR THROUGH FUNCTIONAL INTEGRAL

We limit ourselves for convenience to a system with one degree of freedom and having the Lagrangian of a particle in a potential field, as in Equation 13.124. For the transition amplitude of this system we have expression (13.125). We have the following questions to ask:

1. Is it possible on the basis of functional integration to obtain the formula for the evaluation of the matrix element of a physical operator of the type (13.94)?
2. How can we avoid the difficulty related with the normalization factor in the evaluation of these matrix elements?

In order to answer these questions we define the matrix element of a physical operator using the quantum mechanical definition. Suppose the system is in the Heisenberg state, $|q_a,t_a\rangle$ at the starting moment t_a and $|q_b,t_b\rangle$ is the state of the system at the end moment t_b. The matrix element $\langle F(t)\rangle_{ba}$ of a Heisenberg operator of a physical quantity $\hat{F}(t)$ may be defined as follows:

$$\langle F(t)\rangle_{ba} \equiv \frac{\langle q_b,t_b\,|\,\hat{F}(t)\,|\,q_a,t_a\rangle}{\langle q_b,t_b\,|\,q_a,t_a\rangle} \tag{13.159}$$

If we move to the Schrödinger representation, then considering functional integration for the transition amplitude, the numerator in Equation 13.159 may be represented as follows:

$$\langle q_b,t_b\,|\,\hat{F}(t)\,|\,q_a,t_a\rangle = \langle q_b|\exp\left\{-\frac{i}{\hbar}\hat{H}(t_b-t)\right\}\hat{F}\exp\left\{-\frac{i}{\hbar}\hat{H}(t-t_a)\right\}|q_a\rangle$$

$$= \int dq'dq\langle q_b|\,\exp\left\{-\frac{i}{\hbar}\hat{H}(t_b-t)\right\}|q\rangle\langle q|\,\hat{F}\,|q'\rangle\langle q'|\,\exp\left\{-\frac{i}{\hbar}\hat{H}(t-t_a)\right\}|q_a\rangle$$

$$= \int dq'dq\int Dq(t)\,\exp\left\{\frac{i}{\hbar}S[q_b,t_b;q,t]\right\}\langle q|\,\hat{F}\,|q'\rangle\,\exp\left\{\frac{i}{\hbar}S[q',t;q_a,t_a]\right\} \tag{13.160}$$

For brevity, we consider that the operator $\hat{F}$ is diagonalized in the q-representation:

$$\langle q|\hat{F}|q'\rangle = F(q')\delta(q-q') \tag{13.161}$$

In this case

$$\langle q_b,t_b\,|\,\hat{F}(t)\,|\,q_a,t_a\rangle = \int Dq(t)F[q(t)]\,\exp\left\{\frac{i}{\hbar}S[q_b,t_b;q_a,t_a]\right\} \tag{13.162}$$

Thus, the resultant expression for matrix element through functional integration has the form

$$\langle F\rangle = \frac{\int Dq(t)F[q(t)]\,\exp\left\{\frac{i}{\hbar}S[q_b,t_b;q_a,t_a]\right\}}{\int Dq(t)\exp\left\{\frac{i}{\hbar}S[q_b,t_b;q_a,t_a]\right\}} \tag{13.163}$$

The formula (13.163) may be generalized for the case of a time-ordered product of some Heisenberg operators:

$$
\begin{aligned}
&\frac{\langle q_b,t_b | \hat{T}[\, \hat{F}(t_1)\hat{F}(t_2)\cdots\hat{F}(t_N)\,] | q_a,t_a \rangle}{\langle q_b,t_b \| q_a,t_a \rangle} \\
&= \frac{\displaystyle\int Dq(t)F[q(t_1)]F[q(t_2)]\cdots F[q(t_N)]\exp\left\{\frac{i}{\hbar}S[q_b,t_b;q_a,t_a]\right\}}{\displaystyle\int Dq(t)\exp\left\{\frac{i}{\hbar}S[q_b,t_b;q_a,t_a]\right\}}
\end{aligned}
\tag{13.164}
$$

13.6   PROPERTY OF PATH INTEGRAL DUE TO EVENTS OCCURRING IN SUCCESSION

The action for a path obtained by joining two points (q_a,t_a) and (q_b,t_b) by the intermediate point (q_c,t_c) satisfies the relation:

$$
S = \int_{t_a}^{t_b} L(q,\dot{q},t)\,dt = \int_{t_a}^{t_c} L(q,\dot{q},t)\,dt + \int_{t_c}^{t_b} L(q,\dot{q},t)\,dt
\tag{13.165}
$$

Thanks to this equation, the additivity of S translates into a sum over products of transition amplitudes:

$$
K(q_b,t_b;q_a,t_a) = \int K(q_b,t_b;q_c,t_c)\; K(q_c,t_c;q_a,t_a)\,dq_c
\tag{13.166}
$$

The transition amplitude to go from point q_a to q_b is the sum over all possible intermediate values of q_c, using the transition amplitudes for the particle to go from q_a to q_c and then q_c to q_b.

Suppose between the times t_a and t_b we select two intermediate times t_c and t_d, then the transition amplitude for the particle to go from q_a to q_b may take the form:

$$
K(q_b,t_b;q_a,t_a) = \int\int K(q_b,t_b;q_c,t_c)\; K(q_c,t_c;q_d,t_d)\; K(q_d,t_d;q_a,t_a)\,dq_c dq_d
\tag{13.167}
$$

This process may be continued until the time scale is divided into N intervals, using $N-1$ intermediate times $t_1, t_2, t_3, \ldots t_{N-1}$:

$$
\begin{aligned}
&K(q_b,t_b;q_a,t_a) \\
&= \int\ldots\int K(q_b,t_b;q_{N-1},t_{N-1})K(q_{N-1},t_{N-1};q_{N-2},t_{N-2})\ldots K(q_1,t_1;q_a,t_a)\,dq_{N-1}dq_{N-2}\ldots dq_1
\end{aligned}
$$

$$
\tag{13.168}
$$

This can be interpreted as the transition amplitude of a particle having left a point q_a at time t_a to arrive at a point q_b at time t_b, having passed successively through all the points at the time moments t_i. The quantity (13.168) is understood as the sum (integral) that corresponds to coherent superposition of the transition amplitudes that are associated with all possible paths starting from the point with time t_a and ending at a point with time t_b.

13.7 EIGENVECTORS

In ordinary quantum mechanics the information about a particle is carried by the eigenvector $|q,t\rangle$. The operator that maps such a particle to a given domain of space is

$$P(q,t) = |q,t\rangle\langle q,t| \tag{13.169}$$

Then the state vector $|q,t\rangle$ is an eigenvector of $P(q,t)$, with the eigenvalue equal to the probability density, $\langle q,t|q,t\rangle$.

The expression of the transition amplitude was developed by closely following the motion of a particle in getting to a particular point in a given time. We may consider the amplitude to arrive at a particular point without a special discussion of the previous motion. In quantum mechanics, the wave function $\Psi(q,t) = \langle q|\Psi(t)\rangle$ is that total amplitude for the particle to found at (q,t), arriving from the past in some situation.

Let us examine how the wave function $\Psi(q,t) = \langle q|\Psi(t)\rangle$ relates to the transition amplitude $K(q_b,t_b;q_a,t_a)$, in which a particle comes from (q_a,t_a) and arrives at (q_b,t_b). We saw in Equation 13.50 how the state vector $|\Psi(t)\rangle$ at time $t = t_b$ evolves from an initial one. Forming the scalar product with a position eigenstate, $|q_b\rangle$ we obtain

$$\Psi(q_b,t_b) = \langle q_b|\Psi(t_b)\rangle = \left\langle q_b\left|\int dq\,|q\rangle\,K(q,t_b;q_a,t_a)\right. = \int dq\,\langle q_b|q\rangle\,K(q,t_b;q_a,t_a)\right.$$

$$\tag{13.170}$$

But as the states have delta normalization, there results

$$\Psi(q_b,t_b) = \int dq\,\delta(q_b - q)\,K(q,t_b;q_a,t_a) = K(q_b,t_b;q_a,t_a) \tag{13.171}$$

Therefore, $K(q_b,t_b;q_a,t_a)$ is the amplitude to be found currently in state $|q_b\rangle$.

Consider again a time t_c that falls between the times t_a and t_b. Then using relations (13.45) and (13.47), we have the amplitude $\langle q_b|\Psi(t_b)\rangle$ given by

$$\Psi(q_b,t_b) = \int dq_c\,K(q_b,t_b;q_c,t_c)K(q_c,t_c;q_a,t_a) = \int dq_c\,K(q_b,t_b;q_c,t_c)\Psi(q_c,t_c)$$

$$\tag{13.172}$$

The transition amplitude is also an eigenfunction in which is inherent the history of the arrival of a particle at the point (q_b, t_b).

Usually, in quantum mechanics, an eigenvector gives the probability of finding a particle at a given point but not the answer on where the particle was found earlier. But the transition amplitude K tells us from where and for what time moment will the particle arrive at the point (q_b, t_b). That is, K is a wave function for the present state, in which the initial state is rigorously defined.

The wave function $\langle q_b | \Psi(t_b) \rangle$ corresponds to some distribution. The transition amplitude $K(q_b, t_b; q_a, t_a)$ plays the role of the wave function when the initial condition is rigorously defined:

$$\Psi(q_b, t_b) = \int K(q_b, t_b; q_a', t_a) \delta(q_a' - q_a) dq_a' \tag{13.173}$$

We may see that the effects of all the previous history of a particle may be expressed in terms of the single function $K(q_b, t_b; q_a, t_a)$.

13.8 TRANSITION AMPLITUDE FOR TIME-INDEPENDENT HAMILTONIAN

We assume a system has a time-independent Hamiltonian $\hat{H}$. This corresponds to any situation where the action S does not depend explicitly on time. In this case, the transition amplitude $K(q_b, t_b; q_a, t_a)$ does not depend on the individual times. It is a function only of the time interval $T = t_b - t_a$. This leads to a state vector $|\Psi(t)\rangle$ that periodically depends on time.

The state vector is described by the time-dependent Schrödinger equation:

$$i\hbar \frac{\partial |\Psi(t)\rangle}{\partial t} = \hat{H} |\Psi(t)\rangle \tag{13.174}$$

Supposing $\hat{H}$ is time independent, we can assume $|\Psi(t)\rangle = e^{-iEt/\hbar} |\Phi\rangle$ and look for the energy eigenstates; hence, we have a stationary eigenvalue problem:

$$\hat{H} |\Phi_n\rangle = E_n |\Phi_n\rangle \tag{13.175}$$

From here, we may obtain the energy spectrum E_n and the eigenvectors $|\Phi_n\rangle$, or equivalently, the eigenfunctions $\Phi_n(q) = \langle q|\Phi_n\rangle$ in the coordinate representation. From quantum mechanics, a general state vector $|\Psi(t)\rangle$ may be expressed using a linear combination of the energy eigenvectors $|\Phi_n\rangle$ as a basis:

$$|\Psi(t)\rangle = \sum_n C_n \exp\left\{-i\frac{E_n t}{\hbar}\right\} |\Phi_n\rangle \equiv \sum_n C_n |\Phi_n(t)\rangle \tag{13.176}$$

where the time-evolving energy eigenstates are

$$|\Phi_n(t)\rangle \equiv \exp\left\{-i\frac{E_n t}{\hbar}\right\}|\Phi_n\rangle \tag{13.177}$$

It is possible to express the transition amplitude $K(q_b,t_b;q_a,t_a)$ in terms of the states in Equation 13.177. We suppose the system starts at time t_a in the position eigenstate $|q_a\rangle$. Considering Equation 13.177, we can find the C_n expansion coefficients using a scalar product with $\langle\Phi_m|$ from the left:

$$\langle\Phi_m|\Psi(t_a)\rangle = \sum_n C_n \exp\left\{-i\frac{E_n t_a}{\hbar}\right\}\langle\Phi_m|\Phi_n\rangle = C_m \exp\left\{-i\frac{E_m t_a}{\hbar}\right\} \tag{13.178}$$

where it is assumed that the energy eigenstates are orthonormal. Hence, the expansion coefficients are

$$C_n = \exp\left\{i\frac{E_n t_a}{\hbar}\right\}\langle\Phi_n|\Psi(t_a)\rangle = \exp\left\{i\frac{E_n t_a}{\hbar}\right\}\langle\Phi_n|q_a,t_a\rangle \tag{13.179}$$

Then, from here, the state vector at a later time t_b is found from Equation 13.47 as

$$|\Psi(t_b)\rangle = \sum_n \exp\left\{i\frac{E_n t_a}{\hbar}\right\}\langle\Phi_n|q_a\rangle\exp\left\{-i\frac{E_n t_b}{\hbar}\right\}|\Phi_n\rangle \tag{13.180}$$

This means the transition amplitude is directly found as

$$K(q_b,t_b;q_a,t_a) = \langle q_b|\Psi(t_b)\rangle = \sum_n \exp\left\{i\frac{E_n t_a}{\hbar}\right\}\langle\Phi_n|q_a\rangle\exp\left\{-i\frac{E_n t_b}{\hbar}\right\}\langle q_b|\Phi_n\rangle$$

$$= \sum_n \Phi_n(q_b)\exp\left\{-i\frac{E_n(t_b-t_a)}{\hbar}\right\}\Phi_n^*(q_a) = \sum_n \Phi_n(q_b,t_b)\Phi_n^*(q_a,t_a) \tag{13.181}$$

This can be summarized as

$$K(q_b,t_b;q_a,t_a) = \begin{cases} \sum_n \exp\left\{-\frac{iE_n(t_b-t_a)}{\hbar}\right\}\Phi_n(q_b)\Phi_n^*(q_a), & t_b > t_a \\ 0, & t_b < t_a \end{cases} \tag{13.182}$$

The expression is an interesting contrast to our original definition using q-space functions (13.46). Indeed, it might even be viewed as if the particle moves around in all possible locations in the energy space (acquiring different phase factors) as it propagates from the selected initial position to the final position.

13.9 EIGENVECTORS AND ENERGY SPECTRUM

13.9.1 HARMONIC OSCILLATOR SOLVED VIA TRANSITION AMPLITUDE

The harmonic oscillator is an important problem in quantum mechanics that is easily solvable. The energy spectrum for a quantum harmonic oscillator is known to be

$$E_n = \hbar\omega\left(n + \frac{1}{2}\right) \tag{13.183}$$

The eigenvectors for the oscillator may be obtained with the help of Feynman path integration through the transition amplitude K:

$$K(q_b, t_b; q_a, t_a) = \left(\frac{m\omega}{2\pi i\hbar \sin\omega T}\right)^{1/2} \exp\left\{\frac{im\omega}{2\hbar \sin\omega T}[(q_a^2 + q_b^2)\cos\omega T - 2q_a q_b]\right\} \tag{13.184}$$

This is the Feynman representation of the transition amplitude (a c-number). It is possible to relate the Feynman transition amplitude with the quantum mechanical transition amplitude. The quantum mechanical transition amplitude (13.182) is as follows:

$$K(q_b, t_b; q_a, t_a) = \sum_n \exp\left\{-i\frac{E_n(t_b - t_a)}{\hbar}\right\}\Phi_n(q_b)\Phi_n^*(q_a) \tag{13.185}$$

Comparing Equations 13.184 and 13.185, we may find the spectrum of the oscillator and also the basis wave functions $\Phi_n(q)$. This may be done with the help of the following transformations:

$$i\sin\omega T = \frac{1}{2}\exp\{i\omega T\}(1 - \exp\{-2i\omega T\}), \quad \cos\omega T = \frac{1}{2}\exp\{i\omega T\}(1 + \exp\{-2i\omega T\}) \tag{13.186}$$

From here

$$K(q_b, t_b; q_a, t_a) = \left(\frac{m\omega}{\pi\hbar}\right)^{1/2}\frac{\exp\left\{-\frac{1}{2}i\omega T\right\}}{(1 - \exp\{-2i\omega T\})^{1/2}}$$

$$\times \exp\left\{-\frac{m\omega}{2\hbar}\left[(q_a^2 + q_b^2)\frac{1 + \exp\{-2i\omega T\}}{1 - \exp\{-2i\omega T\}} - 4q_a q_b\frac{\exp\{-i\omega T\}}{1 - \exp\{-2i\omega T\}}\right]\right\} \tag{13.187}$$

This expression has the form in Equation 13.185 if we expand Equation 13.184 in successive powers of $\exp\{-i\omega T\}$. Since the initial factor is $\exp\{-1/2\, i\omega T\}$, then obviously all the terms in this expansion will be of the form $\exp\left\{-1/2\, i\omega T\right\}\exp\left\{-in\omega T\right\}$ for $n = 0, 1, 2, \ldots$. This implies that the energy levels are given by the energy spectrum of Equation 13.183.

How can we find the eigenvectors or eigenstates of the Hamiltonian $\hat{H}$ as represented in some basis? This can be found by doing an expansion in a complete set of states. We demonstrate this method by doing the expansion only up to the term with $n = 2$:

$$
\left(\frac{m\omega}{\pi\hbar}\right)^{1/2} \exp\left\{-\frac{m\omega}{2\hbar}(q_a^2 + q_b^2)\right\} \exp\left\{-\frac{1}{2}i\omega T\right\}\left(1 + \frac{1}{2}\exp\{-2i\omega T\} + \cdots\right)
$$

$$
\times \left\{
\begin{array}{l}
1 + \dfrac{2m\omega}{\hbar} q_a q_b \exp\{-i\omega T\} + \dfrac{4m^2\omega^2}{2\hbar^2} q_a^2 q_b^2 \exp\{-2i\omega T\} \\[2mm]
- \dfrac{m\omega}{\hbar}(q_a^2 + q_b^2)\exp\{-2i\omega T\}\cdots
\end{array}
\right\}
\tag{13.188}
$$

Let us pick from here the coefficient of the lowest term:

$$
\left(\frac{m\omega}{\pi\hbar}\right)^{1/2} \exp\left\{-\frac{m\omega}{2\hbar}(q_a^2 + q_b^2)\right\} \exp\left\{-\frac{1}{2}i\omega T\right\} = \exp\left\{-\frac{i}{\hbar}E_0 T\right\}\Phi_0(q_b)\Phi_0^*(q_a)
\tag{13.189}
$$

This implies that

$$
E_0 = \frac{\hbar\omega}{2}, \quad \Phi_0(q) = \left(\frac{m\omega}{\pi\hbar}\right)^{1/4}\exp\left\{-\frac{m\omega}{2\hbar}q^2\right\}
\tag{13.190}
$$

The function $\Phi_0(q)$ could also be selected to be complex valued:

$$
\Phi_0(q) = \left(\frac{m\omega}{\pi\hbar}\right)^{1/4}\exp\left\{-\frac{m\omega}{2\hbar}q^2 + i\delta\right\}
\tag{13.191}
$$

where δ is a real valued constant. However, this does not give a new physical interpretation and the arbitrary phase is irrelevant.

The next-order term in the expansion is

$$
\exp\left\{-\frac{1}{2}i\omega T\right\}\exp\{-i\omega T\}\left(\frac{m\omega}{\pi\hbar}\right)^{1/2}\exp\left\{-\frac{m\omega}{2\hbar}(q_a^2 + q_b^2)\right\}\frac{2m\omega}{\hbar}q_b q_a
$$

$$
= \exp\left\{-\frac{i}{\hbar}E_1 T\right\}\Phi_1(q_b)\Phi_1^*(q_a)
\tag{13.192}
$$

It follows that

$$
E_1 = \frac{3}{2}\hbar\omega, \quad \Phi_1(q) = \left(\frac{2m\omega}{\hbar}\right)^{1/2}q\,\Phi_0(q)
\tag{13.193}
$$

The next term yields

$$E_2 = \frac{5}{2}\hbar\omega, \quad \Phi_2(q) = \frac{1}{\sqrt{2}}\left(\frac{2m\omega}{\hbar}q^2 - 1\right)\Phi_0(q) \tag{13.194}$$

This may be done in a similar manner for other terms and we obtain the energy eigenvalues and eigenvectors. However, this method makes finding $\Phi_n(q)$ rather tedious.

We may introduce a less direct but less difficult way as follows. We suppose the oscillator when excited moves from the state $|f(q)\rangle$ to the state $|g(q)\rangle$:

$$|f(q)\rangle = \sum_n f_n |\Phi_n(q)\rangle, \quad |g(q)\rangle = \sum_n g_n |\Phi_n(q)\rangle \tag{13.195}$$

where $|\Phi_n(q)\rangle$ are orthogonal basis vectors of the Hamiltonian $\hat{H}$. Considering Equations 13.194 and 13.195, we may compute the expectation value of the transition amplitude for the oscillator from state $|f(q)\rangle$ to $|g(q)\rangle$ as follows:

$$\int \langle f(q_b)|K(q_b,t_b;q_a,t_a)|g(q_a)\rangle dq_a dq_b = \sum_n f_n^* g_n \exp\left\{-i\frac{E_n(t_b - t_a)}{\hbar}\right\} \tag{13.196}$$

We select the special pair of vectors $|f(q)\rangle$ and $|g(q)\rangle$ such that the expansion on the RHS of Equation 13.196 should be simple. When we obtain these functions f_n or g_n, then we may obtain some information about the wave functions $\Phi_n(q)$ from Equation 13.195. Let us select our functions $|f(q)\rangle$ and $|g(q)\rangle$ in the following form:

$$f(q) = \left(\frac{m\omega}{\pi\hbar}\right)^{1/4} \exp\left\{-\frac{m\omega}{2\hbar}(q-\alpha)^2\right\}, \quad g(q) = \left(\frac{m\omega}{\pi\hbar}\right)^{1/4} \exp\left\{-\frac{m\omega}{2\hbar}(q-\beta)^2\right\}$$

$$\tag{13.197}$$

that are Gaussian with the centers, respectively, at α and β. Suppose $t_a = 0$, $T = t_b - t_a$ and $f_n = f_n(\alpha)$, $g_n = g_n(\beta)$, then we substitute Equations 13.184 and 13.197 into the LHS of Equation 13.196 and this yields

$$\sum_n f_n^*(\alpha)g_n(\beta) \exp\left\{-i\frac{E_n}{\hbar}T\right\}$$

$$= \exp\left\{-\frac{i\omega T}{2} - \frac{m\omega}{4\hbar}(\alpha^2 + \beta^2 - 2\alpha\beta \exp\{-i\omega T\})\right\}$$

$$= \exp\left\{-\frac{i\omega T}{2}\right\} \sum_n \exp\left\{-\frac{m\omega}{4\hbar}(\alpha^2 + \beta^2)\right\} \frac{(2\alpha\beta)^n}{n!} \exp\{-in\omega T\}\left(\frac{m\omega}{4\hbar}\right)^n \tag{13.198}$$

From here, it follows that

$$f_n(\alpha) = \frac{\alpha^n}{\sqrt{n!}}\left(\frac{m\omega}{2\hbar}\right)^{n/2}\exp\left\{-\frac{m\omega}{4\hbar}\alpha^2\right\}, \quad g_n(\beta) = \frac{\beta^n}{\sqrt{n!}}\left(\frac{m\omega}{2\hbar}\right)^{n/2}\exp\left\{-\frac{m\omega}{4\hbar}\beta^2\right\}$$

$$(13.199)$$

and

$$E_n = \hbar\omega\left(n + \frac{1}{2}\right) \tag{13.200}$$

Thus, we have been able to find the expansion coefficients in Equation 13.188. We may now find Φ_n from Equation 13.53:

$$f(q) = \left(\frac{m\omega}{\pi\hbar}\right)^{1/4}\exp\left\{-\frac{m\omega}{2\hbar}(q-\alpha)^2\right\} = \sum_n \frac{\alpha^n}{\sqrt{n!}}\left(\frac{m\omega}{2\hbar}\right)^{n/2}\exp\left\{-\frac{m\omega}{4\hbar}\alpha^2\right\}\Phi_n(q)$$

$$(13.201)$$

If we consider the generating function for the Hermite polynomials:

$$\exp\{-q^2 + 2q\xi\} = \sum_{n=0}^{\infty}\frac{q^n}{n!}H_n(\xi) \tag{13.202}$$

where

$$H_n(\xi) = (-1)^n \exp\{\xi^2\}\frac{d^n}{d\xi^n}\exp\{-\xi^2\} \tag{13.203}$$

are the Hermite polynomial then from Equation 13.201, the vector $|\Phi_n\rangle$ takes the form

$$\Phi_n(q) = \left(2^n n!\right)^{-1/2}\left(\frac{m\omega}{\pi\hbar}\right)^{1/4}H_n\left(q\sqrt{\frac{m\omega}{\hbar}}\right)\exp\left\{-\frac{m\omega}{2\hbar}q^2\right\} \tag{13.204}$$

This yields the wave functions (in the q-representation) of an excited state of an oscillator.

13.9.2 COORDINATE REPRESENTATION OF TRANSITION AMPLITUDE OF FORCED HARMONIC OSCILLATOR

Let us write the Lagrangian for an oscillator subjected to a time-dependent force $\gamma(t)$:

$$L = \frac{m\dot{q}^2}{2} - \frac{m\omega^2 q^2}{2} + \gamma(t)q \tag{13.205}$$

We write the equation of motion:

$$m\ddot{q} + m\omega^2 q - \gamma = 0 \tag{13.206}$$

We can now find the classical action. Let q_H and q_{NH} be respectively homogeneous and nonhomogeneous solutions of the equation of motion:

$$q(t) = q_H(t) + q_{NH}(t) \tag{13.207}$$

For a free harmonic oscillator we have

$$m\ddot{q} + m\omega^2 q = 0 \tag{13.208}$$

The general solution is

$$q(t) = A\cos(\omega t + \alpha) \tag{13.209}$$

and if

$$q(t) = \begin{cases} q_a, & t = t_a = 0 \\ q_b, & t = t_b = T \end{cases} \tag{13.210}$$

then

$$q(t = t_a = 0) = A\cos\alpha = q_a \tag{13.211}$$

$$q(t = t_b = T) = A\cos(\omega T + \alpha) = q_b \tag{13.212}$$

From here, it follows that the homogeneous solution $q_H(t)$ is

$$q_H(t) = \frac{1}{\sin\omega T}[q_a \sin\omega(T - t) + q_b \sin\omega t] \tag{13.213}$$

The homogeneous solution is selected in such a way that the initial and end moments give the desired initial and end coordinates.

We find next the nonhomogeneous solution. The nonhomogeneous solution is selected such that at $t = t_a = 0$ and $t = T$, it turns out zero. Our solution is found in the form

$$q(t) = \frac{1}{\sin\omega T}[q_a \sin\omega(T - t) + q_b \sin\omega t] + \frac{1}{m\sin\omega T}\int_0^T G(t,s)\gamma(s)ds \tag{13.214}$$

that substituting into Equation 13.206 yields the equation for the unknown function $G(t,s)$:

$$\frac{1}{m\sin\omega T}\int_0^T (\ddot{G} + \omega^2 G)\gamma(s)ds = \frac{\gamma(t)}{m} \tag{13.215}$$

This is an integral equation that can be solved by assuming

$$\frac{1}{\sin \omega T}(\ddot{G} + \omega^2 G) = \delta(s - t) \tag{13.216}$$

If we solve this equation considering the fact that the nonhomogeneous solution should be selected such that it vanishes at $t = t_a = 0$ and $t = T$, then the nonhomogeneous solution is

$$q_{NH}(t) = \frac{1}{m\omega}\left[\int_0^t \gamma(s)\sin \omega(t - s)ds + \frac{\sin \omega t}{\sin \omega T}\int_0^T \gamma(s)\sin \omega(s - T)ds\right] \tag{13.217}$$

If we consider $q(t) = q_H(t) + q_{NH}(t)$, then the action of a forced harmonic oscillator is

$$S_{FHO} = \frac{m\omega}{2\sin \omega T}\left[\begin{array}{l}(q_a^2 + q_b^2)\cos \omega T - 2q_a q_b + \dfrac{2q_a}{m\omega}\displaystyle\int_0^T \gamma(s)\sin \omega(T - s)ds \\[2mm] + \dfrac{2q_b}{m\omega}\displaystyle\int_0^T \gamma(s)\sin \omega s\, ds \\[2mm] - \dfrac{2}{m^2\omega^2}\displaystyle\int_0^T\int_0^t \gamma(s)\gamma(s')\sin \omega(T - s')\sin \omega s\, ds\, ds'\end{array}\right] \tag{13.218}$$

This is a functional of the force $\gamma(t)$. If $\gamma(t) \equiv 0$, then we have the action of a free harmonic oscillator.

The transition amplitude of a forced harmonic oscillator may be easily evaluated:

$$K_{FHO}(q_b, t_b; q_a, t_a) = \left(\frac{m\omega}{2\pi i\hbar \sin \omega T}\right)^{1/2}\exp\left\{\frac{i}{\hbar}S_{FHO}\right\} \tag{13.219}$$

This corresponds to a purely quantum mechanical result. The absolute square of $K_{FHO}(q_b, t_b; q_a, t_a)$ gives the probability per unit length of finding the mass back to where it started at the beginning of an oscillation with period T. This gives

$$\left|K_{FHO}(q_b, t_b; q_a, t_a)\right|^2 = \frac{m\omega}{2\pi\hbar \sin \omega T} \tag{13.220}$$

13.9.3   MATRIX REPRESENTATION OF TRANSITION AMPLITUDE OF FORCED HARMONIC OSCILLATOR

In this section, we continue with the problem of a forced harmonic oscillator. In physics, such systems include polyatomic molecules in varying external fields, crystals through which an electron passes and excites the oscillator modes, and so on. Our problem in this section is to study one of these problems. We analyze the behavior of a single oscillator disturbed by an external potential. We do this by investigating a single harmonic oscillator that is linearly coupled to some external force or distur-bance $\gamma(t)$. The Lagrangian for such a system was already described:

$$L = \frac{m\dot{q}^2}{2} - \frac{m\omega^2 q^2}{2} + \gamma(t)q \qquad (13.221)$$

We assume for convenience that during a certain time interval $T = t_b - t_a$, the force $\gamma(t)$ is turned on from time $t_a = 0$ and turned off by time $t_b = T$. The oscillator is assumed to be free initially before $t_a = 0$ and finally after $t_b = T$. In the present section, we consider the oscillator to be found initially at time t_a in state n and later at time t_b in state m:

$$\left|\Phi_n(q_a, t_a)\right\rangle \mapsto t_a, \quad \left|\Phi_m(q_b, t_b)\right\rangle \mapsto t_b \qquad (13.222)$$

Here, $|\Phi_n(q,t)\rangle$ and $|\Phi_m(q,t)\rangle$ as seen earlier are energy eigenstates that are solutions of the unperturbed time-independent Schrödinger equation. The force $\gamma(t)$ is consid-ered to be a perturbation.

Since the quantum state of the oscillator changes as a result of a perturbation, it is necessary to find the transition amplitude G_{mn} that the oscillator initially in the state n at time t_a is now found in the state m at time t_b. This is a more convenient represen-tation than the coordinate representation used earlier.

This transition amplitude G_{mn} may be evaluated as follows, as the matrix element of the evolution operator taken between final and initial energy states of interest (see Equation 13.54):

$$G_{mn} = \left\langle \Phi_m \left| \hat{U}(t_b, t_a) \right| \Phi_n \right\rangle = \left\langle \Phi_m \left| \exp\left\{ -\frac{i\hat{H}(t_b - t_a)}{\hbar} \right\} \right| \Phi_n \right\rangle \qquad (13.223)$$

Inserting identity operators as expansions over position states, we have

$$G_{mn} = \int dq_b \int dq_a \left\langle \Phi_m \| q_b \right\rangle \left\langle q_b \left| \exp\left\{ -\frac{it_b \hat{H}}{\hbar} \right\} \exp\left\{ \frac{it_a \hat{H}}{\hbar} \right\} \right| q_a \right\rangle \left\langle q_a \| \Phi_n \right\rangle \qquad (13.224)$$

$$G_{mn} = \int \Phi_m^*(q_b) K_{FHO}(q_b, t_b; q_a, t_a) \Phi_n(q_a) dq_a dq_b \qquad (13.225)$$

This is the transition amplitude from the state n to the state m.

Suppose that $|\Phi_n(t_a)\rangle$ and $|\Phi_m(t_b)\rangle$ are Schrödinger state vectors at the indicated times, for the free harmonic oscillator, then using Equation 13.42 to get these from the Heisenberg states $|\Phi_n\rangle$ and $|\Phi_m\rangle$, we have

$$G_{mn} = \exp\left\{ \frac{i}{\hbar} E_n t_a - \frac{i}{\hbar} E_m t_b \right\} \int \Phi_m^*(q_b, t_b) K_{FHO}(q_b, t_b; q_a, t_a) \Phi_n(q_a, t_a) dq_a dq_b$$

$$\qquad (13.226)$$

If m and n are different from zero this expression of the transition amplitude, G_{mn} becomes a complicated problem to be solved. If $m = n = 0$, then our problem may be conveniently handled as we have to evaluate a Gaussian integral. Thus, to make things easier for us we find the expression of G_{mn} by considering this first situation $m = n = 0$ and we evaluate the amplitude G_{00}. What should be its physical interpretation? The amplitude G_{00} is that where, as a result of an external perturbation, the oscillator does not change its state from the ground state:

$$G_{00} = \exp\left\{-\frac{i\omega T}{2}\right\}\int \Phi_0^*(q_b)K_{FHO}(q_b,t_b;q_a,0)\Phi_0(q_a)dq_a dq_b \qquad (13.227)$$

or

$$G_{00} = F(T)\left(\frac{m\omega}{\pi\hbar}\right)^{1/2}\int dq_a dq_b \exp\left\{\begin{array}{l}-\dfrac{m\omega}{2\hbar}q_a^2 - \dfrac{m\omega}{2\hbar}q_b^2 + \dfrac{im\omega}{2\hbar\sin\omega T}\\ \times\left[(q_a^2 + q_b^2)\cos\omega T - 2q_a q_b + 2q_a f_1 + 2q_b f_2\right]\end{array}\right\}$$

$$(13.228)$$

where

$$f_1 = \frac{1}{m\omega}\int_0^T \gamma(\tau)\sin\omega(T-\tau)d\tau, \quad f_2 = \frac{1}{m\omega}\int_0^T \gamma(\tau)\sin\omega\tau d\tau \qquad (13.229)$$

If we compute the integral in Equation 13.228, then we find amplitude to remain in the ground state

$$G_{00} = \exp\left\{-\frac{1}{2m\hbar\omega}\int_0^T\int_0^\tau \gamma(\tau)\gamma(\sigma)\exp\{-i\omega|\tau-\sigma|\}d\tau d\sigma\right\} \qquad (13.230)$$

Let us now calculate a general case. We make use of the generating function for the Hermite polynomials:

$$\exp\{-q^2 + 2q\xi\} = \sum_{n=0}^\infty \frac{q^n}{n!}H_n(\xi) \qquad (13.231)$$

On this basis we can construct a generating function for the matrix elements G_{mn}. Let us introduce the generating function from definition:

$$g(x,y) = \sum_{n,m}G_{mn}\frac{x^n y^m}{\sqrt{n!m!}} = \sum_{n,m}\frac{x^n y^m}{\sqrt{n!m!}}\exp\left\{-\frac{i}{\hbar}E_m T\right\}$$

$$\times \int \Phi_m^*(q_b,t_b)K_{FHO}(q_b,t_b;q_a,0)\Phi_n(q_a,0)dq_a dq_b \qquad (13.232)$$

Therefore it follows from Equations 13.231 and 13.232 that

$$\sum_m \frac{y^m}{\sqrt{m!}} \exp\left\{-\frac{i}{\hbar}E_m T\right\} \Phi_m(q_b) = \exp\left\{-\frac{i\omega T}{2}\right\}\left(\frac{m\omega}{\pi\hbar}\right)^{1/4}$$

$$\exp\left\{\frac{m\omega}{2\hbar}q_b^2 - \left[\exp\left\{\frac{i\omega T}{\sqrt{2}}y - q_b\sqrt{\frac{m\omega}{\hbar}}\right\}\right]^2\right\} \tag{13.233}$$

$$\sum_n \frac{x^n}{\sqrt{n!}} \Phi_n(q_a) = \left(\frac{m\omega}{\pi\hbar}\right)^{1/4} \exp\left\{\frac{m\omega}{2\hbar}q_a^2 - \left[\exp\left\{\frac{x}{\sqrt{2}} - q_a\sqrt{\frac{m\omega}{\hbar}}\right\}\right]^2\right\} \tag{13.234}$$

From these expressions we have the following:

$$g(x,y) = G_{00} \exp\{xy + ixb_\omega^* + iyb_\omega\} = G_{00}\exp\{b_\omega b_\omega^*\}\exp\{(x + ib_\omega)(y + ib_\omega^*)\}$$

$$= G_{00}\exp\{b_\omega b_\omega^*\}\sum_{p=0}^{\infty}\frac{(x + ib_\omega)^p (y + ib_\omega^*)^p}{p!} \tag{13.235}$$

where

$$b_\omega = \frac{1}{\sqrt{2m\hbar\omega}}\int_0^\tau \gamma(\tau)\exp\{i\omega\tau\}\,d\tau \tag{13.236}$$

If we express $(i(t_2 - t_1)/\hbar) = \lambda$ as a Newton binomial, then

$$g(x,y) = G_{00}\exp\{b_\omega b_\omega^*\}\sum_{p=0}^{\infty}\frac{(x + ib_\omega)^p (y + ib_\omega^*)^p}{p!} =$$

$$= G_{00}\exp\{b_\omega b_\omega^*\}\sum_{p,n,m}\frac{1}{p!}\frac{p!}{n!(p-n)!}\frac{p!}{m!(p-m)!}x^n(ib_\omega)^{p-n}y^m(ib_\omega^*)^{p-m}$$

$$= G_{00}\exp\{b_\omega b_\omega^*\}\sum_{n,m}\frac{1}{\sqrt{m!n!}}\sum_p\frac{x^n y^m}{\sqrt{m!n!}}\frac{(ib_\omega)^{p-n}(ib_\omega^*)^{p-m}p!}{(p-m)!(p-n)!} \tag{13.237}$$

Since, from definition

$$g(x,y) = \sum_{n,m}G_{mn}\frac{x^n y^m}{\sqrt{n!m!}} \tag{13.238}$$

Then

$$G_{mn} = \frac{G_{00}\exp\{b_\omega b_\omega^*\}}{\sqrt{m!n!}}\sum_{p\geq n,m}\frac{(ib_\omega)^{p-n}(ib_\omega^*)^{p-m}p!}{(p-n)!(p-m)!} \tag{13.239}$$

Now that all matrix elements between the energy eigenstates are determined, we can say that these represent the complete solution of the problem of a forced harmonic oscillator. Depending on the particular driving function $\gamma(t)$, the amplitudes for any particular transitions, can, in principle, be found.

13.10   SCHRÖDINGER EQUATION

We show how path integrals relate to the Schrödinger equation. We have seen in previous sections how the wave function $\Psi(q,t) = \langle q \mid \Psi(t) \rangle$ evolves via the transition amplitude $K(q,t + \varepsilon; q',t)$, that is, Equation 13.47 can be expressed as

$$\left| \Psi(t + \varepsilon) \right\rangle = \hat{U}\,(t + \varepsilon, t) \left| \Psi(t) \right\rangle \tag{13.240}$$

Inserting identity operators, one has

$$\left| \Psi(t + \varepsilon) \right\rangle = \int dq \left| q \right\rangle \langle q \left| \hat{U}(t + \varepsilon, t) \right| dq' \left| q' \right\rangle \langle q' \left| \right| \Psi(t) \right\rangle \tag{13.241}$$

Then in the sense of wave functions this gives

$$\Psi(q,t + \varepsilon) = \int dq'\, K(q,t + \varepsilon; q',t)\,\Psi(q',t) \tag{13.242}$$

This expression describes the wave function at time $(t + \varepsilon)$ through the wave functions at time t. Physically, the amplitude to be at position q at time $(t + \varepsilon)$ depends on amplitudes at all locations q' at the earlier time t. We consider two states $\Psi(q',t) \rightarrow \Psi(q,t + \varepsilon)$ separated by an infinitesimal time interval: $(t + \varepsilon) - t = \varepsilon \rightarrow 0$. This permits us to relate the path integration for one time to its value a short time later. In this case we may write the wave function approximately as follows:

$$\Psi(q,t + \varepsilon) = \frac{1}{A} \int \exp\left\{ \frac{i\varepsilon}{\hbar} \left(\frac{m(q - q')^2}{2\varepsilon^2} - U\left(\frac{q + q'}{2}, \frac{(t + \varepsilon) + t}{2} \right) \right) \right\} \Psi(q',t)\,dq' \tag{13.243}$$

Here, U is the potential acting on a mass m in the problem. Let us do the change of variable

$$q = q' + \xi \tag{13.244}$$

then Equation 13.243 becomes

$$\Psi(q,t + \varepsilon) \cong \frac{1}{A} \exp\left\{ -\frac{i\varepsilon}{\hbar} U(q,t) \right\} \int\limits_{-\infty}^{\infty} \exp\left\{ \frac{im\xi^2}{2\hbar\varepsilon} \right\} \Psi(q - \xi, t)\,d\xi \tag{13.245}$$

The term $\varepsilon U\left(\frac{1}{2}(q + q'), \frac{1}{2}(2t + \varepsilon)\right)$ in Equation 13.243 can be replaced by $\varepsilon U(q,t)$ since the error is of a higher order than ε. We consider ξ in Equation 13.245 to be a small parameter. This permits the exponential function to change very slowly. The exponential function $\exp\{im\xi^2/2\hbar\varepsilon\}$ is an oscillatory function. If it oscillates extremely rapidly around a unit circle in a complex plane, it causes almost complete cancellation. It follows that the only substantial contribution to the integral (13.245) comes from values of ξ in the neighborhood of 0. We see that the phase of the function $\exp\{im\xi^2/2\hbar\varepsilon\}$ changes by the order of a radian when ξ is of the order of $(\varepsilon\hbar/m)^{1/2}$. The largest contributions to the integral result from values of ξ of that order. Considering the fact that $q' = q - \xi$, there follows the Taylor series expansion of the functions $\Psi(q, t + \varepsilon)$ and $\Psi(q - \xi, t)$. The first is expanded to first order in ε and the second to second order in ξ:

$$\Psi(q,t) + \frac{\partial\Psi(q,t)}{\partial t}\varepsilon = \frac{1}{A}\exp\left\{-\frac{i\varepsilon}{\hbar}U(q,t)\right\}$$

$$\times \int_{-\infty}^{\infty} \exp\left\{\frac{im\xi^2}{2\hbar\varepsilon}\right\}\left[\Psi(q,t) - \frac{\partial\Psi(q,t)}{\partial q}\xi + \frac{1}{2}\xi^2\frac{\partial^2\Psi(q,t)}{\partial q^2} + \cdots\right]d\xi \quad (13.246)$$

For both sides of Equation 13.246 to agree in the limit as ε approaches 0, the normalization factor A must take the value:

$$A = \left(\frac{2\pi\hbar\varepsilon i}{m}\right)^{1/2} \quad (13.247)$$

To simplify Equation 13.246, we evaluate the following three integrals there:

$$\int_{-\infty}^{\infty} \exp\left\{\frac{im\xi^2}{2\hbar\varepsilon}\right\}d\xi = \left(\frac{2\pi\hbar\varepsilon i}{m}\right)^{1/2}, \quad \int_{-\infty}^{\infty}\exp\left\{\frac{im\xi^2}{2\hbar\varepsilon}\right\}\xi d\xi = 0,$$

$$\int_{-\infty}^{\infty}\exp\left\{\frac{im\xi^2}{2\hbar\varepsilon}\right\}\xi^2 d\xi = \left(\frac{\hbar\varepsilon i}{m}\right)\left(\frac{2\pi\hbar\varepsilon i}{m}\right)^{1/2} \quad (13.248)$$

Thus

$$\Psi(q,t) + \frac{\partial\Psi(q,t)}{\partial t}\varepsilon = \left[\Psi(q,t) + \frac{i\hbar\varepsilon}{2m}\frac{\partial^2\Psi(q,t)}{\partial q^2}\right]\exp\left\{-\frac{i\varepsilon}{\hbar}U(q,t)\right\} \quad (13.249)$$

As $\varepsilon \to 0$, we expand the exponential function in the expression in a series and we have

$$\Psi(q,t) + \frac{\partial\Psi(q,t)}{\partial t}\varepsilon = \left[\Psi(q,t) + \frac{i\hbar\varepsilon}{2m}\frac{\partial^2\Psi(q,t)}{\partial q^2}\right]\left[1 - \frac{i\varepsilon}{\hbar}U(q,t)\right] \quad (13.250)$$

From here, there follows

$$i\hbar \frac{\partial \Psi(q,t)}{\partial t} = -\frac{\hbar^2}{2m}\frac{\partial^2 \Psi(q,t)}{\partial q^2} + U(q,t)\Psi(q,t) \tag{13.251}$$

Of course, this is the usual time-dependent Schrödinger equation. By design, evolution of the wave function in time via the transition amplitude is equivalent to evolution via the Schrödinger equation.

13.11   GREEN FUNCTION FOR SCHRÖDINGER EQUATION

The transition amplitude $K(q_b,t_b;q_a,t_a)$ for $t_b > t_a$, satisfies the Schrödinger equation because it is also a wave function. We know that the transition amplitude is the kernel of the wave function:

$$\Psi(q_b,t_b) = \int K(q_b,t_b;q_a,t_a)\Psi(q_a,t_a)dq_a \tag{13.252}$$

The time-dependent Schrödinger equation controls the state's wave function:

$$i\hbar \frac{\partial \Psi(q_b,t_b)}{\partial t_b} = \hat{H}\,\Psi(q_b,t_b) \tag{13.253}$$

Since we want to use the formula of the type (13.252) only for $t_b > t_a$, then we can set $K = 0$ for $t_b < t_a$. Thus, the exact definition of the transition amplitude should be expressed:

$$\theta(t_b - t_a)\,K(q_b,t_b;q_a,t_a) \tag{13.254}$$

and substituting in Equation 13.252 with that result substituted in Equation 13.253, we arrive at

$$i\hbar \frac{\partial[\theta(t_b - t_a)K]}{\partial t_b} = i\hbar\theta(t_b - t_a)\frac{\partial K}{\partial t_b} + i\hbar K \frac{\partial \theta(t_b - t_a)}{\partial t_b} \tag{13.255}$$

We have for $t_b > t_a$

$$i\hbar \frac{\partial K}{\partial t_b} = \hat{H}K \tag{13.256}$$

On the other hand, in the limit $t_b = t_a$, Equation 13.252 must be an identity operation; hence

$$\lim_{\varepsilon\to 0} K(q_b,t_a + \varepsilon;q_a,t_a) = \begin{cases} \infty, & q_b = q_a \neq 0 \\ 0, & q_b \neq q_a \end{cases} \to \delta(q_b - q_a) \tag{13.257}$$

Then

$$ i\hbar \frac{\partial[\theta K]}{\partial t_b} - \hat{H}[\theta K] = i\hbar\delta(q_b - q_a)\delta(t_b - t_a) \qquad (13.258) $$

The function that satisfies such an equation whose RHS is proportional to a *four-dimensional delta function* is called the *Green function* for the Schrödinger equation. Consequently $K(q_b,t_b;q_a,t_a)$ is a *Green function*. The interpretation in Equation 13.252 is physically simple. The wave function $\Psi(q_a,t_a)$ acts as a source term, and the amplitude $K(q_b,t_b;q_a,t_a)$ propagates the effect of that source to the space–time point (q_b,t_b), where it generates its contribution to $\Psi(q_b,t_b)$. Integration over all such sources gives the total wave function at (q_b,t_b).

13.12 FUNCTIONAL INTEGRATION IN QUANTUM STATISTICAL MECHANICS

It should be noted that the goal of quantum statistical mechanics is basically the study of quantum equilibrium systems having the Gibbs distribution function. These quantum systems are found in the state of thermodynamic equilibrium with an external classical system (thermostat or thermal reservoir). Equilibrium is characterized by fixed values of thermodynamic parameters of the system, say, for example, temperature T, volume V, and number of particles N. In this way, the total energy of the thermostat is assumed to be much greater than the energy of the system and it is quasi-continuous. The behavior of the quantum system itself, however, is assumed to be independent of the nature of the thermostat.

The most convenient object of the quantum theory to describe such nonisolated systems is the density matrix. It is for this reason that we introduced quantum mechanical and quantum statistical density matrices in Sections 1.13 and 1.14. In the present chapter, we are going to represent this density matrix through path integration. This also enables us to make a transition to the quasi-classical approximation of quantum statistical mechanics. We use this to evaluate the first quantum correction to the classical Boltzmann distribution function.

13.13 STATISTICAL PHYSICS IN REPRESENTATION OF PATH INTEGRALS

In thermal equilibrium, the statistical probability of a state with energy E_n is given by

$$ W_n = \frac{1}{Z}\exp\{-\beta E_n\} \qquad (13.259) $$

where $\beta = 1/T$ is the inverse temperature, and the normalization factor

$$ Z = Z(\beta) = \sum_n \exp\{-\beta E_n\} \qquad (13.260) $$

is called the canonical *partition function*, as seen earlier (Chapter 3).

Suppose a physical observable F is defined by the operator $\hat{F}$, and the system's evolution is described by a Hamiltonian $\hat{H}$ with the basis (eigen) vectors $|\Phi_n\rangle$, defined from the eigenvalue equation:

$$\hat{H}|\Phi_n\rangle = E_n|\Phi_n\rangle \tag{13.261}$$

Then a diagonal matrix element F_{nn} is

$$F_{nn} = \langle\Phi_n|\hat{F}|\Phi_n\rangle = \int dq\,\Phi_n^*(q)\,\hat{F}(q)\Phi_n(q) \tag{13.262}$$

with $\Phi_n(q) = \langle q|\Phi_n\rangle$ and $\Phi_n^*(q) = \langle\Phi_n|q\rangle$.

The expectation value of the physical observable F is defined as

$$\begin{aligned}
\bar{F} &= \sum_n F_{nn} W_n = \sum_n F_{nn} \frac{\exp\{-\beta E_n\}}{Z} = \sum_n \int dq\,\Phi_n^*(q)\,\hat{F}(q)\Phi_n(q)\frac{\exp\{-\beta E_n\}}{Z} \\
&= \frac{1}{Z}\int dq'\,dq\,\delta(q'-q)\,\hat{F}(q)\sum_n \Phi_n^*(q')\Phi_n(q)\exp\{-\beta E_n\}
\end{aligned} \tag{13.263}$$

This is a fundamental law that is the summit in statistical physics as we saw originally in Chapter 1.

In Chapter 1, we defined a weighted function of the eigenstates for any operator

$$\rho(q,q') = \sum_n W_n\Psi_n(q)\Psi_n^*(q') = \sum_n \langle q|\Psi_n\rangle W_n\langle\Psi_n|q'\rangle \tag{13.264}$$

the so-called *statistical density matrix* in the coordinate representation. Really, this is a matrix element of the density operator:

$$\hat{\rho} \equiv \sum_i W_i|\Psi_i\rangle\langle\Psi_i| \tag{13.265}$$

Then an expectation value can be found via a trace, for any representation

$$\langle F\rangle \equiv \bar{F} = \int dq'dq\,\hat{F}(q)\delta(q'-q)\rho(q,q') = Tr(\hat{F}\hat{\rho}) \tag{13.266}$$

Let us examine, for comparison, the quantum transition amplitude (13.182), defined using the energy eigenstates:

$$K(q,t_b;q',t_a) = \sum_n \exp\left\{-\frac{iE_n(t_b-t_a)}{\hbar}\right\}\Phi_n(q)\,\Phi_n^*(q') \tag{13.267}$$

If we do the change of variable

$$\frac{i(t_b - t_a)}{\hbar} = \beta \tag{13.268}$$

then K is transformed to $Z\rho\ (q,q')$. Thus, all formulae obtained in path integration in quantum mechanics may be transformed into statistical physics formulae by the change of variables

$$t_b - t_a = -i\hbar\beta \tag{13.269}$$

Further, we know the trace of the density operator is unity:

$$\mathrm{Tr}\{\hat{\rho}\} = \sum_n \langle \Psi_n | \hat{\rho} | \Psi_n \rangle = \sum_n \langle \Psi_n | \left(\sum_i W_i | \Psi_i \rangle \langle \Psi_i | \right) | \Psi_n \rangle = \sum_n \sum_i \delta_{ni} W_i \delta_{in} = 1 \tag{13.270}$$

because the probabilities must sum to 1. So, for example, the trace of the transformed transition amplitude will give the statistical sum, Z.

$$\mathrm{Tr}\{Z(\beta)\hat{\rho}\ (\beta)\} \rightarrow \mathrm{Tr}\{\hat{K}\ (-i\hbar\beta)\} = Z(\beta) \tag{13.271}$$

Here, we suppose that a transition operator can be defined in analogy with the density operator as

$$\hat{K}(t) = \sum_i \exp\left\{-\frac{i}{\hbar} t\,\hat{H}\right\} |\Psi_i\rangle\langle\Psi_i| \tag{13.272}$$

Its matrix elements between the states $|q\rangle$ (at transformed time $t = -i\hbar\beta$) and $|q'\rangle$ (at an initial zero of the transformed time) give back the transition amplitude quoted above. Note, however, what happens when the trace is taken. The trace operation means that the propagation begins at some state $|q\rangle$ and ends at that same state:

$$\mathrm{Tr}\{\hat{K}\} = \int dq \langle q | \hat{K} | q \rangle \tag{13.273}$$

But the transition operator evolves the system. Suppose we insert a set of $N-1$ intermediate states, as identity operators, while partitioning the parameter β into parts

$$N \times (\beta/N) = N\Delta\tau \tag{13.274}$$

Now with

$$\exp(-\beta\hat{H}) = \left[\exp\left(-\frac{\beta\hat{H}}{N}\right)\right]^N \tag{13.275}$$

there results

$$\text{Tr}\{\hat{K}\} = \int dq \int dq_{N-1}\cdots\int dq_1 \langle q|e^{-\frac{\beta}{N}\hat{H}}|q_{N-1}\rangle\langle q_{N-1}|e^{-\frac{\beta}{N}\hat{H}}|q_{N-2}\rangle\cdots\langle q_1|e^{-\frac{\beta}{N}\hat{H}}|q\rangle \tag{13.276}$$

Each matrix element within the integrand involves propagation of the system, but through a small "imaginary time" interval, $\Delta\tau = \beta/N$. The integrations over the intermediate positions clearly become a path integral, in the limit of $N \to \infty$. Most intriguing, the system evolves back to its original state after a total imaginary time interval equal to the inverse temperature, β. The system evolves, and yet we use the amplitude to return to the original state, and, we sum over all possible initial states. Note also that there is really no distinction as to which q_i identifies the "initial" state. The expression is unchanged under a shift of all the names. Starting from any of them, the system evolves and we use the amplitude to return to that same state.

Hence, to go from quantum to statistical physics, we may do the change of variables into an imaginary time τ that has a range limited by the inverse temperature:

$$t = -i\hbar\tau, \quad 0 \le \tau \le \beta \tag{13.277}$$

Because the system returns to any original state, it can be considered a kind of periodic evolution, that is, a periodic boundary condition in the imaginary time.

Let us consider again the transition amplitude (13.94)

$$K(q,t_b;q',t_a) = \int_{q'}^{q}\exp\left\{\frac{i}{\hbar}\int_{t_a}^{t_b}L(\dot{q},q)\,dt\right\}Dq(t) \tag{13.278}$$

Let the Lagrangian $L(\dot{q},q)$ be as follows:

$$L(\dot{q},q) = \frac{m\dot{q}^2}{2} - U(q) \tag{13.279}$$

But

$$\dot{q} = \frac{dq}{dt} = \frac{i}{\hbar}\frac{dq}{d\tau} \tag{13.280}$$

Then, from the above we may evaluate the action

$$S = \int_{t_a}^{t_b} \left(\frac{m\dot{q}^2}{2} - U(q) \right) dt$$

$$= \int_0^{\beta} \left(-\frac{m}{\hbar^2} \left(\frac{dq}{d\tau} \right)^2 - U(q) \right)(-i\hbar)d\tau = -i\hbar \int_0^{\beta} \left(-\frac{m}{\hbar^2} \dot{q}^2 - U(q) \right) d\tau \qquad (13.281)$$

or

$$\frac{i}{\hbar} S = \int_0^{\beta} \left(-\frac{m}{\hbar^2} \dot{q}^2 - U(q) \right) d\tau \equiv \underline{S} \qquad (13.282)$$

where

$$\underline{S} = \underline{S}[q(\tau), -i\hbar\beta] \qquad (13.283)$$

is dimensionless and real. Thus, we may obtain the density matrix expressed through functional integration as follows:

$$Z\rho(q,q') = K(q,q',-i\hbar\beta) = \int_{q'}^{q} \exp\{\underline{S}[q,q',-i\hbar\beta]\} Dq(\tau) \qquad (13.284)$$

We may express the partition function Z through the energy operator $\hat{H}$:

$$Z = \sum_n \exp\{-\beta E_n\} = \sum_n \langle n | \exp\{-\beta \hat{H}\} | n \rangle = \mathrm{Tr}(\exp\{-\beta \hat{H}\}) \qquad (13.285)$$

We may also express the partition function through the transition amplitude:

$$Z = \sum_n \int \langle \Phi_n \| \Phi_n \rangle \exp\{-\beta E_n\} = \int \delta(q - q') dq' dq \sum_n \langle \Phi_n(q) \| \Phi_n(q') \rangle \exp\{-\beta E_n\}$$

$$(13.286)$$

and as

$$\sum_n \langle \Phi_n(q) \| \Phi_n(q') \rangle \exp\{-\beta E_n\} = Z\rho(q,q') = K(q,q',-i\hbar\beta) \qquad (13.287)$$

Then, as stated above, we use a delta function to take the trace, and we arrive at

$$Z = \int \delta(q - q')\,dq\,dq'\,Z\rho(q,q')$$
$$= \int \delta(q - q')\,dq\,dq'\,K(q,q',-i\hbar\beta) = \mathrm{Tr}\{\hat{K}(-i\hbar\beta)\} \qquad (13.288)$$

For example, let us evaluate the partition function for an oscillator:

$$Z = \sum_n \exp\{-\beta E_n\} = \sum_{n=0}^{\infty} \exp\{-\beta\hbar\omega(n + \tfrac{1}{2})\} = \left[2\sinh\left(\frac{\beta\hbar\omega}{2}\right)\right]^{-1} \qquad (13.289)$$

Thus, the partition function of a one-dimensional oscillator is

$$Z = \left[2\sinh\left(\frac{\beta\hbar\omega}{2}\right)\right]^{-1} \qquad (13.290)$$

For comparison, consider using the transition amplitude for a one-dimensional oscillator once more (from expressions (13.155), where $T = t_b - t_a$):

$$K(q_b,t_b;q_a,t_a) = F(T)\exp\left\{\frac{im\omega}{2\hbar\sin\omega T}[(q_a^2 + q_b^2)\cos\omega T - 2q_a q_b]\right\} \qquad (13.291)$$

If we use the Feynman units (translate to imaginary time variable) $t = -i\hbar\tau$, $0 \le \tau \le \beta$, then we have to change the time interval T into $-i\hbar\beta$, and we obtain

$$K(q_b,q_a,-i\hbar\beta) \equiv Z\rho(q_b,q_a) = F(-i\hbar\beta)$$
$$\times \exp\left\{-\frac{m\omega}{2\hbar\sinh(\beta\hbar\omega)}[(q_a^2 + q_b^2)\cosh(\beta\hbar\omega) - 2q_a q_b]\right\} \qquad (13.292)$$

where

$$F(-i\hbar\beta) = \left(\frac{m\omega}{2\pi\hbar\sinh(\beta\hbar\omega)}\right)^{1/2} \qquad (13.293)$$

Then the partition function found from the trace (using $q_b = q_a = q$) is as follows:

$$Z = \int dq\,K(q,q,-i\hbar\beta)$$
$$= \int dq\,F(-i\hbar\beta)\exp\left\{-\frac{2m\omega q^2}{2\hbar\sinh(\beta\hbar\omega)}(\cosh(\beta\hbar\omega) - 1)\right\} = \left[2\sinh\left(\frac{\beta\hbar\omega}{2}\right)\right]^{-1}$$
$$(13.294)$$

This agrees with the result obtained above by more elementary methods.

13.14 PARTITION FUNCTION OF FORCED HARMONIC OSCILLATOR

Suppose in Lagrangian (13.205) for a forced harmonic oscillator, we use Feynman's units (transform to imaginary time) $t = -i\hbar\tau$, $0 \leq \tau \leq \beta$, where the inverse absolute temperature is β:

$$L = -\frac{m\dot{q}^2}{2\hbar^2} - \frac{m\omega^2 q^2}{2} + \gamma(\tau)q \qquad (13.295)$$

By $\gamma(\tau)$ it is understood to take the given time-dependent driving force $\gamma(t)$ and make the replacement, $t \rightarrow -i\hbar\tau$. The dot means derivative with respect to τ. Considering this Lagrangian, the action for a forced harmonic oscillator is

$$S_{PHO} = F(\beta) \begin{bmatrix} (q_\beta^2 + q_0^2)\cosh\beta\,\hbar\omega - 2q_\beta q_0 + \dfrac{2q_0}{m\omega}\displaystyle\int_0^\beta \gamma(\tau)\sinh\hbar\omega\,(\beta - \tau)d\tau \\[2ex] + \dfrac{2q_\beta}{m\omega}\displaystyle\int_0^\beta \gamma(\tau)\sinh\hbar\omega\,\tau\,d\tau \\[2ex] - \dfrac{2}{m^2\omega^2}\displaystyle\int_0^\beta\int_0^\tau \gamma(\tau)\gamma(\tau')\sinh\hbar\omega\,(\beta - \tau')\sinh\hbar\omega\,\tau\,d\tau\,d\tau' \end{bmatrix} \qquad (13.296)$$

where

$$F(\beta) = \frac{m\omega}{2\sinh\beta\,\hbar\omega}, \qquad q_\beta \equiv q(\beta), \quad q_0 \equiv q(0) \qquad (13.297)$$

Let us examine the case of the interaction of an oscillator, say a dispersionless phonon with a particle with position $\vec{r}$, mass m, also affected by some potential $U(\vec{r})$. Then the Lagrangian of the system is

$$L(\vec{r},q) = \frac{m\dot{\vec{r}}^2}{2} - U(\vec{r}) + \frac{\dot{q}^2}{2} - \frac{\omega^2 q^2}{2} + \gamma(\vec{r})q \qquad (13.298)$$

Here, q is the coordinate of the oscillator with a unit mass. The coupling between the oscillator and the particle is carried by the driving term, $\gamma(\vec{r}(t))$. Using Feynman's units yields (dot is derivative with respect to τ here):

$$L(\vec{r}(\tau),q(\tau)) = -\frac{m\dot{\vec{r}}^2}{2\hbar^2} - U(\vec{r}(\tau)) - \frac{\dot{q}^2}{2\hbar^2} - \frac{\omega^2 q^2}{2} + \gamma(\vec{r}(\tau))q \qquad (13.299)$$

The statistical sum of the system described by the Lagrangian in Equation 13.298, by finding the trace of the transition amplitude, is

$$Z = \int dq\, dq'\, d\vec{r}\, d\vec{r}\,' \delta(q - q')\delta(\vec{r} - \vec{r}\,')$$

$$\times \int_{r'}^{r} D\vec{r} \int_{q'}^{q} Dq \exp\left\{ \int_{0}^{\beta} \left[-\frac{m\dot{\vec{r}}^2}{2\hbar^2} - U(\vec{r}) - \frac{\dot{q}^2}{2\hbar^2} - \frac{\omega^2 q^2}{2} + \gamma(\vec{r}(\tau))q \right] d\tau \right\} \quad (13.300)$$

We may exclude first the variable of the oscillator q. To do this, we use the fact that the delta function enforces $q = q'$ and evaluate a Gaussian integral of the type (13.80). Then, in our case we have

$$Z = Z_L \int d\vec{r}\, d\vec{r}\,' \delta(\vec{r} - \vec{r}\,') \int_{r'}^{r} D\vec{r} \exp\left\{ \int_{0}^{\beta} \left[-\frac{m\dot{\vec{r}}^2}{2\hbar^2} - U(\vec{r}) \right] d\tau - \Phi_\omega[\vec{r}(\tau)] \right\} \quad (13.301)$$

where

$$\Phi_\omega[\vec{r}(\tau)] = \frac{\hbar}{4\omega} \int_{0}^{\beta} d\tau \int_{0}^{\tau} d\tau'\, \gamma(\vec{r}(\tau))\gamma(\vec{r}(\tau'))F_\omega\left(|\tau - \tau'| \right) \quad (13.302)$$

$$F_\omega\left(|\tau - \tau'| \right) = \frac{\cosh \hbar\omega\left(|\tau - \tau'| - \dfrac{\beta}{2} \right)}{\sinh \dfrac{\beta\hbar\omega}{2}}, \quad Z_L = \left[2\sinh\left(\frac{\beta\hbar\omega}{2} \right) \right]^{-1} \quad (13.303)$$

It may be seen that in order to evaluate the transition amplitude in Equation 13.300 the coordinates of the oscillator are first eliminated by path integration, then the result is the path integral relative to the coordinate $\vec{r}$ of the particle.

13.15 FEYNMAN VARIATIONAL METHOD

It is possible to form a variational principle based on Feynman's method of functional integration [20,21]. It applies to quantum mechanical systems as well as to statistical systems for a finite temperature. In quantum mechanics, one uses the direct variational method (known as the Ritz method). For $T = 0$, a limit of the Feynman variational method leads to the Ritz variational method.

Let us analyze Feynman's variational method. Consider the partition function in terms of the dimensionless scaled action $\underline{S}[q(\tau)]$:

$$Z = \int dq \int_{q}^{q} \exp\{\underline{S}[q]\}Dq(\tau) = \int \delta(q - q')dq\, dq' \int_{q'}^{q} \exp\{\underline{S}[q]\}Dq(\tau) \quad (13.304)$$

We can introduce the trial action functional $\underline{S}_0[q(\tau)]$ that is hoped to be close to the true action, $\underline{S}[q(\tau)]$, and suppose we can evaluate both of these. Imagine that we cannot evaluate the true partition function, due to whatever mathematical difficulties, but we are able to evaluate a partition function for this trial action. Then the true partition function is

$$Z = \int \delta(q - q') dq dq' \int_{q'}^{q} \exp\{\underline{S} - \underline{S}_0\} \exp\{\underline{S}_0\} Dq(\tau)$$

$$= \int \delta(q - q') dq dq' \int_{q'}^{q} \exp\{\underline{S} - \underline{S}_0\} \exp\{\underline{S}_0\} Dq(\tau)$$

$$\times \frac{\int \delta(q - q') dq dq' \int_{q'}^{q} \exp\{\underline{S}_0\} Dq(\tau)}{\int \delta(q - q') dq dq' \int_{q'}^{q} \exp\{\underline{S}_0\} Dq(\tau)} \qquad (13.305)$$

From here, it follows that

$$Z = Z_0 \left\langle \exp\{\underline{S} - \underline{S}_0\} \right\rangle_{\underline{S}_0} \qquad (13.306)$$

where the trial partition function is

$$Z_0 = \int \delta(q - q') dq dq' \int_{q'}^{q} \exp\{\underline{S}_0\} Dq(\tau) \qquad (13.307)$$

and the correction term is

$$\left\langle \exp\{\underline{S} - \underline{S}_0\} \right\rangle_{\underline{S}_0} = \frac{\int \delta(q - q') dq dq' \int_{q'}^{q} \exp\{\underline{S} - \underline{S}_0\} \exp\{\underline{S}_0\} Dq(\tau)}{\int \delta(q - q') dq dq' \int_{q'}^{q} \exp\{\underline{S}_0\} Dq(\tau)} \qquad (13.308)$$

From definition the Helmholtz free energy F is

$$F = -T \ln Z \qquad (13.309)$$

and also

$$F = E - T\tilde{S} \qquad (13.310)$$

where $\tilde{S}$ is the entropy in this case. Thus, the internal energy is

$$E = -T \ln Z + T\tilde{S} \qquad (13.311)$$

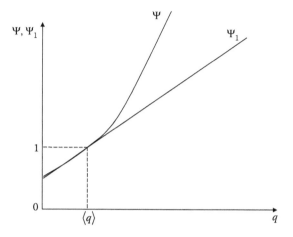

FIGURE 13.5 Comparison of the values of two functions $\Psi(q)$ and $\Psi_1(q)$ at a given point $\langle q \rangle$.

It should be noted that

$$\langle \exp\{q\} \rangle \geq \exp\{\langle q \rangle\} \tag{13.312}$$

This is the Feynman inequality, where q is a random variable and $\langle q \rangle$ is the weighted average of q.

The use of the Feynman units is very advantageous in that we now have an action that is real so that the Feynman inequality in Equation 13.312 is applicable in Equation 13.306. The geometrical interpretation of Equation 13.312 and a proof is given below with the aid of Figure 13.5.

13.15.1 PROOF OF FEYNMAN INEQUALITY

Suppose we have two functions of some random variable q:

$$\Psi = \exp\{q\}, \quad \Psi_1 = \exp\{\langle q \rangle\}(q - \langle q \rangle + 1) \tag{13.313}$$

These we represent in Figure 13.5. If $q = \langle q \rangle$, then

$$\Psi_1 = \Psi = \exp\{\langle q \rangle\}, \quad \langle q \rangle = \int q P(q) \, dq \tag{13.314}$$

and

$$\frac{d\Psi}{dq}\bigg|_{q=\langle q \rangle} = \exp\{\langle q \rangle\}, \quad \Psi(q = \langle q \rangle) = \exp\{\langle q \rangle\}$$

$$\frac{d\Psi_1}{dq}\bigg|_{q=\langle q \rangle} = \exp\{\langle q \rangle\}, \quad \Psi_1(q = \langle q \rangle) = \exp\{\langle q \rangle\} \tag{13.315}$$

where P(q) is the probability distribution of the random variable q.

It may be seen that the functions Ψ_1 and Ψ, together with their derivatives, coincide at the point $q = \langle q \rangle$ and

$$\exp\{q\} \geq \Psi_1 = \exp\{\langle q \rangle\}(q - \langle q \rangle + 1) \tag{13.316}$$

If we multiply both sides of this inequality by the probability P(q) and we integrate with respect to q then the Feynman inequality in Equation 13.312 follows:

$$\langle \exp\{q\} \rangle \geq \exp\{\langle q \rangle\} \tag{13.317}$$

13.15.2 APPLICATION OF FEYNMAN INEQUALITY

We apply the Feynman inequality to estimate the partition function (13.306)

$$\langle \exp\{\underline{S} - \underline{S}_0\} \rangle_{\underline{S}_0} \geq \exp\langle \{\underline{S} - \underline{S}_0\} \rangle_{\underline{S}_0} \tag{13.318}$$

and

$$Z = Z_0 \langle \exp\{\underline{S} - \underline{S}_0\} \rangle_{\underline{S}_0} \geq Z_0 \exp\langle \{\underline{S} - \underline{S}_0\} \rangle_{\underline{S}_0} \tag{13.319}$$

From

$$F = -T \ln Z \tag{13.320}$$

Then

$$F = -T \ln Z_0 - T \ln \langle \exp\{\underline{S} - \underline{S}_0\} \rangle_{\underline{S}_0} \tag{13.321}$$

But

$$\ln \langle \exp\{\underline{S} - \underline{S}_0\} \rangle_{\underline{S}_0} \geq \ln \exp\langle \{\underline{S} - \underline{S}_0\} \rangle_{\underline{S}_0} = \langle \underline{S} - \underline{S}_0 \rangle_{\underline{S}_0} \tag{13.322}$$

Thus

$$F \leq -T \ln Z_0 - T \langle \underline{S} - \underline{S}_0 \rangle_{\underline{S}_0} \equiv F_F \tag{13.323}$$

which defines F_F, the Feynman variational free energy. The relation to the true Helmholtz free energy is

$$F_F \geq F = E - T\tilde{S} \tag{13.324}$$

and as $T \to 0$, then $F = E$ and

$$E \le F_F = -T \ln Z_0 - T \langle \underline{S} - \underline{S}_0 \rangle_{\underline{S}_0} \qquad (13.325)$$

or

$$E \le F_F = -\frac{1}{\beta} \ln Z_0 - \frac{1}{\beta} \langle \underline{S} - \underline{S}_0 \rangle_{\underline{S}_0} \qquad (13.326)$$

Thus, the variational energy (F_F) is an upper limit on the true internal energy of the system at low temperature.

Now we show that for low temperatures, the Feynman variational principle coincides with that of Ritz. Let us consider the stationary Schrödinger equation:

$$\hat{H}|\Phi_n\rangle = E_n|\Phi_n\rangle \qquad (13.327)$$

where E_n is an energy eigenvalue and the eigenstates are $|\Phi_n\rangle$. We suppose that for whatever reason, it is difficult to solve this eigenvalue problem exactly. However, in the Ritz variational method, we want to find the lowest-energy eigenvalue E_0. To do this, one formulates a normalized trial state function, $|\Psi\rangle$, which could (in principle) be expressed in terms of the energy eigenstates:

$$|\Psi\rangle = \sum_n C_n|\Phi_n\rangle, \quad \langle \Psi|\Psi\rangle = \sum_n |C_n|^2 = 1 \qquad (13.328)$$

The trial state is a guess of the true ground state. The expectation value of the energy in this trial state is obtained by averaging the Hamiltonian $\hat{H}$:

$$E_\Psi = \langle \Psi|\hat{H}|\Psi\rangle = \left(\sum_n C_n|\Phi_n\rangle\right)^* \hat{H}\left(\sum_{n'} C_{n'}|\Phi_{n'}\rangle\right) = \sum_{nn'} E_{n'} C_n^* C_{n'} \langle \Phi_n||\Phi_n\rangle$$

$$= \sum_{nn'} E_{n'} C_n^* C_{n'} \delta_{nn'} = \sum_n E_n |C_n|^2 \ge \sum_n E_0 |C_n|^2 \ge E_0, \qquad (13.329)$$

where E_0 is the actual ground-state energy of the system. This means that the energy obtained from the trial state is an upper limit estimate of the ground-state energy:

$$E_0 \le E_\Psi \qquad (13.330)$$

As is well known, any trial state can be used to estimate E_0 this way; however, trial wave functions with the correct symmetry (e.g., no nodes) will give estimates much closer to the true ground-state energy.

13.16 FEYNMAN POLARON ENERGY

Here, we consider an excellent example application of these techniques. We examine a lattice in which is found an electron (or hole). The electron (hole) in an ionic or polar medium polarizes its neighborhood. The electron (hole) "dresses" itself with this polarization and moves in the medium where it interacts with lattices vibrations (phonons). This state of the electron (hole) together with its polarization and interacting with the phonon cloud is called a polaron [20,21,51,52]. It has an effective mass that is higher than that of the bare electron (hole). It should be noted that without the lattice vibrations, there is a periodic potential that acts on the electron (hole). We approximate the effect of this potential by supposing that the electron (hole) behaves as a free particle but with a mass being the electron (hole) band mass. Typically, the electron is coupled to longitudinal optical phonons, for example, in an ionic or polar crystal, that have some long-wavelength frequency $\omega_{\vec{q}}$ only weakly dependent on the wave vector $\vec{q}$. The polaron moves with greater difficulty through the lattice than an undressed electron would move.

The polaron concept is of interest because it describes the particular physical properties of charge carriers in polarizable media. The polaron idea constitutes an interesting field theoretical model consisting of a fermion interacting with a scalar boson (phonon) field. Landau introduced the polaron concept as the self-localized state of a charge carrier in a homogeneous polar medium [53]. Subsequently Landau and Pekar investigated the self-energy and the effective mass of the polaron [54]. This was shown by Fröhlich to correspond to the adiabatic or strong-coupling regime [55]. The work of Fröhlich [55] first pointed out the typical behavior for the weak-coupling polaron. With the help of Fröhlich's work, Feynman found higher orders of the weak-coupling expansion for the polaron energy [56]. The Feynman model has remained the most successful approach to the polaron problem over the years.

In this section, we evaluate the energy of a polaron using the Feynman variational principle. If we know the statistical sum Z, then we may find the free energy F

$$F = -T \ln Z = -\frac{1}{\beta} \ln Z \tag{13.331}$$

and the internal energy E

$$E = -\frac{d}{d\beta} \ln Z \tag{13.332}$$

The internal statistical sum may be expressed in the form of a trace using the scaled dimensionless action:

$$Z = \int \delta(\vec{r}_2 - \vec{r}_1) \, d\vec{r}_2 d\vec{r}_1 \int_{\vec{r}_1}^{\vec{r}_2} \exp\{\underline{S}[\vec{r}]\} D\vec{r} = \mathrm{Tr}\left\{ \int \exp\{\underline{S}[\vec{r}]\} D\vec{r} \right\} \tag{13.333}$$

Let us write the concrete expression of the action

$$\underline{S}[r] = \int L(-i\hbar\tau)\,d\tau \tag{13.334}$$

where

$$L(-i\hbar\tau) = -\frac{m\dot{r}^2}{2\hbar^2} + \sum_{\vec{q}}\left[-\frac{1}{2\hbar^2}\dot{Q}_{\vec{q}}^2 - \frac{1}{2}\omega_{\vec{q}}^2 Q_{\vec{q}}^2\right] - \sum_{\vec{q}}\gamma_{\vec{q}}(\vec{r})Q_{\vec{q}} \tag{13.335}$$

This is the Lagrangian of the electron–phonon interaction. The electron coordinate is $\vec{r}$ and m is its effective mass. The longitudinal optical phonon generalized coordinates are $Q_{\vec{q}}$ for a mode of frequency $\omega_{\vec{q}}$ at wave vector $\vec{q}$. The amplitude of the electron–phonon coupling $\gamma_{\vec{q}}(\vec{r})$, considering a dispersionless phonon frequency $\omega_{\vec{q}} \approx \omega_0$, may be defined as follows:

$$\gamma_{\vec{q}}(\vec{r}) = \sqrt{\frac{2}{V}}V_{\vec{q}}\begin{cases}\sin(\vec{q}\cdot\vec{r}), & q_x > 0 \\ \cos(\vec{q}\cdot\vec{r}), & q_x \le 0\end{cases}, \quad V_{\vec{q}} = \left[\alpha_F\pi\left(\frac{\hbar\omega_0}{q}\right)^2\left(\frac{\hbar}{2m\omega_0}\right)^{1/2}\right]^{1/2}$$

$$\tag{13.336}$$

Here, V_q is the electron–phonon coupling amplitude, V the volume of the region, and α_F is a dimensionless *Fröhlich electron–phonon coupling constant*, defined as follows:

$$\alpha_F = \frac{e^2}{2\hbar\omega_0}\left(\frac{1}{\varepsilon_\infty} - \frac{1}{\varepsilon_0}\right)\left(\frac{2m\omega_0}{\hbar}\right)^{1/2} \tag{13.337}$$

This depends on ε_0 and ε_∞, the static and high-frequency dielectric constants of the crystal. In the present example, we wish to compute the energy and effective mass of the polaron (electron dressed with polarization and moving in the phonon cloud). The parameter α_F is very important in this calculation, as it directly determines how strongly the electron is affected by phonons. Thus, the size of α_F directly determines the scale of the polaron energy.

From Chapter 8, we can recall that the dispersion in a crystal behaves as shown in Figures 8.6 and 8.7. In this polaron example, we will be interested in the region described by the almost constant part of the optical branch near zero wave vector. This is the branch for which positive ions move in a direction opposite to that of the neighboring ions. For convenience of computation of the polaron problem, we consider dispersionless optical phonon modes with frequency ω_0. The value of this frequency then determines the Fröhlich electron–phonon coupling strength.

In order to solve the polaron problem with less effort, it will be better to introduce a Lagrangian that better imitates the physical situation described by the exact Lagrangian in Equation 13.335. This procedure imitates the perturbation theory. A classical potential for the polaron problem might well be expected to be a good approximation when we consider tight binding. Thus, to imitate the polaron problem, a good model is one where instead of the electron being coupled to the lattice, it is coupled by some "spring" to another particle (fictitious particle) and the pair of particles are free to wander about the crystal. Thus, the model Lagrangian may be selected in the one oscillatory approximation:

$$L_0 = -\frac{m\dot{r}^2}{2\hbar^2} - \frac{M\dot{R}^2}{2\hbar^2} - \frac{M\omega_f^2(\vec{R} - \vec{r})^2}{2} \tag{13.338}$$

It is an approximation of the exact Lagrangian of the system, L. Here, M and ω_f are respectively the mass of a fictitious particle and the frequency of the elastic coupling that will serve as variational parameters; R is the coordinate of the fictitious particle. The model system conserves the translational symmetry of the system. A more judicious choice of L_0 to simulate a physical situation may give a better upper bound for the energy.

In order to find the polaron energy, we proceed to find the following quantities:

1. $E_0 = -T \ln Z_0$, the free energy based on the approximate Hamiltonian

$$Z_0 = \int_{-\infty}^{+\infty} \delta(\vec{r} - \vec{r}')\delta(\vec{R} - \vec{R}')d\vec{r}d\vec{r}'d\vec{R}d\vec{R}' \int_{r'}^{r}\int_{R'}^{R} \exp\{\underline{S}_0\}D\vec{r}D\vec{R} \tag{13.339}$$

Here

$$\underline{S}_0 = \int_0^\beta d\tau \left[-\frac{m}{2\hbar^2}\left(\frac{d\vec{r}}{d\tau}\right)^2 - \frac{M}{2\hbar^2}\left(\frac{d\vec{R}}{d\tau}\right)^2 - \frac{M\omega_f^2(\vec{R} - \vec{r})^2}{2} \right] \tag{13.340}$$

is the action functional of the model system. We do a change of variables. We consider the motion of the center of mass and also the relative motion:

$$\vec{\rho}_0 = \frac{\vec{r}m + \vec{R}M}{m + M}, \quad \vec{\rho} = \vec{r} - \vec{R} \tag{13.341}$$

From here, we have

$$\vec{r} = \vec{\rho}_0 + \frac{M}{m + M}\vec{\rho}, \quad \vec{R} = \vec{\rho}_0 - \frac{m}{m + M}\vec{\rho} \tag{13.342}$$

The quantities $\vec{\rho}_0$ and $\vec{\rho}$ are respectively the coordinates of the center of mass and of the relative motion.

The model Lagrangian may be conveniently represented in quadratic form as follows: The first term describes the translational motion of the center of mass; the second and third describe the oscillatory motion.

Let us define the quantities in the Lagrangian as follows:

$$\mu = \frac{Mm}{m+M}, \quad v^2 = u^2\omega_f^2, \quad u^2 = \frac{m+M}{m} \tag{13.343}$$

The parameter μ is a reduced mass and v is a scaled frequency. Considering the change of variable and the following path integral:

$$Z_0 = \int \delta(\vec{\rho}_0 - \vec{\rho}_0')\delta(\vec{\rho} - \vec{\rho}')d\vec{\rho}_0 d\vec{\rho}_0' d\vec{\rho} d\vec{\rho}' \int_{\vec{\rho}_0'}^{\vec{\rho}_0} \int_{\vec{\rho}'}^{\vec{\rho}} \exp\{\underline{S}_0\}D\vec{\rho}_0 D\vec{\rho} \tag{13.344}$$

then we have

$$Z_0 = \frac{Vm^{3/2}u^3}{(2\pi\hbar^2\beta)^{3/2}} \left[2\sinh\left(\frac{\beta\hbar v}{2}\right) \right]^{-3} \tag{13.345}$$

From the Lagrangian of the exact system, we may evaluate the partition function of the exact system:

$$Z = \prod_{\vec{q}} \int \delta(Q_{\vec{q}} - Q_{\vec{q}}')\delta(\vec{r} - \vec{r}')dQ_{\vec{q}}dQ_{\vec{q}}'d\vec{r}d\vec{r}' \int_{\vec{r}'}^{\vec{r}}\int_{Q_{\vec{q}}'}^{Q_{\vec{q}}} \exp\{S\}D\vec{r}DQ_{\vec{q}} \tag{13.346}$$

or

$$Z = \prod_{\vec{q}} Z_L \int \delta(\vec{r} - \vec{r}')d\vec{r}d\vec{r}' \int_{\vec{r}'}^{\vec{r}} D\vec{r} \exp\left\{ \int_0^\beta \left(-\frac{m\dot{\vec{r}}^2}{2\hbar^2} \right)d\tau + \Phi_{\omega_{\vec{q}}}[\vec{r}] \right\} \tag{13.347}$$

Here

$$Z_L = \left[2\sinh\left(\frac{\beta\hbar\omega_{\vec{q}}}{2}\right) \right]^{-3} \tag{13.348}$$

is the lattice sum and

$$\Phi_{\omega_{\vec{q}}}[\vec{r}] = \frac{\hbar}{4\omega_{\vec{q}}} \int_0^\beta\int_0^\beta d\sigma\, d\sigma'\gamma(\vec{r}(\sigma))\gamma(\vec{r}(\sigma'))F_{\omega_{\vec{q}}}(|\sigma - \sigma'|) \tag{13.349}$$

is the functional of the electron–phonon interaction (influence phase). In it is inherent all approximations. We may now write the action of the exact functional of the exact system as follows:

$$\underline{S[\vec{r}]} = -\int_0^\beta \frac{m\dot{\vec{r}}^2}{2\hbar^2} d\tau + \sum_{\vec{q}} \Phi_{\omega_{\vec{q}}}[\vec{r}] \tag{13.350}$$

Now it is possible to evaluate the action functional of the model system. This may be done by calculating the partition function of the model system:

$$Z_0 = \iint \delta(\vec{r} - \vec{r}')\delta(\vec{R} - \vec{R}')d\vec{r}d\vec{r}'d\vec{R}d\vec{R}'$$

$$\times \int_{r'}^{r} D\vec{r} \int_{R'}^{R} D\vec{R} \, \exp\left\{ \int_0^\beta \left[-\frac{m\dot{\vec{r}}^2}{2\hbar^2} - \frac{M\dot{\vec{R}}^2}{2\hbar^2} - \frac{M\omega_f^2}{2}(\vec{r}^2 + \vec{R}^2 - 2\vec{r}\cdot\vec{R}) \right] d\tau \right\} \tag{13.351}$$

Thus, after path integration, we have

$$Z_0 = Z_{\omega_f} \iint \delta(\vec{r} - \vec{r}') \, d\vec{r}d\vec{r}'$$

$$\times \int_{r'}^{r} D\vec{r} \exp\left\{ \int_0^\beta \left[-\frac{m\dot{\vec{r}}^2}{2\hbar^2} - \frac{M\omega_f^2}{2}\vec{r}^2 \right] d\tau - \Phi_{\omega_f}[\vec{r}] \right\} \tag{13.352}$$

Here

$$Z_{\omega_f} = \left[2\sinh\left(\frac{\beta\hbar\omega_f}{2} \right) \right]^{-3} \tag{13.353}$$

and

$$\Phi_{\omega_f}[\vec{r}] = \frac{\hbar M\omega_f^3}{4} \int_0^\beta \int_0^\beta d\sigma \, d\sigma' \vec{r}(\sigma)\vec{r}(\sigma')F_{\omega_f}\left(|\sigma - \sigma'|\right) \tag{13.354}$$

is the influence phase of the interaction of the electron (hole) with the fictitious particle and

$$\underline{S_0} = \int_0^\beta \left[-\frac{m\dot{\vec{r}}^2}{2\hbar^2} - \frac{M\omega_f^2}{2}\vec{r}^2 \right] d\tau + \Phi_{\omega_f}[\vec{r}] - 3\ln\left[2\sinh\left(\frac{\beta\hbar\omega_f}{2} \right) \right] \tag{13.355}$$

$$\underline{S} - \underline{S_0} = \int_0^\beta \frac{M\omega_f^2}{2}\vec{r}^2 d\tau + \sum_{\vec{\kappa}} \Phi_{\omega_{\vec{\kappa}}}[\vec{r}] - \Phi_{\omega_f}[\vec{r}] + 3\ln\left[2\sinh\left(\frac{\beta\hbar\omega_f}{2} \right) \right] \tag{13.356}$$

$$\ln Z \geq \ln Z_0 + \left\langle \underline{S} - \underline{S}_0 \right\rangle_{\underline{S}_0} \tag{13.357}$$

In order to evaluate this, it is necessary to evaluate

$$\left\langle \vec{r}^2(\tau) \right\rangle_{\underline{S}_0}, \quad \left\langle \vec{r}(\tau)\vec{r}(\sigma) \right\rangle_{\underline{S}_0}, \quad \left\langle \cos(\vec{q} \cdot (\vec{r}(\tau) - \vec{r}(\sigma))) \right\rangle_{\underline{S}_0} \tag{13.358}$$

We have to do this with the help of the generating function

$$\Psi_{\vec{q}}(\xi, \eta) = \left\langle \exp\left\{ i(\vec{q} \cdot (\xi \vec{r}(\tau) - \eta \vec{r}(\sigma))) \right\} \right\rangle_{\underline{S}_0} \tag{13.359}$$

From here, we have, for example

$$\left\langle \vec{r}(\tau)\vec{r}(\sigma) \right\rangle_{\underline{S}_0} = \frac{1}{\kappa^2} \frac{\partial^2 \Psi}{\partial \xi \partial \eta}\Big|_{\underline{S}_0}, \quad \left\langle \cos(\vec{q} \cdot (\vec{r}(\tau) - \vec{r}(\sigma))) \right\rangle_{\underline{S}_0} = \frac{1}{2}\left[\Psi_{\vec{q}}(1,1) + \Psi_{\vec{q}}(-1,-1) \right] \tag{13.360}$$

$$\Psi_{\vec{q}}(\xi, \eta) = \left\langle \exp\{\vec{q} \cdot (\xi \vec{r}(\tau) - \eta \vec{r}(\sigma))\} \right\rangle_{\underline{S}_0}$$

$$= \frac{\mathrm{Tr}\left\{ \int\limits_{R'}^{R} \int\limits_{r'}^{r} D\vec{r}D\vec{R} \; \exp\{\vec{q} \cdot (\xi \vec{r}(\tau) - \eta \vec{r}(\sigma)) + \underline{S}_0\} \right\}}{\mathrm{Tr}\left\{ \int\limits_{R'}^{R} \int\limits_{r'}^{r} D\vec{r}D\vec{R} \; \exp\{\underline{S}_0\} \right\}} \tag{13.361}$$

We can find the expression of the generating function by expanding our variables in quadratic form. This may be done by making a change of variables considering the motion of the center of mass and relative motion as above. This enables us to have two forced harmonic oscillators with frequencies $v_0 = 0$, $v_1 = v$. The generating function is conveniently obtained as follows:

$$\Psi_{\vec{q}}(\xi, \eta) = \prod_{i=0,1} \exp\left\{ -\frac{\hbar}{4m_i v_i} \int\limits_0^\beta \int\limits_0^\beta dt\,ds\gamma(t)\gamma(s)F_{v_i}\left(|t-s|\right) \right\} \tag{13.362}$$

where

$$\gamma_0(t) = i\vec{q}[\xi\delta(t-\tau) - \eta\delta(t-\sigma)], \quad \gamma_1(t) = \frac{M}{m+M}i\vec{q}[\xi\delta(t-\tau) - \eta\delta(t-\sigma)] \tag{13.363}$$

$$v_0 = 0, \quad v_1 = v, \quad m_0 = M + m, \quad m_1 = \mu = \frac{mM}{m+M} \tag{13.364}$$

If we consider the fact that

$$F_{v_0}(0) \cong \frac{2}{\beta\hbar v_0} + \frac{\beta\hbar v_0}{6} + \cdots, \quad F_{v_0}\left(|\tau - \sigma|\right) \cong \frac{2}{\beta\hbar v_0}$$

$$+ \frac{\hbar v_0}{\beta}\left[|\tau - \sigma| - \frac{\beta}{2}\right] - \frac{1}{12}\beta\hbar v_0 + \cdots \tag{13.365}$$

$$\Psi_{\vec{q}}(\xi, \eta) = \exp\left\{\begin{array}{l} -\dfrac{\hbar\vec{q}a_2}{4mv}\left[(\xi^2 + \eta^2)F_v(0) - 2\xi\eta F_v\left(|\tau - \sigma|\right)\right] \\[2ex] -\dfrac{\hbar^2\vec{q}a_1\beta}{4m}\left[\dfrac{1}{6}(\xi^2 + \eta^2) - \dfrac{2\xi\eta}{\beta^2}\left(|\tau - \sigma| - \dfrac{\beta}{2}\right)^2 + \dfrac{1}{6}\xi\eta\right] \end{array}\right\} \tag{13.366}$$

$$a_1 = \frac{1}{u^2}, \quad a_2 = 1 - \frac{1}{u^2}, \quad u^2 = \frac{M+m}{m} \tag{13.367}$$

we may then obtain

$$\langle\vec{r}(\tau)\vec{r}(\sigma)\rangle_{S_0} = \frac{3\hbar a_2}{2mv}F_v\left(|\tau - \sigma|\right) - \frac{3\hbar^2 a_1\beta}{4m}\left[-\frac{2}{\beta^2}\left(|\tau - \sigma| - \frac{\beta}{2}\right)^2 + \frac{1}{6}\right] \tag{13.368}$$

$$\langle\vec{r}^2(\tau)\rangle_{S_0} = \frac{3\hbar a_2}{2mv}F_v(0) - \frac{\hbar^2 a_1\beta}{16m} \tag{13.369}$$

These formulae depend on the quantity $|\tau - \sigma|$. This is an indication that our quantities depend on the past. They should be of the type of retarded functions. The significance of the interaction with the past is that the perturbation caused by the moving electron (hole) takes "time" to propagate in the crystal.

We look for an easy way to evaluate the integrals in Equations 13.352 and 13.354. It is observed that the functions in the integrand have a common argument $(|\tau - \sigma| - \beta/2)$. For this reason, let us generalize the function in the integrand in Equations 13.352 and 13.354 to be $G(|\tau - \sigma| - \beta/2)$ to aid in the simplification of the evaluation of the integrals. Thus, we look at the following function that yields the values of the integrals for given $G(|\tau - \sigma| - \beta/2)$ functions:

$$f(\beta) = \int_0^\beta\int_0^\beta G\left(|\tau - \sigma| - \frac{\beta}{2}\right)d\sigma d\tau \tag{13.370}$$

We do the change of variables

$$\tau - \sigma \equiv \rho, \quad 0 \le \sigma \le \tau \tag{13.371}$$

then

$$f(\beta) = 2\int_0^\beta G\left(\rho - \frac{\beta}{2}\right)(\beta - \rho)d\rho \tag{13.372}$$

We consider again the change of variable $\rho = \beta y$, where $0 \le y \le 1$. If $y = 1$, then $\rho = \beta$. Again, if we also do again the change of variable

$$\frac{\beta(1 - y)}{2} \equiv -\tau, \quad d\tau = -\frac{\beta}{2}dy, \quad 0 \le \tau \le \infty \tag{13.373}$$

then from the fact that

$$E = -\frac{d}{d\beta}\ln\left(\frac{Z_F}{Z_L}\right) \tag{13.374}$$

and considering low temperatures $\beta \to \infty$ $(T \to 0)$, the polaron variational energy may be obtained as follows:

$$E = \frac{3}{4}\hbar v\left(1 - \frac{1}{u}\right)^2 - \frac{C}{8\omega_0\pi\sqrt{\pi}}\int_0^\infty d\tau \frac{\exp\{-\omega_0\tau\}}{\sqrt{A(\tau)}} \tag{13.375}$$

$$C = \frac{\alpha_F R_p^{5/2}\pi\omega_0^{5/2}\hbar^{3/2}}{\sqrt{m}}, \quad A(\tau) = \frac{\hbar a_2}{2mv}(1 - \exp\{-v\tau\}) + \frac{\hbar a_1^2\tau}{2m} \tag{13.376}$$

In Equation 13.375, we may now vary $v = u\,\omega_f$ to obtain the lowest upper bound for the polaron energy. The constant C depends on α_F, the dimensionless *Fröhlich electron–phonon coupling constant*. For a crystal such as sodium chloride, $\alpha_F = 5$. For weak coupling, $\alpha_F < 1$, for strong coupling $\alpha_F > 7$, then intermediate coupling is between these ranges. The strong coupling for bulk crystals is not even achieved for alkali halides. It may be achieved for nanostructures. For nanostructures, the strong coupling range is shifted to ones of weak and intermediate coupling ranges. The electron–phonon interaction and the polaron effect in these nanostructures nowadays receive much attention due to their application in optoelectronic and novel generations of electronic devices.

The Feynman method of evaluation is for the entire range of coupling strength. Applying the Feynman variational principle for path integrals results in an upper bound for the polaron self-energy at all α_F, that at weak and strong coupling gives accurate limits.

Feynman obtained the smooth interpolation for the ground-state energy, between weak and strong coupling polarons. We evaluate further the asymptotic expansions of Feynman's polaron energy.

If we consider strong coupling for polarons in our problem, then the fictitious particle is more massive than the electron (hole)

$$M \gg m \tag{13.377}$$

We also know $v = u\,\omega_p$, then for strong coupling, we have

$$E = \frac{3}{4}\hbar v - \frac{C}{\pi^{1/2}R_p^{5/2}\omega_0^2}\left(\frac{mv}{\hbar}\right)^{1/2} \tag{13.378}$$

Considering the fact that for the minimum energy, we have

$$\frac{dE}{dv} = 0 \tag{13.379}$$

then the minimum energy is

$$E = -\frac{1}{3\pi}\alpha_F^2\hbar\omega_0 = -0.1061\alpha_F^2\hbar\omega_0 \tag{13.380}$$

In another strong coupling limit, Feynman found the energy to be

$$E = -0.1061\alpha_F^2\hbar\omega_0 - 2.83\hbar\omega_0 \tag{13.381}$$

The last term in the energy accounts for the weak coupling limit.

Let us consider now the weak coupling limit, $M \ll m$. In this case, the electron (hole) is almost completely free. Here, we may let

$$u^2 = \frac{M+m}{m} = 1 + \frac{M}{m} \equiv 1 + \varepsilon \tag{13.382}$$

where $\varepsilon \equiv M/m$ is a small parameter. Thus

$$u \equiv 1 + \frac{\varepsilon}{2} \tag{13.383}$$

If we minimize the energy considering this, then we have

$$E = -\alpha_F\hbar\omega_0 - 0.0123\alpha_F^2\hbar\omega_0 \tag{13.384}$$

Fröhlich [52] and Lee, Low, and Pines [51] found the weak coupling energy to be

$$E = -\alpha_F\hbar\omega_0 \tag{13.385}$$

Here, Fröhlich provided the first weak-coupling perturbation theory results and Lee, Low, and Pines used two elegant successive canonical transformation formulations. The Lee, Low, and Pines method puts the Fröhlich result on a variational basis.

We may see from our results that a classical potential is not a very good representation of the physical situation except possibly at very large coupling energies. The reasons for this limitation of the model may be twofold:

1. The electron is not confined to any particular part of the crystal. It is free to move in any direction. Any potential obviously will tend to keep the electron near its minimum.
2. From the expression of the exact action functional, the potential that the electron feels at any "time" depends on its position at previous times. That is to say, the effect that the electron has on the crystal propagates at a finite velocity and can make itself felt on the electron at a later time. This is less for a tighter binding. Here, the reaction of the crystal occurs much faster. Thus, a classical potential might well be expected to be a good approximation only for tight binding.

We may conclude that a model that will not suffer from either of the above objections would be one in which, instead of the electron being coupled to the lattice, it is coupled by some "spring" to another particle. The resultant pair of particles would be free to move in the crystal.

References

1. B. K. Agarwal, M. Eisner, *Statistical Mechanics*, Wiley Eastern Ltd., New Delhi, 1988.
2. W. P. Allis, M. A. Herlin, *Thermodynamics and Statistical Mechanics*, McGraw-Hill Book, New York, 1952.
3. Robert P. Bauman, *Modern Thermodynamics with Statistical Mechanics*, Macmillan, New York, 1992.
4. N. N. Bogolubov, N. N. Bogolubov Jr., *An Introduction to Quantum Statistical Mechanics*, World Scientific, Singapore, 1982.
5. J. J. Brehm and W. J. Mullin, *Introduction to the Structure of Matter*, 1 Edition, John Wiley & Sons, New York, 1989.
6. S. G. Brush, *Statistical Physics and the Atomic Theory of Matter: From Boyle and Newton to Landau and Onsager*, Princeton Series in Physics, Princeton University Press, Princeton, New Jersey, 1983.
7. P. Debye, E. Hückel, On the theory of electrolytes. Freezing point depression and related phenomena, *Physikalische Zeitschrift*, 24, 9, 185, 1923.
8. R. P. Feynman, *Statistical Mechanics: A Set of Lectures*, W. A. Benjamin Cummings Pub. Co. Inc., Reading, MA, 1972.
9. R. P. Feynman, *Statistical Mechanics*, Perseus Books, Reading, MA, 1998.
10. Von Harold L. Friedman, *A Course in Statistical Mechanics*, Prentice Hall Inc., Englewood Cliffs, New Jersey, 1985.
11. W. Jones and H. N. March, *Non-Equilibrium and Disorder*, Dover Pub. Inc., New York, 1973.
12. L. P. Kadanoff and G. Baym, *Quantum Statistical Mechanics*, W. A. Benjamin, New York, 1962.
13. R. Kubo, *Statistical Mechanics*, North-Holland, Amsterdam, 1990.
14. L. D. Landau and E. F. Lifshitz, *Statistical Mechanics: Part I*, Pergamon Press, Oxford, 1981.
15. L. D. Landau and E. F. Lifshitz, *Statistical Mechanics: Part II*, Pergamon Press, Oxford, 1981.
16. B. R. Holstein, *Topics in Advanced Quantum Mechanics*, Addison-Wesley Pub. Co., Reading, MA, 1992.
17. K. Huang, *Statistical Mechanics*, John Wiley & Sons, New York, 1987.
18. A. Böhm, *Quantum Mechanics*, Springer-Verlag, New York, 1979.
19. R. H. Dicke and C. Wittke, *Introduction to Quantum Mechanics*, Addison-Wesley Pub. Co., Reading, MA, 1960.
20. R. P. Feynman, *Doctoral Thesis: A New Approach to Quantum Theory* (Editor, Laurie M. Brown), World Scientific Pub. Co. Pte. Ltd., Singapore, 2005.
21. R. P. Feynman and A. R. Hibbs, *Quantum Mechanics and Path Integrals*, McGraw-Hill, New York, 1965.
22. P. Dirac, The Lagrangian in quantum mechanics, *Phys. Zeitschr.d. Sowjetunion* 3, 1, 64, 1933.
23. S. Flügge, *Practical Quantum Mechanics*, Springer-Verlag, Berlin, 1974.
24. L. D. Landau and E. F. Lifshitz, *Quantum Mechanics Non-Relativistic Theory*, 3rd ed., Pergamon Press, Oxford, 1977.
25. E. M. Lifshitz, L. P. Pitaevskii, *Physical Kinetics, Pergamon Press*, Oxford, 1981.
26. S. M. McMurry, *Quantum Mechanics*, Addison-Wesley Pub. Co., Wokingham, 1994.

27. E. H. Sondheimer, A. H. Wilson, The diamagnetism of free electrons, *Proc. R. Soc. London.* A, 210, 1101, 173, 1951.
28. D. L. Goodstein, *States of Matter*, Dover Pub. Inc., New York, 1985.
29. W. Jones and H. N. March, *Perfect Lattice in Equilibrium*, Dover Pub. Inc., New York, 1973.
30. S. K. Ma, *Statistical Mechanics*, World Scientific, Philadelphia, 1985.
31. A. B. Pippard, *The Elements of Classical Thermodynamics*, Cambridge University Press, Cambridge, 1957.
32. L. E. Reichl, *A Modern Course in Statistical Physics*, University of Texas Press, Texas, 1980.
33. F. Reif, *Fundamentals of Statistical and Thermal Physics*, McGraw-Hill, Inc., New York, 1965.
34. P. C. Riedi, *Thermal Physics: An Introduction to Thermodynamics, Statistical Mechanics and Kinetic Theory*, Oxford University Press, Oxford, 1988.
35. L. T. Hill, *Statistical Mechanics*, Dover Publications, Inc., New York, 1987.
36. A. W. Harison, *Solid State Theory*, Dover Publications, Inc., New York, 1979.
37. W. E. Parry, R. E. Turner, Green functions in statistical mechanics, *Rep. Progr. Phys.*, 27, 1, 23, 1964.
38. J. Willard Gibbs, *Elementary Principles of Statistical Mechanics: Developed with Especial Reference to the Rational Foundation of Thermodynamics*, Dover Publications, New York, 1960.
39. I. S. Gradshteyn and I. M. Ryzhik, *Tables of Integrals, Series, and Products*, Academic Press Inc., New York, 1980.
40. Douglas L. Martin, Specific heat of copper, silver and gold below 30 K, *Physical Review* 8, 5357,1973.
41. E. H. Sondheimer, A. H. Wilson, The Diamagnetism of free electrons, *Proc. R. Soc. Lond.* A 210, 1101, 173, 1951.
42. J. Bardeem, L. N. Cooper, J. R. Schrieffer, Theory of superconductivity, *Phys. Rev.* 108, 1175, 1957.
43. L. P. Gorkov, On the energy spectrum of superconductors, *Soviet Physics JETP* 7, 34, 505, 1958.
44. J. R. Schrieffer, *Theory of Superconductivity*, W. A. Benjamin, New York, 1964.
45. R. Meservey, B. B. Schwartz, In: *Superconductivity*, ed. By R. D. Parks, Marcel Dekker Inc., New York, 1969.
46. F. London, On the Bose-Einstein condensation, *Phys. Rev.*, 54, 947, 1938.
47. L. Tisza, On the theory of quantum liquids: Application to liquid helium I & II, *Journal of Physics and Radium, Ser. VIII* 1, 164, 350, 1940.
48. P. L. Kapitza, Viscosity of liquid helium below the lambda-point, *Nature* 141, 74, 1938.
49. L. D. Landau, The theory of superfluidity of helium II. *J. Phys. USSR*, 5, 71, 1941.
50. L. D. Landau. On the theory of superfluidity of Helium II. *J. Phys. (USSR)*, 11, 91, 1947.
51. T. D. Lee, F. E. Low, D. Pines, The motion of slow electrons in a polar crystal, *Phys. Rev.*, 90, 297, 1953.
52. H. Fröhlich, Electrons in lattice fields, *Advances in Physics* 3, 325, 1954.
53. L. D. Landau, Electron motion in crystal lattices, *Phys. Z. Sowjetunion.* 3, 664,1933.
54. S. I. Pekar, *Research on Electron Theory of Crystals, US AEC Transl. AEC-tr-555*, Washington, DC, 1963.
55. H. Fröhlich, Theory of electrical breakdown in ionic crystals, *Proc. Roy. Soc. A* 160, 230, 1937.
56. R. P. Feynman, Slow electrons in a polar crystal, *Phys. Rev.* 97, 660, 1955.

Index